Climate Emergency – Managing, Building, and Delivering the Sustainable Development Goals

Christopher Gorse • Lloyd Scott
Colin Booth • Mohammad Dastbaz
Editors

Climate Emergency – Managing, Building, and Delivering the Sustainable Development Goals

Selected Proceedings from the International Conference of Sustainable Ecological Engineering Design for Society (SEEDS) 2020

Editors
Christopher Gorse
Engineering, Leeds Sustainability Institute
Leeds Beckett University
Leeds, UK

Colin Booth
Architecture and the Built Environment
University of the West of England
Bristol, UK

Lloyd Scott
Technological University Dublin
Dunboyne, Ireland

Mohammad Dastbaz
Architecture and the Built Environment
University of Suffolk
Ipswich, Suffolk, UK

ISBN 978-3-030-79452-1   ISBN 978-3-030-79450-7 (eBook)
https://doi.org/10.1007/978-3-030-79450-7

© The Editor(s) (if applicable) and The Author(s), under exclusive license to Springer Nature Switzerland AG 2022
This work is subject to copyright. All rights are solely and exclusively licensed by the Publisher, whether the whole or part of the material is concerned, specifically the rights of translation, reprinting, reuse of illustrations, recitation, broadcasting, reproduction on microfilms or in any other physical way, and transmission or information storage and retrieval, electronic adaptation, computer software, or by similar or dissimilar methodology now known or hereafter developed.
The use of general descriptive names, registered names, trademarks, service marks, etc. in this publication does not imply, even in the absence of a specific statement, that such names are exempt from the relevant protective laws and regulations and therefore free for general use.
The publisher, the authors, and the editors are safe to assume that the advice and information in this book are believed to be true and accurate at the date of publication. Neither the publisher nor the authors or the editors give a warranty, expressed or implied, with respect to the material contained herein or for any errors or omissions that may have been made. The publisher remains neutral with regard to jurisdictional claims in published maps and institutional affiliations.

This Springer imprint is published by the registered company Springer Nature Switzerland AG
The registered company address is: Gewerbestrasse 11, 6330 Cham, Switzerland

Preface

September 2020 marked the fifth anniversary of the International "Sustainable Ecological Engineering Design for Society" (SEEDS) Conference, which was held in Leeds and brought together researchers from across the world to exchange ideas about our common problems and the challenges our planet faces.

The fifth SEEDS conference was the first conference we had to hold virtually given the terrible COVID-19 pandemic that our planet was going through. The challenges faced by various societies in dealing with the pandemic also highlighted the importance of our collective efforts to work towards UN's "Sustainable Development Goals" (SDG) which significantly include health and education.

Our fifth SEEDS conference is appropriately covering themes around "Social Values and Sustainability", "Education and Sustainability", "Health Safety and Well-being", and "Smart Digital Innovation". The selected proceedings before you also cover other important issues around "Programme Management, Project Management and Supply Chain", "Building Comfort, Performance and Energy", "Retrofit for Energy Efficiency and Comfort", "Sustainable Practice", and "Sustainable Transport".

Three weeks after our first SEEDS conference, in September 2015, world leaders, from 150 different countries, attended the "UN Sustainable Development Summit" in New York to discuss the challenges facing our planet and the fast disappearing natural resources. The Conference set a vision for 2025–2030 to develop a "plan of action for people, planet and prosperity".

It is important to observe that since the 2015 conference we have been through serious challenges and setbacks which include some world leaders denying that there is an environmental emergency and undoing years of works to address the significant challenges our planet faces. We started 2021 with fresh hopes that there might be a possibility of more of united approach by key world players in tackling our climate emergency, thanks to the collective consciousness of people around the world and the weight of public opinion balanced against the importance of "profit" and greed.

We, in the SEEDS community, strongly believe that through research and proven practice, we should foster ideas on how to tackle the serious challenges we all face and aim to reduce negative impacts on the environment while providing for the health and well-being of society. Professions and fields of research required to ensure buildings meet user demands and provide healthy enclosures are many and diverse.

We hope that SEEDS 2020 selected proceedings provide a platform for interested policy makers, researchers, practitioners, and educators to discuss the recognised and important problems affecting sustainable built environment.

Suffolk, UK
Leeds, UK
April 2021

Mohammad Dastbaz
Christopher Gorse

Keywords Building comfort, Building performance, Construction, Education and sustainability, Sustainability, Energy and buildings, Well-being, Project management, Social sustainability, Sustainable transport, Social value

Acknowledgements

We would like to thank all the contributing authors for their tireless efforts in providing us with their valuable research that has made this book possible. We also would like to thank the Springer team for working closely with us and providing advice and support when needed.

Special thanks to Atlanta Blair for her constant and valuable administrative support, for being patient, sending numerous e-mails, as deadlines loomed, and for gathering all the information and material required getting the book to the production stage.

Thanks to all the colleagues at Leeds Sustainability Institute (LSI) at Leeds Beckett University, specially Ellen Glover, who provided us with their valuable support and their research material and helped shape the structure of these selected proceedings.

Christopher Gorse
Lloyd Scott
Colin Booth
Mohammad Dastbaz

Contents

Part I Social Value and Sustainability

1. The Problems of Achieving Social Sustainability: A Cultural Shift 3
   John Sturges and Christopher Gorse

2. A Hammer Only Sees Nails: Why Construction Contractors Monetise Social Value 13
   Greg Watts and Anthony Higham

3. How the Social Value Act (2012) Contributes to Wealth Inequality in the UK Construction Industry 23
   Greg Watts and Anthony Higham

4. Transitioning from a Linear to a Circular Construction Supply Chain 33
   Emmanuel Manu, Nii Amponsah Ankrah, and Jamila Bentrar

Part II Building Comfort, Performance and Energy

5. Efficient Management of Environmental Control Within Electrical Substations 47
   Mark Collett, William Swan, Richard Fitton, and Matthew Tregilgas

6. Nearly Zero-Energy Buildings (nZEB) and Their Effect on Social Housing in Ireland: A Case Study Review 59
   Joseph Glennon, Michael Curran, and John P. Spillane

7. Performance of Distributed Energy Resources in Three Low-Energy Dwellings During the UK Lockdown Period 71
   Rajat Gupta and Matt Gregg

8	**Learnings from the Evolution of the University of Suffolk EcoLab: Adopting People-Centred Design Approaches to Encourage the Mass Uptake of Energy Transition Solutions in the Housing Sector** Benjamin Powell	85
9	**A Review of Sustainable Construction Practices in Ghana** Moses K. Ahiabu, Fidelis A. Emuze, and Dillip Kumar Das	93
10	**Key Factors Influencing Deployment of Photovoltaic Systems: A Case Study of a Public University in South Africa** Nutifafa Geh, Fidelis A. Emuze, and Dillip Kumar Das	105

Part III Retrofit for Energy Efficiency and Comfort

11	**Is It Possible to Develop Lamella and Airey Properties Ecologically?** Elliott Young	121
12	**Implications of a Natural Ventilation Retrofit of an Office Building** Ashvin Manga and Christopher Allen	137

Part IV Education and Sustainability

13	**Developing a Sustainable Urban Environment Through Teaching Asset Management at a Postgraduate Level** David Thorpe and Nasim Aghili	151
14	**The Impact of Department of Construction Management Facebook Environment-Related Posts** John Smallwood	163
15	**'The Breakfast Room Game': A Case of an Innovative Construction Project Management Simulation for Year 6 Children** Ian C. Stewart	175

Part V Health Safety and Wellbeing

16	**Underground Utility Services on Irish Construction Projects: Current Work Practices and the Effectiveness of the Health and Safety Authority (HSA) Code of Practice** Shane Carmody, Michael Curran, and John P. Spillane	191
17	**Health and Safety Practices and Performance on Public Sector Projects: Site Managers' Perceptions** Nomakhwezi Mafuya and John Smallwood	203

18	**Understanding the Concept of Resilience in Construction Safety Management Systems**	217
	Isaac Aidoo, Frank Fugar, Emmanuel Adinyira, and Nana Benyi Ansah	
19	**Optimisation of the Process for Generation, Delivery and Impact Assessment of Toolbox Talks on a Construction Site with Multiple Cultures**	235
	Máire Feely, James G. Bradley, and John Spillane	

Part VI   Programme Management, Project Management and Supply Chains

20	**Bagging a Bargain Begets Amnesia: Insights of Integrating Responsible Sourcing into Building Information Modelling**	249
	Sophie Ball and Colin A. Booth	
21	**An Analysis of Adversarial/Cooperative Attitudes in Construction Contracting: How Approaches to Adversarial Procurement Might Have a Lasting Effect on Project Culture**	265
	Emily Harrison and John Heathcote	
22	**"Megaprojects to Mega-Uncertainty" Is About Risk Management to Perform**	277
	Charlene Chatelier, Adekunle S. Oyegoke, Saheed Ajayi, and John Heathcote	
23	**How Calls for New Theory Might Address Contemporary Issues Affecting the Management of Projects**	297
	John Heathcote	
24	**Can Hard Paradigm Artefacts Support Soft Paradigm Imperatives? An Unpaired Comparative Experiment to Determine whether Visualisation of Data Is an Effective Collaboration and Communication Tool in Project Problem-Solving**	313
	Alison Davies and John Heathcote	
25	**Collusion within the UK Construction Industry, An Ethical Dilemma**	327
	Joseph Thorp and Hadi Kazemi	
26	**The Myth of the Post Project Review**	337
	Amelia Maya Guinness and John Heathcote	
27	**The Tendency Towards Suboptimal Operational Planning**	349
	Thomas Price and John Heathcote	

Part VII Smart Digital Innovation

28 Challenges of Projects Supporting Smart Cities' Development 371
Zahran Al-Hinai, John Heathcote, and Hadi Kazemi

29 I Spy with My Little Eye: Improving User Involvement in Elderly Care Facility Design through Virtual Reality 385
Abdul-Majeed Mahamadu, Udonna Okeke, Abhinesh Prabhakaran, Colin A. Booth, and Paul Olomolaiye

30 Smart Enterprise Asset Management 395
Mark Lenton, Dave Lister, Jim Garside, Richard Pleace, Gary Shuckford, Simon Roberts, Tim Platts, Paul Redmond, Christopher Gorse, Bashar Alhnaity, and Ah-Lian Kor

31 Strategic Management of Assets and Compliance through the Application of BIM and Digital Twins: A Platform for Innovation in Building Management 415
Gav Roberts, Lee Reevell, Richard Pleace, Ah-Lian Kor, and Christopher Gorse

32 BIM Education Through Problem-Based Learning Exercise: Challenges and Opportunities in an Inter-Professional Module 431
Shariful Shikder

33 Investigating the Trinity Between Sustainability and BIM-Lean Synergy: A Systematic Review of Existing Studies 441
Hafize Büşra Bostancı, Onur Behzat Tokdemir, and Ali Murat Tanyer

Part VIII Sustainable Practice

34 Management Policy for Laboratory Electronic Waste from Grave to Cradle 461
Mercy Chinenye Iroegbu, Samantha Chinyoka, Tariro Ncube, and Salma M. S. Alarefi

35 The Place of Urban Forestry in our Viable Urban Futures: A Cosmetic or a Metaphysic? 475
Alan Simson

36 Illicit Crops, Planning of Substitution with Sustainable Crops Based on Remote Sensing: Application in the Sierra Nevada of Santa Marta, Colombia 483
Hector Leonel Afanador Suárez, Gina Paola González Angarita, Leila Nayibe Ramírez Castañeda, and Pedro Pablo Cardoso Castro

37	**Bibliometric Study on Particle Emissions of Natural and Alternative Building Materials**............................	495
	Nana Benyi Ansah, Emmanuel Adinyira, Kofi Agyekum, and Isaac Aidoo	

Part IX Sustainable Transport

38	**Social Acceptance and Societal Readiness to EVs**	511
	Khalid Kamal Abdelgader Mohamed, Henok K. Wolde, and Salma M. S. Alarefi	
39	**Transport and Waste: Killing Two Birds with One Stone—The Sustainable Energy**	525
	Hina Akram, Shoaib Hussain, and Talib E. Butt	

Index... 533

About the Editors

Christopher Gorse is the director for the Leeds Sustainability Institute, located at Leeds Beckett University. The building performance and sustainability research unit that Professor Gorse leads has amassed one of the most comprehensive sets of actual building thermal performance data in the UK. The work has informed government policy and regulation, is used extensively by industry, and is now embedded in the work of the International Energy Agency's programme to inform whole building testing and performance measures. The research team has extensive knowledge and expertise in whole building performance tests in the field and laboratories, hygrothermal and thermal modelling, building simulation, elemental building component testing, energy and behaviour monitoring.Work extends across domestic and industrial developments, refurbishment, and retrofit projects and looks at the impact and effectiveness of interventions and changes. The efficiencies of building fabric, services, and renewable energy technologies are a major focus of the group's work. The more extensive work of the Sustainability Institute includes energy, waste, business and social governance, information and communication technology, corporate social responsibility, planning, ecology, process management, and project management, all engaged in reducing negative impact on the natural environment.Professor Gorse leads a sub-task group for the International Energy Agency, Annex 58, on whole-scale building testing, working with over 30 international partners. More recently, Professor Gorse led a major contract project funded by the Department of Energy and Climate Change to undertake research on the Core Cities Green Deal Go Early project. The work involved the intensive and extensive monitoring and evaluation of buildings that are benefiting from Green Deal and Eco funding in the Leeds area.

Lloyd Scott is a professor of Academic Advisor and Partnership Co-Ordinator in the School of Surveying and Construction Management at Dublin Institute of Technology (DIT). He joined the DIT as a lecturer in Construction Management and Technology in 2000. He is currently supervising 7 PhD researchers. Apart from his lecturing, supervision, research, and academic administrative duties, Lloyd has completed a PhD in the field of Built Environment Education and has developed a framework for assessment led learning strategies for Built Environment education. In 2014 he accepted the position of "Professor of Practice" at the Haskell and Irene Lemon Construction Science Division in the College of Architecture at the University of Oklahoma. Along with this he has produced many peer-reviewed conference and journal papers. He serves as the Associate Editor of the *International Journal of Construction Education and Research*. His research interests include modern approaches to thermal performance in domestic construction, development of sustainable energy sources and their practical application in Ireland, and project delivery methods for a sustainable environment in Ireland. He also serves on the editorial board of *Structural Survey*, an academic journal that publishes contemporary and original research in building pathology and building forensics, refurbishment, and adaptation. In 2016 Lloyd accepted the position of Research Fellow at the Sustainability Institute at Leeds Beckett University.

Colin A. Booth is Associate Head of Research and Scholarship in the Faculty of Environment and Technology at the University of the West of England (UWE), Bristol, UK. He holds the distinguished titles of both Visiting Professor of Civil Engineering and Visiting Professor of Sustainability at prestigious international universities. He is the author/co-author of eight books and ~190 scientific articles. His research interests include sustainability in the built environment, environmental management in construction, property level flooding, sustainable drainage systems (SuDS), climate change adaptation strategies, and urban pollution.

About the Editors

Mohammad Dastbaz graduated in Electrical and Electronic Engineering and went on to do a PhD in "Design, Development and Evaluation of Multimedia Systems" at Kingston University. In 1989, he established one of UK's first multimedia computer companies called "Systems 2000 Ltd." and was one of the only companies alongside Philips Corporation to participate in the UK's first Multimedia Systems Exhibition at London Olympia in 1990. He Joined Kingston University in 1994 as a research lead for developing multimedia-aided learning packages and has since worked in a number of UK universities, progressing to become the Dean of the School of Computing and Engineering at the University of East London. In 2011, Professor Dastbaz joined Leeds Metropolitan University (now Leeds Beckett University) as their dean and pro-vice-chancellor for Faculty of Arts, Environment and Technology.Professor Dastbaz remains research active and has published over 60 refereed journal and conference papers, books and book chapters. He is on a number of editorial boards of international journals, has been chair of a number of international conferences (including IEEE's Information Visualisation), and remains a member on a number of international conference committees. His latest publications include an edited volume on *Green Information Technologies*; *Building Sustainable Futures—Design and the Built Environment*, and a series of four edited volumes on *Technology and Sustainable Futures* published by Springer.Professor Dastbaz is a Fellow of the BCS—The Chartered Institute for IT, Fellow of the Higher Education Academy, and a Fellow of the Royal Society of Arts (RSA).

About the Authors

Nasim Aghili graduated in BSc Rural Reclamation Technology Engineering from the Faculty of Technology and Engineering, Hamadan, Iran, in 2007, then she succeeds to take a place in the major of Urban Geography and Planning at Islamic Azad University as a master's student in 2007. She received a Ph.D. in Facilities Management from Universiti Teknologi Malaysia in 2017. She is a former Research Fellow at the University of Southern Queensland, Tutor at Deakin University, and Teaching Fellow at Universiti Sains Malaysia. Her areas of research interests include Facility Management, Asset Management, Building Information Modelling (BIM), Green Building, Energy Management, and Sustainable Environment.

Moses K. Ahiabu is a Lecturer in the Department of Building Technology, Ho Technical University, Ho—Ghana for over 15 years and the Departmental Examination Officer (DEO). He holds a Master's Degree (MSc) in Engineering Project Management from Coventry University, UK. Mr. Ahiabu is currently pursuing a Doctoral Studies at Central University of Technology, Bloemfontein, South Africa. Sustainability, health and safety, and project management are the primary research interests of Mr. Ahiabu. His paper presented at the SEEDS Conference 2020 won Highly Recommended Award. Central University of Technology, Bloemfontein, South Africa

Ho Technical University, Ho, Ghana

Isaac Aidoo has an M.Phil. in Construction Management and a B.Sc. in Building Technology He is a member of Ghana Institution of Construction (MGIOC), Institution of Engineering and Technology (IET). He is a Building Technologist. He has over 20 years of experience in construction site management. He is currently a Full-Time Lecturer with the Building Technology Department of Accra Technical University, Accra-Ghana, and a PhD (Construction Management) candidate at the Department of Construction Technology and Management, Kwame Nkrumah University of Science and Technology, Kumasi, Ghana. His research area is Resilience in Construction Safety Management systems in the construction Industry.

Hina Akram has an MPhil and an MSc, whose research career cuts through the five spheres of environment—the lithosphere, atmosphere, hydrosphere, biosphere, and anthroposphere/technosphere. She always seeks sustainable development and climate change to be cross-cutting themes of her research endeavours.

Salma M. S. Alarefi holds a BEng in Telecommunication Engineering and a PhD in Electronics and Computing Systems from the University of Essex in 2013 and 2018 respectively. Dr. Al Arefi is an award-winning author in sustainable electronic and electrical engineering. Her research and teaching interests focus largely on the application and uptake acceleration of renewable energy systems.

Zahran Al-Hinai has a master's degree in Strategic Project Management from Leeds Beckett University in 2018. He has experience in Strategic Management and Project Management through the experience of working at the Supreme Council for Planning in the Sultanate of Oman and Oman 2040 Vision Office. He has worked on important national projects such as Oman Vision 2040 and the ninth and tenth Five Year Plan for the Sultanate. He has also participated in establishing the Strategic Management Office (SMO) in the SCP. Zahran has established several companies, one of them being Value Investment.

Amelia has always had a fascination in the realm of social sciences. Having the opportunity to study the profound and valuable connection that social theory has with the practical world of her profession in project management has been a stimulating and vital part of the overarching progression of the management world, hopefully continuing the research to further comprehend the practical applications and benefits that social input may have on cross-sectoral careers. She is forever thankful for the support of likeminded individuals who share the passion for continued learning in order to excel in the future of ideas and practical knowledge.

Nii A. Ankrah is a Senior Lecturer and Director of construction programmes at the Department of Civil Engineering, Aston University. He has over 13 years of research experience in the areas of organisational behaviour and project management. He is well published in the areas of health and safety, supply chain management, dispute resolution, cradle to cradle in the built environment, and sustainability, with over 57 refereed publications comprising journal papers, books, and conference papers.

Nana Benyi Ansah is a civil engineering professional with over 15 years site construction and teaching experience in many related subjects. He was an adjunct lecturer in strength of materials and theory of structures at the Pentecost University (Ghana, 2015–2020). Currently, he is a lecturer at the Department of Building Technology, Accra Technical University (Accra, Ghana).

Sophie Ball is a Design Manager for Kier Construction Ltd., where she works full-time on public funded projects. Before this she spent almost a decade working in the real estate sector, where she managed a large property portfolio. She was an undergraduate at the University of Gloucestershire where she studied BSc Geography, and she was a postgraduate at the University of the West of England (UWE), Bristol, where she studied MSc Sustainable Development in Practice. Her research interests include sustainability in the built environment, environmental management in construction, climate change mitigation and adaptation strategies.

Jamila Bentrar is an Urban Project Manager and also a futurist in Lille Metropole (Urban Authority in North of France). She has a master's degree in Environmental Geosciences and undertakes project management of various development activities, such as natural risk management and brownfield rehabilitation. Her interests are in foresight, innovation, and circular models. In the exercise of her profession, she participates in many innovative projects aimed at the development of demonstrators to preserve and valorise the territory's resources.

Colin A. Booth is Associate Head of Research and Scholarship in the Faculty of Environment and Technology at the University of the West of England (UWE), Bristol, UK. He holds the distinguished titles of both Visiting Professor of Civil Engineering and Visiting Professor of Sustainability at prestigious international universities. He is the author/co-author of eight books and ~190 scientific articles. His research interests include sustainability in the built environment, environmental management in construction, property level flooding, sustainable drainage systems (SuDS), climate change adaptation strategies, and urban pollution.

Hafize Büşra Bostancı after graduating from the Department of Architecture of İzmir Institute of Technology in 2013, completed her master's studies at the Department of Architecture of Gazi University in 2019 in information technologies in building science and construction technologies. She is continuing her PhD studies at the Department of Architecture of Middle East Technical University (METU) as a full-time research assistant since 2019.

James G. Bradley is currently a lecturer on the Construction Management and Engineering programme in the School of Engineering at the University of Limerick, teaching undergraduates in the areas of compliance and project management. As a senior member within the research group at the Irish Construction Management Research Centre (ICMRC), his research interests lie in the areas of training and development, mental health and well-being, and Big Data in the construction sector. Prior to taking up his role at the University of Limerick, Jim had more than 25 years industrial experience, holding senior management positions and acting as a consultant across a range of sectors.

Talib E. Butt has a PhD and an MSc. Currently, he is working as Senior Lecturer at Northumbria University, Newcastle Upon Tyne, UK.

Charlene Chatelier 1. Association for Project Management. (2005). *Project management body of knowledge*. Association for Project Management. 2. Dvir, D., Raz, T., & Shenhar, A. J. (2003). An empirical study of the relationship between project planning and project success. *International Journal of Project Management, 21*(2), 89–95. 3. Flyvbjerg, B. (2016). The fallacy of beneficial ignorance: A test of Hirschman's hiding hand. *World Development, 84*, 176–189. 4. Liu, Z., et al. (2016). Handling social risks in government-driven mega project: An empirical case study from West China. *International Journal of Project Management, 34*(2), 202–218. 5. Oyegoke, A. S., & AI Kiyumi, N. (2017). The causes, impacts and mitigations of delay in megaprojects in the Sultanate of Oman. *Journal of Financial Management of Property and Construction, 22*(3), 286–302.

Samantha Chinyoka received her BSc Hons degree in Applied Physics from National University of Science and Technology, Bulawayo, Zimbabwe, in 2015, where she was awarded the NUST book prize for academic excellence. She also received her MSc degree in Electrical Engineering and Renewable Energy Systems from the University of Leeds, Leeds, UK, in 2020. Her research interests are in integration of renewable energy systems at both distribution and transmission levels as well as sustainable industrial practices.

Mark Collett is an MPhil student at the University of Salford and conducting research with Electricity North West Ltd. through a knowledge exchange partnership. Mark has a background in energy engineering with research interests in building performance and low carbon buildings. Mark will shortly be commencing a PhD study focusing on building performance evaluation.

Michael Curran is an experienced Researcher and Lecturer in Construction Management at many universities across Ireland and the UK. His research interests include Stakeholder Management and Engagement, Corporate Social Responsibility, Health and Safety, Sustainability, and Site Management. He has lectured and supervised many student dissertations on both Undergraduate and Postgraduate Construction Management courses at the University of Limerick, Queen's University Belfast, University College London, and the Institute of Technology Sligo. Michael completed his PhD at Queen's University Belfast, and this is complemented with an MSc in Construction Project Management and a BEng (Hons) in Civil Engineering, also from Queen's.

Dillip Kumar Das has a PhD in Urban and Regional planning with Civil Engineering and City Planning background. Currently he is engaged in teaching, research, and community engagement activities at the University of KwaZulu Natal, South Africa. His research and consulting interests under sustainable urban and regional development include systems analysis, system dynamics modelling, infrastructure planning and management, smart cities, and transportation planning. He has co-authored two books as the lead author and published several peer-reviewed research articles.

About the Authors

Alison Kay Davies graduated with distinction from the MSc Strategic Project Management course at Leeds Beckett University and is a programme manager with NHS England and Improvement's People Directorate.

Fidelis A. Emuze is Professor and Head of the Department of Built Environment at the Central University of Technology, Free State (CUT), South Africa. Lean construction, health, safety, and well-being and sustainability constitute the primary research interest of Dr. Emuze, who is a National Research Foundation (NRF) rated researcher. Dr. Emuze is the editor of *Value and Waste in Lean Construction* (published by Routledge), *Valuing People in Construction* (published by Routledge), and co-editor of *Construction Health and Safety in Developing Countries* (published by Routledge). Dr. Emuze is the International Coordinator of CIB W123—People in Construction Working Commission.

Máire Feely graduated from the University of Limerick in 2020, completing her BSc. (Hons) in Construction Engineering and Management. On completion, she joined Linesight, where she took up the role of Graduate Project Manager, working on a variety of projects.

Nutifafa Geh is currently a doctoral candidate at the Central University of Technology, Free State. He holds a master's degree in Environmental Management from Liverpool Hope University, UK, and a bachelor's degree in Building Technology from Kwame Nkrumah University of Science & Technology, Ghana. Prior to enrolling for his doctoral studies, he worked in the construction industry and the oil and gas sector in Ghana. His research interests include green building, construction procurement, modular construction, and campus sustainability. Nutifafa is the corresponding author and can be contacted via ngehonline@gmail.com.*Web Profile*ORCID iD: https://orcid.org/0000-0002-6934-0359

Christopher Gorse is Professor of Project and Construction Management and Director of the Leeds Sustainability Institute. Visiting Professor of Lean Construction CUT South Africa and Visiting Professor of the University of Suffolk. He is Chair of the International Conference for Sustainable Ecological Engineering Design for Society (SEEDS), immediate past Chair of the Association of Researchers in Construction Management (ARCOM), and a founding member of the Building Performance Network. He is a Chartered Builder, a member of the Engineering Professors' Council, and holds Principal Investigator positions for major construction, environment, and energy research projects. Chris's work is interdisciplinary and centred on the management of construction, performance of buildings, and transformation of the built environment. Related fields of expertise include the built environment, building technology, policy, governance and management of construction; project, programme, and process management; supply chain management; social value, partnering, group behaviour and interaction: building performance and evaluation; building characterisation, buildings and energy flexibility; smart buildings and energy integration; and sustainability in the built environment.

Amelia Maya Guinness graduated with distinction from the MSc Project Management course at Leeds Beckett University. She is now a practicing project manager in IT development.

Rajat Gupta is *Director* of the multi-disciplinary *Oxford Institute for Sustainable Development* and *Low Carbon Building Research Group* at Oxford Brookes University, where he also holds senior professorial chair in *sustainable architecture and climate change*. His research interests lie in evaluating building performance, carbon mapping buildings and neighbourhoods, smart energy systems, and scaling up energy retrofits. As Principal Investigator, he has won over £13 million in research grants from UK Research Councils, Innovate UK, EU, United Nations, and Industry. He has published widely, including journal papers on future direction of energy research and evaluation of an innovative retrofit programme.

Emily Harrison graduated with distinction from the MSc Project Management course at Leeds Beckett University. She is an experienced project manager with a Quantity Surveying practice in Leeds, UK.

About the Authors

Thomas Haskins is a postgraduate student on the MSc Project Management course at Leeds Beckett University and a project management intern with Hermes Distribution Leeds, UK.

John Heathcote trained as a civil engineer and delivered a broad range of engineering and business improvement projects in a 22-year career in practice. He holds an HNC in civil engineering, and an honours degree in project management and an MBA from Sheffield Hallam University's Sheffield Business School. He now lectures and researches into the management of projects, based at Leeds Beckett University, where he has been for 17 years. John has been a visiting lecturer at the University of Applied Sciences Wuerzburg-Schweinfurt. Chairing the APM's Value Management Specific Interest Group for 5 years promoting an interest in projects as an open system concept, John's research takes an interest in reconceptualising normative project management to exploit projects' latent value, something that might be key to creating a more sustainable future. His PhD (to be published in 2022) explores the tensions between modernist organisations and project approaches proposing a reinterpretation of the theory of project management based on the lived experiences of professionals and associated research studies that work with live projects.

Anthony Higham is a Chartered Quantity Surveyor and Chartered Construction Manager with over 20 years of experience, ten of which he spent in industry. He is currently a Senior Lecturer at the University of Salford, UK.

Halton Housing is no ordinary Housing Association. Established in 2005, we have experienced tremendous success. We own and manage over 7000 homes across Runcorn and Widnes. We have got ambitious growth plans and aim to build and acquire over 1000 homes in the next 5 years! Our customers are at the heart of our business and our vision is Improving People's Lives.For us, innovation is much more than technology. We aim to embed creativity and fresh thinking into everything we do, providing real solutions to real problems—from how we could enhance the way we deliver services to our customers, to how we develop our neighbourhoods, improve our properties, and support our people.

Shoaib Hussain has a PhD and an MSc. His research interest covers environmental pollution, nanotechnology, wastewater, and plant science. Currently, he is working as a Head of Analytical Lab of a multi-national pharma industry.

Mercy Chinenye Iroegbu holds a B.Eng. from Imo State University, Owerri, Nigeria (2011–2016) and a Master of Science (Eng) from the University of Leeds, UK (2019–2020), both degrees in Electronic and Electrical Engineering. Her research interests focus largely on sustainable electronic and electrical generation. She is the recipient of the Women's Engineering Society postgraduate award of the International Women in Engineering Day 2020 (INWED20).

Hadi Kazemi is the Course Director for Construction Management in the School of Built Environment, Engineering and Computing at Leeds Beckett University. His research focuses on the application of smart functionalities including AI, Digital Twins, Machine Learning, and Construction 4.0 in construction and project management contexts. He is also an advocate of promoting sustainable practices within the construction industry. Prior to academia, he worked as a project and operations manager in construction and manufacturing sectors.

Ah-Lian Kor is part of Leeds Beckett MSc Sustainable Computing Curriculum Development Team. She has been involved in several EU projects for Green Computing, Innovative Training Model for Social Enterprises Professional Qualifications, and Integrated System for Learning and Education Services. She has published work on ontology, semantic web, web services, portal, and semantics for GIS. She is active in AI research and has developed an intelligent map understanding system and reasoning system.

Nomakhwezi Mafuya is a registered Environmental Health Practitioner with the Health Professions Council of South Africa (HPCSA), specialising in Health, Safety, and Environmental (HSE) management. She recently completed an MSc in the Built Environment (Construction Health and Safety Management specialisation) as offered by the Nelson Mandela University. She has 7 years of experience in the construction industry and is currently responsible for HSE management at the Eastern Cape Department of Public Works and Infrastructure within the Sarah Baartman Region.

Abdul-Majeed Mahamadu is a Senior Lecturer in Project Management and Building Information Modelling (BIM) in the Faculty of Environment and Technology at the University of the West of England (UWE), Bristol, UK. He is a research active academic with expertise in the areas of BIM and emerging digital and immersive (Virtual Reality) technologies as well as their applications in design, construction, and operation of buildings. He is the author of over 60 scientific articles.

Ashvin Manga is a contract lecturer and MSc candidate at Nelson Mandela University. A former medical student, sustainability and occupant well-being became his overarching research area. He presents the module Environment and Services which is focused on sustainability within the built environment and green technologies. His BSc Honour research titled "Implications of a natural ventilation retrofit" was presented at SEEDS 2020 and awarded the highly commended award in Building Performance. His masters research involved the development of a computer vision application that counts the number of students in a lecture room using CCTV cameras for improved air quality and facility management.

Emmanuel Manu is an Associate Professor in Quantity Surveying and Project Management at the School of Architecture, Design, and the Built Environment. He has over 8 years of experience undertaking research activities in the areas of construction supply chain management, circular economy in the built environment, sustainable procurement and social value, and smart and digital processes for performance improvement in the built environment. He has collaborated with industry partners to deliver research and consultancy projects that are all aimed at improving the performance of construction supply chains and driving the transition towards a smart and circular built environment.

Khalid Kamal Abdelgader Mohamed holds a BEng in Electronic Engineering from the University of Medical Sciences and Technology, Sudan, and an MSc in Electrical Engineering and Renewable Energy Systems from the University of Leeds, UK, in 2017 and 2020 respectively. Khalid has worked in industrial automation and renewable energy industries in Sudan. The award-winning author has published multiple papers in the field of power systems, electric vehicles, and renewable energy.

Tariro Ncube received her BSc Hons in Electrical and Electronic Engineering from the University of Zimbabwe in 2013. She also completed an MSc in Electrical and Renewable Energy Systems from the University of Leeds in 2020. Her research interests include Power Systems Dynamics, Protection and Control of power systems with the integration of Renewable Energy Resources.

Udonna Okeke is a Senior Lecturer in Engineering Management in the Faculty of Environment and Technology at the University of the West of England (UWE), Bristol, UK. He is a Programme Leader in Aerospace Engineering. His research expertise is in the areas of innovation and innovation management including their application in healthcare settings.

Paul Olomolaiye is a Professor of Construction Engineering and Management and currently Pro Vice-Chancellor for Equalities and Civic Engagement at the University of the West of England (UWE), Bristol. He is Chair of the Cabot Learning Federation, a Multi-Academy Trust of 21 Schools in and around Bristol, and a Non-Executive Director of Avon and Wiltshire NHS Mental Health Trust. He has widely published, with over 200 journal and conference publications and authorship of two major books on Construction Productivity and Stakeholder Management.

Tim Platts has worked continuously in his consultancy for a wide range of clients, both blue chip and SME, including some of the world's leading tech companies and a list of educational institutions at Higher Education level (including being commissioned by the University of Salford to author a paper on BIM in 2012) as well as working in a social business promoting a BIM curriculum in schools across the UK (Class Of Your Own).

Benjamin Powell is a qualified architect who runs his own practice which he established in the late 2017—studiomanifest.co.uk. He is a member of the Architects

Registration Board (ARB) and Royal Institute of British Architects (RIBA). Alongside this, he is a lecturer in Architecture at the University of Suffolk.His practice takes a people-centred approach and seeks more meaningful collaboration between disciplines. His work at the university looks to encourage greater dissemination of academic theory through professional practice, with the aim of improving the long-term impact that architects have on addressing some of the major issues we face as a society.

Abhinesh Prabhakaran is a doctoral researcher (PhD) in the Faculty of Environment and Technology at the University of the West of England (UWE), Bristol, UK. His research interests include Virtual Reality, Artificial Intelligence, Building Information Modelling (BIM), Construction Safety, Eye Tracking, and Visual Search Behaviour.

Thomas Price graduated with distinction from the MSc Project Management course at Leeds Beckett University, with vast experience within both construction and business change projects, with a passion for research into reconceptualising live projects to assess the potential of latent value.

Gav Roberts is a Disruption Officer at Halton housing. He works at the forefront of innovation, developing radical technology changes across a number of business areas in the social housing sector.Gavin now drives innovation strategy at Halton Housing by changing how people think, behave, do business, and learn using cutting-edge disruptive techniques.Gavin has a flare and passion for creative technologies and embraces innovation in all aspects of his life.

Jessica Schips graduated with distinction from the MSc Project Management course at Leeds Beckett University, after completing action research into independent planning and protest projects.

Shariful Shikder is involved in teaching and research activities related to sustainability and BIM in the School of Built Environment, Engineering and Computing at Leeds Beckett University. Before joining Leeds Beckett University, he worked in the UK construction industry and developed expertise in BIM and project management. Previously, Shariful also worked in the EPSRC-funded research projects at Loughborough University, where his research area included the innovative use of building simulation tools in delivering optimised facility design, environmental and energy performance of healthcare buildings, and sustainable design strategies to mitigate the impact of climate change. Shariful obtained his PhD from Loughborough University, which looked at the thermal performance of residential buildings accommodating older people.

Alan Simson is a landscape architect, an urban forester, and an urban designer. Currently, he is Emeritus Professor of Landscape Architecture + Urban Forestry at Leeds Beckett University. Prior to working in academia, he worked in the UK New

Towns, particularly Telford, where he ran the Urban Forestry Programme for over 10 years. He has also worked in private practices, including his own, both in the UK and abroad. He is involved in a number of overseas and national activities, including the European Forum on Urban Forestry and is Chair of the White Rose Community Forest.

John Smallwood is the Professor of Construction Management in the Department of Construction Management at Nelson Mandela University and the Principal of Construction Research Education and Training Enterprises (CREATE). Both his MSc and PhD (Construction Management) addressed construction health and safety (H&S). He has conducted extensive research and published in the areas of construction H&S, ergonomics, and occupational health (OH), but also in the areas of environment, health and well-being, primary health promotion, quality management, and risk management.

John P. Spillane is an experienced research-focused lecturer in Construction Management at the University of Limerick, with a demonstrated history of working in both higher education and the construction industry. As a Chartered Construction Manager, working within the remit of Digital Construction, Sustainability, Dispute Resolution, and Confined Site Construction, John is currently working on a number of research projects throughout Ireland and the UK, while also the Director of the Irish Construction Management Research Centre, supervising a number of PhD students and co-ordinating a number of industry-based knowledge transfer projects.

Ian C. Stewart is a Reader at the University of Manchester, Faculty of Science and Engineering. He is Programme Director of the MSc in Management of Projects and sits on the steering committee of the University College for Interdisciplinary Learning. He also teaches around business strategy, change and commercial management at AMBS, Warwick and EIGSI, La Rochelle. He is interested in the engagement of industry practitioners in project management higher education, the human costs of experiential learning in very large class situations, and the influence of higher education on graduate identity as project professionals.

Ali Murat Tanyer after graduating from the Department of Architecture of Middle East Technical University in 1997, started his master's studies at the same department. He completed his MSc degree in 1999 in the field of construction management. He gained his PhD degree from Salford University in 2005. His main areas of expertise are construction information technology and sustainability and how they support built environment. He has been a full-time instructor since 2006 and currently working as a full-time professor at METU.

David Thorpe is Associate Professor in Engineering and Technology Management at the University of Southern Queensland, which he joined following an extensive career as a practising civil engineer. He is a Fellow of the Institution of Engineers, Australia, in which he is a Chartered Civil Engineer and Engineering Executive.

David's career objective is to be a world leading professor in strategically focused sustainable engineering, construction, and technology management, a goal that is supported by his research interests in sustainable engineering and built environment management. He has authored approximately 100 publications, has supervised 12 Research and Higher Degree students to completion, and is a Fellow of the Leeds Sustainability Institute.

Joseph Thorp has recently graduated with a degree in Quantity Surveying from Leeds Beckett University and looking to build a career in Construction Industry.

Onur Behzat Tokdemir after graduating from the Civil Engineering Department of Middle East Technical University, gained his MSc and PhD degrees from the Civil Engineering Department of Illinois Institute of Technology. His main areas of expertise are construction and project management, lean construction, occupational health, sustainability, and green buildings. He has been a full-time instructor and currently working as a full-time professor at METU.

Greg Watts is the Director of the Quantity Surveying Programme and a lecturer at the University of Salford. He is also a Chartered Quantity Surveyor and prior to moving into academia gained extensive experience in the construction industry working for contractors delivering multi-million-pound projects. Greg's research interests include Social Value and Corporate Social Responsibility in the Construction Industry.

Henok Kebede Wolde holds an MSC in Electrical Engineering and Renewable Energy Systems from the University of Leeds which he pursued as Chevening scholar student. Henok has 7 years of industrial experience working as Energy Engineer with international telecommunication companies, including Huawei and Ericsson. He also has experience working as a research assistant and has authored and co-authored a number of conference publications.

Elliott Young is an Associate Architectural Technologist. He has recently graduated from Leeds Beckett University, studying a 4-year BSc Architectural Technology course where he achieved a first class honours degree. During his 4-year course at Leeds Beckett University, he took a placement year in industry. For this, he worked for an architectural practice, based in Huddersfield. The company focused on the development of non-traditional housing across the UK; from his experience during his placement year, he expanded on the practical knowledge and created the following research paper. He is currently working in a small practice expanding his practical knowledge and working towards becoming chartered through CIAT.

Part I
Social Value and Sustainability

Chapter 1
The Problems of Achieving Social Sustainability: A Cultural Shift

John Sturges and Christopher Gorse

1.1 Introduction

The importance of social sustainability has been emphasised in the 2019 SEEDS conference (Sturges, 2019). The only truly sustainable societies are those that have achieved social sustainability, whereby every member of the society subscribes to the way that it is organised. At the time when our thoughts on this paper commenced Autumn, 2019, the Extinction Rebellion protests were in progress in Europe, the USA and other various parts of the world (BBC, 2019). These protests were and continue to be based on the idea of directly influencing governments to take action on the current environmental crisis, particularly climate change. The premise of the protests would bring about media coverage, evoke public support and persuade governments and communities to adopt sustainability. Most movements, including appeals by icons, such as David Attenborough and Greta Thunberg, address those in leading position to change but also engage through public address whole communities. For such messages to resonate and result in action, all ages, sectors communities and counties will need to commit, a cultural shift is required.

Diamond (2006) has described how some societies that have been sustainable have achieved social sustainability usually by bottom-up approaches and in one case by a top-down approach (Tokugawa Japan). These societies were physically cut off from the rest of the world, or in the case of Japan, they were cut off by an edict of the ruling Shogun. However, the twenty-first century world is a globalised world, whereby all societies are linked by trade, transport and communication systems of various kinds. The links between our nations and international connectedness are being emphasised by the outbreak and global spread of the Corona 19 virus.

J. Sturges · C. Gorse (✉)
Leeds Sustainability Institute, Leeds Beckett University, Leeds, UK
e-mail: C.Gorse@leedsbeckett.ac.uk

© The Author(s), under exclusive license to Springer Nature Switzerland AG 2022
C. Gorse et al. (eds.), *Climate Emergency – Managing, Building, and Delivering the Sustainable Development Goals*, https://doi.org/10.1007/978-3-030-79450-7_1

Within 3 months of the virus being detected, it was considered a pandemic by the World Health Organization (Liu, et al., 2020; WHO, 2021). The international response to the pandemic may lead many to think that an international change to social sustainability is achievable in the face of the existential threat of climate change.

However, economies of the developed countries present the greatest influence on globalisation, and our international connectedness currently fails to offer the degree of social cohesion required to respond to sustainability crisis. Indeed, when addressing future sustainability, most policies give priority to economic sustainability before environmental sustainability and with little consideration for social sustainability (Toli & Murtagh, 2020). For the climate emergency to be acted on with sufficient haste commerce, governments and communities must act together or the impact will be felt.

The tipping point for climate change is rapidly approaching, and the loss of diversity in the ecosystem is so critical that humans will be witness to irreversible change in the next decade (Attenborough & Hughes, 2020). The recently reported loss of 219 billion tonnes of ice per year and sea levels rising considerably faster than previously predicted are set to displace millions of people, and, with the loss of the ice's cooling effect, a further acceleration to global warming is anticipated (Selley et al., 2021; Shepherd et al., 2020).

Environmentalist are increasing the pressure on governments, but so too are those industries likely to be adversely affected by the sustainability agenda and those changes proposed under the Sustainable Development Goals.

Governments usually wish to maintain the impression that they can solve all or most of society's problems, and so 'Extinction Rebellion' aims their protests at governments and also at The United Nations Assembly in New York. Unfortunately, they are not the only ones seeking to influence governments. The worlds of industry and big business also seek government influence via lobbying and via substantial contributions to political parties and campaign funds. Over the past 40 years or so, the number of lobbyists in Washington, USA, has increased by a factor of 10, and the money spent on lobbying has increased from millions to billions of dollars. Clearly this money is not donated altruistically, and the lobbyists are not in Washington for the 'high life'. Considerable efforts also go into producing misleading information about certain products where the producers have an interest in maintaining or increasing sale levels (Monbiot, 2007). Notable past examples include the lobbies for tobacco and for the production and use of chlorofluorocarbons (CFCs) as refrigerants.

Extinction Rebellion is a group of open, highly visible influencers, whereas the lobbyists operate covertly. Everyone is aware of Extinction Rebellion, whereas the lobbyists prefer to remain unseen. Money and campaign contributions are powerful influences, and while street protests have some effect, it appears to be limited to eliciting sympathetic utterances by some government officials. Extinction Rebellion has staged demonstrations in various capitals including London and at the United

Nations in New York. Are they making their case to the right people? Are there others who need to hear the message? If progress is to be achieved at the necessary scale and speed, then everyone needs to be involved. What are the barriers to getting the message and its urgency to everyone to get them on side?

1.2 Problems of Social Sustainability

The foregoing account illustrates both sides of the problem. On the one hand, experience teaches us that sustainability can only be achieved when everyone in society subscribes to its achievement. This means that everyone realises that sustainability is the only way forward long term and is willing to accept the various limitations that are imposed by it. On the other hand, an industrialised form of consumer capitalism is now widely perceived as the only way forward, even in what are or were communist regimes, China being a notable example. Furthermore, the West is wedded to democratic systems of the government, where representatives are usually elected by a simple majority vote. In some peoples' minds, this is the 'only' way to do things. People living under autocratic regimes often express an understandable longing for democratic government. However, democracy does have limitations. If we look again at those societies that achieved sustainability, none of them had a formal system of democratic government as we understand it. The bottom-up societies in Tikopia, Ladakh and New Guinea operated by consensus, and the top-down society in Japan followed the edicts of the Shogun, trusting him to make wise decisions in the interests of the common good of his people. For two and a half centuries, this system worked, and it only broke down under intense pressures, initially from outside Japan. These pressures eventually took root among part of the Japanese population who began to urge changes, so eventually the Shogun returned to the Emperor the power to rule in the 1868 Meiji Restoration. The requisite situation for sustainability had been lost once Japanese society became divided, showing once again how important social sustainability really is.

However, the problem goes deeper than this. The western world is wedded to notions of personal liberty and the assertion of individual rights with little or no limitations on an individual's choices. Politicians rarely mention individual responsibilities, a notable exception being John Kennedy's inaugural address in 1961, where he famously said: 'ask not what your country can do for you, ask rather what you can do for your country'. Europe, America and the western world hold to democracy and to the ideals of individual liberty, and while China does not, they both pursue consumer capitalism.

Two current news items illustrate the problems we face. Sales of electric cars are beginning to increase, but at the same time sales of less economical SUV types are increasing, and many car manufacturers say the production of SUV types amount to 40% of their output. Even the top-end luxury car makers such as Rolls Royce, Bentley and Aston Martin now market SUV models as part of their ranges.

Consumption of plastics by supermarkets has increased because consumers treat the heavier plastic bags for which they are charged in the same way that they treated the 'free' ones. If supermarkets were serious about this problem, the price of a plastic reusable bag would reflect its environmental cost if wasted. This would make plastic reusable bags relatively expensive and may induce consumers to change their behaviour, reusing rather than disposing of the bags.

The UK is increasingly experiencing problems with heavier rainfall and flooding, impacting heavily on communities across the country. Yet in towns and cities, people continue to pave over their gardens to make parking space for their ever-larger and more numerous vehicles. Hundreds of hectares of rain-absorbent land have been covered over in this way in the past two decades. The same amount of rainfall occurs but run-off is increased. This flooding may in large part be ascribed to climate change, since warm air holds more water vapour than cold air. Despite this fact, many people are willing to ascribe increased rainfall to 'freak weather' conditions. Why is this so?

Turning again to our western notions of democracy, people view it through 'rose-tinted' glasses. People are given a vote at election time and vote for their MP in the UK or their Congressman or Senator in the USA. However, in between elections, they can write to their MP over any matters of concern, and most MPs do a good job of representing their constituents. In between elections, individual MPs can exert some influence on their behalf. This is not the only influence brought to bear on governments in the UK or in the USA. The ease with which senior figures move between the private and public sectors is an indication of close relationships that exist between the private sector, big business and government. Reference has already been made to the number of lobbyists in Washington and to the sums of money expended on influence.

The ideals of western democracy – life, liberty and the pursuit of happiness – have been given, perhaps most memorably, by some former US presidents, especially Jefferson, Lincoln and Kennedy. Words of high-flown rhetoric that they uttered are remembered by people all over the world. In his Gettysburg address, President Lincoln paid tribute to those who had given their lives in the cause of the Union. The civil war had another 2 years to run, but he encouraged his listeners to honour the dead by re-dedicating themselves to the task of preserving the Union, with its ideals of equality and freedom. He concluded the last sentence of his address with the following words: 'and that the government of the people, by the people, for the people shall not perish from the earth'.

The words *for the people, by the people* imply that a truly democratic form of government is the future for the USA after the war and henceforth. No reference is made to very large campaign donations or support for political parties now made. President Lincoln will not have foreseen how his country's government would evolve in the twentieth century. The industrial and 'big business' sector had not developed into the major centre of power that it now is. President Eisenhower's farewell speech was also memorable for his warning about the power of 'the military industrial complex'. In the USA, since the closing of World War II, the military has also become a huge centre of power, with the huge industry sectors of aerospace, electronics, weapons, etc. now closely involved in Washington. The need for a highly educated population to staff expanding government and industry has led to the growth in importance and size of academia in America, Europe and beyond. Another group with influence that cannot be ignored is 'the media', the press, the broadcasters and the internet.

In many fields of human society, what is apparent is not always the complete picture. Business organizations have management structures, but sometimes the person with the most influence is not the one in the top seat. American management theorists saw the mismatch between the formal organization chart and called this 'the informal management structure'. Big business exerts influence not only by financial contributions but also by issuing 'spoiling publicity' designed to mislead, as referred to above (Monbiot, 2007). When so much power and influence is in such a small number of hands, how easy is it to operate in a truly democratic way? People eagerly accept and take up all the latest products put out by the industrial economy, but they never ask about the price. 'I can order it on-line', 'I can have it delivered to the office', 'I can get my hands on it tomorrow', etc. No paperwork, no trips to the shops and so on. Some complain about the vast sums of money made by the entrepreneurs who market these things. These superrich people dispose of greater wealth than some of the poorest nations of the world. If California was an independent country, it would be among the ten richest nations on Earth, and yet there are thousands of homeless people in Los Angeles. Such social inequality is the mark of a nonsustainable form of society.

The streets in towns and cities are full of white delivery vans, and the townscape now consists of fast-food outlets, betting shops, estate agents, charity shops, etc. This is part of the visible price. The invisible price is the coming into being of huge data centres to serve the ever-expanding internet. Each centre is filled with file servers and consumes as much power as a city the size of large UK cities. Is it to be wondered at that the CO_2 emissions continue to climb? In the meantime, everything is gradually being connected to the internet – the 'internet of things'. If these trends continue, a future power outage could bring civilisation to a halt.

It is interesting to note that, as lockdown measures were taken and movement restrictions introduced to reduce the Corona 19 virus transmissions, global daily energy use drastically changed (Le Quere et al., 2020; Liu et al., 2020) and emissions from personal travel and pollution in the major cities fell (Gorse & Scott, 2021). However, this was countered by domestic energy and media use as home-based activities increased the internet's carbon footprint (Obringer et al., 2021). We need to recognise and be responsible for our actions.

Social sustainability requires everyone in society to share the ideals and to work together to achieve them. We now have a good understanding of how we have brought the world to its present unsustainable (as far as human existence is concerned) state. We have 'globalised' the problems, but we have not globalised this knowledge, and we have certainly given no thought as to how we might globalise any solutions that we may devise.

This discussion has shown that when considering political influence, it is not a duopoly of government and people (electorate) but that other centres of power have emerged in the modern world. These are the industry sectors including manufacturing, services and agriculture, the military and academia. The media play an important role in providing information transmission between the centres of power. Millions of people switch on TV for the news every day from CNN, BBC, etc., and the content of the broadcast depends on the news editors. The news that we receive is what the broadcast editor deems to be most newsworthy, not always what is really news or all that has occurred.

The way that societies have developed in Europe, America and the wider world illustrates how we have moved away from social sustainability and raised the barriers to its achievement. The largest international industrial organizations have grown to the point where their size and power rivals or even exceeds that of many national governments.

1.3 Likely Barriers to Achieving Social Sustainability

The foregoing sections have attempted to set out the scale of the problems we face. We live in a 'globalised' world, i.e. a world where the same social forces are at work in many parts of the world, a world with a population of over 7.5 billion. One hundred years ago, people in remote parts of Indonesia lived in complete ignorance of conditions in Europe or America. Because of the internet and the ubiquity of the mobile phone, this is no longer the case, and this has undoubtedly helped to drive the great migration crisis being felt in many places throughout the world today.

One hundred years ago, there were societies living in some remote parts of the world that lived sustainably (Diamond, 2006; Norberg-Hodge, 2000). These societies lived apparently happy lives but without the physical trappings of the modern world. Furthermore, they had stable population numbers, because everyone in them knew that there was an upper limit to how much food they could produce and that their existence and survival depended upon not generating too many mouths to feed. Interestingly, when they did begin to interact with and to absorb western ways, they began to consume more energy and to produce more children. A population explosion inevitably followed the adoption of western ways. In the west, we recognise population pressure as a big factor in bringing about adverse effects in our world, but we do not recognise that our adoption of consumer capitalism has driven the population explosion that is now seen as a problem.

The achievement of sustainability must be an objective shared by everyone, or it will not be attainable. We have seen that, besides the government and the people, other centres of power have emerged in our highly developed societies. These centres of power are interconnected via the media, but there is not complete transparency, which is yet another difficulty.

1.4 The Importance of Social Sustainability

We are led to believe that social sustainability occurs when relationships come together to sustain an ecosystem generating the capacity to support healthy communities now and for future generations (Barron & Gunlett, 2002). At its simplest form, social sustainability is the relational effort to ensure the health and wellbeing of people now and in the future. Although often identified as one of the three pillars of sustainability, within the field of sustainability, the 'social' element receives least attention (Spangenberg, 2006; McKenzie, 2004, p. 14) and is largely overlooked in debates and political discourse (Woodcraft, 2015). The concept of social sustainability is diverse, possibly the most difficult to measure, and receives less political attention when compared with economic and environmental sustainability (Toli & Murtagh, 2020). Notwithstanding this imposition, respect for our fellow citizens and their wellbeing does exist. The recent COVID-19 pandemic demonstrates the scale of action nations are prepared to make but also exposes the challenges of introducing consistent measure across a nation's action. While restrictions were found to be effective in reducing transmission, measure taken and response rates varied as countries attempted to balance social and economic impact (Alfano & Ercolano, 2020; IMF, 2021). However, as the reality of the pandemic was made evident, country-wide measures were introduced to protect the public, and with each national restriction, a sharp economic decline was experienced (IMF, 2021).

The pandemic has demonstrated that nations are prepared to take action necessary to protect their fellow citizens. Notwithstanding the action taken across the globe, the social unrest and resistance experienced in some countries show just how difficult it is to change social and cultural behaviour as economies suffer. The projection of climate change having a devastating impact on the wellbeing of fellow citizens 'in years to come' appears insufficiently imminent for a consistent response at this moment in time. Unfortunately, our anthropogenic actions are having a devastating impact on the ecosystem and ultimately the niche climate conditions upon which humans rely to exist (Xu et al., 2020). Social sustainability and our ability to protect humans and other species now have to become a central tenet to reduce the impact of climate change and ensure human survival.

Humans clearly evolved and outdid most other living animals as a result of their ability to communicate and work together. As humans, we are recognised as the life form that has exploited the ability to work together. Human relationship skills have been used to build resource, sustain health and wellbeing and protect ourselves and offspring from threats. While in the past humans have done this extremely successfully, recent events may raise questions about our ability to perform social sustainability at a global scale.

One of the key questions raised is can social sustainability exist within the commercial world? Allmahmound and Doloi (2015) argue that sustainability outcomes are best achieved when taking account the satisfaction of stakeholders. Within the global context, the stakeholder perspective seems short-sighted. However, the aim is to enhance the importance of the social context and overall benefits. There is an argument that as we continue to develop economically and indeed sustainably, we neglect the life—the social dimension within the context of sustainability has been treated with apprehension (Hill & Bowen, 1997; Edum-Fotwe & Price, 2009).

While Ducker would have us believe that 'culture eats strategy for breakfast', we have yet to experience the cultural change that can engender social sustainability.

1.5 Concluding Remarks

One proposed solution to the climate crisis was that of geoengineering the planet (Royal Society, 2009). The Stockholm Resilience Centre Report and the Royal Society's geoengineering report were both published in 2009, and a decade has since elapsed without international concerted action. This clearly illustrates the lack of united political will to agree a common policy. This is frustrating, because we understand the essence of the problem and we have many technical solutions to hand, and while nothing on the right scale is done, the problem becomes more urgently in need of solution. Unfortunately, it remains true that until everyone sees the lack of sustainable operation as an existential threat, many people will disregard it as mere background noise. At all levels of society worldwide, people are focused on the need for economic growth. Those societies that were sustainable knew very clearly that they faced an existential threat if they did not maintain a stable population, and every member of these societies was well aware of this fact. The challenge we face today is that of presenting the information to everyone in a way that puts emphasis on the fact that we face an existential threat.

A recent genre of book has come into being dealing with 'sustainable companies', of which the one by Laszlo (2003) is an early example. The book contains some very interesting case studies of companies that have made great efforts to reduce their environmental impacts while remaining profitable enterprises. Laszlo rightly differentiates between 'shareholder value' and 'stakeholder value', and he highlights various methodologies for getting company workforces onside in the quest for improved company performance—in other words, how to achieve social sustainability within the company. The book is written with a deep knowledge of American management theory but with no knowledge of the physical and biological sciences that underpin how our planet functions. Furthermore, the author mistakes environmental impact mitigation for sustainability, and sustainability is much more than environmental protection. Without an appreciation of system thermodynamics and evolutionary biology, authors are writing blind to the problem. To gain an understanding of the problems that we face requires a good deal of effort and a degree of scientific knowledge that many do not possess. Many people are familiar with the first law of thermodynamics, but very few understand the second law and especially its implications, and this despite that fact that the consequences of ignoring the second law are all around us. It is true that thermodynamics is a recondite subject, but that does not diminish its importance. Later works in this same genre of management and sustainability (e.g. Zokaei et al., 2013) show that management thinkers still do not understand the ramifications of system thermodynamics. They still assume that impact mitigation is the same thing as sustainability, and like Laszlo (2003), they offer yet more interesting case studies and more well-thought-out methods for promoting social sustainability within organisations. However, the whole of society must be engaged, and these authors have nothing to say about this.

A point made by another author, Speth (2008), is that we face 'apathy, ignorance and greed', and not just CO_2 emissions, global warming, biodiversity loss, etc. This is not universally true, but it certainly is true for large sections of society. Most people are too focused on the problems of their day-to-day existence to become concerned about a problem that they think belongs to someone else. To gain some understanding of the present state of our world takes considerable effort, but most people want to carry on as normal and adopt a head-in-the-sand approach to the unmistakable and mounting evidence of the reality of global warming and its effects. It is difficult to avoid the conclusion that the world is so committed to the status quo and business as usual that governments national and local, industry, academia, the media, etc. are not prepared to begin thinking seriously about confronting the changes that need to be made.

All the strategies discussed, apart from an immediate cessation of fossil fuel burning, will take time to be developed to an effective level. This is time that we do not have.

References

Alfano, V., & Ercolano, S. (2020). The efficacy of lockdown against COVID-19: A cross-country panel analysis. *Applied Health Economics and Health Policy, 18*(4), 509–517. https://doi.org/10.1007/s40258-020-00596-3.

Allmahmound, E., & Doloi, H. K. (2015). Assessment of social sustainability in construction projects using social network analysis. *Facilities, 33*(3–4), 152–176.

Attenborough, D., & Hughes, J. (2020). *A life on our planet: My witness statement and a vision for the future.* Ebury Publishing.

Barron, L., & Gunlett, E. (2002). *wacoss housing and sustainable communities indicators project.* The Regional Institute Online Publishing. http://www.regional.org.au/au/soc/2002/4/barron_gauntlett.htm

BBC (2019). Extinction rebellion protests: What happened? BBC News, 25th April. https://www.bbc.co.uk/news/uk-england-48051776

Diamond, J. (2006). *Collapse. How societies choose to fail or succeed.* Penguin Books.

Edum-Fotwe, F. T., & Price, A. D. F. (2009). A social ontology for appraising sustainability of construction projects and developments. *International Journal of Project Management, 27*(4), 313–322.

Gorse, C., & Scott, L. (2021). Smart integrated and sustainable development in changing times. *Built Environment, Frontiers.* https://www.frontiersin.org/articles/10.3389/fbuil.2021.670559/full

Hill, R. C., & Bowen, P. A. (1997). Sustainable construction: Principles and a framework for attainment. *Construction Management and Economics, 15*(3), 223–239.

IMF. (2021). *Policy tracker.* International Monetary Fund. https://www.imf.org/en/Topics/imf-and-covid19/Policy-Responses-to-COVID-19

Laszlo, C. (2003). *The sustainable company. How to create lasting value through social and environmental performance.* Island Press.

Le Quere, C., Jackson, R. B., Jones, M. W., Smith, A. J. P., Abernethy, S., Andrew, R. M., De-Glo, A. J., Willis, D. R., Shan, D. R., Canadell, J. G., Friedlinsgstein, P., Creutzig, F., & Peters, G. P. (2020). Temporary reduction in daily global CO_2 emissions during the COVID-19 forced confinement. *Nature, Climate Change, 10*, 647–653.

Liu, Z., Ciais, P., Deng, Z., et al. (2020). Near-real-time monitoring of global CO_2 emissions reveals the effects of the COVID-19 pandemic. *Nature Communications, 11*, 5172.

Mckenzie, S. (2004). *Social sustainability: Towards some definitions.* Hawke Research Institute Working Paper Series, No. 27, p. 29.

Monbiot, G. (2007). *How to stop the planet burning.* Penguin Books.

Norberg-Hodge, H. (2000). *Ancient futures.* Revised Edition. Rider, Ebury Press.

Obringer, R., Rachunok, B., Maia-Silva, D., Arbabzadeh, M., Nateghi, R., & Mandani, K. (2021). The overlooked environmental footprint of increasing internet use. *Resources, Conservation and Recycling, 167*, 105389.

Royal Society. (2009). *Geoengineering the climate, governance and uncertainty.* The Royal Society.

Selley, H. L., et al. (2021). Widespread increase in dynamic imbalance in the Getz region of Antarctica from 1994 to 2018. *Nature Communications, 12*, 1133. https://doi.org/10.1038/s41467-021-21321-1.

Shepherd, A., et al. (2020). Mass balance of the Greenland Ice Sheet from 1992 to 2018. *Nature, 579*, 233–239. https://doi.org/10.1038/s41586-019-1855-2.

Spangenberg, J. H. (2006). Assessing social sustainability: Social sustainability and its multicriteria assessment in a sustainability scenario for Germany. *International Journal of Innovation and Sustainable Development, 1*(4), 318–348. https://www.researchgate.net/publication/233990880_Assessing_social_sustainability_Social_sustainability_and_its_multicriteria_assessment_in_a_sustainability_scenario_for_Germany.

Speth, J. G. (2008). *The bridge at the edge of the world. Capitalism, the environment and crossing from crisis to sustainability.* Yale University Press.

Sturges, J. L. (2019). The significance of social sustainability. In: L. Scott, & C. Gorse (Eds.), *Proceedings 5th annual SEEDS conference*, 11–12 September 2019.

Toli, A. M., & Murtagh, N. (2020). The concept of sustainability in smart city definitions. *Built Environment, Frontiers.* https://www.frontiersin.org/articles/10.3389/fbuil.2020.00077/full

Woodcraft, S. (2015). Understanding and measuring social sustainability. *Journal of Urban Regeneration and Renewal, 8*(2), 133–144. https://www.researchgate.net/publication/286595877_Understanding_and_measuring_social_sustainability.

WHO (2021). Timeline: WHO's COVID-19 response, World Health Organization. https://www.who.int/emergencies/diseases/novel-coronavirus-2019/interactive-timeline#!

Xu, C., Kohler, T. A., Lenton, T. A., Svenning, J., & Scheffer, M. (2020). Future of the human climate niche. *Proceedings of the National Academy of Sciences, 117*(21), 11350–11355. https://doi.org/10.1073/pnas.1910114117.

Zokaei, K., Lovins, H., Wood, A., & Hines, P. (2013). *Creating a lean and green business system. Techniques for improving profits and sustainability.* CRC Press.

Chapter 2
A Hammer Only Sees Nails: Why Construction Contractors Monetise Social Value

Greg Watts and Anthony Higham

2.1 Introduction

Construction is a significant industry for the UK economy and represents approximately 9% of economic output (BEIS, 2018). The public sector represents around 26% of this workload and offers a degree of certainty in uncertain economic times. For any construction contractor wanting to secure public sector projects, the Social Value Act (2012) is now used throughout public procurement. The Act can be described as the step forward on the organisational CSR journey that reappeared in the 1950s. This places obligations on private sector companies to adopt socially responsible practices and participate in activities that increase social value. Through placing a legal obligation on the public sector, the Social Value Act formalised and legitimised the use of social value as a procurement criterion to match that of cost, time and quality; contracts would no longer be let on a lowest upfront cost basis. This places a requirement on construction contractors to measure and communicate their social value in order to be successful in public sector procurement.

As social value is arguably a subjective concept, measurement and communication could prove difficult. To circumvent any potential problems that may arise because of this subjectivity, contractors adopt objective measurement methods. Why contractors engage with SV measurement has been explored, but how this engagement occurs, and the ramifications of any measurement practices adopted, is yet to be fully understood. Through interviews with ten leading UK construction contractors, this paper seeks to address this gap in current knowledge.

G. Watts (✉) · A. Higham
School of Science, Engineering and Environment, University of Salford, Salford, UK
e-mail: G.N.Watts@salford.ac.uk

© The Author(s), under exclusive license to Springer Nature Switzerland AG 2022
C. Gorse et al. (eds.), *Climate Emergency – Managing, Building, and Delivering the Sustainable Development Goals*, https://doi.org/10.1007/978-3-030-79450-7_2

2.2 From CSR to SV

Amongst the plethora of definitions published, the 'CSR pyramid' first proposed by Carroll in 1979, then further developed in 1991, has served as somewhat of a seminal base for CSR research with its wide-ranging encompassing approach to the subject. Carroll (1991) argues that a company's CSR journey starts with their economic responsibilities and then moves to their legal responsibilities and then ethical responsibilities before finally arriving (at the top of the pyramid) at philanthropic responsibility. This approach to defining CSR served as an all-encompassing umbrella concept under which numerous interpretations and approaches exist (Barthorpe, 2010).

The book *Social Responsibilities of the Businessman* by Bowen (1953) is credited with bringing to the forefront of modern business actions the age-old concept of general goodwill by those with the ability to make a positive difference. In the book, it is wealthy business owners who are challenged to increase their social responsibility due to increasing post-war prosperity (Bowen, 1953). The concept then grew throughout the 1960s and 1970s with the advent of social movements and the increased realisation that business behaviour could be shaped by stakeholder mobilisation (Carroll, 1999). This journey then took CSR and wider business responsibility through an evolution of sorts from social responsibility to more environmentally focussed strategies. Perhaps fuel for those who link business intentions with only superficial action to illustrate their responsibility for work winning purposes, a decline of CSR activity was witnessed in the 1980s which has been linked to the economic issues of the time. However, in defence of business, it can be argued that without economic certainty a company may not be able to operate in the long term let alone commit resources to socially responsible actions where they are not directly linked to business operations. This perhaps adds further credibility to the CSR pyramid definition proposed by Carroll (1991) as CSR is a philanthropic business priority only after economic, legal and ethical targets have been achieved. The prosperous 1990s brought a return of CSR to the forefront of business activity but with a continued evolution of understanding and expectation. As business wealth and stakeholder involvement and demands increased over the millennium and into the last decade the focus of organisational CSR expanded (or perhaps arguably went full circle) to include all manner of social, economic and environmental factors that business may be directly or indirectly involved in, or may be completely separate but judged of such societal importance, there is an expectation upon all businesses to play their part. A recent industrially focussed global business survey confirms that CSR is now embraced by the majority of large-scale organisations as a central part of their business identity (KPMG, 2017). Whilst the motivations of the companies involved were not explored in great detail, it can be argued that their increasing adoption of socially responsible practices, in the UK at least, can be in part attributed to legislation increasing socially focussed criterion in public sector procurement practices.

One example of such legislation is The Public Services (Social Value) Act (2012) (SVA). The SVA governs public body procurement behaviour by legitimising the use of criterion in procurement other than purely financial (Watts et al., 2019a). The SVA allows public bodies to take additional social value achieved (as put forward by the tenderer) into consideration when awarding contracts so that the successful contractor is not necessarily the one who put forward the lowest immediate cost (Loosemore & Higgon, 2016).

2.3 The Construction Industry

The construction industry includes the design, construction, maintenance and demolition of built assets, engineering and infrastructure works. According to a UK Government briefing document in 2018, the construction industry contributed £117 billion to the UK economy, some 6% of the total, and is responsible for over 2.4 million jobs across 343,000 different businesses (Rhodes, 2019). Construction therefore has a significant impact upon the UK economic output. The public sector accounts for approximately 26% of UK construction work, and historic economic data has shown that public sector workload remains fairly buoyant during times of economic uncertainty (Rhodes, 2015).

This could help explain why contractors engage with the public sector despite the additional requirements imposed by the SVA. Management philanthropic values and succumbing to the pressures imposed by wider stakeholder expectations cannot be underestimated as drivers of social value behaviour. Nevertheless, the SVA plays a key role in imposing social value requirements upon contractors who engage with public sector work. The introduction of the SVA therefore ultimately placed an obligation on any company wanting to win public sector work to be able to effectively measure and communicate their social value (Watson et al., 2016).

2.4 Measuring Social Value: The Battle Between Subjectivity and Objectivity

In order to successfully communicate such SV, an agreement must be reached on a definition, at a minimum between the company and client but also ideally between a wide cohort of stakeholders (Watts et al., 2019a). However, with no widely agreed definition between multiple stakeholders who each hold potentially unique interpretations, accurate measurement and clear communication are difficult to achieve (Loosemore & Higgon, 2016). Attempts to effectively communicate with wider cohorts simultaneously are arguably the problem at the heart of the CSR and SV debates, in that these concepts mean different things to different people (Watts et al., 2019a). Approaches can be categorised as objective or subjective. Objective is where a fact-based natural science approach is adopted that implies a phenomenon has an existence independent of social actors (Robson & McCartan, 2017). Subjective is where meanings are socially constructed and therefore subject to change (Bryman, 2016). Objective attempts to define CSR and SV are adopted by some parties but have ultimately led to competing and often conflicting definitions proposed that serve to further exacerbate attempts to reach a consensus amongst stakeholders (Zhao et al., 2012). Subjective attempts to define CSR allow each stakeholder when faced with the same communication to arrive at their own interpretation (Griffith, 2011). However, one problem that exists with such subjectivity is that it is open to potential abuse by some companies who mask their failure to effectively engage with CSR and SV with ambiguous and opaque terminology (Watts et al., 2019b). Whilst numerous different methods of measuring SV exist (Wood & Leighton, 2010; Higham et al., 2018), many that are currently adopted by construction contractors objectively quantify SV so that it can be easily communicated as part of tender returns. Indeed, it has been argued that social value has objective requirements in the need to be measured, communicated and widely understood (Loosemore & Higgon, 2016)—a perspective that has been largely perpetuated on the view that social value is delivered through "investments that intentionally target specific social objectives along with a financial return and measure the achievement of both" (Social Impact Investment Taskforce, 2014, p. 1).

Literature is dominated by an array of tools, metrics, frameworks and models that have been developed with the sole aim of predicting and measuring social value attainment. Proponents of these techniques such as Ding (2008), Carter and Fortune (2007), Rees (2009) and Higham et al. (2018) argue that such evaluation frameworks provide fundamental building blocks for comprehensive change, by providing practical, transparent and simple to understand criteria to which the industry can respond in manageable steps, thereby empowering construction professionals to think about sustainability in an experiential way, with the safety net of expert guidance, checks and balances (Schweber, 2013). Yet Haapio and Viitaniemi (2008), Ding (2008) and more recently Higham et al. (2018) have questioned the validity of monetising sustainability, a theoretical construct far removed from the operation of the market mechanism. At the core of their objection is the assertion that monetary units are likely to limit the validity of any sustainability evaluation produced.

The most contentious issue in the adoption of such objective frameworks is the quantification and monetisation of intangible social outcomes using financial proxies (Arudson et al., 2013; Krley et al., 2013) which can lead analysts to take some extremely imaginative and adventurous pathways when appraising social return (Krley et al., 2013) or lead contractors to apply retrospective social value justifications to achieve client requirements (Russel, 2013). Yet it is this objective need to quantify such a subjective concept that perpetuates the tension at the heart of SV debates. Contractors need to measure and communicate the subjectivity of SV and arguably choose an objective, monetary-derived method by which to do this. Although this decision in itself is presumed yet not explored in any great detail in the literature, the ramifications of this process are also not fully understood. This paper seeks to explore how and why contractors engage with SV and understand any impact objectively measuring and communicating SV has on the concept's subjectivity.

2.5 Research Methodology

Ontology is questioning the nature of reality and consists of two dominant ontological positions: objectivism and constructivism (Bryman, 2016). As objectivism derives from the fact-based natural sciences, in the case of social value, this approach is ultimately of a positivist epistemological position and therefore requires to be expressed through quantitative data, i.e. many of the measurement tools currently used in the construction industry. Constructivism refers to a belief that meanings are socially constructed and agreed, and epistemologically is of an interpretivist position. This is where meanings are subjective and so best expressed through qualitative data. Social value measurements ultimately find qualitative data difficult to quantify and so tend towards a quantitative approach.

The packaging of social value as quantitative data driven from an objectivist standpoint by many social value measurement tools adopted in the UK construction industry continues to dominate practice. This paper, however, subscribes to the argument that social value is best understood from a constructivist viewpoint. Indeed, it is through this approach that the views, perceptions, actions and motivations of construction contractors can be best understood (Cresswell, 2013). Therefore, qualitative data was gained from semi-structured interviews due to their ability to explore a depth of participant understanding and allowed topics to be built upon and interesting avenues that arose during the course of the interview to be pursued further (Byrne, 2012).

Ten semi-structured interviews were conducted with individuals involved in work winning from ten different UK construction contractors. The websites of the top 20 construction contractors by turnover in 2019 were viewed, and introductory emails were sent to individuals identified online as being appropriate to the research. Ten positive responses were received. This included three procurement and work winning directors, four bid managers and three senior estimators. The interviews were then conducted over the phone lasting between 45 and 90 min. All interviews were recorded and then transcribed, and a process of narrative analysis was conducted. Narrative analysis involves data in the form of stories. It is the process of summarising participant responses to reveal their deeper understandings without reducing responses to quantitative variables (Loosemore & Bridgeman, 2018). Such stories can be compared or even grouped together to reveal both similarities and differences of opinions and interpretations and are increasingly popular as a social science analysis method (Griffin & May, 2012).

2.6 Findings and Discussion

Analysis of the interviews revealed many consistent findings across all contractors. Firstly, reinforcing trends identified in the industrial survey conducted by KPMG (2017), all interviewees argued that CSR and SV are embraced as a core part of all contractors' business models. There was also a general consensus across all interviewees that they have increased their social value focus due to the requirements of the SVA. However, all contractors also stipulated that even without the SVA in place during procurement they would still continue with their social value behaviours due to management values. Yet all contractors interviewed also won an element of their annual work from the public sector so this research could not determine the validity of this further. Although this area wasn't the focus of this paper, it was an interesting insight and one that is worth further exploration in future research.

When the topic of social value measurement was discussed, all contractors were involved with measurement to different degrees. Again, there was a consensus of all interviewees which revealed that measurement of social value was utilised to support and enhance social value communications, supporting findings in the literature that social value has objective requirements and needs to be measured and communicated (Loosemore & Higgon, 2016). However, differences in understandings were witnessed in each contractor's perceptions of social value measurement. Whilst each contractor discussed terms and practices that overlapped in places, they all had somewhat different perceptions of social value that were driven in part from the needs and perceptions of their numerous clients.

In total, six different types of social value measurement tools were utilised by the ten contractors, with all but one using a recognised third-party measurement tool. This illustrates how there appears to be no single market leading social value measurement method, and with each method communicating social value in a different way, this will arguably contribute to the difficulties and confusions that persist in reaching an agreed definition of social value. However, all the measurement tools currently in use attributed monetary metrics to social value. All the measurement tools were of an objective and qualitative nature and so arguably reduced social value down to the figures. Interestingly there was no consistency reported across the six different measurement tools, even when measuring the same social value behaviour. For instance, one contractor reported using two different social value measurement tools across their business operations and adopting a 'pick and mix' approach when it came time to compiling tenders for procurement or even when reporting on progress to existing clients. The results of whichever tool had the highest monetary output were used, even if this meant using one tool to measure one practice and a different tool to measure a different practice. The interviews explored this further and revealed the contractor saw no contradiction over the legitimacy of either social value measurement tool. It appears the tools were not questioned or rigorously explored, simply used and exploited for the benefits they brought.

Analysis of the interviews also revealed that seven of the contractors interviewed felt the social value measurement method they used fully measured and communicated social value. There was a consensus across these seven contractors that social value could be fully measured, and they saw no issues in the reduction of benefits to financial metrics. They were all of the opinion that social value was objective and, once agreed between stakeholders, could be easily measured and communicated using financial metrics. This is despite arguments in the literature that social value means different things to different people (Watts et al., 2019a) and a lack of widely agreed definitions results in social value being difficult to measure and communicate (Loosemore & Higgon, 2016). The three remaining contractors also all believed that social value was objective and that, despite different stakeholder understandings and perceptions, agreements could be reached as to what social value is, and then measurement proxies should be used. Two contractors felt that the majority of social value could be captured in some form of financial metrics but admitted there was some elements of social value that arose from the practices they undertook which could not be accurately captured in monetary terms—but other metrics could be used—the tools to measure such social value just were not available yet. These two contractors believed such social value was impossible to accurately capture at present and so made no effort to do so. Only one contractor interviewed believed there was a need to measure social value using both financial and nonfinancial metrics, and they were currently considering how best to do this but had no solution at the time of interviewing. All contractors, therefore, only participated in social value practices they were able to accurately measure, and that resulted in a set of financial variables that could be easily communicated to numerous stakeholders. Any social value practices that couldn't be easily measured using currently adopted measurement tools were likely to be ignored or marginalised, with focus shown to social value practices that easily leant themselves to financial measurement.

2.7 Conclusion

Construction contractors engage with social value measurement in order to be able to communicate their social value practices and results to stakeholders such as public sector clients, as this will increase the likelihood of successful procurement. However, social value is a subjective concept and so not one that lends itself easily to measurement. Nevertheless, contractors are increasingly adopting measurement tools that take an objective approach and reduce social value to monetary figures. Through interviews with ten UK construction contractors, this paper reveals that the majority of contractors interviewed see social value only as that which can be measured with financial metrics; it seems a hammer can only see nails. Whilst some contractors do see the subjective nature of social value posing a problem for measurement in that every aspect cannot be easily measured with financial metrics, at present amongst the vast majority of contractors, no effort is being made to measure these practices. In fact, the social value practices that do not result in easy-to-measure-and-communicate financial metrics are being largely ignored. The findings of this paper reveal that social value practices which result in a higher amount of social value generated are potentially being ignored for practices which are easier to measure and communicate. Such findings contribute to understanding the ramifications of contractor decisions to measure and communicate social value in objective terms. In practice this may mean public sector clients need to consider the procurement requirements they are imposing on contractors in order to achieve the highest amount of social value possible and not just that which can be easily condensed to financial terms. Further research could be conducted around understanding what other drivers are behind contractor reporting practices, and also further research could be aimed at establishing how social value practices that do not lend themselves to be financial metrics can be effectively measured and communicated.

References

Arudson, M., Lyon, F., McKay, S., & Moro, D. (2013). Valuing social value? The nature and controversies of measuring social return on investment. *Voluntary Sector Review, 4*(1), 3–18.

Barthorpe, S. (2010). Implementing corporate social responsibility in the UK construction industry. *Property Management, 28*(1), 4–17.

BEIS. (2018). *Industrial strategy - Construction sector deal.* https://assets.publishing.service.gov.uk/government/uploads/system/uploads/attachment_data/file/731871/construction-sector-deal-print-single.pdf

Bowen, H. (1953). *Social responsibilities of the businessman.* Harper & Row.

Bryman, A. (2016). *Social research methods* (5th ed.). Oxford University Press.

Byrne, B. (2012). Qualitative interviewing. In C. Seale (Ed.), *Researching society and culture* (3rd ed.). Sage.

Carroll, A. B. (1991). The pyramid of corporate social responsibility: Toward the moral management of organizational stakeholders. *Business Horizons, 4*, 39–48.

Carroll, A. (1999). CSR: Evolution of a definitional construct. *Business & Society, 38*, 268–295.

Carter, K., & Fortune, C. (2007). Sustainable development policy perceptions and practice in the UK social housing sector. *Construction Management and Economics, 25*(4), 399–408.

Cresswell, J. (2013). *Research design: Qualitative, quantitative, and mixed methods approaches.* Sage.

Ding, G. K. C. (2008). Sustainable construction–The role of environmental assessment tools. *Journal of Environmental Management, 86*, 451–464.

Griffin, A., & May, V. (2012). Narrative analysis and interpretative phenomenological analysis. In C. Seale (Ed.), *Researching society and culture* (3rd ed.). Sage.

Griffith, A. (2011). Fulfilling contractors' corporate social responsibilities using standards-based management systems. *International Journal of Construction Management, 11*(2), 37–47.

Haapio, A., & Viitaniemi, P. (2008). A critical review of building environmental assessment tools. *Environmental Impact Assessment Review., 28*(7), 469–482.

Higham, A. P., Barlow, C., Bichard, E., & Richards, A. (2018). Valuing sustainable change in the built environment: Using SuROI to appraise built environment projects. *Journal of Facilities Management., 16*(3), 315–355.

Krley, G., Münscher, R., & Mülbert, K. (2013). *Social return on investment (SROI): State-of the-art perspectives.* Universtat Heidelberg.

KPMG. (2017). *The road ahead; The KPMG survey of corporate responsibility reporting 2017.* KPMG Global Sustainability Services 2017.

Loosemore, M., & Bridgeman, J. (2018). The social impact of construction industry schools-based corporate volunteering. *Construction Management and Economics, 36*(5), 243–258.

Loosemore, M., & Higgon, D. (2016). *Social enterprise in the construction industry*. Routledge.

Rees, W. E. (2009). The ecological crisis and self delusion: Implications for the building sector. *Building Research and Information, 37*(3), 300–311.

Rhodes, C. (2015). *Construction industry: Statistics and policy*. Nr 01432. House of Commons Library. www.parliament.uk/briefing-papers/sn01432.pdf

Rhodes, C. (2019). *Construction industry: Statistics and policy*. Nr 01432. House of Commons Library. https://commonslibrary.parliament.uk/research-briefings/sn01432/

Robson, C., & McCartan, K. (2017). *Real world research* (4th ed.). Wiley.

Russel, S. (2013). *Journey to impact: A practitioner perspective on measuring social impact*. Midland Heart.

Schweber, L. (2013). The effect of BREEAM on clients and construction professionals. *Building Research and Information., 41*(2), 129–145.

Social Impact Investment Taskforce. (2014) *Impact investment: The invisible heart of markets* [Online]. http://www.socialimpactinvestment.org/reports/Impact%20Investment%20Report%20FINAL[3].pdf

Watson, K. J., Evans, J., Karvonen, A., & Whitley, T. (2016). Capturing the social value of buildings: The promise of Social Return on Investment (SROI). *Building and Environment, 103*, 289–301.

Watts, G., Fernie, S., & Dainty, A. (2019a). Paradox and legitimacy in construction: How CSR reports restrict CSR practice. *International Journal of Building Pathology and Adaptation, 37*(2), 231–246.

Watts, G., Ferne, S., & Dainty, A. (2019b). Measuring social value in construction. In: *ARCOM 2019 (35th Annual Conference)*, 2nd–4th September 2019, Leeds Beckett University.

Wood, C., & Leighton, D. (2010). *Measuring social value: The gap between theory and practice*. https://www.demos.co.uk/files/Measuring_social_value_-_web.pdf

Zhao, Z., Zhao, X., Davidson, K., & ZUO, J. (2012). A corporate social responsibility indicator system for construction enterprises. *Journal of Cleaner Production, 29-30*, 277–289.

Chapter 3
How the Social Value Act (2012) Contributes to Wealth Inequality in the UK Construction Industry

Greg Watts and Anthony Higham

3.1 Introduction

The United Kingdom (UK) construction industry has an annual value of over £99bn, accounts for 9% of the UK economic output and is responsible for creating over two million jobs (Rhodes, 2019). Despite such significant economic benefits, society remains highly critical of the industry on account of its widely publicised negative impacts on society and the environment (Barthorpe, 2010). Consequently, the many positives to the industry are often overlooked and taken for granted such as the responsible behaviour of organisations and the infrastructure and buildings we depend on every day. Attempts to change the negative societal opinions of the construction industry have resulted in construction organisations embracing corporate social responsibility (CSR). CSR can be described as organisations adopting practices to protect and improve the environment and society as part of their business activities (Carroll, 2015). Whilst organisations embrace CSR for a variety of reasons, it is arguably the public sector that pushes the CSR agenda forward.

The public sector itself accounts for approximately 26% of the construction industry, and in addition to procuring goods and services for public use, public sector organisations such as local authorities (LAs) are now also able to maximise expenditure for additional societal benefit with the use of The Public Services (Social Value) Act (2012) (SVA). Using the SVA when awarding construction contracts allows LAs to consider the wider value contractors offer and not just the lowest cost (Watson et al., 2016). The SVA helps LAs maximise the societal benefits they experience in times of austerity. Whilst the focus of the SVA and contractor CSR are not directly aimed at wealth inequality, if such concepts are delivered

G. Watts (✉) · A. Higham
School of Science, Engineering and Environment, University of Salford, Salford, UK
e-mail: G.N.Watts@salford.ac.uk

successfully, it could be presumed that wealth inequity would reduce. However, despite the increasing use of the SVA, wealth inequality in the UK is increasing. This area is currently unexplored and so the aim of this deductive research is to analyse this emerging area in more detail. By conducting interviews with both contractors and LAs, and utilising the theoretical lens of legitimacy theory, this paper seeks to understand the impact of the SVA and funding cuts on LAs, the ramifications for construction contractors and their CSR practices and the impact such legislation and practices have upon the levels of wealth inequality in society.

3.2 The Changing Landscape for Local Authorities

LAs are a substantial public sector client for the construction industry. However, the funding of LAs is directly influenced by government rules and regulations. During the previous decade, the government has engaged in a process of austerity. This can be defined as a shared feeling of hardship and reduced expectations, or perhaps more optimistically as a shared feeling of hope that such joint suffering will ultimately lead to a prosperous economy and increased living standards (Coleman, 2016). However, it is argued austerity impacts the most vulnerable groups in society the hardest (Horridge et al., 2019).

In a recent article, Bulman (2018) reported that government austerity has manifested itself in budget cuts that have gone so deep they resulted in the bankruptcy of Northamptonshire council. A report by the Local Government Association (LGA) (2017) confirms this could be the first of many LAs to fail as a reduction in government grants will mean almost half all LAs will receive no central government funding by 2020. LAs were instead primed to retain more of their own business rates raised (75% raising to 100% instead of the current 50%) (LGA, 2017). It is now government policy that LAs are to have responsibility for their own funding decisions (Bulman, 2018). However, it is argued that this increase in business rates would not cover the funds lost from grants and will leave LAs facing a cliff edge in funding reduction as the business rate increase will not be in place during the phasing out of the current grant system (LGA, 2017). Therefore, by 2024, it is forecasted that majority of LA funding will come from council tax receipts and retained business rates and not from central government grants as is does currently (LGA, 2018).

LAs will therefore have to be self-sufficient, with money received equal or greater than money spent (presuming deficits will try to be avoided). This will potentially result in a more aggressive attitude to LA spending and further spending cuts. Whilst some may argue this may be a fairer system, a report from the Institute of Fiscal Studies found that those LAs who would receive the most from retained business rates would not necessarily be those LAs who had the most spending needs (Smith et al., 2018). The study models the impact of what the business rate reduction would have looked like between 2006 and 2014, finding 25% of the LAs worst hit having 13% less spending power relative to their needs compared to the 25% of LAs who would have experienced the least cuts (Smith et al., 2018). Some LAs are therefore set to be worse off financially after the government strategy is fully implemented. However, it is argued that one way LAs can seek to maximise expenditure, thereby offsetting some of the negative impacts of budget cuts, is by utilising the SVA during procurement (Cabinet Office, 2015).

3.3 The Social Value Act (2012)

The SVA places a legislative duty on public sector bodies in England and Wales to consider the wider value that can be achieved during procurement and not just the lowest cost tender returned (Watson et al., 2016). Social value has been described as something that will add benefit to both immediate stakeholders and wider society (Kuratko et al., 2017). It is an actionable concept (Watson et al., 2016) that results in a positive contribution to communities (Raiden et al., 2019). At a more nuanced level, it is argued that only the social value created above and beyond the actual goods and services of the transaction can be considered as actual social value. However, there are currently no widely agreed metrics to measure the social value created through procurement, and the metrics that are used are subject to disagreement and confusion (Loosemore & Higgon, 2016). It is argued, however, that such ambiguity is purposeful, as a government review of the SVA reported that it allows LAs to identify and focus on the social value most important to them (Cabinet Office, 2015). Social value can therefore include practices that aim to have either short-term or long-term benefits. However, it is an assumption made by all stakeholders that the social value practices undertaken ultimately result in some sort of benefit for the intended recipients. The SVA can therefore perhaps be considered as a positive tool in the arsenal of LAs in maximising spending, arguably helping achieve 'more for less' in their procurement (Watts et al., 2019).

However, whilst the use of SV is growing amongst LAs due to its legal requirement and the increased legitimacy it affords to LAs, it is argued the SVA is not widely utilised by all LAs due to low awareness, slow reactions in adopting new procurement strategies, a lack of leadership and a fear of legal challenges from its incorrect application (Cabinet Office, 2015). Where the SVA has been adopted, benefits have been realised, with 'success stories' including an increase in fair trade products specified and more employment opportunities provided to disadvantaged groups (Loosemore & Higgon, 2016). Contractors, however, may argue that they have engaged in such behaviour long before the requirements of the SVA. Indeed, a review of contractor CSR reports revealed reports, published in 2007 some 5 years before the introduction of the SVA, contain examples of contractors engaging in practices intending to improve society (Watts et al., 2019). It could therefore be argued that such socially responsible contractor behaviours were undertaken before the introduction of the SVA. It could be argued that the recent push towards SV forms part of the CSR movement and agenda which has been increasing and evolving since the 1950s.

3.4 Corporate Social Responsibility

The modern advent of organisational CSR behaviour can be traced back to the publication of the book *Social Responsibilities of the Businessman* by Bowen in 1953. Bowen argued for a giving back to society from business leaders due to the increasing industrial prosperity (Bowen, 1953). The concept has evolved since this time, focusing upon civil rights movements, woman's rights, consumer protection and environmental concerns as well as important social issues (Carroll, 2015). This evolution and increasing demand placed on organisations can be attributed to each generation expecting more responsibility from business in general. Such holistic expectations have made the concept of CSR diverse. CSR can therefore be described as organisations protecting and improving the welfare of the environment and wider society as part of their business activities (Carroll, 2015). Although like the concept of SV, CSR is ambiguous and means different things to different people (Barthorpe, 2010).

Requests from clients during procurement for additional metrics to be met, such as those considered as part of an organisation's CSR, have also grown in importance. Now the criteria of time, cost and quality are no longer considered a triumvirate, and contractors are required to engage with and evidence their CSR in order to be successful during procurement. The benefits such CSR activity brings to stakeholders and intended recipients are also widely reported (Cabinet Office, 2015). Therefore, setting aside organisational motivations for CSR, it can be argued that the impacts of CSR activity are positive on both the environment and wider society. Indeed, it has been shown that one of the reasons contractors engage in CSR is for the positive difference it makes (Barthorpe, 2010). Arguably, the SVA has been introduced to harness the benefits offered by CSR and encourage more private sector organisations to adopt CSR practices. Therefore, some may believe that if the SVA was widely adopted by all LAs in all geographical regions, such CSR benefits would become more common and could alleviate negative issues experienced in society at the same time as alleviating some of the constraints LAs face due to funding cuts. It is with such thinking in mind that the widespread use of the SVA is encouraged (Cabinet Office, 2015). However, the relationship between the SVA and societal inequality is yet to be explored in the literature nor has any potential undesirable consequences of the SVA. This research seeks to offer one of the first explorations of the unintended negative impacts of the SVA and adopts the theoretical lens of legitimacy theory to assist in understanding key actor decisions.

3.5 Legitimacy Theory

Legitimacy can be described as the perception that an organisation conforms to social norms and expectations and therefore has a social licence to operate (Bachmann & Ingenhoff, 2016). Legitimacy theory provides a theoretical

Table 3.1 Legitimacy classifications as derived from Duff (2017)

Type of legitimacy	Sub-category
Pragmatic—where practical and logic consequences arise from organisational exchanges with stakeholders	Exchange—organisations embrace practices they hope will result in legitimacy
	Influence—stakeholders believe organisations consider societal interests
	Dispositional—stakeholders believe organisations have societal concern
Moral—where stakeholders believe an organisation is doing the 'right thing'	Consequential—stakeholders judge organisations on what they achieve
	Procedural—organisations adopt socially accepted practices
	Structural—stakeholders perceive an organisation is structured to achieve its advertised aims
	Personal—stakeholders believe those in charge of the organisation have high morals
Cognitive—where stakeholders believe an organisation's motivation reflects their own	Comprehensibility—where an organisation purposefully structures itself to be understandable to stakeholders
	Taken for granted—stakeholders perceive the organisation to be one of the only ones who can meet their needs

framework to understand and explain how the decisions of individuals and organisations are governed and motivated by legitimacy seeking behaviours (Duff, 2017). It is argued there are three main categories of legitimacy, each with nuanced sub-categories as can be seen in Table 3.1.

Both individuals and organisations seek different types of legitimacy at different stages through their actions and communications and can attempt to progress along a continuum from pragmatic to cognitive legitimacy in the eyes of others (Belal & Owen, 2015). Legitimacy theory explains organisational decisions as being motivated by seeking legitimacy from stakeholders (Duff, 2017). Indeed, it could be argued that contractors embrace CSR to increase legitimacy perceptions amongst LAs which is achieved when LAs in turn procure only contractors who embrace CSR. This study adopts legitimacy theory as a lens to analyse and understand LA and contractor behaviour in regard to their adoption, encouragement and use of the SVA and wider CSR practices.

3.6 Research Method

In this research, the view is adopted that social value is fundamentally a subjective concept as numerous arguments throughout the literature highlight how different stakeholders have different social value interpretations (Watts et al., 2019). This view therefore dictates a constructivist ontological position, with the understanding that meanings are socially constructed between different actors and are therefore best understood through a qualitative research strategy (Bryman, 2016).

This research is concerned with the interpretation, enactment and ramifications of the SVA by both LAs and construction contractors, and so purposive sampling was undertaken. This is where participants are identified based on their ability to satisfy the research requirements (Robson & McCartan, 2017). In this research, the top 50 construction contractors by turnover were identified and 10 picked at random. An online search for LAs who had an advertised use of the SVA was also undertaken, and from the 22 identified, 10 were picked at random. From the respective organisation websites, key actors who have responsibility for procurement of construction works, and those who have responsibility for carrying out SV practices, were selected. Email introductions were then sent to each individual outlining the research and requesting interviews. Six LAs and five contractors replied positively. In total, 11 interviews were conducted. Galvin (2015) conducts an extensive review of 54 previous studies and concludes that in order for a researcher to have a high confidence in their findings 11 to 15 interviews are optimum. Semi-structured interviews were then conducted which allowed the core topics of interest to the research to be covered whilst allowing flexibility for the interviews to pursue interesting and unexpected lines of enquiry (Bryman, 2016). The interviews were conducted by telephone due to the wide geographical spread of the participants whilst also allowing the participants to respond to questions from comfortable and familiar surroundings (Creswell, 2013).

Narrative analysis was employed to encourage participants to discuss relevant topics and also as a method of analysing responses. Narrative analysis encourages interviewees to respond to questions asked by telling stories of their experience, with the researcher then extracting relevant information from these stories (Sandelowski, 1991). Such stories and responses are summarised and grouped together so detailed insights and understandings can be revealed (Loosemore & Bridgeman, 2017). In this research, as participant understandings of the SVA were sought, in addition to examples of the SVA use, participants were requested to tell stories of how they have used or experienced the SVA, why it was used in those ways, in what contexts and any benefits and drawbacks experienced. Any wider reflection shown by the interviewees when using the SVA, and potential consideration of ramifications, both positive and negative, were also discussed. The categories of legitimacy theory were used to both structure the interview questions and as themes by which to categorise responses. For example, LAs were asked if they judge contractors on their social value achievements (consequential legitimacy), and contractors were asked why they engaged with social value practices and if they engaged with such practices to be viewed as legitimate (exchange legitimacy). Contractors were also asked the amount of LA work undertaken, their motivation for conducting SV practices and if such SV practices were undertaken across all projects regardless of client. The LAs were also asked about the changes to their funding, procurement practices and their previous and current social value requirements from contractors. The types of legitimacy then became headings under which the summarised responses gained through the narrative analysis were grouped. Any stories that shared characteristics with legitimacy theory categories were then grouped together under the appropriate heading, thereby revealing if any aims or actions by either LAs or contractors were driven by legitimacy motivations and, if so, which type of legitimacy. This allowed a deeper consideration of stakeholder actions to be undertaken and allowed the theoretical lens of legitimacy to reveal deeper insights into stakeholder action and intention.

3.7 Findings and Discussion

The interviews with LAs revealed that changes to their funding are indeed underway, and are influencing procurement decisions, including impacting the way they procure construction contracts, confirming arguments in the literature. All LAs reported it was the introduction of the SVA that encouraged them to consider the wider benefits they could achieve from expenditure. All LAs also reported that prior to the introduction of the SVA, simply considering the lowest priced tender was acceptable and regularly undertaken. Although the majority of the LAs revealed that cost is still the most important factor, they reported the SVA now enables them to make further requests from each contractor. The interviews also reinforced arguments in the literature that LAs are attempting to ensure SV resulting from procurement occurs within their own geographical remit. This was to illustrate spending on issues of local importance to satisfy the majority of their stakeholders. All LAs were largely unapologetic about this and discussed how it was for the long-term good of their own area. One LA even stated how they want local spend to occur within ten miles from the city centre and not ten miles from the location of the site, as sometimes the latter will cross the boundary of another LA and they want to retain all the spend benefits themselves (Table 3.2).

The interviews also revealed that LAs acted this way in their search for moral and cognitive legitimacy in the eyes of local stakeholders such as community groups, residents and local business leaders. This raises the prospect of those LAs with lower socio-economic communities failing to raise enough funds to invest in construction projects and so therefore unable to experience the advantages the SVA offers. LAs who have the financial means can therefore maximise the benefits their local communities experience. However, analysis of the interviews reveals that being motivated to achieve legitimacy in the eyes of immediate stakeholders comes at the expense of wider stakeholders, with those LAs interviewed inadvertently serving to widen the wealth gap between affluent and poorer communities. When this was raised, LAs cited concerns about their own communities. Although no client wanted inequality to increase elsewhere, they all discussed what their responsibility was to their own populations first and foremost. Interestingly, the contractors did not acknowledge how the focus of their SV activities could perhaps contribute to wealth inequality and instead, like LAs, argued that such SV need would be picked up by LAs and contractors in those areas.

Table 3.2 Summary of research findings

Salient findings	Relationship with literature
Changes to LA funding is influencing construction procurement decisions	This builds on findings by Bulman (2018) and LGA (2018) that LA funding is changing with the potential ramifications for the construction industry illustrated. The paper provides the insight that LAs are aware of, and preparing for, changes to their funding by maximising use of the SVA
The SVA is increasingly used and does help LAs achieve more from procurement	The findings reinforce those in the literature of Raiden et al. (2019) and illustrate that recommendations made by the Cabinet Office (2015) are occurring in industry
LAs use the SVA to ensure contractor spending is within their own geographical remit to maximise local benefits experienced. This is conducted to increase LA legitimacy in the eyes of local communities	Building on the use of legitimacy theory as a theoretical lens to understand behaviour (Duff, 2017), this paper contributes to understanding how legitimacy theory can govern client actions and this can influence contractor behaviour
LAs without the funds to use for construction projects may lose out on experiencing any SV-related benefits within the areas they cover	Such findings build on existing research by LGA (2018) and Smith et al. (2018) and extend current findings to reveal possible negative connotations to current and planned government policy
Construction contractors engage with CSR and SV activity for the wider societal good and not simply to win work	Research by Watson et al. (2016) and Raiden et al. (2019) explore CSR and SV activity, and these findings contribute to, reinforce and further our understandings of contractor CSR and SV motivations

It was perhaps unsurprising to learn contractors discussed the procurement benefits CSR activity brings to their organisation. However, all those interviewed also strongly believed the benefits to society and the environment are the main motivations behind their CSR strategies and actions. Contractors generally believed they would still conduct SV practices even without the requirements imposed by the SVA and discussed current examples of CSR activity they were undertaking where the SVA had not been used, including for private sector clients who were not overly concerned with SV during procurement. Contractors also often structured their organisations with specially hired staff to oversee the social value practices undertaken (structural legitimacy) and perceive they had to adhere to LA practices expected (procedural legitimacy). It was also interesting that every contractor interviewed felt authentic support from top level management and business owners to deliver high-impact social value practices (personal legitimacy). The use of legitimacy theory also revealed contractors were either motivated by, or at least aware of, the moral legitimacy benefits achieved by embracing CSR activity. Contractors are therefore willing to conduct CSR activity in the locations specified by LAs if it meant they could be both successful in procurement and engage with CSR activity.

3.8 Conclusion

This research offers one of the first explorations of the unintended negative impacts of the SVA and the potential scenarios that could arise if the SVA is increasingly adopted by LAs. This increasing use is both encouraged by central government and somewhat driven by the government's austerity measures and the upcoming changes to how LAs receive funding resulting in future budget cuts. The research findings reveal that whilst contractors are motivated to embrace CSR for the benefit of society, current legislative and procurement practices dictate the focus of such CSR. If this focus is within a LAs' geographical remit, the contractor will ultimately be more successful in public sector procurement. The most affluent LAs will procure construction works and receive SVA benefits, further enhancing their communities and so increasing the funds they receive to invest in construction work, whereas LAs without enough funds to invest will not experience the same SV benefits from contractor CSR activity. The focus of such CSR activities solely in areas at the discretion of affluent LAs utilising the SVA could inadvertently contribute to increasing social inequality.

The limitations of this study include the number of interviewees conducted, as they cannot be said to be representative of the hundreds of LAs and contractors operating in the UK. Therefore, the generalisability of the research findings is limited. It would also be of interest to compare the actions of LAs who have elected mayors and those who do not, to reveal if the shifting focus of local politics impacts how LAs spend their budgets. It is recommended that further research will need to be conducted with a wider sample of LAs, including those of different socio-economic levels and those who do and do not use the SVA in procurement of construction works to increase understanding of the wider SVA ramifications. This research is of particular importance to contractors tendering for LA work and for those contractors motivated to engage in CSR by a desire to positively contribute to society. The research is also of importance to LAs and government policy makers in their consideration of the successes and failures of the SVA. The findings of this research add to the current understanding surrounding SV procurement in the UK construction industry and highlight previously unexplored areas of importance regarding the use of the SVA. It is therefore recommended that LAs think carefully before they use the SVA, and perhaps a longer-term, nationwide approach to harnessing the power of contractor CSR should be discussed amongst all LAs to ensure the benefits can be experienced by all geographical regions.

References

Barthorpe, S. (2010). Implementing corporate social responsibility in the UK construction industry. *Property Management, 28*(1), 4–17.
Bachmann, P., & Ingenhoff, D. (2016). Legitimacy through CSR disclosures? The advantage outweighs the disadvantages. *Public Relations Review, 42*, 386–394.

Belal, A., & Owen, D. (2015). The rise and fall of stand-alone social reporting in a multinational subsidiary in Bangladesh. *Accounting Auditing and Accountability Journal, 27*(8).

Bowen, H. (1953). *Social responsibilities of the businessman.* Harper & Row.

Bryman, A. (2016). *Social research methods* (5th ed.). Oxford University Press.

Bulman, M. (2018). English councils brace for biggest government cuts since 2010 despite 'unprecedented' budget pressures. *The Independent.* Published Monday 1 Oct 2018.

Cabinet Office. (2015). *Social value act review.*

Carroll, A. (2015). Corporate social responsibility: The centrepiece of competing and complementary frameworks. *Organisational Dynamics., 44*, 87–96.

Coleman, R. (2016). Austerity futures: Debt, temporality and (hopeful) pessimism as an austerity mood. *New Formations, 87*, 83–101.

Creswell, J. (2013). *Qualitative inquiry and research design: Choosing among five approaches.* Sage.

Duff, A. (2017). Corporate social responsibility as a legitimacy maintenance strategy in the professional accountancy firm. *The British Accounting Review., 49*(6), 1–19.

Galvin, R. (2015). How many interviews are enough? Do interviews in energy consumption research produce reliable knowledge? *Journal of Building Engineering, 1*, 2–12.

Horridge, K., Dew, R., Chatelin, A., Seal, A., Macias, L., Cioni, G., Kachmar, O., & Wilkes, S. (2019). Austerity and families with disabled children: A European survey. *Developmental Medicine & Child Neurology, 61*(3), 329–336.

Kuratko, D., McMullen, J., & Hornsby, C. (2017). Is your organisation conducive to the continuous creation of social value? Toward a social corporate entrepreneurship scale. *Business Horizons., 60*, 271–283.

Local Government Association. (2017). *Council funding to be further cut in half over next two years.*

Local Government Association. (2018). *Local government funding. Moving the conversation on.*

Loosemore, M., & Bridgeman, J. (2017). Corporate volunteering in the construction industry: Costs and benefits. *Construction Management and Economics, 35*(10).

Loosemore, M., & Higgon D. (2016). *Social enterprise in the construction industry.* Routledge, Cabinet Office.

Raiden, A., Loosemore, M., King, A., & Gorse, C. (2019). *Social value in construction.* Routledge.

Rhodes, C. (2019). *Construction industry: Statistics and policy.* Nr 01432. House of Commons Library. www.parliament.uk/briefing-papers/sn01432.pdf.

Robson, C., & McCartan, K. (2017). *Real world research* (4th ed.). Wiley.

Sandelowski, M. (1991). Telling stories: Narrative approaches in qualitative research. *The Journal of Nursing Scholarship, 23*(3), 161–166.

Smith, N., Phillips, D., & Simpson, P. (2018). *100% business rates retention may lead to divergences in English councils' funding without growth.* Institute for Fiscal Studies.

Watson, K. J., Evans, J., Karvonen, A., & Whitley, T. (2016). Capturing the social value of buildings: The promise of social return on investment (SROI). *Building and Environment, 103*, 289–301.

Watts, G., Dainty, A., & Fernie, S. (2019). Paradox and legitimacy in construction: How CSR reports restrict CSR practice. *International Journal or Building Pathology and Adaptation, 37*(2), 231–246.

Chapter 4
Transitioning from a Linear to a Circular Construction Supply Chain

Emmanuel Manu, Nii Amponsah Ankrah, and Jamila Bentrar

4.1 Introduction

It is estimated that the construction and operation of the built assets account for about 60% of materials consumption in the UK (BAM and ARUP, 2017). Construction, demolition and excavation waste is also estimated to account for approximately 20–30% of waste in the EU (BAM and ARUP, 2017). This substantial resource consumption in the construction sector makes it a priority industry for promoting a circular economy. Circular economy is a sustainable economic paradigm that seeks to bridge the gap between production and consumption activities (Witjes & Lozano, 2016) by maintaining materials as valuable and useful resources for production. Products, components and materials in a circular economy have to be kept at their highest possible utility and value at all times (Ellen MacArthur Foundation, 2015a) to be useful resources for production. The Ellen MacArthur Foundation (2015b, p. 2) has defined circular economy as a 'new economic model that seeks to ultimately decouple global economic development from finite resource consumption'. There is already evidence that adoption of circular business models can promote sustainability. By applying a life cycle assessment methodology, Nasir et al. (2017) compared a circular building insulation supply chain with the linear equivalent and found 40% reduction in carbon emissions. Other sustainability

---

E. Manu (✉)
School of Architecture, Design and Built Environment, Nottingham Trent University, Nottingham, UK
e-mail: emmanuel.manu@ntu.ac.uk

N. A. Ankrah
School of Engineering and Applied Science, Aston University, Birmingham, UK

J. Bentrar
Métropole Européenne de Lille, Lille Cedex, France

benefits of circular economy such as waste material reduction and reuse (Hossain et al., 2020) have also been reported. Given the high resource consumption and waste from construction and operation of built assets, circular economy adoption in construction has continued to attract research interest in the literature.

Successful implementation of circular economy in the construction industry will be dependent on construction supply chains adopting circular business models (circular construction supply chains). Complex networks of supply chain firms are involved throughout the life cycle of built assets, from design, construction and operation, to demolition at the end of life phase. The target outcomes for promoting circular economy in the construction supply chain will be to achieve material efficiency through reuse and recycling of construction waste (Ghisellini et al., 2018) and to design out waste (Ellen MacArthur Foundation, 2015b). For such circular economy ambitions to be realised in the construction industry, construction supply chain firms will have to rethink traditional business models, which are more aligned to the linear take-make-dispose economy. Recent studies have highlighted the difficulty in implementing circular economy in the supply chain (Schraven et al., 2019). Schraven et al. (2019) investigated the Dutch stony materials supply chain and found that the key challenges inhibiting circular economy transition in the supply chain are the lack of incentives for the supply chain to make changes towards circularity, lack of mutual interest between supply chain actors, high uncertainties and risks and clashes of perceptions across the supply chain.

From the existing literature, limited studies have addressed this issue of construction supply chains transitioning towards a circular economy, particularly from the perspectives of organisations that have already been through this transition journey. The aim of this study, as part of a much wider study on circular economy transition in the construction supply chain, was to explore the factors that are critical to enabling the transition from a linear to a circular construction supply chain.

In the sections that follow, a literature review is presented on linear and circular supply chains, including circular business models that can be adopted by firms in the construction sector. This is followed by a methodology section, the research findings, results and discussions and the conclusion.

4.2 Linear and Circular Supply Chains

Supply chains have been defined by Christopher (2011, p. 13) as "the network of organisations that are involved through upstream and downstream linkages, in the different processes and activities that contribute value in the form of a product or service delivered to an ultimate consumer". Supply chains that operate in any industry are therefore responsible for the flow of products, information and financial resources within complex production and consumption systems. Traditional supply chains operate business models that are embedded in the linear economy. The linear economy is driven by a take-make-dispose model of production and consumption which is wasteful and unsustainable (Ellen MacArthur Foundation, 2015a; Ruiz et al., 2020). Construction supply chains operating within the linear economy use virgin resources to construct built assets, and these assets are not designed to facilitate disassembly or easy adaptability to changing user requirements. They soon become derelict waste liabilities that are demolished rather than deconstructed, hence resulting in high loss of material quality and eventually waste. Construction supply chains that design, construct, operate and recover built assets have predominantly operated within this linear economy (linear supply chains).

Circular economy as a sustainable economic model (Virlanuta et al., 2020) offers an alternative vision for production and consumption activities. It has been defined by Kirchherr et al. (2017, pp. 224–225) as 'an economic system that is based on business models which replace the "end-of-life" concept with reducing, alternatively reusing, and recycling materials in production/distribution and consumption processes, with the aim to accomplish sustainable development, which implies creating environmental quality, economic prosperity and social equity, to the benefit of current and future generations'. Within the circular economy, production and consumption activities are intentionally designed to maintain material value and integrity over time as useful resources in a circular flow. Kirchherr et al. (2017) found that circular economy definitions frequently congregate around the 4Rs of reduction, reuse, recycling and recovery activities, all of which create different loops in the circular flow. However, Morseletto (2020) proposed ten circular economy strategies, which are recover, reuse, recycle, repurpose, refurbish, remanufacture, repair, reduce, rethink and refuse. In Fig. 4.1, the ten circular economy strategies proposed by Morseletto (2020) have been linked to circular business models that are applicable to the built environment. For circular economy to be successfully realised in the construction industry, supply chains that operate in the industry need to transition by adopting these circular business models (circular supply chains).

A business model is an articulation of the logic that organisations use to identify, capture and deliver value to their consumers within viable structures of revenues and cost (Teece, 2010). BAM and ARUP (2017) categorised circular business models in the built environment into circular design, circular use and circular recovery. However, through synthesis from a range of sources (BAM and ARUP, 2017; Esposito et al., 2018; EIT Climate-KIC, 2019), eight circular business models are provided in Fig. 4.1, i.e. circular design, circular construction, product as a service, life cycle extensions, sharing platforms, end-of-life management, circular material production and supply and energy recovery models. These circular business models are driven by value identification, capture and delivery throughout the life cycle of built assets. Since the considerable amount of waste generated by the construction industry can be traced back to the early stages of design (Esposito et al., 2018), circular design models focus on designing out waste in a structure by paying attention to the circularity and reusability of the individual materials and components within it. Circular designs should also facilitate the use of circular construction techniques (e.g. design for manufacture, assembly and disassembly). Circular construction techniques and processes (e.g. construction assembly and installation) should achieve circularity of the structure, with claims that prefabrication and off-site construction are techniques that can help achieve circularity (Esposito et al.,

Project lifecycle	Circular economy strategy		Description	Circular business models
Smarter design and construction	R0	Refuse	Phase-out materials (e.g. virgin materials) or environmentally damaging production processes by abandoning its function or offering the same function with a different material.	Circular design – design of product, projects and process for the circular economy

Circular construction – construction for easy disassembly |
| | R1 | Rethink | Design and construct built assets that facilitate other R strategies so that intensity of use and efficiency is increased (e.g. design for disassembly, design for flexibility and adaptability) | |
| | R2 | Reduce | Design and construct built assets so that they consume fewer natural resources. | |
| Extend lifespan of built assets | R3 | Reuse | Reuse built assets or their component parts which are still in good condition to fulfil original function. | Product as a service

Digital sharing platforms and marketplaces

Life cycle extension – repurpose, refurbish and maintenance |
	R4	Repair	Repair and maintenance of defective products so that is can be used with its original function.	
	R5	Refurbish	Restore built assets and their components to bring them up to date.	
	R6	Remanufacture	Use parts of discarded components or parts from built assets in the construction of new projects with the same function.	
	R7	Repurpose	Adapt built assets for different purposes or use parts of discarded components from built assets in the construction of new projects with different functions.	
Useful application of materials at end of life	R8	Recycle and Upcycle	Process materials from built assets to obtain the same or higher-grade quality raw materials.	End-of-life management – deconstruction of built assets

Circular material (re)production and supply – production and supply of recycled/upcycled materials |
| | R9 | Recovery | Incineration of material with energy recovery. | Energy recovery - waste-to-energy |

Fig. 4.1 Circular economy strategies and business models (Adapted from Morseletto, 2020)

2018). The extension of the lifespan of built assets through active maintenance regimes, repairs, upgrades and refurbishments all fits within the life cycle extension business model. Sharing platforms are also circular business models that enable multiple clients/customers to use the same resources, hence an increase in resource utilisation and reduction in demand for new products. A typical example of such circular business models is platforms like FLOOW2, which allows contractors with idle construction equipment (e.g. earth diggers and excavators) to maximise equipment utilisation through sharing and exchanges (Esposito et al., 2018).

Circular material production and supply business models are based on recycling or upcycling of used (non-virgin) materials that would otherwise have resulted in waste. Through secondary production processes, these materials are re-produced as resources for the same product (recycling) or higher-grade products (upcycling). Circular material production and supply business models also capture, store and recycle waste energy (e.g. waste heat). Energy recovery business models can be compatible and complimentary to recycling in the circular economy (Van Caneghem et al., 2019; Morseletto, 2020) although this is the least sustainable of the circular business models because it destroys materials forever and encourages waste (see Vilella, 2019). Construction supply chain firms that adopt any of these circular business models will be paramount for achieving circular economy transition in the construction industry. The factors that are critical for driving such a transition in the construction supply chain are largely unexplored.

4.3 Research Methodology

The study adopted a qualitative methodology that is underpinned by an interpretivist epistemology. Due to this interpretivist epistemology, the researchers had to constantly engage with participants as 'insiders' (Creswell, 2013) to uncover deeper meanings through interactive dialogue and interpretation (Ponterotto, 2005). Data was gathered through face-to-face semi-structured interviews conducted with participants in both the UK and France. The focus on these two countries (UK and France) was as a result of existing collaborations with organisations that had adopted or promoted circular business models. Altogether, a total of 14 participants (3 in UK and 11 in France) were interviewed across 12 organisations (3 in the UK and 9 in France). All the participants were purposefully selected because they had engaged in or championed the implementation of circular economy in the construction supply chain. However, despite the range of circular business models summarised in Fig. 4.1, all the participants in this study had only implemented or promoted two of the circular business models in Fig. 4.1, i.e. end-of-life management and the returns, recycling and upcycling business models. All the participants were senior-level representatives of the various organisations summarised in Table 4.1.

Apart from Participant 4 who was interviewed individually, all the other French participants were interviewed using focus group sessions. Participants 5—7 in Table 4.1 were interviewed together as focus group 1, participants 8–9 as focus

Table 4.1 Semi-structured interview participants and their organisations

Organisation	Label	Participant(s)	Location
Asbestos denaturing and recycling firm	RC-01	Participant 1	UK
Demolition and aggregate recycling firm	DC-RC -01	Participant 2	UK
Supplier of recycled insulation-social enterprise	SCP	Participant 3	UK
Ceramic thermal energy capture, storage recycling firm	RC-02	Participant 4	France
Federation of concrete manufacturers	FCM	Participant 5	France
Manufacturer of prefabricated concrete products	MCMP-01	Participant 6	France
Manufacturer of architectural masonry products	MCMP-02	Participant 7	France
Real estate developer	RED	Participant 8	France
Developer and social landlord	DSL	Participant 9	France
Demolition, aggregate recycling and recycling equipment manufacturer	DC-RC-REM	Participant 10	France
Concrete producer—Ready mix	CP	Participant 11 Participant 12 Participant 13	France
Waste valorisation consultancy	WVC	Participant 14	France

CP concrete producer, *DC* demolition company, *RC* recycling company, *DSL* developer and social landlord, *REM* recycling equipment manufacturer, *MCMP* manufacturer of concrete and masonry products, *WVC* waste valorisation consultancy, *SCP* supplier of circular products, *FCM* federation of concrete manufacturers, *RED* real estate developer

group 2 and participants 10–14 as focus group 3. The focus group interviews were used for interviewing the majority of the French participants because it offered the logistical benefit of interviewing multiple individuals simultaneously (Onwuegbuzie et al., 2009), as compared to the individual interviews. The individual interviews lasted approximately 50 min, whilst the focus group interviews lasted approximately 120 min. These semi-structured interviews were very exploratory in nature. Participants were asked to share their experiences of implementing circular business models, any challenges they faced and advice for construction supply chain firms that are interested in transitioning towards the circular economy.

The data was analysed using a thematic analysis approach (Braun & Clarke, 2006; Peterson, 2017) which started with transcription of the audio interviews. Thematic analysis is an interpretive data analytic process that requires the immersion of one's self in the data to identify common ideas or themes about what is being studied and the research question(s) being addressed (Peterson, 2017). The transcripts were analysed manually by coding segments of text (Miles & Huberman, 1994) based on an inductive approach, with the initial codes assigned after familiarisation and interpretation of meanings in the text. These initial codes were then organised and re-organised until a coherent pattern that was consistent with the main research question emerged. The output from the analysis is therefore a reflection of the researchers' interpretation of participants' views, rather than an objective and quantifiable generalisation. Nonetheless, this approach aligns with the interpretivist ontology (Peterson, 2017) that was chosen for the study.

4.4 Results and Discussion

Participants provided their perspectives on actions and factors that were needed to successfully drive circular economy transition in the construction supply chain. The thematic analysis results are summarised in Fig. 4.2. The achievement of design for deconstruction visions in the built environment was a critical success factor which, according to participants, will accelerate transition towards circular economy in the supply chain. Instances were cited of difficulties in the recycling process that could be traced back to the way structures had been designed. Some participants also linked a regime where structures are designed for deconstruction to minimised recycling costs. Previous studies have recognised the importance of designing buildings as material banks, with material passports that facilitate end-of-life decisions (Honic et al., 2019). This finding has implications for design professionals in construction who will have to develop more circular design expertise. On the issue of flexibility of regulations and standards to accommodate circular innovations, participants in France were optimistic that the soon-to-be-passed French 'License to Innovate' bill could promote the level of flexibility that can foster innovation. In the UK, this regulatory flexibility was more about achieving a better balance between energy efficiency targets and end-of-life utility of built assets. This view is captured in the interview extract from the demolition and aggregate recycling firm in the UK:

> Policy at the minute has been entirely driven by erm, energy usage. That's it. We're absolutely driven by efficiencies of buildings when they finish. Well, they've got to stop looking at that, policy's got to change to buildings end of life and whether it can be deconstructed, or whether it can be re-purposed.—DC-RC-01 [Participant 2]

Another critical success factor was to minimise the cost of recycling (secondary production costs) to make circular economy transition in the construction supply chain more profitable. Both participants in France and the UK linked this minimisation of secondary production costs to the creation of shorter resource flow loops. Creating shorter resource flow loops by recycling materials on site will minimise transportation and storage costs and, as a positive consequence, also reduce whole life carbon from secondary production activities. When it was not possible to operate within a specified proximity of the recycling facility, participant firms in both the UK and France deployed mobile recycling plants to shorten the material flow loop and generate cost-efficiencies. A participant in France remarked:

> Transportation has a great impact…the opportunity to re-use materials right on the site can eradicate the transportation costs. It also minimises the need for storage space for re-processed materials. If the mobile unit cannot operate on site, then the platform machine is used, which increases the cost of transportation and storage.—DC-RC-REM [Participant 10]

Tingley et al. (2017) also recognised the need for cost minimisation strategies with regard to structural steel reuse. Some participants suggested that economic incentives like EU subsidies and higher taxes on virgin natural resources will strengthen the case for circular economy transition in the construction supply chain. A participant in France (CP—Participant 12) cited an example where some local authorities

Main theme	Higher level Codes	Lower level Codes	Indicated by
Critical success factors for circular supply chain transition	Maturity of markets for circular products		CP – [Participant 11, Participant 12 Participant 13]
	Flexible business models that allow for innovation	Smaller companies are more agile so easily adopt new processes and circular innovations.	FCM- [Participant 5] WVC – [Participant 14] RC-02 – [Participant 4]
	Change in mindset and communication strategy	Change mindset from waste management to resource management	DC-RC-REM - [Participant 10] DSL – [Participant 9] DC-RC -01 – [Participant 2] DSL – [Participant 9]
		Label recycled products as eco-products rather than products from waste.	WVC – [Participant 14] FCM- [Participant 5] DC-RC-REM - [Participant 10]
	Minimizing costs of secondary production	Creation of shorter loops	FCM- [Participant 5] DSL – [Participant 9] DC-RC -01 – [Participant 2]
		Tax breaks, carbon credits and other economic incentives	RC-01 – [Participant 1] CP – [Participant 11, Participant 12 Participant 13]
	Collaboration amongst industry stakeholders		DC-RC -01 – [Participant 2] DC-RC-REM - [Participant 10] CP – [Participant 11; Participant 12; Participant 13; Participant 14] RED – [Participant 8] DSL- [Participant 9]
	Flexible regulations and standards that embrace innovation	Regulations and standards that keep pace with circular economy innovations	DC-RC-REM - [Participant 10] CP – [Participant 11; Participant 12; Participant 13; Participant 14]
		Regulations that achieve a balance between energy efficiency and design for deconstruction	DC-RC -01 – [Participant 2]
	Industry leadership		FCM- [Participant 5] MCMP-01 – [Participant 6] MCMP-01- [Participant 6] MCMP-02- [Participant 7]
	Design for deconstruction	Demountable building techniques that facilitate quick, easy, and cheap deconstruction at the end-of-life phase	DC-RC -01 – [Participant 2] FCM- [Participant 5] DC-RC-REM - [Participant 10] CP – [Participant 11; Participant 12; Participant 13; Participant 14]
	Availability of resources in commercial quantities		DC-RC -01 – [Participant 2] FCM- [Participant 5] RED – [Participant 8] DSL- [Participant 9]

Fig. 4.2 Results from the thematic analysis process on critical success factors

now allow a developer to build 30% more on the land if they could demonstrate a decrease in their CO_2 emissions through use of low-carbon cement with recycled content. With the high cost of resources in Europe, it is evident that some economic incentives will enable circular economy transition in the construction supply chain. Nasir et al. (2017) argued that reaching higher levels of circularity will require such policy interventions due to higher economic costs. Another critical success factor expressed by most participants was the need for strong industry collaborations to promote circular economy in the supply chain. Some of the French participants had been beneficiaries of such collaborations between circular supply chain firms (e.g. deconstruction specialists), trade federations, client organisations, designers (architects, engineers) and regulators (local and central government authorities) working together to drive circular economy transition.

The expertise of specialist deconstruction supply chains will be needed at the early stages of design to advise on how choice of building systems, materials and design can facilitate easy deconstruction at the end-of-life phase of built assets. The added costs from such early engagements could be justified by the potentially higher values of the assets at their end of life. The demolition and aggregate recycling firm in the UK explained:

> …when a project team get together to build a structure, they should be talking to the people that might remove that structure at the end of its life. From a waste point of view, and a deconstruct point of view.—DC-RC-01 [Participant 2]

Collaborations with material certification agencies are also important in a post-Grenfell era to ensure that new circular products are rigorously tested and certified to meet stringent safety and performance standards. The findings indicate a need for much wider collaborations than what Witjes and Lozano (2016) claimed should exist between procurers and suppliers. The results also indicate that strong industry leadership is required to facilitate circular economy transition in the construction supply chain. This could take the form of major clients specifying substantial amount of recycled content for their projects and constructing circular buildings as demonstration projects, which, according to Hossain et al. (2020), is still lacking, and even trade federations that support their member firms to build capacity towards circular economy transition. The Federation of Concrete Manufacturers (FCM-Participant 5) exhibited this strong leadership by providing support for member firms (including MCMP-01 and MCMP-02) to adopt circular business models. This support resulted in one of the participants (MCMP-02-Participant 7) developing a new drainage and paving product using waste seashells from the food industry supply chain.

Another critical success factor was the use of positive communication to challenge negative perceptions and mindsets about recycled and upcycled products, even amongst the supply chain firms themselves. Materials and products that are recycled from demolition waste will need to be branded for the benefits they offer as eco-materials and eco-products. This re-branding was beginning to yield some benefits as one participant explained:

> In Paris, eco-products are now beginning to be sold for more than natural products due to new generation of mindsets that are more environmentally conscious. The approach is to talk about a new eco-material and eco-product and not recycled waste materials. The wider public do not trust materials that are recycled from waste. The assumption is that since it is not a natural material, I will pay less for it. It is a matter of communication.—DC-RC-REM [Participant 10]

Demolition firms will also need a mindset shift from demolition and waste management firms to deconstruction specialists, resource managers and urban miners. Virlanuta et al. (2020) revealed that behavioural challenges can affect circular economy transition and will have to be overcome. From the results, positive communication strategies and a change in mindsets will be needed to overcome some of the psychological and behavioural drawbacks. Other critical success factors that were expressed by participants were the availability of secondary resources in commercial quantities within the local geographical area, the need to be flexible in outlook to continuously develop circular innovations and the need for maturity of the market for circular products to achieve a critical mass for greater demand. Since the commercial availability of secondary resources will continue to be critical for construction supply chain firms that are into recycling, there is scope to utilise digital and AI tools that help to map and predict secondary resources availability in terms of volumes and time. The use of digital and AI platforms to drive circular economy has already been acknowledged in previous studies (Ellen MacArthur Foundation, 2019; Honic et al., 2019). However, the findings from this study imply that such tools will be needed not only for tracking materials for recycling but also for supporting the high levels of collaboration that is needed to drive circular economy transition within the construction supply chain context. A typical example is the Upcyclea platform in France, which is an AI-powered digital platform that connects different partners with material flow data to facilitate the efficient management of circular resources.

4.5 Conclusion

The above study provides insight into the views of organisations that have already embraced circular economy transition in the construction supply chain. The findings have revealed nine critical success factors for circular economy transition: (1) maturity of markets for circular products; (2) flexible business models that allow for innovation; (3) change in mindset and communication strategy; (4) minimising costs of secondary production; (5) collaboration amongst industry stakeholders; (6) flexible regulations and standards that embrace innovation; (7) industry leadership; (8) design for deconstruction; and (9) availability of resources in commercial quantities. It is also evident that the collaboration needed amongst industry stakeholders to drive circular economy transition in the construction supply chain is enormous. These collaborations will need to extend beyond the conventional construction supply chain identified by Cox et al. (2006), to include wider stakeholders like regulatory and certification bodies and insurers (for instance, where new circular products are being used) and lawyers (to draft contracts that incorporate circular economy requirements). There is the need for major clients with high purchasing power to take lead in driving such collaborations by specifying circular economy requirements and targets as part of their tenders. These tender requirements will most likely drive collaborations between the clients, deconstruction specialists and designers at the early stages of projects. The findings in this study also imply that economic incentives that minimise secondary production costs (e.g. tax breaks and subsidies for circular adoption and higher taxes for virgin resources) can be used by policymakers to encourage circular economy transition in the construction supply chain.

The present study is not without limitations. First, despite the range of circular business models that exist, all the participants in this study had only implemented two of the circular business models, which relate to recycling and upcycling of waste. Secondly, whilst the study sought to generate views from construction supply chains across two countries (UK and France), there were only three participant organisations from the UK. Further transnational studies can be undertaken to understand how construction supply chains in other countries are transitioning towards the circular economy. Finally, the output from the analysis cannot be generalised given the qualitative exploratory nature of the study. There is the need for further research on critical success factors for circular economy transition in the supply chain that is based on quantitative and generalisable methods.

References

BAM, & ARUP. (2017). *Circular business models for the built environment.* https://www.arup.com/-/media/arup/files/publications/a/8436_business-models-low-res.pdf

Braun, V., & Clarke, V. (2006). Using thematic analysis in psychology. *Qualitative Research in Psychology, 3*(2), 77–101.

Christopher, M. (2011). *Logistics and supply chain management* (4th ed.). Edinburgh Pearson Education Ltd.

Cox, A.W., Ireland, P., & Townsend, M. (2006). Managing in construction supply chains and markets: reactive and proactive options for improving performance and relationship management. Thomas Telford Services Ltd.

Creswell, J. W. (2013). *Qualitative inquiry and research design: Choosing among five approaches.* Sage.

EIT Climate-KIC. (2019). *The challenges and potential of circular procurements in public construction projects.*

Ellen MacArthur Foundation. (2015a). *Growth within: A circular economy for a competitive Europe.*

Ellen MacArthur Foundation. (2015b) *Towards a circular economy: Business rationale for an accelerated transition.*

Ellen MacArthur Foundation. (2019). *Artificial intelligence and the circular economy - AI as a tool to accelerate the transition.*

Esposito, M., Tse, T., & Soufani, K. (2018). Introducing a circular economy: New thinking with new managerial and policy implications. *California Management Review, 60*(3), 5–19.

Ghisellini, P., Ripa, M., & Ulgiati, S. (2018). Exploring environmental and economic costs and benefits of a circular economy approach to the construction and demolition sector. A literature review. *Journal of Cleaner Production, 178*, 618–643.

Honic, M., Kovacic, I., & Rechberger, H. (2019). Improving the recycling potential of buildings through Material Passports (MP): An Austrian case study. *Journal of Cleaner Production, 217*, 787–797.

Hossain, M. U., Ng, S. T., Antwi-Afari, P., & Amor, B. (2020). Circular economy and the construction industry: Existing trends, challenges and prospective framework for sustainable construction. *Renewable and Sustainable Energy Reviews, 130*, 109948.

Kirchherr, J., Reike, D., & Hekkert, M. (2017). Conceptualizing the circular economy: An analysis of 114 definitions. *Resources, Conservation and Recycling, 127*, 221–232.

Miles, M. B., & Huberman, A. M. (1994). *Qualitative data analysis: An expanded sourcebook*. Sage.

Morseletto, P. (2020). Targets for a circular economy. *Resources, Conservation and Recycling, 153*, 104553.

Nasir, M. H. A., Genovese, A., Acquaye, A. A., Koh, S., & Yamoah, F. (2017). Comparing linear and circular supply chains: A case study from the construction industry. *International Journal of Production Economics, 183*, 443–457.

Onwuegbuzie, A. J., Dickinson, W. B., Leech, N. L., & Zoran, A. G. (2009). A qualitative framework for collecting and analyzing data in focus group research. *International Journal of Qualitative Methods, 8*(3), 1–21.

Peterson, B. L. (2017). Thematic analysis/interpretive thematic analysis. In *The international encyclopedia of communication research methods* (pp. 1–9). Wiley.

Ponterotto, J. G. (2005). Qualitative research in counseling psychology: A primer on research paradigms and philosophy of science. *Journal of Counseling Psychology, 52*(2), 126.

Ruiz, L. A. L., Ramón, X. R., & Domingo, S. G. (2020). The circular economy in the construction and demolition waste sector–a review and an integrative model approach. *Journal of Cleaner Production, 248*, 119238.

Schraven, D., Bukvić, U., Di Maio, F., & Hertogh, M. (2019). Circular transition: Changes and responsibilities in the Dutch stony material supply chain. *Resources, Conservation and Recycling, 150*, 104359.

Teece, D. J. (2010). Business models, business strategy and innovation. *Long Range Planning, 43*(2–3), 172–194.

Tingley, D. D., Cooper, S., & Cullen, J. (2017). Understanding and overcoming the barriers to structural steel reuse, a UK perspective. *Journal of Cleaner Production, 148*, 642–652.

Van Caneghem, J., Van Acker, K., De Greef, J., Wauters, G., & Vandecasteele, C. (2019). Waste-to-energy is compatible and complementary with recycling in the circular economy. *Clean Technologies and Environmental Policy, 21*(5), 925–939.

Vilella, M. (2019). *Waste-to-energy is not a sustainable business the EU says*, Zerowaste Europe. https://zerowasteeurope.eu/wp-content/uploads/2019/09/zero_waste_europe_policy_briefing_sustainable_finance_en.pdf

Virlanuta, F. O., David, S., & Manea, L. D. (2020). The transition from linear economy to circular economy: A behavioral change. *FORCE: Focus on Research in Contemporary Economics, 1*(1), 4–18.

Witjes, S., & Lozano, R. (2016). Towards a more Circular Economy: Proposing a framework linking sustainable public procurement and sustainable business models. *Resources, Conservation and Recycling, 112*, 37–44.

Part II
Building Comfort, Performance and Energy

Chapter 5
Efficient Management of Environmental Control Within Electrical Substations

Mark Collett, William Swan, Richard Fitton, and Matthew Tregilgas

5.1 Introduction

Electrical substations are critical infrastructure assets enabling the transmission and distribution of electricity. There is a requirement to maintain specific environmental conditions of humidity and temperature to protect the equipment housed within the substations. To better understand how to efficiently obtain these conditions, distribution network operator (DNO) Electricity North West Limited (ENWL) have partnered with the University of Salford to undertake this study. This research aims to evaluate the existing methods of environmental control in electrical substations and identify, implement and measure the effectiveness of interventions to improve the efficient management of environmental control within electrical substations.

ENWL own and operate 558 substations where this environmental control is required. The associated energy consumption with these assets is 10.5 million kWh/annum. In the UK it is estimated there are over 5000 substations where this form of environmental control is applicable (Energy Networks Association, 2015). ENWL wish to improve the efficiency of environmental control systems to reduce this energy consumption as well as the associated CO_2 emissions. Furthermore, a more critical benefit is that obtaining more compliant environmental conditions will reduce the risk of damage to electrical distribution equipment from occurring, the effects of which can be catastrophic, including damage to the electricity network, loss of electricity supply to thousands of customers or serious injury.

M. Collett (✉) · W. Swan · R. Fitton
School of Science, Engineering and Environment, University of Salford, The Crescent, Salford, UK
e-mail: m.collett2827@student.leedsbeckett.ac.uk

M. Tregilgas
Electricity North West, Stockport, UK

5.2 Literature Review

5.2.1 Substations

An electrical substation is a subsidiary station of a distribution system where voltage is transformed from high to low, or the reverse, using transformers and/or where circuit switching takes place (Ofgem, 2017). There are multiple categories of substation defined by the voltage levels at which they operate. The categories relevant to this research are grid and primary assets. In a primary substation, both the primary voltage (33 kV) and secondary voltage (11 kV or 6.6 kV) are high voltage (Ofgem, 2017). A grid substation operates a primary voltage of 132 kV to a secondary voltage of 33 kV. For the purposes of this paper, substations refer to grid and primary assets unless stated otherwise.

To understand what is typical of the substation population, asset management data from ENWL was analysed to identify the most popular criterion for various building aspects of the substation population (Electricity North West Limited, 2018a). This is summarised in Table 5.1.

A substation that embodies the criteria listed in Table 5.1 is Pendleton primary. An aerial image of this substation is shown in Fig. 5.1.

5.3 Required Environmental Conditions and Approach to Environmental Control

Within substations there is a need to maintain a low humidity environment to inhibit corrosion of the electrical assets and prevent partial discharges occurring. Partial discharges (PD) are electrical discharges occurring inside or on the surface of electrical insulation materials caused by high-voltage electrical stressing of the insulation system when equipment is energised. When PD occurs it emits energy as light, heat, sound and gaseous discharge, such as ozone and nitrous oxides (Byrne, 2014). Research completed from EA Technology (Byrne, 2014) and Chongqing University (Meng et al., 2013) has demonstrated that in high humidity environments partial discharge is more likely to occur. The effects of partial discharge can be significant,

Table 5.1 Analysis of ENWL substation population

Building aspect	Most popular criterion	Percentage of the substation population that this applies
Building type	Standalone	96
Building construction	Stone/brick	95
Roof type	Flat	72
Roof construction	Asphalt	67
Substation commissioned	1960s	41

5 Efficient Management of Environmental Control Within Electrical Substations

Fig. 5.1 Pendleton Primary Substation

including equipment failures, danger to staff and the public and network failures. As such, it is in the interest of the substation operators to manage environmental conditions accordingly.

In ENWL substations, a target of 50% relative humidity (RH) is required in switch rooms and control rooms (areas where distribution equipment is housed). In these areas a minimum temperature of 10 °C is required to enable dehumidification systems to function. Further conditions to be obtained in the substation are a boosted temperature of 15 °C to provide thermal comfort for operatives working in the building and a minimum temperature of 5 °C in WC areas to prevent freezing (Electricity North West Limited, 2007).

These conditions are obtained through a combination of dehumidification and heating. Condensate dehumidifiers such as the EBAC CD30e or CD100e are used with electric convector heaters (Electricity North West Limited, 2007). The heaters are controlled through a Sangamo Power Saver PSB 01 thermostat (Electricity North West Limited, 2018b). A temperature of 12 °C is set as a safety factor to avoid the minimum of 10 °C being reached.

The EBAC dehumidification units operate in a condensing cycle to remove moisture from the air. In this process moist air is passed over cooling coils to lower its temperature past its dewpoint (step 1), at this stage the water content in the air is condensed, and it is drained away from the process. The cooled air is then heated to lower the specific humidity, ω and RH to the desired level (step 2) (Struchtrup, 1988). The conditioned air is then returned to the room (step 3). This process is visualised in the psychrometric chart shown in Fig. 5.2.

The warmer the air, the greater the cooling that occurs and hence the more effective removal of moisture can occur. Whilst no data on the effect of temperature is available for the EBAC units utilised, similar systems state a reduced moisture

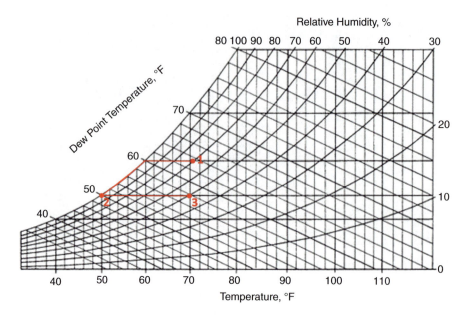

Fig. 5.2 Condensate Dehumidification

extraction rate with temperature by approximately 60% from 30 °C to 12 °C (Dantherm, 2016).

The EBAC units state their extraction rate as 10 and 30 litres/day for the CD30e and CD100e units, respectively. This extraction rate is given at a 30 °C, 80% RH, a condition that will give a maximum extraction rate due to the high RH and temperature allowing for significant cooling of the air in the condensate cycle. This is a condition that rarely occurs in the UK so is likely overstating the systems' extraction rate in regular conditions. They also state effective volumes of 85 m³ (CD30e) and 300 m³ (CD100e)(EBAC, 2019).

Dehumidification demand, expressed in mass of moisture to be removed in a given time, can be calculated through use of Eq. (5.1) (Narayan et al., 2019):

$$D = \rho V n (\omega_1 - \omega_2) \tag{5.1}$$

Where D is dehumidification demand (kgh⁻¹), ρ is air density (kgm⁻³), V is volume of space (m³), n is air change rate (h⁻¹), ω_1 is the specific humidity of the ambient unconditioned air and ω_2 is the required specific humidity of the conditioned air (kg$_{water}$ kg$_{air}$⁻¹).

The air change rate, n, is not easily measured. However, it can be estimated through an air permeability test where a building is pressurised and the flow rate of the fan recorded to determine the air change rate at 50 pascals pressure difference. Whilst this figure is not representative of the air change rate at normal conditions, it can be extrapolated to estimate an average air change rate through analytical means (Sherman & Grimsrud, 1980) or through computational modelling.

The stated effective volumes of the EBAC units are used when selecting systems for deployment in substations. It is not clear what air change rate has been assumed from the manufacturer in the determination of these volumes nor what target RH is applied. Unlike other commercial buildings, air permeability is not a specified design criterion of substations; hence prior to this research, air permeability testing is not a procedure carried out in substations.

There is often a difference between predicted energy performance and actual energy use; this phenomenon is known as a performance gap and is well known to occur in non-domestic buildings (van Dronkelaar et al., 2016). Causes of performance gap can be the limitations of measurement systems caused by placement of sensors and calibration issues such as sensor drift, the change of measured value from a given input over time (Sawhney, 1985).

This approach to environmental control and its effectiveness in substations is to be evaluated through the following method to understand its effectiveness, where efficiencies can be gained and if any performance gap is present.

5.4 Research Review and Methodology

The efficient management of environmental control is researched through a case study method. Case studies are beneficial in exploring subjects and issues where relationships are ambiguous or uncertain (Gray, 2018). The relevant engineering concepts such as building performance, building fabric and efficiency of mechanical/electrical systems are well established; however there is no known research in their application to substations.

5.5 Research Method

Three substations that are archetypal of the substation population displaying the criteria identified in Table 5.1 were selected for investigation as case studies. The substations identified were Pendleton, South East Macclesfield and Windermere.

5.5.1 Environmental Monitoring Campaign

To understand the effectiveness and energy demand of the existing environmental control systems, an environmental and energy monitoring campaign was undertaken prior to interventions being delivered. A Monnit ™ wireless sensor system was installed in each substation. Temperature and RH sensors were fixed in each room of the substation at both high and low levels to account for stratification. Current meters were installed on all relevant electrical circuits for the environmental

Fig. 5.3 Remote Environmental and CT Sensors

control systems. On the exterior of the substations, a weather station was installed to capture the external environmental conditions. These sensors are pictured in Fig. 5.3. All control settings for heaters and dehumidifiers were checked as being in line with ENWL's policy when the monitoring equipment was installed.

For the sites where a full year's data was not available, the monthly average was used to estimate a full year's data for dehumidification demand. For heating demand, data was extrapolated against heating degree day data to estimate a full year's demand (CIBSE, 2006).

To determine the approximate heating demand required for the sites, a computational thermal model was produced using design builder software. U values of the building elements were taken through U value measurement in line with ISO 9869 (BSI, 2014) where possible, but estimations based on the construction of the building fabric and age of the asset were made. Air permeability was calculated via a blower door test in line with ATTMA TSL2 technical standard (ATTMA, 2010).

Dehumidification demand was calculated by using Eq. (5.1). The following variables were used:

- V is the total conditioned volume of the substation in m^3.
- n is the average air change rate calculated through the computation model in hour^{-1}.
- ω_1 is 0.00670 kg$_{water}$ kg$_{air}^{-1}$ the associated specific humidity with 12.0 °C, 77%RH (determined as average ambient climatic conditions through 1 years monitoring).
- ω_2 is 0.00434 kg$_{water}$ kg$_{air}^{-1}$ the specific humidity achieved when lowering the RH of ambient air to the target 50%.
- ρ is the density of air is taken as 1.21 kgm^{-3}.

To determine the associated annual energy demand, this was equated to the power demand of 0.46 kW and extraction rate stated in the EBAC data sheets. The extraction rate was reduced by 60% to 0.17 kg h⁻¹ accounting for the reduced temperature at which they are deployed.

Example calculation for Pendleton substation:

$$D = \rho V n(\omega_1 - \omega_2) = 1.21 * 253.476 * 0.502 * (0.00670 - 0.00434) = 0.363 \text{ kg h}^{-1}$$

$$\text{Annual moisture removal rate} = 0.363 * 24 * 365 = 3{,}183.04 \text{ kg year}^{-1}$$

Associated annual energy demand
$$= 3{,}183.04 \text{ kg year}^{-1} \div 0.17 \text{ kg h}^{-1} * 0.46 \text{ kW} = 8{,}612.93 \text{ kWh}$$

5.5.2 Interventions

To improve the performance of the buildings, a series of interventions were derived from the following principles: reducing undesirable air infiltration, segregation of conditioned and non-conditioned spaces, lowering the thermal conductivity of building fabric and replacing heating control systems. Consideration to the cost of interventions was necessary to comply with budget constraints. These are listed as follows and pictured in Fig. 5.4.

- **Pendleton**

Existing roof covering is removed and replaced with insulated warm roof.

Remedial work to the interior to improve air permeability: filling unused cable ducts/penetrations, sealing around door frames and repointing brickwork.

Rigid insulation boards were used as a trench splitter system acting as a thermal and moisture barrier between the cable trench and the conditioned space above.

Replacement thermostats installed.

- **Windermere**

An insulated partition was constructed between the conditioned control room and a now decommissioned area that is no longer heated or humidity controlled.

Insulated Roof

Trench Splitter System

Sealed unused cable penetrations

Insulated Partition

Blanked Ventilation Points

Replacement Thermostat

Fig. 5.4 Installed interventions pictured

Existing ventilation points were blanked with plywood and sealed with a silicone bead. These are thought to be installed as part of the original construction prior to the requirement of humidity control in substations and hence no longer required.

Remedial work to improve air permeability: sealing of door frames and cable trench space.

Replacement thermostats installed.

- **South East Macclesfield**

Existing ventilation points were blanked with plywood and sealed with a silicone bead. These are thought to be installed as part of the original construction prior to the requirement of humidity control in substations and hence no longer required.

Remedial work to improve air permeability: sealing of cable ducts and door frames.

Replacement thermostats installed.

The effectiveness of these measures in terms of reducing heat demand was calculated through inputting the interventions into the thermal models where possible. Follow-up air permeability tests were undertaken and the results used to determine reduced dehumidification demand.

5.6 Research Results

Table 5.2 shows the heating demand for the three substations. The demand is shown as calculated through the computational modelling for the untouched substation, as measured through the CT sensors and as calculated with interventions applied.

Table 5.3 shows the dehumidification demand for the three substations. The demand is shown as calculated through Eq. (5.1), as measured through CT sensors and as calculated with interventions applied.

Table 5.4 shows the volume weighted average environmental conditions recorded within the conditioned areas of the three substations. Where cells are blank monitoring equipment was yet to be installed.

Table 5.2 Substation heating demand

Substation	Baseline calculated heating demand (kWh)	Measured heating demand (kWh) (% ± against calculated baseline)	Calculated enhanced heating demand (kWh) (% ± against calculated baseline)
Pendleton	2,740	4,433 (+62%)	904 (−67%)
Windermere	1,746	13,018 (+646%)	1,296 (−26%)
South East Macclesfield	9,277	27,188 (+193%)	8,483 (−9%)

5 Efficient Management of Environmental Control Within Electrical Substations

Table 5.3 Substation dehumidification demand

Substation	Baseline calculated dehumidification demand (kWh)	Measured dehumidification demand (kWh) (% ± against calculated baseline)	Calculated enhanced dehumidification demand (kWh) (% ± against calculated baseline)
Pendleton	8,613	7,381 (−16%)	5,971 (−31%)
Windermere	11,382	3,125 (−73%)	6,112 (−46%)
South East Macclesfield	10,542	6,255 (−42%)	5,304 (−50%)

Table 5.4 Substation environmental conditions

	Pendleton				Windermere				South East Macclesfield			
	High		Low		High		Low		High		Low	
Month	°C	% RH	°C	% RH	°C	% RH	°C	% RH	°C	% RH	°C	% RH
Apr-19	19.7	41.7	18.1	49.2								
May-19	19.5	47.1	18.3	53.0								
Jun-19	20.6	49.6	19.5	56.4								
Jul-19	24.9	48.1	23.4	55.1					24.3	57.1	22.8	62.9
Aug-19	23.6	49.8	22.3	57.7	20.3	60.9	19.0	66.4	23.5	55.3	22.0	61.2
Sep-19	20.5	51.1	19.5	57.6	17.8	62.3	16.4	68.5	22.5	52.8	20.9	58.8
Oct-19	16.8	56.8	15.7	65.1	17.0	58.4	14.4	68.9	20.5	52.0	18.8	58.1
Nov-19	13.5	58.3	12.1	67.4	15.4	50.3	11.8	63.3	18.7	49.0	16.7	56.0
Dec-19	12.4	56.9	10.8	67.5	14.8	50.7	11.0	65.5	18.2	47.8	16.1	55.1
Jan-20	12.2	56.6	10.8	67.5	14.9	51.8	11.1	67.0	18.8	45.7	16.5	53.3
Feb-20	11.8	58.9	10.6	69.5	14.5	53.0	10.8	68.1	18.7	43.7	16.4	51.1
Mar-20	13.5	57.4	12.2	70.6	15.6	51.5	12.5	64.1	19.4	40.5	16.8	48.1

Table 5.5 Total savings and simple payback analysis

Substation	Potential savings—heating and dehumidification (kWh)	Cost of interventions	Payback period (years)
Pendleton	4,939	£7,260	9.4
Windermere	8,735	£3,220	2.4
South East Macclesfield	19,656	£4,674	1.5

Table 5.5 shows the overall potential savings from the interventions as well as associated costs and payback period.

The main theme from the results is a performance gap between the heating and dehumidification demand calculated and as measured. This is leading to over and under-conditioning of temperature and humidity levels. Interventions are shown to decrease the calculated demand of environmental control systems.

5.7 Discussion

Table 5.2 shows the recorded heating demand for all three sites is significantly greater than that calculated through the thermal models. This is an example of performance gap occurring driven by sensor drift in the heating controls, meaning that they are unable to maintain a specified set point. As such despite a set point of 12 °C set on the thermostats, Table 5.4 demonstrates that through the winter months both Windermere and South East Macclesfield have been heated beyond this with average temperatures as high as 18.2 °C recorded in December 2019. Additionally, the WC area in Pendleton is noted to be heated excessively; as this is not a conditioned area, this is not accounted for in Table 5.4. However, the average temperature in December 2019 was 23.3 °C. The sensor drift causing this performance gap can be caused by deterioration of components inside the sensor over time.

Table 5.3 shows the substation dehumidification demand is significantly less than that calculated analytically. The corresponding RH values in Table 5.4 show that for substantial periods of the monitoring period, the target of 50% RH is not met. This again shows a performance gap occurring. This is contributed to by placement of control humidistats for the dehumidification systems. In several cases the control humidistat is positioned directly adjacent to the dehumidifier. This causes the space immediately surrounding the unit to be conditioned, but the remainder of the room does not reach the required RH. This can occur both longitudinally and latitudinally, possibly contributing to the stratification of RH demonstrated in Table 5.4. This could be resolved by improved placement of the humidistat and improved distribution of the conditioned air. It is noted that newly commissioned substations are fitted with multiple humidistats to combat this although the dehumidification in the three substations analysed have a single humidistat only.

Further exacerbating the under-conditioning and performance gap is the effect of temperature and moisture removal capacity of condensate dehumidifiers. Pendleton shows higher RH in the winter months than those recorded in summer. Alternative dehumidification systems that use a desiccant material to extract moisture are not affected as significantly as condensate systems and are to be investigated as further work.

Table 5.2 shows a reduction in calculated heating demand across all three substations with interventions applied. It is most significant proportionally at Pendleton where significant investment was made in the enhancement of the roof with insulation reducing conductive heat loss through the building fabric. For Windermere and South East Macclesfield, significant reductions against the calculated baseline are recorded through mostly enhancing the air permeability through low-cost remedial works. However, the savings associated with avoiding excessive heating by replacing existing controls with accurate controls is significant and proportionally would be greater than the savings associated with the enhancement measures across all three sites.

Table 5.3 shows the interventions installed have a significant effect on reducing the calculated dehumidification demand. This is due to air permeability improvements reducing the air change rate that is proportional to the dehumidification demand. These benefits would result in more compliant environmental conditions as infiltration of unconditioned air is reduced. However, to consistently obtain said conditions, the humidity controls should be enhanced to ensure the entirety of a space is conditioned; this would in turn increase consumption but is necessary for the purposes of obtaining the required environmental conditions.

Table 5.5 shows that significant savings are possible in all three substations from the interventions derived and applied. It is noted that the costs of the replacement roof covering at Pendleton are excluded. This is as the works were primarily required as a maintenance activity and, although the addition of insulation is enhancing the performance of the building, splitting the costs of the improvement from the maintenance activity is complex. All sites demonstrate a payback period of under 10 years.

5.8 Conclusions

The environmental condition data obtained shows that the three substations analysed are largely being over-conditioned in temperature and under-conditioned in respect of RH creating a performance gap. The overheating is caused by sensor drift in the heating controls resulting in significant unnecessary energy consumption. The under-conditioning of humidity means the electrical equipment within the substations analysed is at a higher risk of PD occurring as the target of 50% RH is not met. The environmental monitoring of the sites has provided a calibration check on the existing environmental control systems and demonstrated that they are not operating efficiently.

The interventions delivered show that significant reduction in heating and dehumidification demand is achievable and can result in savings of over 19,000kWh/annum and are able to achieve payback periods of under 10 years. These interventions and subsequent energy savings are demonstrated as effective in improving the efficient management of environmental control within substations.

A limitation of the work is that monitoring results analysed are currently pre-intervention only. As further works, continued energy and environmental monitoring is to be conducted to verify the benefits of the installed interventions. Additionally, an alternative dehumidification system is to be investigated, and the research will expand to cover the assessment of a Net Zero Carbon substation currently being constructed by ENWL.

References

ATTMA. (2010). *Measuring air permeability of building envelopes (non-dwellings)*. Northampton.

BSI. (2014). *ISO 9869-1:2014- Thermal insulation — Building elements — In situ measurement of thermal resistance and thermal transmittance; Part 1: Heat flow meter method*. BSI Standards Limited.

Byrne, T. (2014). Humidity effects in substations. In *11th petroleum and chemical industry conference Europe electrical and instrumentation applications* (pp. 1–10). PCIC. https://doi.org/10.1109/PCICEurope.2014.6900056

CIBSE. (2006). *TM41: Degree days: Theory & application*. Author.

Dantherm. (2016). *Mobile dehumidifier selection guide*. https://www.dantherm.com/media/1975455/mobile-dehumidifier-cdt-selection-guide.pdf

EBAC. (2019). *CD30 & CD30e*. http://www.eipl.co.uk/downloads/brochures/CD30.pdf

Electricity North West Limited. (2007). *Code of practice 351 civil design aspects of primary substations*. Author.

Electricity North West Limited. (2018a). *Ellipse asset management database*. Ellipse.

Electricity North West Limited. (2018b). *Proposed revised procedures for environmental control within primary and BSP substations*.

Energy Networks Association. (2015). *Climate change adaptation reporting power second round*.

Gray, D. E. (2018). *Doing research in the real world*. Sage.

Meng, F., Zhang, X., Wu, X., & Xu, B. (2013). Experimental studies on air humidity affecting partial discharge in switchgear. In *Annual report - Conference on electrical insulation and dielectric phenomena, CEIDP* (pp. 1237–1241). https://doi.org/10.1109/CEIDP.2013.6747094.

Narayan, G. P., Sharqawy, M. H., Lienhard V, J. H., & Zubair, S. M. (2019). Thermodynamic analysis of humidification dehumidification desalination cycles. *Desalination and Water Treatment, 16*(1–3), 339–353. https://doi.org/10.5004/dwt.2010.1078.

Ofgem. (2017). Guide to the RIIO-ED1 electricity distribution price control.

Sawhney, A. (1985). *A course in electrical and electronic measurements and instrumentation*. J.C Kapur.

Sherman, M., & Grimsrud, D. (1980). *Measurement of infiltration using fan pressurization and weather data*. https://eetd.lbl.gov/sites/all/files/publications/lbnl-10852.pdf

Struchtrup, H. (1988). *Thermodynamics and energy relations* (Vol. 38). Springer.

van Dronkelaar, C., Dowson, M., Spataru, C., & Mumovic, D. (2016). A review of the regulatory energy performance gap and its underlying causes in non-domestic buildings. *Frontiers in Mechanical Engineering, 1*, 1–14. https://doi.org/10.3389/fmech.2015.00017.

Chapter 6
Nearly Zero-Energy Buildings (nZEB) and Their Effect on Social Housing in Ireland: A Case Study Review

Joseph Glennon, Michael Curran, and John P. Spillane

6.1 Introduction

Housing is vital to everyone, to our economy and environment, whilst representing so much more than just a place to live (Hills, 2007). For more than a century, social housing has offered individuals and families a home that they need, and the foundations of social housing provision in Ireland can be traced back to the mid-1800s (Norris, 2005). Statistics from the Irish Council for Social Housing (2013) identify that social housing accounts for almost 10% of Irish homes. Furthermore, 75% of these dwellings are procured and managed directly by local authorities (Central Statistics Office, 2006), with the majority of funding derived from direct government grants (Norris & Fahey, 2011). Scanlon et al. (2015) support that social housing has been an important part in housing provision for many decades, regarding both new build and regeneration but also in providing adequate affordable housing for a wide range of citizens. However, Reeves et al. (2010) argue that, in the coming decades, there will be considerable challenges in achieving deep cuts in carbon emissions from existing housing stock, as part of the global effort to combat climate change. Climate change is predicted to have a negative effect on the world unless proper mitigation measures are implemented (IPCC, 2014). To combat these issues, Moran et al. (2020) note that buildings in the European Union (EU) are now being designed and constructed to tighter building standards and codes. One key measure aimed at reducing energy consumption in buildings is the mandatory introduction of nearly zero-energy buildings (nZEB) (Attia et al., 2017).

J. Glennon · J. P. Spillane
School of Engineering, University of Limerick, Castletroy, Limerick, Ireland

M. Curran (✉)
School of Natural and Built Environment, Queens University Belfast, Belfast, UK
e-mail: mcurran23@qub.ac.uk

An nZEB is a building with a very high energy performance, and the nearly zero or very low amount of energy required should be covered to a very significant extent by energy from renewable sources produced on site or nearby (D'Agostino & Mazzarella, 2019). Kurnitski et al. (2011) support that an nZEB is estimated to use at least two to three times less energy compared to standard buildings. The Irish Green Building Council (2018) states that EU legislation requires all new buildings to be nZEB by 31 December 2020 and all buildings acquired by public bodies by 31 December 2018 (EPBD, 2010). Goggins et al. (2016) argue that there is a great deal of research available in literature on reducing the impact of the energy associated with the operational stage of a building's lifecycle. However, on review, the literature fails to acknowledge and highlight the integration of nZEB and social housing, particularly within an Irish context. Therefore, it is necessary to identify and evaluate the emergence of nZEB standards and regulations in Ireland and their cost-effectiveness and compliance with social housing construction projects. To address these issues and to fulfil a succinct but prevailing gap in an emerging research area, it is imperative to provide results based on actual events that emerge, when studying a complex environment such as the construction of social housing in Ireland. Thus, concentrating on a relevant area of interest, the aim of this study is to evaluate if social housing in Ireland can comply with new nZEB standards, without increasing the cost of construction. This is achieved by undertaking a sequential mixed method approach, incorporating a combination of both qualitative and quantitative techniques for analysis, including a literature review, semi-structured individual interviews, a focus group seminar and a questionnaire survey. In addressing this aim, it is anticipated that this study will assist both industry practitioners and policymakers in identifying how nZEB standards can positively and financially assist in the construction of social housing projects in Ireland.

6.2 Nearly Zero-Energy Buildings and Social Housing

Across Europe, Salem et al. (2019) identify that various studies have considered whole building retrofit on new and existing case studies to reach nZEB standards. Lindkvist et al. (2014) investigate the barriers and challenges to nZEB projects in Sweden and Norway; Bahadori-Jahromi et al. (2018) explore the retrofit of a UK residential property to achieve nZEB standard; and Cortiços (2018) researches social housing and nZEB within a Portuguese context. In Ireland, there have been some publications investigating the potential for the voluntary Passive House (pH) standard (Colclough et al., 2011; Clarke et al., 2014), which is another recent energy-efficient methodology; however, few have considered nZEB standards and the costs associated with social housing in particular. When constructing a standard residential dwelling to the minimum Irish current building regulations compared to an nZEB standard built using pH methods, Colclough et al. (2018) identify a saving of just €130 by constructing to nZEB standard (€114,862 vs. €114,992). Attia et al. (2017) argue that nZEB require high-quality construction through new specialised construction technologies, and the use of energy efficient technologies and materials is necessary. Silva et al. (2015) support that the know-how of professionals that are able to deal with new technologies and standards is a barrier, and the construction quality is a serious challenge (Ford et al., 2007). In an N. Irish study, Colclough and McWilliams (2019) identify that a three-bed social house (compliant with UK building regulations) costs £7536 to upgrade to energy efficiency multiplier (EEM) standards but costs £10,374 when upgrading to pH standard. Extra costs include increased air tightness and insulation levels, a heat recovery and ventilation system, high-performing windows and doors and using gas central heating. Cost reductions are achieved via the elimination of a traditional heating system, chimney stack and reduced site overheads.

However, Dalla Mora et al. (2018) argue that higher initial investment costs will equate to lower primary energy use, thus, providing a cost saving over time. Moreover, barriers to the implementation of nZEB do exist, even though governments are promoting energy-efficient buildings (Ralph, 2012). Blomsterberg (2012) notes that there are ambiguities, even though legislation is moving towards implementation of nZEB targets, and Lindkvist et al. (2014) substantiate that nZEB only address new buildings, and there are no clear plans for nZEB renovations. Aelenei et al. (2015) discuss many barriers to nZEB including high initial costs of investments, long payback time of investments, limited technical skill in the decision process, lack of awareness concerning economic benefits and uncertainties considering the measurement and verification of nZEB. Attia et al. (2017) believe that countries are poorly prepared for nZEB, and from an economic standpoint, renovation of existing buildings to nZEB is far more expensive than a regular build (Dalla Mora et al., 2015). In their study of Irish retrofit industry professionals and nZEB, Zuhaib et al. (2017) found that 90% of respondents recorded 'cost involved' as the driving factor for their choices in nZEB construction. Furthermore, Curtin (2009) concurs that high upfront costs and homeowners' reluctance for long-term cost savings over short-term expenditures are key barriers to nZEB in Ireland. Nevertheless, Aelenei et al. (2015) argue that nZEB bring many benefits, including energy cost saving, lower dependence on energy suppliers, improved comfort, tax deductions, low interest loans and best practices related to renovation to nZEB. Colclough et al. (2018) outline the potential of nZEB and social housing in Ireland through their study on nZEB and PH standards, providing a foundation to expand the knowledge area further within the context of this research. As building standards continue to improve in the EU, the introduction of statutory nZEB legislation will not only benefit Ireland but many countries in Europe and beyond.

6.3 Research Method

This study is part of a primary investigation which aims to contribute to both industry and academia. A mixed method is adopted, as Creswell (2014) believes that when an inquirer combines both quantitative and qualitative methods, it provides a better understanding of the problem than using either method alone. On completion of an informative literature review, the research method adopted consists of case study analysis, including two exploratory individual interviews and one focus group seminar, with a variety of construction professionals based on social housing development projects situated throughout Ireland. The selection of the three case studies was based on a combination of criterion and convenience sampling strategies: firstly, by identifying construction projects that were social housing developments, and, secondly, by arranging interviews and focus groups depending on the participants' availability at the time of each visit. Similarly, the case study approach is chosen as Yin (2014) suggests it is the most suitable for the 'how' and 'why' research questions. McIntosh and Morse (2015) recommend a semi-structured interview format, as it determines people's subjective reactions to situations, thus, extending the researcher's knowledge on the topic.

From an ethical perspective, the participants are informed of the nature of the research, its purpose and what the resultant data will be used for, prior to commencement of interviews. The identities of those involved remain anonymous, and confidential information such as company names, addresses, client details, etc. are not disclosed. Case A consists of a new social housing development scheme with 18 dwellings in the Leinster region, and the interviewees include a site manager with 12 years of experience, a contracts manager with 15 years of experience and a quantity surveyor with 9 years of experience. Case B is a new social housing development scheme with 22 dwellings also in the Leinster region, and the interviewee is a graduate engineer with 2 years of experience. Case C is a refurbishment project of a social housing development in the Munster region, and the interviewee is a project manager with 20 years of experience. In this case study, the refurbishment entails the transition of open fires to central heating systems. Cases A and B are live projects, and Case C is recently completed. Following the interviews, a questionnaire survey was distributed to various construction companies to further consolidate the findings. Questionnaires are a widely used means of collecting data, and it is an easy way to get responses from many people (Rowley, 2014).

6.4 Results and Analysis

The interviews and focus groups commenced by gaining general background information about each participant and their relevant case study, followed by a discussion on nZEB and social housing in Ireland. Findings from the interviews, focus group and literature review were then combined to generate the questionnaire survey, and this was circulated out to industry. All the resultant data from each research method was amalgamated and thematically analysed, identifying keywords, topics and themes for discussion. From the data gathered through the various research methods, the analysis concentrates on a range of key themes including materials, superstructure, mechanical and electrical (M&E) services, subcontractors,

6 Nearly Zero-Energy Buildings (nZEB) and Their Effect on Social Housing...

Table 6.1 Social houses built to various standards

		Standard build with heat pump	Standard build with solid fuel heating	NZEB build
Thermal envelope	Type I	(W/m²k)	(W/m²k)	(W/m²k)
Insulation U value	Roof	0.12	0.12	0.10
	Ground floor	0.14	0.14	0.12
	Walls	0.13	0.13	0.11
	Windows and doors	1.20	1.20	1.00
Ventilation system		MVHR	Natural ventilation	MVHR
Heating system		Heat pump	Solid fuel	Heat pump
DHW (hot water)		Heat pump	Solid fuel	Heat pump

Table 6.2 Cost comparison

	NZEB build	Standard build with heat pump	Standard build with solid fuel heating
Superstructure	€ 6400	€ 6400	€ 6700
Windows	€ 8512	€ 3735	€ 3735
Insulation (wall)	€ 3168	€ 1825	€ 1825
Internal plastering	€ 10,000	€ 8400	€ 8400
Mechanical	€ 8400	€ 8400	€ 13,100
Electrical	€ 5900	€ 5900	€ 6500
Heating	€ 4400	€ 4400	€ 7000
Site overheads	€ 23,888	€ 22,333	€ 22,333
Insulation (roof)	€ 840	€ 500	€ 500
Attic hatch	€ 125	€ 80	€ 80
Total cost	€ 71,633	€ 61,973	€ 70,173
Cost difference		€ 9660	€ 1460

internal plastering, preliminaries and time. In particular, three main types of social housing were discussed: standard houses with a heat pump, standard houses with solid fuel heating and new houses built to NZEB standard. Due to space limitations, a concise summary of their properties is illustrated in Table 6.1.

A cost comparison is undertaken between the three types of houses, and Table 6.2 identifies where the cost differentiates between each type. It is worth documenting that the findings from the individual interviews, focus group seminar and questionnaire survey are specific to this research – thus, not a generalised view. Nevertheless, this study provides a foundation to advance and expand further, supporting continuous research into nZEB and social housing construction projects in Ireland.

6.5 Discussion

6.5.1 Materials

The most prominent theme that emerged from both the individual interviews and focus groups seminar was the costs and sourcing of various materials needed to comply with nZEB standards. Goggins et al. (2016) agree that more attention is being paid to the energy related to the production of the materials used to create and maintain buildings. Some key elements that were highlighted in the focus group seminar included the insulation of walls and floors, and a debate surrounding double- and triple-glazed windows. The quantity surveyor was able to explain in detail the costs involved on new houses being built to nZEB standard. In a standard build, insulation with partial fill of 100 mm in a 140 mm cavity costs approximately €1825, whereas in an nZEB, insulation fully filled in a 150 mm cavity costs €3170, resulting in an extra €1345 when building to nZEB standard. Regarding the dimensions discussed, the nZEB uses 150 mm insulation in its substructure at a cost of €1650, compared to the standard build which uses just 100 mm insulation at cost of €1100, resulting in an extra €550 when building to nZEB standard.

Attic insulation uses three layers of 200 mm quilt insulation in an nZEB, costing €840 to supply and fit. In a standard build, two layers of 200 mm quilt insulation are used, costing €500 to supply, resulting in an extra €340 when building to nZEB standard. Triple-glazed windows fitted in an nZEB cost approximately €8500, and double-glazed windows in a standard build cost approximately €3750, resulting in a massive difference of €4750. Additionally, attic hatches of the dwelling units were also discussed. An attic hatch for an nZEB must be air-tight and costs €125 extra to install, whereas a standard attic hatch which is sufficient in a traditional build costs €80, resulting in an extra €45 when considering nZEB. In Case A, this amounts to an extra cost of €810 when installing across 18 units. The questionnaire supports the premise that materials for nZEB construction is more expensive than standard builds, and 80% of the respondents concur that materials associated with nZEB will ultimately increase the overall cost of the project.

6.5.2 Superstructure

Another prominent theme shared by the interviewees was the superstructure. The interviewees stated potential savings can be made on the superstructure of a build when considering the necessity of a chimney. The site manager discussed how the provision of a chimney increases overall costs as a result of additional materials, time and labour. Colclough et al. (2018) support that extra costs are incurred on a standard dwelling compared to an nZEB dwelling, with the addition of a chimney. The contracts manager stated that including a heat pump to nZEB standard would cost €6400, whereas including a chimney in the superstructure for solid fuel would cost €6700, thus, leading to a saving of €300 on an nZEB dwelling. Furthermore, 87% of the questionnaire respondents agreed that overall the construction cost of a social house to nZEB standard would be less than a social house built to current standards.

6.5.3 Mechanical and Electrical (M&E) Services

The contracts manager argued that mechanical and electrical (M&E) services are one of the biggest costs to consider when constructing social housing and across all construction projects in general. In this instance, complying with nZEB standards resulted in cheaper M&E costs compared to a traditional build. The supply and fitting of a mechanical ventilation system incorporating mechanical ventilation with heat recovery (MVHR) costs €8400. Also, the price of buying and commissioning a heat pump costs €4400, and the electrical services of supplying and fitting cost €5900, totalling €18,700 in an nZEB build. On the other hand, in a traditional build with solid fuel heating, costs amount to around €13,100 for radiators and the supply and fitting of a solid fuel heater and cylinder. The electrical services cost €6500, and mechanical ventilation fans are also required at an extra cost of €800. This results in a total cost of €20,400 for M&E services in a traditional build and at a cost of €1700 extra overall. These findings support the research of Colclough and McWilliams (2019), where M&E costs on social houses in Northern Ireland are considerably lower, when building to pH standard.

6.5.4 Subcontractors

The capabilities of subcontractors when complying to nZEB standards were another topic of discussion identified and highlighted. The graduate engineer remarked that some of the current subcontractors lack the skills and knowledge needed to implement nZEB standards. The Project Manager on the refurbishment scheme echoed these sentiments but added that training and educational programmes will have to be introduced if subcontractors are to fully comply with the statutory requirements of nZEB guidelines. Frappé-Sénéclauze (2015) identifies similar issues, stating difficulty in finding trained trades and subcontractors for pH construction. Regarding cost, 75% of the questionnaire respondents believe that the price and costs associated with subcontractors would rise when constructing an nZEB. However, the interviews identified that, in some cases, the price of subcontractors can be less in certain areas when considering an nZEB. The quantity surveyor suggested that the price of block layers and M&E subcontractors are less in an nZEB than they are in standard build with oil heating. On the other hand, subcontractors such as window fitters and attic insulators almost always cost more when it is an nZEB. The contracts manager believed that choosing the correct subcontractor can have an effect on overall costs, supporting Clarke and Herrmann (2004) who state that construction firms can profit from subcontractors.

6.5.5 Internal Plastering

Plastering of the internal walls was highlighted by some of the interview participants, and they claim that this can affect the outcome of pricing between an nZEB and a standard build. In a standard build, the internal studs are covered with a 12.5 mm slab, and the walls are taped, jointed and skimmed for €8400. In an nZEB, all internal studs have insulation at the higher cost of €10,000; however in the long term, this provides a better insulated wall. Evola et al. (2014) found that the type of plaster along with the type of brick can provide a cost-effective design solution in residential nZEB buildings.

6.5.6 Preliminaries

All of the interviewees noted that the cost of preliminaries would be higher when introducing nZEB standards, complementing the work of Colclough et al. (2019). Preliminaries on a social house with nZEB standard cost approximately €23,900 per unit, compared to approximately €22,300 for a house built to traditional standards, resulting in the nZEB being €1600 more expensive. 80% of the questionnaire respondents also agreed that preliminaries of an nZEB would be higher than a traditional build.

6.5.7 Time

Along with increased costs, some of the interviewees noted that increased time taken to complete projects is associated with nZEB construction, compared to standard builds. The site manager cited extra work involved in the attic insulation, cavity insulation and time taken to produce and install triple-glazed windows compared to double-glazed. Colclough and McWilliams (2019) support that, in pH standard construction, extra time is required by experienced teams to provide extra detailing on certain elements of construction. Furthermore, 80% of the questionnaire respondents agreed that building social housing to nZEB standard will increase both time and costs, compared to traditional construction.

6.6 Conclusion and Recommendations

Essentially, this exploratory study focuses on the emergence of nearly zero-energy building (nZEB) standards and regulations and their potential cost-effectiveness and compliance with social housing construction projects in Ireland. EU legislation requires all new buildings to be nZEB by the end of 2020 and all buildings acquired by public bodies by the end of 2018. Therefore, it is the responsibility of construction companies, governments and all associated stakeholders that nZEB standards are adhered to in all future construction projects. This study concentrates on social housing projects in particular and to establish if houses can be constructed to nZEB standard without increased costs. The research compares a new house to nZEB standard against a traditional standard build with a heat pump and another traditional standard build with solid fuel heating. Considering the results captured from the case studies and data analysis, a range of cost-effective solutions is identified. Regarding the superstructure, construction of an nZEB dwelling saves €300 as there is no need for a chimney, €1700 is saved for M&E services, and €2600 is saved for heating, as less work is required fitting heat pumps compared to fitting solid fuel appliances. However, in other areas of construction, complying to nZEB standards costs considerably more, including windows (€4750), insulation (€550), internal plastering (€1600) and attic hatches (€45). Overall, the price of a new build social house per unit to nZEB standard is approximately €71,633, compared to the cheaper price of €61,973 for a standard build with a heat pump, and €70,173 for a standard build with solid fuel heating. Even though the price of an nZEB social house is more expensive in the short term at the construction stage, it has the potential to render financial savings in the long term, due to the quality of the overall build and use of energy efficient methods and materials.

However, the findings from the interviews and focus group are case study specific, and only a concise, subjective view of the topic is produced, not a generalised one. Nevertheless, this study provides a foundation to advance and expand into more detailed research and supports continuous investigation into the impact nZEB standards have on social housing construction projects in Ireland. As the new nZEB requirements are bound by EU legislation, the results of this study are beneficial to countries all over Europe, not just Ireland. The findings in this paper can be developed further, where a broader analytical context can be addressed in a subsequent journal publication, and additional theoretical points of departure can be articulated. As there were only three case studies in review, it is recommended that more individual interviews and focus group seminars are considered for qualitative analysis, and a sequential selection strategy is incorporated using criterion selection, such as quota and random sampling. A detailed questionnaire survey should also be composed and distributed to a larger sample to further strengthen the research, introducing a quantitative aspect to the study. Therefore, this provides the basis for informing and verifying the validity and necessity of the research and subsequent investigation going forward. The key contribution of this research identifies that new nZEB buildings have many advantages and disadvantages; nevertheless, in the short term, they are more expensive to construct, but in the long term, they have potential financial savings for new social housing construction projects in Ireland.

References

Aelenei, L., Petran, H., Tarrés, J., Riva, G., Ferreira, A., Camelo, S., Corrado, V., Šijanec-Zavrl, M., Stegnar, G., Gonçalves, H., & Magyar, Z. (2015). New challenge of the public buildings: nZEB findings from IEE RePublic-ZEB project. *Energy Procedia, 78*, 2016–2021.

Attia, S., Eleftheriou, P., Xeni, F., Morlot, R., Ménézo, C., Kostopoulos, V., Betsi, M., Kalaitzoglou, I., Pagliano, L., Cellura, M., & Almeida, M. (2017). Overview and future challenges of nearly zero energy buildings (nZEB) design in Southern Europe. *Energy and Buildings, 155*, 439–458.

Bahadori-Jahromi, A., Salem, R., Mylona, A., Godfrey, P., & Cook, D. (2018). Retrofit of a UK residential property to achieve nearly zero energy building standard. *Advances in Environmental Research, 7*(1), 13–28.

Blomsterberg, Å. (2012). *Barriers to implementation of very low energy residential building and how to overcome them.* North Pass, Lund University.

Central Statistics Office (CSO). (2006). *Census of population – Housing volume.* Central Statistics Office.

Clarke, J., Colclough, S., Griffiths, P., & McLeskey, J. T., Jr. (2014). A passive house with seasonal solar energy store: In situ data and numerical modelling. *International Journal of Ambient Energy, 35*(1), 37–50.

Clarke, L., & Herrmann, G. (2004). Cost vs. production: Disparities in social housing construction in Britain and Germany. *Construction Management and Economics, 22*(5), 521–532.

Colclough, S., Kinnane, O., Hewitt, N., & Griffiths, P. (2018). Investigation of nZEB social housing built to the Passive House standard. *Energy and Buildings, 179*, 344.

Colclough, S., & McWilliams, M. (2019). Cost optimal UK deployment of the passive house standard. In *The international conference on innovative applied energy (IAPE'19)* (p. 261).

Colclough, S., Mernagh, J., Sinnott, D., Hewitt, N. J., & Griffiths, P. (2019). The cost of building to the nearly-zero energy building standard: A financial case study. In *Sustainable building for a cleaner environment* (pp. 71–78). Springer.

Colclough, S. M., Griffiths, P. W., & Hewitt, N. J. (2011). A year in the life of a passive house with solar energy store. In *Energy storage conference*, IC-SES.

Cortiços, N. D. (2018). Social housing to nZEB-Portuguese context. In *Proceedings of the creative construction conference*, 30 June–3 July 2018.

Creswell, J. W. (2014). *Research design: Qualitative, quantitative, and mixed methods approaches* (4th ed.). Sage.

Curtin, J. J. (2009). *Jobs, growth and reduced energy costs: Green print for a national energy efficiency retrofit programme.* The Institute of International and European Affairs.

D'Agostino, D., & Mazzarella, L. (2019). What is a nearly zero energy building? Overview, implementation and comparison of definitions. *Journal of Building Engineering, 21*, 200–212.

Dalla Mora, T. D., Cappelletti, F., Peron, F., Romagnoni, P., & Bauman, F. (2015). Retrofit of an historical building toward NZEB. *Energy Procedia, 78*, 1359–1364.

Dalla Mora, T. D., Pinamonti, M., Teso, L., Boscato, G., Peron, F., & Romagnoni, P. (2018). Renovation of a school building: Energy retrofit and seismic upgrade in a school building in Motta Di Livenza. *Sustainability, 10*(4), 969.

EPBD (Energy Performance Building Directive). (2010). Directive 2010/75/EU of the European parliament and of the council. *Official Journal of the European Union, 334*, 17–119.

Evola, G., Margani, G., & Marletta, L. (2014). Cost-effective design solutions for low-rise residential net ZEBs in Mediterranean climate. *Energy and Buildings, 68*, 7–18.

Ford, B., Schiano-Phan, R., & Zhongcheng, D. (2007). The 4. passivhaus standard in European warm climates, Design guidelines for comfortable low energy homes – Section 2 and 3. Passive-On Project report. In *School of the built environment* (pp. 305–317). University of Nottingham.

Frappé-Sénéclauze, T. P. (2015). *Programs or policies to encourage Passive House in North America*. Pembina Institute.

Goggins, J., Moran, P., Armstrong, A., & Hajdukiewicz, M. (2016). Lifecycle environmental and economic performance of nearly zero energy buildings (NZEB) in Ireland. *Energy and Buildings, 116*, 622–637.

Hills, J. (2007). *Ends and means: The future roles of social housing in England*. ESRC Research Centre for Analysis of Social Exclusion, CASE report 34 ISSN 1465-3001.

ICSH (Irish Council for Social Housing). (2013). *ICSH pre budget submission 2014*. https://www.icsh.ie/content/publications/icsh-pre-budget-submission-2014

IGBC (Irish Green Building Council). (2018). *Nearly zero energy building standard*. https://www.igbc.ie/nzeb/

IPCC. (2014). Climate change 2014, The mitigation of climate change. In *Contribution of working group III to the fifth assessment report of the intergovernmental panel on climate change*, The Intergovernmental Panel on Climate Change.

Kurnitski, J., Allard, F., Braham, D., Goeders, G., Heiselberg, P., Jagemar, L., Kosonen, R., Lebrun, J., Mazzarella, L., Railio, J., & Seppänen, O. (2011). How to define nearly net zero energy buildings nZEB. *REHVA Journal, 48*(3), 6–12.

Lindkvist, C., Karlsson, A., Sørnes, K., & Wyckmans, A. (2014). Barriers and challenges in nZEB projects in Sweden and Norway. *Energy Procedia, 58*, 199–206.

Norris, M. (2005). Social housing. In M. Norris & D. Redmond (Eds.), *Housing contemporary Ireland: Policy, society and shelter* (pp. 160–182). Institute of Public Administration.

Norris, M., & Fahey, T. (2011). From asset based welfare to welfare housing? The changing function of social housing in Ireland. *Housing Studies, 26*(3), 459–469.

McIntosh, M. J., & Morse, J. M. (2015). Situating and constructing diversity in semi- structured interviews. *Global Qualitative Nursing Research, 2*, 1–10.

Moran, P., O'Connell, J., & Goggins, J. (2020). Sustainable energy efficiency retrofits as residential buildings move towards nearly zero energy building (NZEB) standards. *Energy and Buildings, 211*, 109816.

Ralph, G. (2012). *Energy efficiency buildings public private partnership project review*.

Reeves, A., Taylor, S., & Fleming, P. (2010). Modelling the potential to achieve deep carbon emission cuts in existing UK social housing: The case of Peabody. *Energy Policy, 38*(8), 4241–4251.

Rowley, J. (2014). Designing and using research questionnaires. *Management Research Review, 37*(3), 308–330.

Salem, R., Bahadori-Jahromi, A., Mylona, A., Godfrey, P., & Cook, D. (2019). Investigating the potential impact of energy-efficient measures for retrofitting existing UK hotels to reach the nearly zero energy building (nZEB) standard. *Energy Efficiency, 12*(6), 1577–1594.

Scanlon, K., Fernández Arrigoitia, M., & Whitehead, C. M. (2015). Social housing in Europe. *European Policy Analysis, 17*, 1–12.

Silva, S. M., Almeida, M. G. D., Bragança, L., & Carvalho, M. (2015). nZEB training needs in the southern EU countries-South ZEB project. *Latin-American and European Encounter on Sustainable Building and Communities-Connecting People and Ideas, 3*, 2469–2478.

Yin, R. K. (2014). *Case study research: Design and methods* (5th ed.). Sage.

Zuhaib, S., Manton, R., Hajdukiewicz, M., Keane, M. M., & Goggins, J. (2017). Attitudes and approaches of Irish retrofit industry professionals towards achieving nearly zero- energy buildings. *International Journal of Building Pathology and Adaptation, 35*(1), 16–40.

Chapter 7
Performance of Distributed Energy Resources in Three Low-Energy Dwellings During the UK Lockdown Period

Rajat Gupta and Matt Gregg

7.1 Introduction

In 2019 the UK parliament passed legislation requiring the government to achieve net zero greenhouse gas emissions status by 2050; previously the target was 80% of 1990 levels by 2050 (UK Government, 2019). Unfortunately, there have been roadblocks along the way, namely, the significant, almost five-year void in guidance left after Zero-Carbon Homes and Code for Sustainable Homes (CSH) were withdrawn. As a result, many new homes have been built to only minimum standards (CCC, 2019). Ideally, the way forward involves a change to those minimum standards. This Future Homes Standard set for 2025 is expected to create an average home that would produce 75–80% less CO_2 emissions than one built to the 2013 UK Building Regulations (Ministry of Housing Communities and Local Government, 2019). In order to meet this target, the Committee on Climate Change (CCC) (2019) recommends that from 2025 no new homes should be connected to the gas grid. Instead these homes should use low-carbon systems like heat pumps or be connected to community heat networks.

At the same time the domestic sector is behind the other sectors on reducing emissions (BEIS, 2018), there has been an increase in the number of people working from home. The number of people working from home has nearly quadrupled in the last 20 years (CCC, 2019). This statistic, of course, does not take into consideration the Covid-19 lockdown and cultural shift which may occur as a result of employees and employers becoming accustomed to staff working from home (Kelly, 2020). Except for the self-employed, this means that more energy is being used

---

R. Gupta (✉) · M. Gregg
Low Carbon Building Research Group, Oxford Institute for Sustainable Development, School of Architecture, Oxford Brookes University, Oxford, UK
e-mail: rgupta@brookes.ac.uk

during the day at home instead of in the office. One positive outcome from this would be a higher rate of self-consumption (SC) in renewable energy, namely, photovoltaics, especially if coupled with a heat pump.

As the recommendations for meeting the low-carbon targets include a significant shift from gas to electrification of heat and transport, several future scenarios forecast an increase in peak demand. To help offset this demand, there will need to be a large increase in low-carbon generation which tends to be variable and intermittent. Balancing the system, especially in the summer, will also be a challenge, wherein a projected 30 GW of solar connected to the distribution network would mean the difference of almost 20 GW transmission demand depending on whether it is a cloudy or sunny day. Solutions to these problems include energy storage, demand side response (DSR), smart networks and increased interconnection (DECC, 2015). Batteries that charge when there is surplus PV generation and discharge during a dwelling's peak demand, or charge based on time-of-use (TOU) settings to coincide with TOU tariffs, are considered smart technologies in this paper.

In response to these current challenges, this paper empirically examines the effectiveness of distributed energy resources (DERs) comprising smart home batteries coupled with rooftop solar PV on actual energy use and peak demand in three dwellings designed to high thermal standards. The paper also explores the change in daily energy use and performance of DERs during the Covid-19 lockdown period (23 March to 31 May 2020). All three dwellings are located in an eco-development in York (UK), occupied continuously by families, and have identical heating systems (district heating), rooftop solar PV systems (4 kWp) and home batteries (14 kWh). The dwellings were constructed as part of a 5-year (2015–2020) research project called Zero Plus, funded by the European Union's Horizon 2020 Research and Innovation programme. The overall aim of the project was to reduce net regulated energy consumption of the dwellings to 20 kWh/m^2/year and achieve renewable energy generation of 50 kWh/m^2/year by deploying advanced energy technologies.

7.2 Literature Review on Solar PVs and Batteries

There are several ways to benefit from photovoltaic (PV) generation and storage in dwellings depending on country or local policy; these include PV with no self-consumption (direct feed-in to the grid), PV self-consumption, PV self-consumption with active load management (i.e. DSR) and PV self-consumption with battery storage (Johann & Madlener, 2014). The right approach depends on the policy and incentives in an area. As an example, as there is currently no feed-in tariff (FiT) in the UK, the first option would not be viable. Grid-export rates through FiTs have been diminishing in several other regions like they have in the UK. Therefore, maximising self-consumption (SC) from electricity production is becoming an increasingly important consideration for standalone electricity generating systems and also those with connected batteries (O'Shaughnessy et al., 2018). For an average UK household with electricity demand of 4000 kWh/year and 2.9 kWp PV system, the SC has been estimated to be about 35–38% (McKenna et al., 2018). One UK study of PV and battery system combinations showed that over the summer months, a sample of 44 dwellings in Oxford had SC ratios which ranged widely from 19% to 70% (average of 43%) (Gupta et al., 2019). For an average household in Germany, SC of electricity using a PV and battery combination was found to be rarely higher than 50% (Johann & Madlener, 2014).

Self-consumption can be a helpful factor in judging the efficient use of a PV system, but one misleading factor in SC can be the influence of an oversized PV system on the ratio. Even with large batteries, an oversized PV system can be more disadvantageous in jurisdictions where no FiT is paid. Self-sufficiency (SS) ratios can be more telling about the ability of the equipment and timing of use to meet household demand. However, the mismatch between heating load and the solar production profile limits the SS ratio. This is true even when coupled with batteries; to increase SS ratio on annual scale would require seasonal storage (Zhang et al., 2016). One study evaluating the return on investment between lead acid, NaNiCl (sodium-nickel-chloride) and lithium ion (Li-ion), in combination with the PV system, found that the Li-ion battery system is superior in achieving a higher SS ratio with the same life cycle cost (Zhang et al., 2016). As batteries help to increase SS, they have the potential to work against FiTs as an economic driver for electricity generating systems (Truong et al., 2016). Currently, in the case of the UK, however, this is helpful as the FiT is no longer available for new electricity generating installations (Jones et al., 2017).

As is shown, batteries help with respect to SC and SS. In the same way, batteries are also beneficial in shifting or alleviating peak grid consumption. As the grid pressure increases, the UK are trying to relieve this pressure focussing on efforts to reduce the peak demand. In the UK study cited above, aggregating solar generation and storage at a community level showed that peak grid electricity demand between 17:00 and 19:00 was reduced by 8% through the use of smart batteries across 74 dwellings (Gupta et al., 2019). Shifting PV production is helpful in achieving this, but also setting up batteries to take advantage of TOU tariffs is also modelled to be beneficial. One study calculated that with 2 kWh of battery storage per household, the peak demand at low voltage substations could be halved. With homes heated with heat pumps, 3 kWh battery storage per household would avoid increasing the peak demand (Pimm et al., 2018).

As it is important that energy-related claims of smart home technologies are scrutinised (Hargreaves et al., 2018), it is also important to understand the capability of batteries in improving self-consumption and self-sufficiency. There is relatively less empirical research on the actual performance of distributed energy resources in homes especially solar PV coupled with home batteries that can meet the daily energy needs of dwelling. This is what this study investigates.

7.3 The Zero Plus Project Case Study Dwellings

There are three Zero Plus (ZP) dwellings (ZP1, ZP2 and ZP3) in a development located in York, England. Figure 7.1 shows the dwellings and the PV panels located on the southwest – south-facing roofs. On the left are ZP1 and ZP2, both two-bed semi-detached dwellings, and on the right is ZP3, a three-bed detached dwelling. Table 7.1 lists various household characteristics for the three dwellings including PV and battery specifications.

7.4 Methodology

The major focus of this paper is on the statistical analysis of the electricity balance in the dwellings, i.e. consumption, generation and storage. Also included is a brief overview of space heating and domestic hot water (DHW) (disaggregated from total hot water supply) to provide a complete view of total energy consumption in the dwellings. A significant limitation of the dataset is that the space heating data logger in ZP1 logged no data for the entire heating season. So that as much data are aligned as possible, the analysis covers the period from 1 February 2020 to 31 May 2020. Also, with respect to electricity consumption in the dwellings, peak grid demand times are taken as 16:00–20:00 (DECC, 2015).

The analysis approach first reports on the overall view of energy consumption, followed by electricity consumed and generated in the homes with a focus on self-consumption (SC) and self-sufficiency (SS) and times when the batteries charge and discharge. The impact of Covid-19 lockdown on energy use in the dwellings is also explored, using 'pre-lockdown' dates 1 February 2020–22 March 2020 and 'during lockdown' dates 23 March 2020–31 May 2020. Table 7.2 shows the data sources and data gathered in the study and at what frequency these data were gathered. The monitoring was remote; data gathered were transmitted through the wireless network to an online repository.

Fig. 7.1 Zero Plus dwellings: from left to right, ZP1, ZP2 and ZP3

7 Performance of Distributed Energy Resources in Three Low-Energy Dwellings...

Table 7.1 Household characteristics

	ZP1	ZP2	ZP3
Form	2-storey right-side semi-detached	2-storey left-side semi-detached	2-storey detached with adjacent attached garage
Total floor area (m²)	84.4	84.4	129.6
Measured air permeability (design 4.0 m³ h⁻¹ m⁻²@50pa)	5.39 m³ h⁻¹ m⁻²	5.44 m³ h⁻¹ m⁻²	7.53 m³ h⁻¹ m⁻²
No. of bedrooms	2	2	3
First full month in dwelling	August 2019	February 2019	September 2019
Period heated	Mid September to mid May	Start February to early April	Mid September to early April
Number of occupants	3	3	4
Household	1 adult/2 children	2 adults/1 child	2 adults/2 children
Occupancy	Always at home	Always at home; 1 adult works away from home	Always at home
Occupancy pattern as viewed through energy consumption	Most active morning hours: 6–8 am Most active evening hours 6–10 pm Most energy consumed in evening	Most active morning hours: 6–8 am Most active evening hours 7–11 pm Most energy consumed in evening	Most active morning hours: 5–9 am Most active evening hours 6–10 pm Most energy consumed in evening
Building fabric specification	U-values (W/m².K): Exterior wall 0.17, party wall 0.23, roof 0.16, floor 0.14, windows 1.33, air permeability 4.0 m³/(h.m²)@50pa		
Heating	District heating (gas)		
Renewables	Total 4.34 kWp per dwelling: 14×–310 Wp PV panels 50% tilt, 236 SW-W azimuth		
Battery	14 kWh (total), 5 kW continuous power, Li-ion, fully integrated inverter		

Table 7.2 Data sources and details

Variable	Resolution/details	Source
Indoor temperature	Temp.: 0.1 °C at 30 min interval	Orsis data loggers installed in dwellings
Heat energy monitoring, DHW energy monitoring	1 Wh at 30 min interval	Orsis data loggers installed in dwellings
Fans and lighting electricity consumption	1 Wh at 30 min interval	Orsis data loggers installed in dwellings
Total electricity, battery and PV monitoring	0.1 kW at 5 min interval	Battery inverter
Heating degree days (HDD)	0.1 degree day, 15.5 °C base temperature, daily basis	www.degreedays.net

7.5 Results

7.5.1 Total Energy Consumption of the Dwellings

For the period reported, ZP3, the largest in both size and number of occupants, is consuming the most energy overall. Though ZP3 does have more occupants and is a larger dwelling, per area, it is still consuming more heating (gas) and electricity. Space heating is three times of what is used in ZP2 after being normalised by area. With respect to electricity, however, ZP3 has a wheelchair lift installed as one of the occupants uses a wheelchair; they also have a fish tank. For this reason, electricity consumption is expected to be much higher. Table 7.3 lists the overall consumption values in the dwellings.

7.5.2 Electricity Balance in the Dwellings

As ZP3 uses more electricity than the other dwellings, it not only has the highest instantaneous SC of PV electricity, it also consumes more electricity directly from the grid and the battery to offset higher daily power demands. Table 7.4 shows the proportioned breakdown of in- and out-going electricity in the dwellings. The PV and battery combination contributed to a SS range of 76–95%. This, based on a simple annual projection and an electricity rate of 15 p/kWh, could result in annual savings of between £260 and £438.

All dwellings have peak electricity consumption between 16:00 and 18:00. This falls within the typical UK peak grid demand times of 16:00–20:00. Figures 7.2, 7.3, and 7.4 show the average hourly PV generation and source of electricity to meet total HEC. As can be seen in the graphs, both instant consumption of PV generation

Table 7.3 Energy data from 1 February 2020 to 31 May 2020

	ZP1	ZP2	ZP3
Total energy consumption (kWh)	1535[a]	1838	2938
Total heating energy consumption (kWh)	206[a]	243	1144
Total domestic hot water (kWh)	674	863	504
Total household electricity consumption (HEC) (kWh)	655	732	1290
Total PV generation (kWh)	1682	1683	1699
Net electricity consumption	−1027	−951	−409
Net total consumption (kWh)	−147[a]	155	1239
Net total consumption (kWh/m^2)	−1.7[a]	1.8	9.6

[a]ZP1 total heating energy consumption was not monitored due to a faulty data logger. The total heating energy consumption value is roughly estimated normalising ZP2's hourly space heating consumption with hourly living room temperature data and applying this to ZP1's hourly living room temperature data. This is done because ZP1 and ZP2 are identical in form, though slightly different in occupancy. As a result, all values that include space heating energy consumption for ZP1 are estimated

7 Performance of Distributed Energy Resources in Three Low-Energy Dwellings...

Table 7.4 Energy data from 1 February 2020 to 31 May 2020

	ZP1	ZP2	ZP3
Total household electricity consumption (kWh)	655	732	1290
% of HEC from PV (instantaneous SS)	49%	51%	42%
% of HEC from battery (PV and grid mix)	49%	42%	50%
% of HEC from grid (direct consumption)	2%	7%	8%
Total PV generation (kWh)	1682	1683	1699
% of PV exported to grid	53%	57%	32%
% of PV self-consumed (instant and battery)	47%	43%	68%
% of PV SC (instantaneous only)	20%	22%	34%
Total grid consumption (kWh)	30	153	316
% direct from grid	27%	35%	30%
% grid consumption delayed through battery	73%	65%	70%

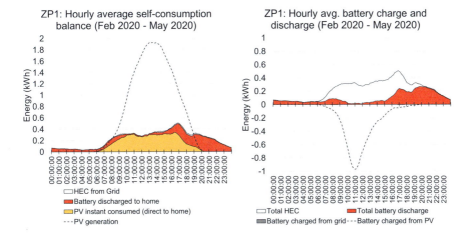

Fig. 7.2 ZP1: Average hourly HEC breakdown (left) and battery balance (right)

and battery discharge are greatly offsetting this peak demand. Between 16:00 and 20:00:

ZP1 average total of 0.8 kWh of direct grid consumption (PV and battery making up 17.8 kWh)—reducing this home grid pressure during peak by 96% on average.

ZP2 average total of 0.4 kWh of direct grid consumption (PV and battery making up 24.2 kWh)—reducing this home grid pressure during peak by 98% on average.

ZP3 average total of 1.7 kWh of direct grid consumption (PV and battery making up 31.6 kWh)—reducing this home grid pressure during peak by 95% on average.

Figures 7.2, 7.3, and 7.4 show the battery charging and discharging times throughout a day over the period. Notably ZP3 performs a significant amount of off-peak charging from the grid overnight; 70% of their total grid consumption is shifted through the battery. ZP1 does not regularly charge the battery from the grid overnight. As can be seen, ZP3 has larger baseline energy consumption than the

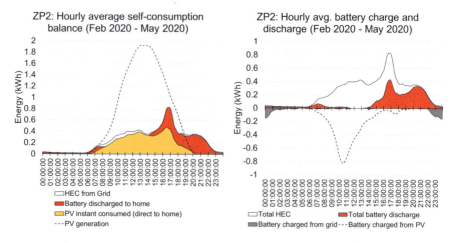

Fig. 7.3 ZP2: Average hourly HEC breakdown (left) and battery balance (right)

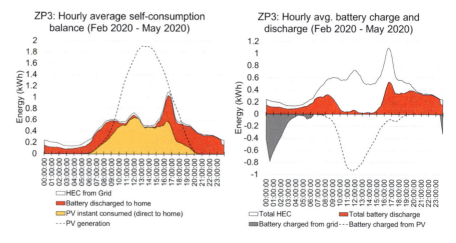

Fig. 7.4 ZP3: Average hourly HEC breakdown (left) and battery balance (right)

other dwellings. Contributors to this are likely their higher use of fans and lights and possibly their fish tank.

7.5.3 Impact of Covid-19 Lockdown on Energy Use in the Dwellings

Overall, in looking at total dwelling electricity consumption, there is an increase in electricity consumption across all dwellings from before Covid-19 lockdown (1 Feb–22 Mar 2020) to during lockdown (23 Mar–31 May 2020). The following presents the investigation of overall daily electricity use, changes to peak demand,

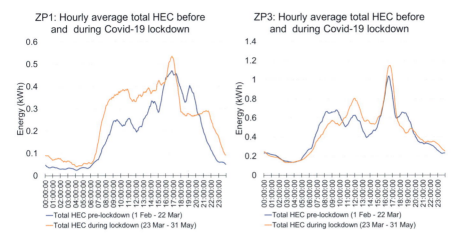

Fig. 7.5 Hourly electricity consumption contrast before and during Covid-19 lockdown (ZP1 left, ZP3 right)

detailing of end uses to investigate the above changes further and finally space heating.

ZP1 increased total daily HEC by 30%, ZP2 increased by 9% and ZP3 increased by only 2%. Figure 7.5 contrasts the shift in average hourly total HEC for each period in ZP1 and ZP3. In ZP1 the increase is notable, where there is more energy consumption in the morning to afternoon and a sharper peak at peak demand times. ZP3 has had a sharper peak during peak demand but no other significant change overall. ZP2, not shown, also had higher noon to peak consumption and a higher peak demand very similar to ZP3. Though during the initial questionnaire assessment both ZP1 and ZP3 stated that they are 'home all the time', it is expected that the occupants in ZP1, in reality, before the lockdown left the house more often to visit friends or relatives albeit possibly not on a regular schedule.

In looking deeper at electricity use in the dwellings, electricity consumption for fans and lighting (sub-metered) was removed from the total electricity use to isolate all remaining uses, called 'appliances' here. The hypothesis is that this 'appliance' consumption should increase as occupants are stuck at home in lockdown; however, this increase is not expected to be overly large, but noticeable. This is because most of the extra use is expected to be in low-power devices like televisions and computers. It is recognised that not being able to isolate cooking energy is a limitation since a shift to cooking at home could be a significant indicator of the occupancy shift. Though lighting consumption could also increase, it is best removed as the lockdown period progressed through days which are increasing in daylight hours. Figure 7.6 shows the fan, lighting and appliance consumption pre-lockdown and during lockdown for ZP1. The only dwelling showing a notable impact of lockdown in appliance use is ZP1, where a slight downward trend before lockdown becomes a significant upward trend. ZP2 and ZP3 (not shown) demonstrated downward trends in consumption during lockdown; however, in ZP2 this is barely noticeable

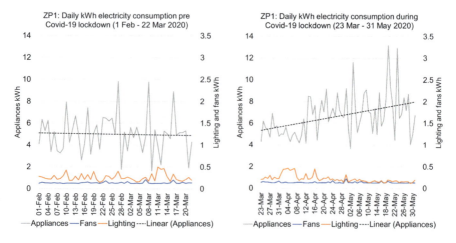

Fig. 7.6 ZP1: Sub-metered electricity use (pre-lockdown left, during lockdown right)

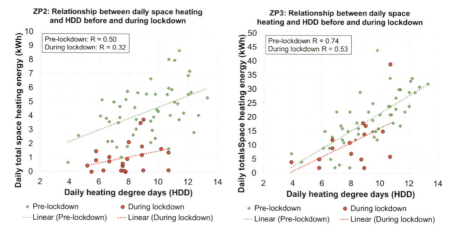

Fig. 7.7 ZP2 (left) and ZP3 (right) correlation between space heating and HDD pre- and during Covid-19 lockdown

and is a little higher than pre-lockdown, and in ZP3 the downward trend also appeared to be slowing during lockdown.

Space heating is another way to observe the impact of lockdown on energy consumption; however, a limitation of this method is that lockdown has occurred as the heating season is ending. For this reason, correlation with heating degree days (HDD) is used to assess impact. Figure 7.7 shows ZP2 and ZP3 daily total space heating as it correlates to the daily HDD for the same period. One aspect to note immediately is that the correlation between space heating and HDD is much weaker during lockdown as opposed to before. This, however, could just be an aspect of the end of the heating season and not necessarily an increase of user heating.

In theory, if there is more use of heating as a result of being at home more, the proportion of total heating consumption above the best fit line will be higher during lockdown. Both dwellings are showing a small increase in proportion of total heating above the best fit line during lockdown (from 63% to 65% in ZP2 and from 61% to 62% in ZP3). This slight increase is, however, not considered high enough to suggest that there is a significant increase in heating in these dwellings as a result of lockdown. As it is the end of the heating season, the number of instances assessed was limited. It is also possible that the warmer than average temperatures had an impact on space heating results (daily measured average 2.3 °C above CIBSE test reference year data for LEEDS—closest location).

7.6 Discussion

The analysis of the Zero Plus dwellings has shown that batteries are 'smart' devices that are helpful in shifting the renewable energy or overnight (off-peak) grid charging to peak demand times in dwellings. As the batteries were able to double the SC of the PV systems and the combined PV and battery combinations were able to reduce average peak load by at least 95%, this technology will undoubtedly help the UK government meet its goals to develop a *smart energy system* and reduce peak electricity demand pressure on the electricity grid. Under several different scenarios, adding storage to the energy system in combination with decarbonisation efforts is considered to increase overall efficiency and resilience in the system (DECC, 2015).

As was demonstrated, based on demand, ZP3 may benefit from an Economy 7 tariff or a dynamic pricing tariff as about 17% of their total HEC was through overnight battery charging. If these homes did not have PV but did have the large batteries, TOU tariffs would be an effective money-saving and peak load shifting option. DSR via appliance timing is perhaps more needed in homes without such large batteries. Future research to explore potential improvement of the dwellings' use of PV and battery storage would include demand side response (DSR) experimentation and evaluation. This is shifting the times when electricity is consumed to take further peak demand pressure off the grid. This can be done in response to a signal (DECC, 2015), perhaps through in-home energy monitoring devices or, as the technology has progressed, through smart phone application linked to smart metering. Obvious examples include timing laundry or dishwasher activation during peak PV generation times or overnight by setting timers on the appliances. If DSR through shifting energy consumption to overnight is recommended, Economy 7 tariff – referring to 7 h overnight where electricity is offered at a cheaper rate—is a potential incentive.

With the Covid-19 lockdown and its potential impact on a larger shift to working from home, generation and storage arrangements such as those exhibited through the case studies would be greatly beneficial. This is particularly true of PV, as there would be more energy consumption throughout the middle of the day particularly in the winter. Furthermore, as there is a recommendation to shift to electrified heating,

PV and batteries would be even more relevant. Working from home contributing to greater instantaneous SC of PV generation would provide a greater return on investment. This is important now more than ever as there is no longer FiT for new PV installations. However, incentives are an effective policy tool. Storage will benefit the overall system and, therefore, should be rewarded for its impact (HM Government and Ofgem, 2017). Like the previous FiT for solar renewable technology, there should be a *peak demand shift tariff* that would incentivise PV (again), batteries or even well-managed DSR. This would pay householders a tiered tariff rate based on the proportion of electricity reduced during peak demand hours. The progress could be judged based on a baseline year for the household and paid on a monthly basis. This example, however, would only work for retrofits. For newly built dwellings, the baseline would likely need to be a local-, perhaps, postcode-level average from which to base improved performance.

7.7 Conclusion

To meet the UK's net zero greenhouse gas emissions target and reduce peak demand pressure from the electricity grid as electrification of heating becomes a more prominent solution to reduced emissions, the combination of electricity generation and storage is shown to be smart and resilient solutions. Though the case study dwellings appear to be typical in their peak-time consumption, the PV and battery combination has been shown to reduce their average direct grid consumption during peak hours by over 95%. Because the PV system on each dwelling is large, instantaneous SC was low, between 20% and 30%, but the batteries helped to increase SC up to around 50%. The estimated cost benefit of this ranged from annual savings of between £260 and £438. Overall, with respect to the distributed energy resources installed, the dwellings resulted in 80% reduction in net total energy consumption during 4 months of the year. The demonstrated systems show benefits for householders through a net reduction of total energy consumption and benefits for the overall energy system through a reduction in peak energy demand.

Another aspect of the study briefly looked at the impact of the Covid-19 lockdown on energy consumption in the dwellings. Though the occupants in two of the dwellings claimed to 'always be at home', there was a slight increase in electricity consumption for the two of them and a notable increase in the other. Space heating consumption did not reflect the same impact; however, it did show poor seasonal responsiveness at times. Such empirical studies are vital for providing the learning that is necessary for future scalability and replicability of distributed energy resources to move towards a smart and flexible energy system in the UK. Future research could explore the extent and magnitude of demand side response (DSR) that could be offered by PV systems and smart batteries, with and without time-of-use tariffs.

Acknowledgement The research study is part of the Zero Plus research project, which has received funding from the European Union's Horizon 2020 Research and Innovation programme under Grant Agreement No. 678407.

References

BEIS. (2018). *Clean growth strategy: Leading the way to a low carbon future.* HM Government https://assets.publishing.service.gov.uk/government/uploads/system/uploads/attachment_data/file/700496/clean-growth-strategy-correction-april-2018.pdf. Accessed 20 July 2020

CCC. (2019). *UK housing: Fit for the future?* C. o. C. Change. https://www.theccc.org.uk/wp-content/uploads/2019/02/UK-housing-Fit-for-the-future-CCC-2019.pdf. Accessed 20 July 2020

DECC. (2015). *Towards a smart energy system.* https://assets.publishing.service.gov.uk/government/uploads/system/uploads/attachment_data/file/486362/Towards_a_smart_energy_system.pdf. Accessed 20 July 2020

Gupta, R., Bruce-Konuah, A., & Howard, A. (2019). Achieving energy resilience through smart storage of solar electricity at dwelling and community level. *Energy and Buildings, 195,* 1–15.

Hargreaves, T., Wilson, C., & Hauxwell-Baldwin, R. (2018). Learning to live in a smart home. *Building Research & Information, 46*(1), 127–139.

HM Government, & Ofgem. (2017). *Upgrading our energy system smart systems and flexibility plan.* UK Government https://assets.publishing.service.gov.uk/government/uploads/system/uploads/attachment_data/file/631656/smart-energy-systems-summaries-responses.pdf. Accessed 20 July 2020

Johann, A., & Madlener, R. (2014). Profitability of energy storage for raising self-consumption of solar power: Analysis of different household types in Germany. *Energy Procedia, 61,* 2206–2210.

Jones, C., Peshev, V., Gilbert, P., & Mander, S. (2017). Battery storage for post-incentive PV uptake? A financial and life cycle carbon assessment of a non-domestic building. *Journal of Cleaner Production, 167,* 447–458.

Kelly, J. (2020). Here are the companies leading the work-from-home revolution. *Forbes.* https://www.forbes.com/sites/jackkelly/2020/05/24/the-work-from-home-revolution-is-quickly-gaining-momentum/?sh=15b6f7871848. Accessed 20 July 2020

McKenna, E., Pless, J., & Darby, S. J. (2018). Solar photovoltaic self-consumption in the UK residential sector: New estimates from a smart grid demonstration project. *Energy Policy, 118,* 482–491.

Ministry of Housing Communities and Local Government. (2019). *The future homes standard 2019: Consultation on changes to part L (conservation of fuel and power) and part F (ventilation) of the building regulations for new dwellings.* https://assets.publishing.service.gov.uk/government/uploads/system/uploads/attachment_data/file/843757/Future_Homes_Standard_Consultation_Oct_2019.pdf. Accessed 20 July 2020

O'Shaughnessy, E., Cutler, D., Ardani, K., & Margolis, R. (2018). Solar plus: Optimization of distributed solar PV through battery storage and dispatchable load in residential buildings. *Applied Energy, 213,* 11–21.

Pimm, A. J., Cockerill, T. T., & Taylor, P. G. (2018). The potential for peak shaving on low voltage distribution networks using electricity storage. *Journal of Energy Storage, 16,* 231–242.

Truong, C. N., Naumann, M., Karl, R. C., Müller, M., Jossen, A., & Hesse, H. C. (2016). Economics of residential photovoltaic battery systems in Germany: The case of Tesla's Powerwall. *Batteries, 2*(2), 14.

UK Government. (2019). *The climate change act 2008 (2050 target amendment) order 2019.* http://www.Legislation.gov.uk, http://www.legislation.gov.uk/uksi/2019/1056/introduction/made. Accessed 20 July 2020

Zhang, Y., Lundblad, A., Campana, P. E., & Yan, J. (2016). Employing battery storage to increase photovoltaic self-sufficiency in a residential building of Sweden. *Energy Procedia, 88,* 455–461.

Chapter 8
Learnings from the Evolution of the University of Suffolk EcoLab: Adopting People-Centred Design Approaches to Encourage the Mass Uptake of Energy Transition Solutions in the Housing Sector

Benjamin Powell

8.1 Introduction

The EcoLab project comes as part of a wider scheme, currently being constructed, to build a digital skills and innovation hub at BT's Adastral Park site in Martlesham. This initiative sees the founding of a satellite campus at the park, BT's main centre for research and development, with the intent to provide closer ties between academia and industry, in order to enhance capacity for the growth and implementation of new sustainable technologies.

At the time of writing, the proposals for the EcoLab scheme (Fig. 8.1) have just been submitted for planning permission, with construction due to commence towards the end of the year. Whilst the purpose of the facility is to be used as a research centre and laboratory, the building takes the form of a two-bedroom house and to this extent complies with national planning polices including space standards.

Working to a limited budget of approximately £200,000, the initial targets for the project are to achieve a zero carbon occupancy and a reduction in embodied carbon of around 50% compared to a typical brick and block house and to construct the building using the latest construction methods, such as offsite fabrication, so as to create the potential for the model to be replicated at scale. Longer term, the completed build will utilise SMART systems to evaluate the performance of sustainable materials, low carbon energy and water consumption, coupled with critical consideration of the wider landscape through ecological design.

B. Powell (✉)
Architect, studiomanifest.co.uk, London, UK
e-mail: B.Powell@uos.ac.uk

Fig. 8.1 Initial concept design proposal for the EcoLab

The project brief is therefore to address two of the most pressing concerns that we face: climate change, and the impact the construction industry has on this, and the UK housing crisis, which sees the chronic shortage of affordable housing.

Whilst we hope that the research from the project will be a worthwhile addition to this extensive body of work, we are acutely aware of the urgency of implementing these outcomes at scale. Like many local authorities, Suffolk County Council declared a climate emergency in March of 2019, with an ambition to move the county towards carbon neutrality by 2030, ahead of national targets for 2050. To get close to achieving this ambitious goal, robust demonstrator outputs, with affordable carbon reduction solutions, will require immediate uptake on a mass scale. So how can we achieve this?

8.2 The Consumer

During early conversations to develop the project brief, the importance of the public in the implementation of sustainable design was highlighted in a conversation with a local planning officer. Discussing rainwater harvesting technology, they stated 'the feedback we regularly receive from developers is that they are open to greater uptake of the [low energy] technology, but currently there isn't the demand from their customers' (M. Williams 2020, personal communication, 16 June).

In the UK, roughly 60% of the new homes built each year are delivered by just ten companies (Great Britain & Department for Communities and Local Government,

2017). Significantly influenced by the market, the importance of these large companies highlights that just because the technology is available does not necessarily mean it will be utilised in a build. Whilst government policy looks to address this (Great Britain & Department for Energy and Climate Change, 2013), if demand from the consumer does not exist, then uptake from the market will be slow (Murphy, 2012; Roy et al., 2007). This is not a new idea. In Roy et al. (1998), identified that 'Designers and manufacturers recognise that environmental factors generally only enter consumer purchasing decisions after product performance, quality, reliability, and value for money' (pp. 268–269).

More recent research suggests that, whilst customers' environmental consciousness has increased, there is still a shortfall in converting this to actual purchasing (Environmental Change Institute, 2001; Farber, 2019). 'Few consumers who report positive attitudes toward eco-friendly products and services follow through with their wallets. In one recent survey 65% said they want to buy purpose-driven brands that advocate sustainability, yet only about 26% actually do so' (White et al., 2019).

Addressing construction specifically, Jackson expands on this: 'The design and construction of LZC (low and zero carbon) eco housing…., while showing what can be achieved, remain as largely experimental 'green niches' created by highly committed individuals and organizations.'

> They tend to be viewed by designers and policy makers as purely functional, technical devices, without sufficient regard for their aesthetic and ergonomic design and brand image, which can have a crucial role in adoption and effective use…material goods are important to us, not just for their functional uses, but because they play vital symbolic roles in our lives (2005, p. 6).

In 2019, the government published *Building a Market for Energy Efficiency: Call for Evidence Summary of Responses (Great Britain & Department for Business, Energy and Industrial Strategy, 2019)*. This came in recognition of the particular challenges of driving demand and addressing supply side barriers to market growth in the owner occupier sector. The document sets out what the government believed to be the current state of the market, what the barriers are to growth and potential policies for addressing those barriers. The findings from this research included six key areas for targeted focus:

- Landlords and their properties within the private rented sector (PRS)
- All audiences in the context of broader advice versus energy efficiency alone
- The need to build a mass market to encourage the supply chain to invest, innovate and scale up, as well as avoid complexity in targeting and search costs
- Building a market for energy efficiency
- Identifying the properties that would benefit most from retrofitting energy efficiency measures
- Considering early adopters of smart technologies and generational age groups of young and old millennials, generation X-ers and baby boomers

In addition, the findings also highlighted a potential lack of low carbon building construction knowledge, coupled with a need to establish social norms and conformity in terms of energy efficiency investments and the impact that decreasing home ownership levels have on reducing the incentive to improve properties or buy an efficient home from the outset.

Understanding these broader constraints, and the financial challenges to home buyers, the Government Advisory Committee on Climate Change undertook a tangent review (2019). Recognising the market's resilience to rapid change, the review presented the urgency for parallel regulatory reform in order to meet the government's climate objectives, establishing that all new homes must be built to be low carbon, energy and water efficient and climate resilient, from 2025 at the latest.

Moving into the next half of this decade, it will be necessary for policy and planning to lead where the market cannot. Nonetheless, the importance of society in the design and implementation of sustainable construction must be recognised as an essential component in achieving the sea change required to overcome the climate emergency.

8.3 Collaborative Environment

To this effect, a focus of our work on the EcoLab so far has been to put in place a collaborative and open methodology that encourages dialogue and which places people at the centre of the development of the scheme.

At an institutional level, the project team blends academics from the traditional sciences, including from Information Systems Engineering and Wildlife and Ecology and Conservation Science, with those more closely associated with the social sciences, most notably the architecture department.

This cross-disciplinary approach has extended to partnerships outside of the university. Figure 8.2 documents the range of organisations that informed the project development in its infancy. This includes a broad church, ranging from BT to the Suffolk Climate Change partnership.

Dialogue is ongoing with the local planning authority from East Suffolk Council (ESC). As well as using the scheme as an exemplar for sustainable design, it has been agreed to share information on the project's costs. It is hoped this information can be used to demonstrate that low-energy, hi-tech, sustainable construction can be used to build affordable homes and therefore provide local authorities with evidence to counter frequent arguments from volume housing providers that these approaches are prohibitively expensive (R. Bishop 2020, personal communication, 9 July).

Project links have also been established with the BRE. As well as advising on the technical performance of the building, with a focus on occupational health, the long-term outcomes from the project will be shared with their smart homes and innovation department. This will feed into research being undertaken at a national level and inform British Standards surrounding sustainable construction best practice, and subsequently, the legislation of this area.

8 Learnings from the Evolution of the University of Suffolk EcoLab: Adopting… 89

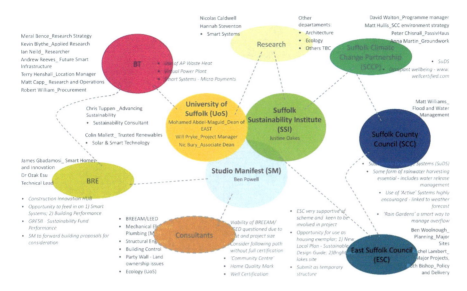

Fig. 8.2 Early project network

Happening in tandem to ongoing dialogue, perhaps one of the most enriching parts of the project so far has been the collaborative design process.

8.4 Hackathon

Taking place virtually during the COVID-19 national lockdown, the 'EcoLab Architecture Hackathon' invited architecture students from around the East of England to work collaboratively to conceive concept designs for the project.

In the first of two sessions, the co-creators, a group of around 12 participants, put together a database of precedents that were discussed and analysed. Students were encouraged to think about the space as somewhere that they would want to live and, as such, look for precedents that were not only sustainable but which looked good and were suited to their context. Participants were then challenged to begin sketching their ideas through a series of design games, including a form of 'speed drawing' in which individuals shared thoughts with one another in a round robin format.

In the second session, participants bought together the ideas from the first week into a coherent design. This exercise followed BIM stage 3 protocol with individuals working in the cloud using a shared digital model (Revit). By the end of the session, the site was occupied with a collection of models, not just showing a diverse range of ideas for low energy design but which could also be used to test the site itself for the ideal location.

As well as being a great opportunity for architecture students to gain experience working on a live project and to see their designs develop into a built scheme, the

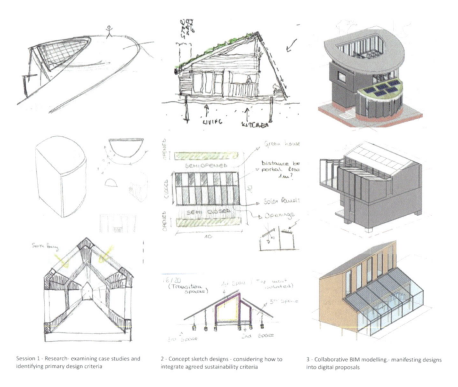

Fig. 8.3 Some examples of ideas generated at each stage of the Hackathon

incredible energy and enthusiasm from participants have provided the project with a unique platform from which to grow. The ideas that were generated ranged from forming a hierarchy of spaces, with the most frequently occupied in the centre, to the use of moveable parts and shared common spaces (Fig. 8.3).

Feedback from the session was positive, with the all respondents (6 out of 12) rating the event either very good or excellent, feeling they had learnt a significant amount about sustainable design and finding working in the online environment a positive experience.

Activities such as this are an essential component in ensuring that all actors involved in the development of the built environment are equipped with the requisite low carbon building construction knowledge. It also outlines the type of approach to training young architects being developed at UoS, 'these events are at the heart of what we do at Suffolk….we are advocates of Architecture as a tool for social and ecological change; an agent for changing how people live for the better and how they make decisions about the built environment. Our students are trained early to approach ways of designing and building with a critical eye and curiosity; skills much needed by all future built environment professionals' (L. Psarologaki (Course Leader) 2020, personal communication, 09 July).

8.5 Next Steps

Further collaborative design initiatives are planned throughout the duration of the development of the scheme and beyond. UoS architecture students will play an active role in the detail design and construction of the scheme, including the potential for some live construction workshops, which will allow students to witness the construction process. We also plan to include a form of post-occupancy evaluation.

The collaborative approach is already being adopted elsewhere. A similar collaborative design event is being planned with Wildlife, Ecology and Conservation Science students that is hoped will explore a range of ambitions, including the implementation of a demonstration garden that links with work being done by Transition Woodbridge to create wildlife corridors, as well as implementing some form of heathland restoration to assist in the protection of the struggling silver-studded butterfly population.

Following the completion of the building, regular studies are expected to be undertaken with the building in occupation, including inviting members of the public to use the space to test how intuitive and user friendly are the various SMART technologies being studied.

But in the immediate future, the priority is to raise the public profile of the project, not just within academic spheres but also to the wider public. Habib, Hardisty and White argue that 'Harnessing the power of social influence is one of the most effective ways to elicit pro-environmental behaviours in consumption' (White et al., 2019). They outline three ways in which this can be done by making sustainable behaviours more evident; making people's commitments to eco-friendly behaviour public; and using healthy competition between social groups.

To this effect, the design of the building is intended to be eye catching (if not necessarily to everyone's tastes) to aid with a media campaign. As well as using the individual and institutional social media accounts that are available to us, press releases are being issued to the local and national press, including architectural and lifestyle publications. It is hoped that attention in these areas will enhance the longer-term relevance and reach of the scheme.

8.6 Summary

Within this collaborative framework, we are working collectively to realise the wider social and affordability aspects of greening the built environment and develop a scheme that is not only technically excellent but which is designed for people. A big, but valuable research ask from what will be a compact, modern and comfortable building footprint, at a time when land availability, resource usage and quality affordable housing represent a national challenge.

References

Committee on Climate Change. (2019). *UK housing: Fit for the future?* https://www.theccc.org.uk/publication/uk-housing-fit-for-the-future/

Consumer behaviour: How people make buying decisions [WWW Document]. (n.d.). https://2012books.lardbucket.org/books/marketing-principles-v1.0/s06-consumer-behavior-how-people-m.html

Environmental Change Institute. (2001). *Retail therapy: Increasing the sales of CFLs*. ECI.

Farber, D. (2019). *The 10 reasons people buy new products [WWW document]*. Medium. https://medium.com/new-markets-insights/the-10-reasons-people-buy-new-products-1489aad9b1c9

Great Britain, & Department for Business, Energy and Industrial Strategy. (2019). *Building a market for energy efficiency: Call for evidence*. https://assets.publishing.service.gov.uk/government/uploads/system/uploads/attachment_data/file/ 813488/building-market-for-energy-efficiency-summary-of-responses.pdf

Great Britain, & Department for Communities and Local Government. (2017). *Fixing our broken housing market*. https://assets.publishing.service.gov.uk/government/uploads/system/uploads/attachment_data/file/ 590464/Fixing_our_broken_housing_market_-_print_ready_version.pdf

Great Britain, & Department for Energy and Climate Change. (2013). *The carbon plan—Reducing greenhouse gas emissions*. https://assets.publishing.service.gov.uk/government/uploads/system/uploads/attachment_data/file/ 47613/3702-the-carbon-plan-delivering-our-low-carbon-future.pdf

Jackson, T. (2005). *Motivating sustainable consumption: A review of the evidence on consumer behaviour and behavioural change*. Report to the Sustainable Development Research Network, January. www.sd-research.org.uk/

Murphy, J. (2012). *Governing technology for sustainability*. Routledge.

Roy, R., Smith, M. T., & Potter, S. (1998). Green product development—Factors in competition. In T. Barker & J. Köhler (Eds.), *International competitiveness and environmental policies* (pp. 265–275). Edward Elgar.

Roy, R., Caird, S., & Potter, S. (2007). People centred eco-design: Consumer adoption of low and zero carbon products and systems. In J. Murphy (Ed.), *Governing technology for sustainability* (pp. 41–62). Earthscan.

White, K., Hardisty, D. J., & Habib, R. (2019). The elusive green consumer. Harvard Business Review. July–August 2019. https://hbr.org/2019/07/theelusive-green-consumer

Chapter 9
A Review of Sustainable Construction Practices in Ghana

Moses K. Ahiabu, Fidelis A. Emuze, and Dillip Kumar Das

9.1 Introduction

The importance of the construction industry to economic development cannot be underestimated. For instance, the construction industry contributes at least 7% to jobs and accounts for about 11% of the world's gross domestic product (GDP) (Roumeliotis & Fenton, 2011). Rapid urbanisation and industrialisation have necessitated an increase in the development of infrastructure projects. Such projects contribute to national economic development. The construction industry consumes a lot of natural resources that lead to climate change to meet the increased infrastructural demand of nations. The construction and maintenance of buildings account for about 40% of the world's natural resources and energy use (Edeoja & Edeoja, 2015, p. 112). The sector is generating unacceptable levels of material waste (Hussin et al., 2013), contributes a quarter of the global total carbon emissions and is responsible for 40–50% of the greenhouse gas emissions (GHG) (Huang et al., 2017).

The global movement against the effect of climate change on the environment has led many countries to focus on sustainable construction. Sustainable construction is believed to be the way for the built environment to contribute to the achievement of sustainable development (Abidin, 2010). The built environment in Ghana is responsible for land degradation, loss of habitat, air and water pollution and high

---

M. K. Ahiabu (✉)
Department of Civil Engineering, Central University of Technology, Free State, Bloemfontein, Republic of South Africa

F. A. Emuze
Department of Built Environment, Central University of Technology, Free State, Bloemfontein, Republic of South Africa

D. K. Das
School of Engineering, University of KwaZulu-Natal, Durban, Republic of South Africa

energy usage (Ofori, 2012; Djokoto et al., 2014). However, if the built environment adopts sustainable construction practices, it can help mitigate these negative impacts. The stakeholders in Ghanaian construction are aware of the destructive nature of conventional construction practices (Ayarkwa et al., 2017). The desire for sustainable construction practices has been on the increase among clients, sponsors, construction professionals, government agencies and regulatory bodies (Ogungbile & Oke, 2019). As a result of the benefits linked to sustainable construction, the built environment is increasingly incorporating sustainability practices into new buildings and existing buildings (Ahn et al., 2013).

Notwithstanding the benefits of sustainable construction, research has shown that implementation of sustainable construction in Ghana is at the infant stage. Unsustainable design and construction practices as well as constant degradation of the environment for construction purposes still exist (Ayarkwa et al., 2017). However, barriers such as perceived initial cost, lack of political commitment, lack of legislation, lack of professional knowledge and technological difficulties must be removed. There is a growing interest in analysing the extent to which the barriers hinder the implementation of sustainable construction (Ayarkwa et al., 2010). Due to the importance of sustainable construction, this research aims to examine the level of implementation of sustainable construction practices in Ghana to reduce the carbon and ecological footprint with emphasis on the barriers and drivers for a successful application.

9.2 Research Method

The study method adopts a conceptual approach to review related literature. A traditional literature review is a comprehensive, critical and objective analysis of the current knowledge on a topic (Baker, 2016). The traditional review identifies central issues and research gaps. According to Jesson et al. (2011), a traditional literature review is a written appraisal of what is already known with no prescribed methodology (cited in Li & Wang, 2018). It is a research method in its own right. It presents a summary of published research to a particular topic of interest in a way that contributes to a better understanding of issues (Jesson et al., 2011). The selection of materials was based on a purposive selection with significant contributions to context-specific knowledge taking precedence over citation metrics. The researcher logically put the publications together to develop a coherent argument on the subject matter. However, there may be essential contributions and knowledge transfer in both a systematic review and a traditional review (Jesson et al., 2011) to identify gaps or inconsistencies in a body of knowledge. The next section is therefore based on the review of the related corpus.

9.3 Literature Review

9.3.1 The Construction Industry in Ghana

The construction industry contributes to the socio-economic development of Ghana. According to Osei (2013), the construction industry of Ghana can be likened to the UK construction industry of 1983. The industry is responsible for the construction projects covering key stages including feasibility, design, construction, operation, decommissioning, demolition and disposal. It delivers infrastructure and construction projects, which is defined by Du Plessis et al. (2002) as a broad process or mechanism for the realisation of human settlements and the creation of infrastructure that supports development. Two ministries supervise such projects in Ghana. The Ministry of Roads and Highways (MRH) supervises the road infrastructure, which is the backbone of the transportation systems in the country. The built environment falls under the Ministry of Works and Housing (MWH), which is the Government of Ghana's central management agency responsible for formulating policies and programmes for the housing and work sub-sectors of the economy. The ministry is mandated in line with Sections 11 and 13 of the Civil Service Act, 1993 (PNDCL 327). It also relies on Executive Instrument (EI. 28, 2017), to initiate and formulate policies for the works and housing sector, as well as coordinate, monitor and evaluate the implementation of policies, plans and programmes for the sustainable management of public land properties, drainage and coastal protection work and operational hydrology as well as safe, secure, decent and affordable housing using technical expertise and innovative methods for all people living in the country (Ministry of Works and Housing, 2017).

The agencies and departments responsible for the built environment include Public Works Department (PWD), Hydrological Services Department (HSD), Engineering Council (EC), Architects and Engineering Services Limited (AESL), Rent Control Department (RCD), Department of Rural Housing (DRH), Public Servants Housing Loan Scheme Board (PSHLSB), Architect Registration Council (ARC), State Housing Company (SHC) and Tema Development Company Limited (TDC). Akinradewo et al. (2019) categorised the industry's stakeholders into four main groupings as users and consumers, demand-side operators, regulators and supply-side operators. The construction activities in the built environment include the extraction and beneficiation of raw materials, the manufacturing of construction materials and components, the construction project cycle from feasibility to deconstruction and the management and operation of the built environment. Despite its significance to the overall national economic development, the built environment is arguably one of the most resource-intensive industries and considered as a significant contributor to environmental pollution (Zimmermann et al., 2005).

9.3.2 Description of Sustainable Construction

Kibert proposed the first definition of sustainable construction which appeared in Tampa (1994) as the creation and responsible management of a healthy built environment based on resource-efficient and ecological principles (cited in Bourdeau, 1999). According to Huovila (1998), sustainable construction aims at minimising the use of energy and emissions that are harmful to the environment and health. Lanting (1998) also conclude that sustainable construction is a process of the building which aims at reducing the adverse health and environmental impact caused by the construction process. Sustainable construction according to the International Council for Research and Innovation in Building and Construction (CIB) is the sustainable production, use, maintenance, demolition and reuse of buildings and construction and their components (CIB, 2004). However, Du Plessis et al. (2002) take sustainable construction further by introducing the idea of restoring the environment as well as highlight the social and economic aspect of sustainability. She defined sustainable construction as a complete process aiming at restoring and maintaining harmony between the natural and the built environments to create settlements that affirm human dignity and encourage economic equity.

In essence, sustainable construction is the application of sustainability principles to the construction life cycle from planning, constructing, mining raw materials to produce construction materials, using low embedded-energy materials, saving water and energy, deconstruction and managing waste (Mustafa & Bakis, 2015). The concept of sustainable construction is hinged on three main pillars, namely, environmental protection (1), social well-being (2) and economic prosperity (3) (Abidin, 2010). Striking the right balance between these factors is what is required to support true sustainability. For constructions to be sustainable, firms should, therefore, have an all-inclusive approach to the projects to achieve balance and consistent synergy among the three tripods of sustainable development (environment, economy and society).

9.3.3 Implementation of Sustainable Construction

There is a need for a sustainable development plan and strategy that should not be confused with the development plan of nations. A sustainability development strategy that creates jobs at the expense of the environment and social impacts and displaces thousands of people and reduces biodiversity defeats its purpose (Du Plessis, 2007). Sustainability lies in the relationship between social, economic and environmental issues. The main challenge for the construction sector lies in finding an approach to the physical, economic and human development to meet the requirement of sustainable development and construction defined by locally identified needs and value system. Du Plessis (2007) identified two main paths for the implementation of sustainable construction as the creation of capable and viable local

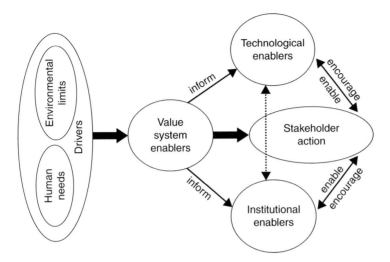

Fig. 9.1 A strategy for enabling sustainable construction (Source: Du Plessis, 2007, p. 72)

construction sector and the ability of the sector to respond to the demands that sustainable development and construction places on its activities. Adjarko et al. (2016) say that the minimisation of resource consumption, improvement of indoor air quality, avoidance of environmental health problems, use of recycled materials, waste reduction and maximisation of reuse are the critical sustainable construction principles applicable to Ghanaian construction. Sustainability can be achieved through strategic sustainable construction enablers. The enablers include technological, institutional and enablers related to a valued system such as social, spiritual or moral values that guide decisions as indicated in Fig. 9.1.

9.3.4 Barriers and Drivers of Sustainable Construction in Ghana

Barriers (please see Table 9.1) refer to attributes and conditions that can prevent and hinder actions or impede progress towards achieving sustainable construction practices (Vandierendonck et al., 2010; Ayarkwa et al., 2017). Barriers have adverse effects on the attainment of sustainable construction. According to Connell (2010), knowledge, awareness and attitude are internal barriers, while the external barriers include the availability of technology, finance and green technology. According to Ayarkwa et al. (2017), the financial incentive is identified as the main barrier to the implementation of sustainable construction in Ghana.

In contrast to barriers, drivers of sustainable construction are various elements that trigger, sustain and expand the uptake and implementation of required practices (Ayarkwa et al., 2017). Some drivers promote the implementation of sustainable

Table 9.1 Barriers to sustainable construction

Barriers	Research approach	Sample	Reference
Lack of financial incentives (high taxes and low-profit margin) Lack of building code and regulation Lack of investment Initial cost (high initial cost) Lack of client demand High cost of environmental services and technology Insufficient research Lack of public awareness Competitive pressure Lack of database and information Lack of green products Lack of professional knowledge and expertise Lack of green technology Tendering and contract requirements High level of perceived risks	Quantitative	Architects in Ghana	Ayarkwa et al. (2017)
Cost considerations and implications Lack of sustainable practices knowledge Attitude of professionals Absence of a sustainability rating tool Total cost of alternative energy sources Total client control on the design	Quantitative	Finland	Bash and Hakkinen (2015)
Lack of building codes and regulations Lack of incentives Higher investment cost Investment risks Higher final cost Lack of public awareness Lack of demand Lack of strategy for promotion Lack of expertise Lack of cooperation Lack of database and information Resistance to change Lack of training Lack of technology Lack of government support Lack of a measurement tool	Quantitative	Construction professionals in Ghana	Djokoto et al. (2014)

(continued)

Table 9.1 (continued)

Barriers	Research approach	Sample	Reference
Lack of building codes and regulations Lack of incentives Higher investment cost Investment risks Higher final cost Unavailability of design and construction team Lack of technology Lack of training Lack of cooperation	Mixed methods	Contractors and consultants in the UK	Opoku and Ahmed (2014)
Increased capital cost contract requirements Large size and diverse company activities Perception that sustainability costs more Managing competing and conflicting targets with other business aims Sustainability is down on the boards' priority list	Quantitative	Architects in Ghana	Bangdome-Dery and Kootin-Sanwu (2013)

construction practices. Revel et al. (2010) categorised the potential benefit to be accrued in terms of jobs, right corporate image and retention of quality staff as pull factors. Push factors are reactions to regulatory demands and financial incentives. Change in legislation, building codes and bylaws are essential moderators that could trigger, promote and sustain sustainable construction (Ayarkwa et al., 2017). According to Ayarkwa et al. (2017), actions such as a change in legislation, building code and bylaws, tax rebates/subsidies on green products and availability of integrated technology can promote and enhance sustainable construction in Ghana (Table 9.2).

9.4 Lessons Learnt

The construction industry in Ghana contributes to the socio-economic development of the country. It is patterned on the UK model. The ministry of works and housing is the agency responsible for construction works in Ghana. Despite the reported challenges, construction professionals are aware of the benefits of sustainable construction in Ghana. However, the implementation of sustainable construction is at

Table 9.2 Drivers of sustainable construction

Drivers	Research approach	Sample	Reference
Client demand and requirements Stakeholder influence Cost efficiency Competitive advantage Legislation and regulations Awareness and knowledge by top management Clear and consistent guidelines for measuring sustainable construction Win more contracts to remain in business Financial incentives (tax rebates, high-profit margin) Company's reputation and brand image to attract and retain the right staff Availability of life cycle cost analysis Moral obligation to protect the environment Investment	Quantitative	Architects in Ghana	Ayarkwa et al. (2017)
Regulations Costs, risks and market value Demand and the role of clients Tendering and procurement processes Process phases and scheduling of tasks Cooperation and networking Knowledge and common terminology Availability of integrated methods Innovation	Quantitative	Construction professionals in Finland	Bash and Hakkinen, (2015)
Demand by stakeholders Financial benefits Need for corporate/social responsibility Environmental sustainability	Qualitative	South African construction professionals	Windapo (2014)
Financial incentives Building regulations Client awareness Client demand Planning policy Taxes/levies Labelling/measurement Investment		Construction professionals in the UK	Pitt et al. (2009)

(continued)

Table 9.2 (continued)

Drivers	Research approach	Sample	Reference
The imposition of stricter regulations Establishment of the longer customer-supplier relationship Awareness of the environmental, social and economic impact Implementation of an environmental management system Push from the top management Implementation of ISO14,000 kind of certifications	Quantitative	Construction professionals in India	Arif et al. (2009)

the infant stage, where the accrual of benefits is minimal. The construction industry, and by implication the built environment, is a significant contributor to global climate change and other environmental threats. It consumes three billion tons of global raw materials annually and produces an enormous amount of waste. Climate change may continue to be a global threat for generations to come if practical actions are not expedited in the built environment. The sector needs to take practical action to mitigate the impact of various threats. The research on the carbon and ecological footprint is of particular significance in dealing with climate change to restore the ecological balance.

Sustainable construction practices are not frequently used in Ghana. The main barriers to the implementation of sustainable construction include lack of financial incentives, lack of demand, lack of governmental commitment, lack of legislation and lack of building codes and regulations to promote sustainable construction. The client demand and requirements appear to be the main drivers of sustainable construction in Ghana. However, there is no distinctive difference between barriers, drivers and enablers of sustainable construction because various researchers use the terms interchangeably. There is a need for further research to define these terms. This will help increase the awareness level for implementing sustainability.

9.5 Conclusion and Way Forward

The construction industry in Ghana has significantly improved, and there is a growing awareness for adopting sustainability practices in the sector. However, the literature on sustainable construction in Ghana is minimal. As such, this study contributes to the literature on sustainable construction in general and Ghana in particular on the current state of implementing sustainable construction practices in Ghana. It is concluded that sustainable construction practices are marginal in Ghana (Ahiabu et al., 2019), although the industry is a significant consumer of natural resources that contribute to climate change. In effect, there should be an increased level of awareness of the benefits of sustainability to ensure that more projects adopt sustainable construction measures to reduce carbon and ecological footprint, which can reportedly combat climate change (Garcia-Olivares, 2015). The government of Ghana should introduce financial incentives to construction companies and professionals in the built environment that adopt sustainability to lower the negative environmental impact of construction activities. The following actions could be taken to reduce the effects of construction activities on the environment:

Minimisation of resources consumption.
Improvement of indoor air quality.
Avoidance of environmental health problems.
Use of recycled materials.
Waste reduction and maximisation of reuse.
Increased sustainability literacy and education.

References

Abidin, N. Z. (2010). Investigating the awareness and application of sustainable construction concept by Malaysian developers. *Habitat International, 34*(4), 421–426.

Adjarko, H., Agyekum, K., Ayarkwa, J., & Amoah, P. (2016). Implementation of environmental sustainable construction principles in the Ghanaian construction industry. *International Journal of Engineering and Management, 6*(2), 585–593.

Ahiabu, M. K., Emuze, F., & Das D. (2019). A Framework towards the Reduction of the Ecological and Carbon Footprint of Construction Activity in Ghana. In *Proceedings of 14th international postgraduate research conference: Contemporary and future directions in the built environment* (pp. 525–537). University of Salford.

Ahn, Y. H., Pearce, A. R., & Wang, G. (2013). Drivers and barriers of sustainable design and construction: The perception of green building experience. *International Journal of Sustainable Building Technology and Urban Development, 4*(1), 35–45.

Akinradewo, O., Aigbavboa, A., Oke, A., & Coffie, H. (2019). The Ghanaian construction industry and road infrastructure development: A review. In *Proceedings of 14th international postgraduate research conference: Contemporary and future directions in the built environment* (pp. 340–349). University of Salford.

Arif, M., Egbu, C., Haleem, A., Kulonda, D., & Khalfan, M. (2009). State of green construction in India: drivers and challenges. *Journal of Engineering Design and Technology 7*(2), 223–234.

Ayarkwa, J., Acheampong, A., Wiafe, F., & Boateng B. E. (2017). Factors affecting the implementation of sustainable construction in Ghana: The Architect's perspective. In: *Proceedings of international conference on infrastructure development in Africa* (Vol. 6, pp. 377–386).

Ayarkwa, J., Ayirebi-Danso, & Amoah, P. (2010). Barriers to implementation of ESCPS in the construction industry in Ghana. *International Journal of Engineering Science, 2*(4).

Baker, D. J. (2016). The purpose, process, and methods of writing a literature review. *AORN Journal, 103*(3), 265–269.

Bangdome-Dery, A., & Kootin-Sanwu, V. (2013). Analysis of barriers (factors) affecting architects in the use of sustainable strategies in building design in Ghana. *Research Journal in Engineering and Applied Sciences, 2*(6), 418–426.

Bash, E., & Hakkinen, T. (2015). *Barriers and drivers for sustainable building*. PhD Proposal, 1.

Bourdeau, L. (1999). Sustainable development and the future of construction: A comparison of visions from various countries. *Building Research and Information, 27*(6), 354–366.
CIB. (2004). *50 years of international cooperation to build a better world.* CIB.
Connell, K. Y. H. (2010). International and external barriers to eco-conscious apparel acquisition. *International Journal of Consumer Studies, 34*(3), 279–286.
Djokoto, S., Dadzie, J., & Ohemeng-Ababio, E. (2014). Barriers to sustainable construction in the ghanaian construction industry: Consultants' perspective. *Journal of Sustainable Development, 7*, 134–143.
Du Plessis, C. (2007). A strategic framework for sustainable construction in developing countries. *Construction Management and Economics, 25*, 67–76.
Du Plessis, C., et al. (2002). *Agenda 21 for sustainable construction in developing countries.* CSIR Report BOU/E0204CSIR CIB & UNEP-IETC, Pretoria.
Edeoja, J. A., & Edeoja, A. O. (2015). Carbon emission management in the construction industry-case studies of the nigerian construction industry. *American Journal of Engineering Research, 4*, 112–122.
Garcia-Olivares, A. (2015). Sustainability of electricity and renewable materials for fossil fuels in a post-carbon economy. *Energies, 8*(12), 13308–13343.
Huang, W., Li, F., Cui, S., Li, F., Huang, L., & Lin, J. (2017). Carbon footprint and carbon emission reduction of urban buildings: A cases in Xiamen city. In *Proceedings of Urban transitions conference,* Shanghai, China, *Procedia engineering* (Vol. 198, pp. 1007–1017).
Huovila, P. (1998). *Sustainable construction in Finland 2010,* Report 2 in CIB, Sustainable Development and the Future of Construction. A comparison of vision from various countries, CIB Report Publication 225, CIB.
Hussin, J. M., Rahman, I. A., & Memon, A. H. (2013). The way forward in sustainable construction: Issues and challenges. *Institute of Advance Engineering and Science, 2*, 15–24.
Jesson, J., Matheson, L., & Lacey, F. (2011). *Doing your literature review: Traditional and systematic approaches.* Sage.
Kibert, C. J. (1994). Establishing principles and a model for sustainable construction. In *Proceedings of the 1st international conference of CIB task group 16 on sustainable construction* (pp. 3–12), Tampa, 6–9 November.
Lanting, R. (1998). *Sustainable construction in the Netherlands,* Report 9 in CIB Sustainable Development and the Future of Construction. A Comparison of Vision from various countries, CIB Report Publication 225, CIB.
Li, S., & Wang, H. (2018). Traditional literature review and research synthesis. In *The Palgrave handbook of applied linguistics research methodology* (pp. 123–144).
Ministry of Works and Housing. (2017). *Sector strategic medium term development plan for 2018–2021.*
Mustafa, Y., & Bakis, A. (2015). Sustainability in the construction sector. In *Proceedings of world conference on technology, innovation and entrepreneurship, procedia – social and behavioural sciences* (Vol. 195, pp. 2253–262).
Ofori, G. (2012). *Developing a construction industry in Ghana: The case for the National University of Singapore* (pp. 1–19).
Opoku, A., & Ahmed, V. (2014). Embracing sustainability practices in UK construction organisations: Challenges facing intra-organizational leadership. *Built Environment Project and Asset Management, 4*(1), 90–107.
Osei, V. (2013). The construction industry and its leakages to the Ghanaian economic-policy to improve the sector's performance. *European Centre for Research Training and Development, International Journal of Development and Economic Sustainability, 1*, 56–72.
Ogungbile, A. J., & Oke, A. E. (2019). Sustainable construction practices in West African countries. In E. Motoasca, A. Agarwal, & H. Breesch (Eds.), *Energy sustainability in built environment. Energy, environment, and sustainability.* Springer.
Pitt, M., Tucker, M., Riley, M., & Longden, J. (2009). Towards sustainable construction: promotion and best practices. *Construction Innovation, 9*(2), 201–224.

Revel, A., Stokes, D., & Chen, H. (2010). Small businesses and the environment: Turning over a new leaf? *Business Strategy and the Environment, 19*(5), 273–288.

Roumeliotis, G., & Fenton, S. (2011). *Global construction growth to outpace DGP*. This decade.

Vandierendonck, A., Liefooghe, B., & Verbruggen, F. (2010). Task stitching: Interplay of reconfiguration and interference control. *Psychological Bulletin, 136*(4), 601.

Windapo, A. O. (2014). Examination of green building drivers in south African construction industry: Economics versus ecology. *Sustainability, 6*(9), 6088–6106.

Zimmermann, M., Althaus, H. J., & Haas, A. (2005). Benchmarks for sustainable construction: A contribution, environmental management systems adoption by Australian organisation: Part 1: Reasons, benefits to develop a standard. *Energy and Buildings, 37*, 1147–1157.

Chapter 10
Key Factors Influencing Deployment of Photovoltaic Systems: A Case Study of a Public University in South Africa

Nutifafa Geh, Fidelis A. Emuze, and Dillip Kumar Das

10.1 Introduction

Globally, universities are on a quest to transform their campuses by implementing numerous sustainable initiatives. By so doing, universities are ensuring that their activities are ecologically sound, socially and culturally just and economically viable (Orr, 2002). Since, universities are similar to small towns having several buildings which need energy to operate, the adoption of photovoltaic (PV) system presents enormous potential for improving the energy performance of the vast portfolio of buildings. In fact, it is reported that the limited availability of green buildings, where the principles of sustainability are demonstrated, is a significant barrier to sustainable development implementation and innovation in universities (Ávila et al., 2017). According to the categorisation provided by Ávila et al. (2017), adoption of PV system belongs to operational innovation, which is the introduction of tools to enhance and maximise the operations of an institution.

Regarding the operationalisation of sustainability within universities in Africa, the study of Ulmer and Wydra (2019) involving 29 different universities in 16 African countries confirmed that there appears to be ongoing implementation of all 4 sustainability strategies—research, teaching, campus operations and community outreach—in the universities. Although greenness variation exists across universities, generally, the uptake of PV systems among South African public universities appears to remain rather marginal and far below expectation. Similar to diffusion of

N. Geh (✉) · F. A. Emuze
Central University of Technology, Free State, Faculty of Engineering, Built Environment and Information Technology, Bloemfontein, South Africa
e-mail: 219000670@stud.cut.ac.za; femuze@cut.ac.za

D. K. Das
University of KwaZulu-Natal, School of Engineering, Durban, South Africa

other innovations, deployment of PV systems is influenced by several factors. For example, high investment cost, lack of financial resources, lack of support from institution administrators, lack of planning and focus and unavailability of space for installation can inhibit adoption of PV systems (Ansari et al., 2013; Ávila et al., 2019; Mah et al., 2018; Shah et al., 2019). Although a significant number of research projects examined the factors which influence the adoption of PV systems, it is evident that studies were largely focused on individual adoption in households and communities. Research exploring the phenomenon from organisational perspective is rather limited. In addition, studies examining the phenomenon exclusively in the context of South African universities (SAUs) seem to be lacking. However, the study of Awuzie and Emuze (2017) provided insight on the phenomenon when they investigated the drivers for implementation of sustainable development (SD) in a SAU. The authors found that the case study institution pursued alternative sources of energy (renewable energy) because it was reckoned as 'cheaper' than conventional sources of electricity (Awuzie & Emuze, 2017). Within this context, a comprehensive analysis of the factors influencing upscaling of PV systems within SAU context is worthwhile. Thus, the aim of this study, through review of extant literature and interview of experts in a SAU, is to (1) ascertain progress on deployment of PV systems in a SAU, (2) identify the various factors which adversely influence upscaling of the technology and (3) determine the appropriate strategies to enhance deployment.

10.2 Literature Review

The quest by universities to pursue sustainable development is bound to face challenges. The literature on diffusion of PV systems has shown that barriers exist in both developed and developing countries. The search for literature revealed that there is limited published research focused exclusively on barriers within university context. However, there are research findings that give insight on barriers faced by universities. For instance, literature on sustainable development implementation or campus sustainability execution gives indication of barriers that bear on adoption of PV systems. Therefore, literature from such areas can augment literature that specifically addresses adoption of PV systems.

Photovoltaic systems are the second most deployed renewable energy technology in the world by installed capacity after wind. This success is greatly attributed to the technology's falling cost (International Renewable Energy Agency [IRENA], 2019). Nevertheless, it is still expensive to install PV systems. A significant body of literature confirms that the high initial cost and unavailability of funds to invest in projects are the prime barriers to adoption (Balcombe et al., 2014; King et al., 2014; Kurata et al., 2018; Mah et al., 2018; Nygaard et al., 2017). The payback period for investments in PV systems is also normally long, and this can foster non-adoption because it takes long to recoup capital investment. In a similar manner, when the benefits to be gained from using PV system is unknown or unappreciated, it can cause people not to adopt, and so is lack of incentives (Ansari et al., 2013; Balcombe et al., 2014; Garlet et al., 2019; Sindhu et al., 2016). Unavailability or complexity of lease or power purchase agreement (PPA) contracts can hinder adoption (Rai et al., 2016; Rosales-Asensio et al., 2019). Likewise, lack of established dealer network can be a barrier to uptake of PV systems (Pode, 2010). Unavailability of space could also trigger non-adoption. In some cases, existing regulations require high-rise public housing blocks to reserve rooftop area as a refuge for occupants in case of fire, and this could place limitation on the space that is available for PV installation (Mah et al., 2018). It can also be due to poor solar orientation of homes and roof architecture (Garlet et al., 2019). Other times, it can result from structural challenges, fear of damage to roof which can cause leaks or shades produced by tall adjacent buildings or tress (Walters et al., 2018; Zhang et al., 2012). For PV systems to function efficiently, it is necessary to have a systematic maintenance culture in place. This added responsibility and cost can also trigger non-adoption (Sindhu et al., 2016).

Concerning SD operationalisation within higher education institutions (HEIs), it is suggested that universities in developed countries are leading while those in developing countries are laggards (Ávila et al., 2019). Ávila et al. (2017) through a survey, involving 301 respondents from 172 universities around the world, examined the barriers to innovation and SD implementation. The authors found that lack of support from university administration seems to be the biggest hurdle. Some other important factors identified were lack of policies which promote green building practices, lack of incentives, lack of environmental committee, lack or inadequate planning and focus and resistance to change. The barriers identified by Ávila et al. (2017) were further corroborated by Ávila et al. (2019) when the authors undertook comparative analysis from continental perspective—North and South America, Africa, Asia, Europe and Oceania. The work of Richardson and Lynes (2007) provided insight on factors which could inhibit the execution of green building practices on university campuses. This study found lack of incentives (resulting in low motivation), lack of support and commitment from stakeholders with decision-making power and lack of quantifiable green building targets as key barriers. In a similar manner, Kasai and Jabbour (2014) found that lack of financial incentive is negatively correlated to adoption of green buildings by universities.

An analysis of facility directors' perception of the challenges to sustainable procurement (SP) in HEIs in Nigeria and South Africa revealed that lack of commitment and support from organisational leadership, lack of funding, non-specification of SD-based requirements in contract documents and lack of commitment from HEIs were hampering progress. Regarding initiatives occurring on campuses to foster SD implementation, Mawonde and Togo (2019) reported that the University of South Africa (UNISA) is undertaking several projects, including installation of solar panels. The university is, however, said to be having funding challenges (Mawonde & Togo, 2019). Green movements, in the form of student-led organisations which advocates for greenness within universities, are also emerging in SAUs. For example, Green Campus Initiative is a popular action group launched in 2012, and its presence is reported in several SAUs (South African Government News Agency [SANews], 2012). The relevance of sustainability offices in fostering SD in universities is also emphasised in the literature, and their lack can leave a gap within a university community (Adomßent et al., 2019; Filho et al., 2019). Another important key player in the energy sector is government, and the literature is unambiguous about the significant role of governments. For example, legislations and fiscal incentives are one way governments can foster investments in renewable energy. Therefore the lack of these support systems from government can slow the rate of diffusion of PV systems in a locality (Mah et al., 2018; Shah et al., 2019).

10.3 Research Methodology

This study adopted a case study design approach. This approach was preferred because it offered the opportunity to conduct a detailed and intensive analysis of issues exclusively pertinent to the case study (Bryman, 2004; Kumar, 2011). The selected case is Central University of Technology, Free State (CUT), in South Africa. CUT is one of the 26 public universities with two campuses; the main campus is located in Bloemfontein, and the second campus is in Welkom. The institution was established in 1981 as Technikon Free State, and it was in 2004, due to the re-branding and merging of universities in South Africa, that the institution received its new name and unique identity as a University of Technology (UoT). The university provides a wide range of programmes at undergraduate, graduate and postgraduate levels. CUT has a student population of 20,676 as of January 2020.

A two-step approach was adopted in collecting the data. Firstly, information was gathered from institution documents and website to ascertain the level of deployment of PV systems at the institution, including site visits. In addition, information gathered at this stage was verified from the institution's Estates and Infrastructure Department. Secondly, semi-structured interview with key stakeholders was conducted to identify key factors which adversely influence deployment and to determine the strategies to enhance upscale of PV systems. The interview protocol was structured into two parts—the first section sought for background information, and the second section comprised three questions regarding participants' views on barriers, solution to the barriers and general comments. The interview protocol was submitted to the university research committee, and approval was received before the data was collected.

Purposive sampling technique was used in selecting participants for the study (Bryman, 2004). This approach was deemed most suitable due to the objective of selecting only participants who were well informed about deployment of PV systems. Participants were selected and invited to participate in the study if they were involved in infrastructure development and building maintenance and/or SD implementation in the university. Due to COVID-19 restrictions, communication with the selected participants was via email—the interview protocol was sent to all 15 selected staff requesting them to grant an interview. Follow-up emails were sent to people who did not respond to the email after 2 weeks. A total of seven staff participated in the study (referred to as P1, P2, P3, P4, P5, P6 and P7), and

10 Key Factors Influencing Deployment of Photovoltaic Systems: A Case Study...

Table 10.1 Information on interviewees

No	Stakeholder group	Reference code	Years of employment at CUT
1	Academic staff	P1	5
2	General staff	P2	20
3	General staff	P3	15
4	Academic staff	P4	25
5	General staff	P5	8
6	General staff	P6	10
7	Academic staff	P7	13

the average year of employment with CUT is 13.7 years (Table 10.1). The participants consisted of four general staff and three academic staff. Because the focus of the study is to elicit information on factors hampering progress, the authors were of the view that respondents will be more open to share their views when they have surety of confidentiality and anonymity in the reportage of their views on the phenomenon. Therefore, the interviewees were given this assurance prior to engaging them in the study. This was also the reason for the limited information provided on interviewees' demographics in Table 10.1. In all, five participants were interviewed via Zoom in May–June 2020 when university staff were working remotely due to COVID-19 restrictions (P1, P2, P3, P4, P5). After restrictions were partly lifted, one face-to-face interview was conducted at the Bloemfontein campus in July 2020 (P6). Interview sessions lasted for an average of 30 min, and sessions were recorded by permission of the interviewees. Prior to ending the interview sessions, the interviewer reiterated the points highlighted by the interviewees before requesting general comments from them at the end. This ensured re-affirmation of the views of the respondents and also enhanced accuracy in reportage of their views. One participant could not be interviewed, hence the interview questions were emailed to the participant, and it was filled and returned (P7).

The recordings from the interviews were manually transcribed verbatim, and the textual data was thereafter thematically analysed. The analysis was manually carried out in line with the guidelines recommended by Kumar (2011). The analysis process included identifying the main themes which emerged from the data, organising responses under the main themes and integrating themes and responses in the text in the report (Kumar, 2011, p. 278). The seven main themes which emerged from the data are presented and discussed in the next section.

10.4 Results and Discussion

In this section, the findings are presented and discussed in accordance with the objectives. First, the progress on PV installation at the case study university is presented. Second, the challenges facing further deployment of the technology are

Table 10.2 Factoring influencing upscaling of photovoltaic systems

No	Description	P1	P2	P3	P4	P5	P6	P7
1	Funding challenges	x	x	x	x	x	x	x
2	Installation space challenges	x			x		x	x
3	Limited commitment to operationalising policies and practices	x		x	x			
4	Limited knowledge regarding the benefits	x						
5	Lack of sustainability office		x	x				
6	Lack of schemes which support PV development	x		x		x		
7	Lack of external pressure or motivation	x				x		

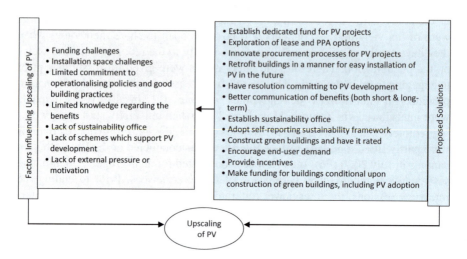

Fig. 10.1 Photovoltaic deployment framework

presented and discussed, and the strategies which were deemed necessary to aid upscaling are set forth (Table 10.2 and Fig. 10.1).

10.4.1 Photovoltaic Systems Deployment at CUT

The institution's first solar project was launched in 2013, and this was the installation of ten 40 W cell phone solar-powered charging stations on Bloemfontein and Welkom campuses for use by students. The charging stations were called solar-flower, and each of the ten stations can charge up to four devices simultaneously. The solar-flower was designed, fabricated and installed by CUT's internal experts (Central University of Technology [CUT], 2014). Furthermore, in 2018, the institution completed a 153 kW capacity solar project—12 dual-axis solar tracking PV systems mounted on 8-m-high masts. The project was executed in collaboration with a private partner—Karah Assets (Pty) Limited—and is located at the Bloemfontein campus (CUT, 2018). Energy storage systems were not installed; hence, the power generated is not stored but is directly fed into the local grid of the university. The 153 kW system was funded by Karah Assets (Pty), and the private partner is also responsible for maintenance of the system; the university pays for the power consumed at an agreed price which is lower than the market price. The 153-kW system is also used for educational and research purposes. Also, since 2014, the Faculty of Engineering, Built Environment and Information Technology has been carrying out PV installations in phases on BHP Billiton Building at Bloemfontein campus. This project has been mainly executed for educational and research purposes. Currently, installation on the rooftop of BHP Billiton Building has reached a total capacity of 20 kW, and the electricity generated is used in the building.

10.4.2 Factoring Influencing Upscaling of PV Systems and Remedial Measures

Interviewees identified various factors which they perceive are adversely influencing the upscaling of PV system at CUT. Overall, seven main factors emerged from the data and are discussed below (Table 10.2 and Fig. 10.1).

Funding Challenges

It is widely acknowledged that PV systems are still quite expensive, and the high initial cost is a major factor which can hamper deployment (Kurata et al., 2018). Apart from the panels, there are other expensive components, such as cabling and energy storage systems (P2, P4, P6). All the interviewees perceived that a major challenge facing upscaling of PV at CUT is related to funding. According to P2, 'obviously the first one is financial', and P1 opined that, 'funding is a major challenge'. Mainly, the respondents were of the view that PV projects were not receiving the needed funding because of two major reasons. Firstly, it was believed there is lack of funds, and secondly because the university has other pressing needs which require financial resources, hence funds were allocated to these needs which were perceived to be of higher priority. Respondent P5 expounded on the latter phenomenon by saying,

> When you bring solar power and other requirements you find that if you try to categorise, it doesn't rise to the level where it is funded in the categories (P5).

Additionally, one of the respondents spoke concerning the same subject by articulating that although money is being made available for infrastructure development by the Department of Higher Education and Training (DHET), the funds are not invested in PV systems when decisions are taken, because it is considered as being too expensive (P3).

In offering solutions to the funding challenges, it was suggested there should be long-term planning to make sure there is dedicated funding for PV projects (P5). It was also advocated that general procurement systems are not best suited for PV procurement, especially when only PV is being procured and not a building project, hence the need to innovate procurement processes that give more value for money (P2, P5). The university's current 153 kW PV system is based on a public private partnership, where the university pays for the power that is generated, and this option is seen by some of the respondents as a viable acquisition option that the university can continue to explore in upscaling PV systems on campus, as opposed to acquiring the system on outright purchase basis (P1, P2, P6).

Challenges Relating to Installation Space

Unavailability or limited space could also trigger non-adoption or hamper upscaling of PV projects (Mah et al., 2018; Walters et al., 2018; Zhang et al., 2012). This study identified one of the challenges CUT is tackling, that is, finding suitable spaces to install the PV systems at affordable price and also to ensure that it performs at optimal level without affecting campus aesthetic. Specifically, it was confirmed by interviewees P4 and P6 that the university is currently planning to expand the 153-kW installation, and this requires installation of additional solar trackers like the existing systems at convenient positions on campus or installing fixed systems. The spatial challenge pointed out by the interviewees is only peculiar to Bloemfontein campus (P1, P4, P6, P7). However, it was wholly acknowledged that Welkom, as a developing campus, has a lot of space for PV installations.

In overcoming the challenge relating to space, interviewees recommended careful planning and assessment of spaces that can be used to install more PV systems, both within and outside the university campus. For instance, interviewee P4 confirmed the university is in consultation with owners of an adjoining private property to secure space for expansion of the PV systems.

Limited Commitment to Operationalising Policies and Practices

Universities in their quest to implement SD tenets have realised the need to set out clear visions and policies that will guide action and development. It is therefore laudable and not surprising that CUT adopted its *Vision 2020* in 2010. Consequently, the institution's commitment to its transformation agenda is affirmed and revealed in several policy documents, such as: (1) Transformation Plan 2016–2020, (2) the 2016–2020 Strategic Plan, (3) the special project for the period 2018–2022 approved at the Management Lekgotla held on 28–29 September 2017 and (4) the ongoing development of Vision 2030 and Strategic Plan 2021–2025 (CUT, 2010, 2017a, b, 2020).

According to Richardson and Lynes (2007) and Shriberg (2003), top leadership commitment and support is a significant driver for delivering sustainable practices in universities. Also, the study of Ávila et al. (2017, 2019) established that limited commitment from decision-makers can significantly impair innovation and SD operationalisation within universities. Similarly, the findings from this study suggests that limited commitment to operationalising university policies and green building practices, such as PV deployment during new construction/refurbishment projects, is a major factor which influences PV deployment rate at CUT (P1, P3, P4). In providing insight on this institution-related barrier, respondent P3 opined by stating:

> The barrier is commitment as far as I am concerned. The university does have sustainable development policy and it's got a commitment. So the commitment [has to go] beyond paper. The major barrier is operationalising your commitment. Because by operationalising you need to commit resources, financial resources (P3).

In overcoming the commitment barrier, it was suggested there should be a system in place which in a way 'forces' the university to ensure that new buildings incorporate PV systems at the time of construction or make provision for easy installation at a later time. Refurbishment of existing buildings was also suggested to be executed with similar principles to afford the same opportunities (P1, P3, P4).

Limited Knowledge or Appreciation of Benefits from Existing Projects

There is a general consensus among scholars that people or organisations adopt innovations because of the perceived benefits (Balcombe et al., 2014; Davis, 1989; Iacovou et al., 1995). However, it has also been recognised that, when people perceive that the benefits are too small, it could trigger non-adoption. For example, Balcombe et al. (2014) found that some people would reject PV because of the perception that the environmental benefits are too small. In like manner, a respondent postulated that, although the university is planning to upscale, the motivation is very low and the university has not attained the level it could have in terms of deployment because there is limited buy-in from stakeholders. According to this respondent, there is limited buy-in because the benefits accrued from existing projects may not be satisfactory or acknowledged enough to motivate a speedy upscaling of the institution's PV portfolio (P1). Reiterating the importance of knowing the benefits of existing projects in order to engender further deployment, interviewee P1 stated that:

> Before you go in to say you want to undertake a bigger one, a large scale one, what is the impact this one has made now on the balance sheet of the university, on occupant comfort in the building … do we see light from the solar being used during load-shedding? … so limited knowledge concerning the impact of what we have now, or the phase one we have now can serve as a barrier because you can't secure buy ins (P1).

The main evidence from the data points out that better communication of the impact of existing projects is very essential in motivating and securing maximum support from decision-makers for scaling up. It was also suggested there is the need to compute the life cycle cost of projects and showcase the benefits that are accruable across the whole of PV lifecycle.

Lack of Sustainability Office

Evidence from the literature alludes that sustainability offices (SOs) are very key in strengthening collaboration among stakeholders and an effective tool in driving and supporting the implementation of sustainable initiatives within universities (Adomßent et al., 2019; Filho et al., 2019). SOs can also be useful in fostering awareness among students and staff on matters relating to sustainable development (ibid). In congruence with the literature, another factor pointed out by some interviewees was the lack of sustainability office (SO) in CUT. According to some interviewees, there used to be a SO in CUT, but it was shut down for unknown reasons (P2, P3).

Although the lack of SO in CUT is not a direct barrier to deployment of PV, the lack of it indirectly creates an impact due to the vacuum it creates in terms of advocacy roles. The findings therefore suggest that perhaps the presence of SO could further enhance implementation of sustainable initiatives at CUT, including deployment of PV systems.

Lack of Schemes Which Support Photovoltaic Development

The scores allocated to energy in green building rating systems are very substantial. Score are awarded for energy efficiency as well as for renewable energy generation. Green certification of university buildings can therefore trigger the adoption of PVs as a green energy generation instrument in buildings. According to P1, green building rating can be a subtle way of encouraging universities to adopt PVs. Similarly, self-reporting frameworks, for example, Sustainability Tracking, Assessment and Rating System (STARS), can encourage universities to implement sustainable initiatives and report their sustainability performance. Therefore, the absence of such schemes could diminish sustainability drive on campus. The lack of green-rated buildings and the lack of campus sustainability self-reporting framework were affirmed by interviewees as factors which may have diminished the drive to upscale PV in CUT (P1, P3, P5). Interviewee P5 articulated this view by saying:

> So if you look at all these public institutions in South Africa and you ask which ones have any kind of certification, there is nothing, simply because that is not a requirement and the second nobody is demanding that. So again, if you certify it is a luxury, it is you who want to pursue that one (P5).

Although universities are not mandated to certify their building, the interviewees were in agreement that it would be impactful if universities are encouraged to construct green buildings and have it certified. Also, according to P5, the adoption of a campus sustainability self-reporting framework can boost PV deployment, and it was confirmed by the same respondent that the university is in the process of filling this gap.

Lack of External Motivation and Incentives

The lack of incentives to universities also emerged as a relevant factor which lessens the determination for PV deployment at CUT. Other scholars, such as Kasai and Jabbour (2014) and Richardson and Lynes (2007), also found incentives to be an essential driver for construction of green buildings in universities. Furthermore, the absence of demand for solar electricity and pressure from institution stakeholders, such as students and staff, were cited as factors that indirectly influence PV deployment decisions. A comment from an interviewee was:

> Because we are educating people for sustainability and they always know that using solar is more sustainable … so now if there is pressure from users that we want more of solar electricity … that pressure group will make the university dedicate more funding for it (P5).

The literature affirms that government legislations and policy are very instrumental in renewable energy development (Mah et al., 2018; Shah et al., 2019). This study identified that there is a belief that the lack of appropriate legislation which promotes renewable energy development in universities affects PV deployment decisions at CUT (P1). In addition, it was acknowledged that Department of Higher Education and Training (DHET) which is the financier of most university buildings does not make construction of green buildings (by extension PV adoption) mandatory (P5). It is therefore the view of some interviewees that incentives, legislation, end-user demand and DHET 'motivation' could help in installing more PVs on campus.

10.5 Conclusion

Universities can partly succeed in the transition to green campus by adopting sustainable building practices such as the installation of PV systems. Understanding the dynamics of PV deployment on campuses is important to make the transition happen effectively. By adopting a qualitative case study research design, this study examined the views of university staff at a SAU and contributed to the existing body of knowledge in two key ways. First, it provided insight into seven factors that could negatively impact the upscaling of solar PV. This includes funding challenges, installation space challenges, limited commitment to operationalising policies and good building practices, limited knowledge regarding benefits, lack of sustainability office, lack of schemes which support PV development and lack of external motivation and incentives. Second, it produced valuable insight from the experts on proposed measures to enhance diffusion of PVs. The findings of the study are limited to the case study and do not offer generalisation of the results across public universities in South Africa. Also, the findings are solely based on the perception of the informants, and although the participants were purposefully selected due to their role and expertise, the authors acknowledge the possibility that the views of the participants may not be absolute interpretation of the phenomenon on campus.

Given the scope of the study, the findings are expected to be useful to relevant university stakeholders in devising strategies to strengthen the institution's capacity to transform the built environment to serve better as a model in society. Furthermore, valuable insight is provided into strategic areas that government agencies, especially those in charge of infrastructure development, could look into to further support universities to transform into sustainable campuses.

Acknowledgements This paper reports part of the findings of a doctoral research project which was supported by the National Research Foundation of South Africa (Grant Number: 116085) and Central University of Technology, Free State.

References

Adomßent, M., Grahl, A., & Spira, F. (2019). Putting sustainable campuses into force: Empowering students, staff and academics by the self-efficacy Green Office Model. *International Journal of Sustainability in Higher Education, 20*(3), 470–481.
Ansari, M. F., Kharb, R. K., Luthra, S., Shimmi, S. L., & Chatterji, S. (2013). Analysis of barriers to implement solar power installations in India using interpretive structural modeling technique. *Renewable and Sustainable Energy Reviews, 27*, 163–174.
Ávila, L. V., Beuron, T. A., Brandli, L. L., Damke, L. I., Pereira, R. S., & Klein, L. L. (2019). Barriers to innovation and sustainability in universities: An international comparison. *International Journal of Sustainability in Higher Education, 20*(5), 805–821.
Ávila, L. V., Leal Filho, W., Brandli, L., Macgregor, C. J., Molthan-Hill, P., Özuyar, P. G., & Moreira, R. M. (2017). Barriers to innovation and sustainability at universities around the world. *Journal of Cleaner Production, 164*, 1268–1278.
Awuzie, B., & Emuze, F. (2017). Promoting sustainable development implementation in higher education: Universities in South Africa. *International Journal of Sustainability in Higher Education, 18*(7), 1176–1190.
Balcombe, P., Rigby, D., & Azapagic, A. (2014). Investigating the importance of motivations and barriers related to microgeneration uptake in the UK. *Applied Energy, 130*, 403–418.
Bryman, A. (2004). *Social research methods* (2nd ed.). Open University Press.
Central University of Technology. (2010). *Vision 2020.* https://www.cut.ac.za/vision-2020
Central University of Technology. (2014). *CUT innovation lead the way towards a greener future.* https://www.cut.ac.za/news/cut-innovation-lead-the-way-towards-a-greener
Central University of Technology. (2017a). Vision 2020 and beyond [online]. *CUT Website.* https://www.cut.ac.za/vision-2020-and-beyond
Central University of Technology. (2017b). *Reimagining cut as a transformative, transformational, entrepreneurial, engaged university and "model" University of Technology in Africa.*

Central University of Technology. (2018). *CUT solar project: The sun is the limit*. https://www.cut.ac.za/news/cut-solar-project-the-sun-is-the-limit

Central University of Technology. (2020). *Communiqué to CUT community: Update on the Vision 2030 Engagement Platform*. https://www.cut.ac.za/announcements/72

Davis, F. D. (1989). Perceived usefulness, perceived ease of use, and user acceptance of information technology. *MIS Quarterly, 13*(3), 319–340.

Filho, W. L., Will, M., Salvia, A. L., Adomßent, M., Grahl, A., & Spira, F. (2019). The role of green and sustainability offices in fostering sustainability efforts at higher education institutions. *Journal of Cleaner Production, 232*, 1394–1401.

Garlet, T. B., Ribeiro, J. L. D., de Souza Savian, F., & Mairesse Siluk, J. C. (2019). Paths and barriers to the diffusion of distributed generation of photovoltaic energy in southern Brazil. *Renewable and Sustainable Energy Reviews, 111*, 157–169.

Iacovou, C. L., Benbasat, I., & Dexter, A. S. (1995). Electronic data interchange and small organizations: Adoption and impact of technology. *MIS Quarterly, 19*(4), 465–485.

International Renewable Energy Agency. (2019). *Future of Solar Photovoltaic: Deployment, investment, technology, grid integration and socio-economic aspects (A global energy transformation paper)*. Abu Dhabi.

Kasai, N., & Jabbour, C. J. C. (2014). Barriers to green buildings at two Brazilian engineering schools. *International Journal of Sustainable Built Environment, 3*(1), 87–95.

King, G., Stephenson, J., & Ford, R. (2014). *PV in Blueskin: Drivers, barriers and enablers of uptake of household photovoltaic systems in the Blueskin communities, Otago, New Zealand*. Centre for Sustainability, University of Otago.

Kumar, R. (2011). *Research methodology: A step-by-step guide for beginners* (3rd ed.). Sage.

Kurata, M., Matsui, N., Ikemoto, Y., & Tsuboi, H. (2018). Do determinants of adopting solar home systems differ between households and micro-enterprises ? Evidence from rural Bangladesh. *Renewable Energy, 129*, 309–316.

Mah, D. N., Wang, G., Lo, K., Leung, M. K. H., Hills, P., & Lo, A. Y. (2018). Barriers and policy enablers for solar photovoltaics (PV) in cities: Perspectives of potential adopters in Hong Kong. *Renewable and Sustainable Energy Reviews, 92*, 921–936.

Mawonde, A., & Togo, M. (2019). Implementation of SDGs at the University of South Africa. *International Journal of Sustainability in Higher Education, 20*(5), 932–950.

Nygaard, I., Hansen, U. E., Mackenzie, G., & Pedersen, M. B. (2017). Measures for diffusion of solar PV in selected African countries. *International Journal of Sustainable Energy, 36*(7), 707–721.

Orr, D. W. (2002). *The nature of design: Ecology, culture and human intention*. Oxford University Press.

Pode, R. (2010). Solution to enhance the acceptability of solar-powered LED lighting technology. *Renewable and Sustainable Energy Reviews, 14*(3), 1096–1103.

Rai, V., Reeves, D. C., & Margolis, R. (2016). Overcoming barriers and uncertainties in the adoption of residential solar PV. *Renewable Energy, 89*, 498–505.

Richardson, G. R. A., & Lynes, J. K. (2007). Institutional motivations and barriers to the construction of green buildings on campus: A case study of the University of Waterloo, Ontario. *International Journal of Sustainability in Higher Education, 8*(3), 339–354.

Rosales-Asensio, E., de Simón-Martín, M., Borge-Diez, D., Pérez-Hoyos, A., & Comenar Santos, A. (2019). An expert judgement approach to determine measures to remove institutional barriers and economic non-market failures that restrict photovoltaic self-consumption deployment in Spain. *Solar Energy, 180*, 307–323.

South African Government News Agency. (2012). *African Green campus initiative launched*. https://www.sanews.gov.za/south-africa/african-green-campus-initiative-launched

Shah, S. A. A., Solangi, Y. A., & Ikram, M. (2019). Analysis of barriers to the adoption of cleaner energy technologies in Pakistan using Modified Delphi and Fuzzy Analytical Hierarchy Process. *Journal of Cleaner Production, 235*, 1037–1050.

Shriberg, M. P. (2003). Sustainability in US Higher Education: Organizational factors influencing campus environmental performance and leadership. *International Journal of Sustainability in Higher Education, 4*(1), 1–335.

Sindhu, S., Nehra, V., & Luthra, S. (2016). Recognition and prioritization of challenges in growth of solar energy using analytical hierarchy process: Indian outlook. *Energy, 100*, 332–348.

Ulmer, N., & Wydra, K. (2019). Sustainability in African higher education institutions (HEIs) existing activities. *International Journal of Sustainability in Higher Education, 21*(1), 18–33.

Walters, J. P., Kaminsky, J., & Huepe, C. (2018). Factors influencing household solar adoption in Santiago, Chile. *Journal of Construction Engineering and Management, 144*(6), 72.

Zhang, X., Shen, L., & Chan, S. Y. (2012). The diffusion of solar energy use in HK: What are the barriers? *Energy Policy, 41*, 241–249.

Part III
Retrofit for Energy Efficiency and Comfort

Chapter 11
Is It Possible to Develop Lamella and Airey Properties Ecologically?

Elliott Young

11.1 Introduction

This research paper investigates and draws conclusions about the most eco-friendly way in which to redevelop "non-traditional" properties to suit current building regulations. The term "non-traditional" generally refers to prefabricated building systems where frames and construction methods are significantly different from those used for more traditional masonry construction. "1970 was the peak of the non-traditional housing phase with over 55,000 homes built and by 1975 around 420,000 had been erected by local authorities in England and Wales alone" (Williams, n.d.). The two different types of non-traditional dwellings this paper has focused on are Airey, a common PRC frame structure, and Lamella, a timber-framed structure. These have been chosen from the wide selection of different dwellings, in order to provide a manageable research focus, based on a comparative approach leading to an appropriate common conclusion.

Consideration has been taken of other factors, such as cost, which might hinder the development of these dwellings. However, as public concern for global warming increases, can we afford not to update these uninsulated dwellings if we are to hit our government's proposal that the UK will be the "first major economy to pass new laws to reduce emissions to net zero by 2050" (GOV.UK, 2019). "If you really think that the environment is less important than the economy, try holding your breath while you count your money" (McPherson, 2019).

E. Young (✉)
Leeds Beckett University, School of the Built Environment, Engineering and Computing, Northern Terrace, Queens Square Court, Leeds, UK

11.2 Literature Review

There is currently very limited impartial literature on non-traditional dwellings. Due to this, the main background information on this topic comes from the BRE. This is supplemented throughout using private lucrative companies' information; however, this may be biased, so less academically robust. Sources include:

11.2.1 BRE

The BRE have outlined the building structures and their inherent defects, with over 400 non-traditional dwelling types in their publication, *Non-traditional Houses: Identifying Non-traditional Houses in the UK 1918–75*. They provide a more in-depth analysis into Airey properties in their "Airey houses: technical information and guidance" publication. These publications have been used as a base point from which to find out possible development options.

11.2.2 Structherm

This company has done extensive research into the solution for developing Airey properties with structural external wall insulation. Their work in this field has been considered as part of this research paper to rectify the inherent defects of Airey properties.

11.2.3 Mapei

Mapei is a company that has researched into the structural strengthening of concrete structures, and from this, it has designed products to strengthen and repair a wide range of concrete structure failures. Its research has also been considered to derive a method of rectifying the Airey concrete structure defects.

11.3 Research Review and Methodology

The design of the methodology for the research used qualitative and quantitative mixed methods (Creswell & Creswell, 2017). This rationalised the triangular relationship between philosophical worldviews, appropriate research designs and research methods and how each of these can be used to help format and structure this paper. The research review considered the following:

11.3.1 Philosophical Worldview

Of the four main philosophical worldviews, Post-positivism, Constructivism, Transformative Paradigm and Pragmatism, the best matches for this topic of research were considered to be Constructivism and Pragmatism, because Constructivism is the act of "relying as much as possible on the participants' views of the situation being studied. The questions become broad and general so that participants can construct the meaning of a situation, typically forged in discussions or interactions with other persons" (Creswell & Creswell, 2017). The justification for using this philosophy as a basis for this research is the need to construct a general view from expert practitioners. Similarly, Pragmatism is where "researchers emphasize the research problem and question and use all approaches available to understand the problem" (Creswell & Creswell, 2017). This philosophy was therefore considered appropriate to use to identify the current defects within the two property types and the ways in which to rectify them.

11.3.2 Research Designs

The three main research designs for this type of research are quantitative, qualitative and mixed methods. Due to the nature of this topic, the research process has adopted a mixed method design, because of the need for exploratory sequential procedures of review and data comparison. For example, "the researcher first begins with a qualitative research phase and explores the views of participants. The data is then analysed, and the information used to build into a second, quantitative phase" (Creswell & Creswell, 2017). It was considered that this research design would help to extrapolate the participants' data into usable information and see if it matches up with existing literature.

11.3.3 Research Methods

In order to conduct the research, the process incorporated comparative analysis of the reported current defects with the Lamella and Airey house types using existing literature and data from a structured questionnaire undertaken with senior architectural technologists, who specialise in the field under research. This comparison helped to identify and clarify where explanations were incomplete or unavailable. Questions included the analysis of the consistency or peculiarity of these defects across the construction and performance of these house types; for example: are the defects common to both types or peculiar to each?

11.3.4 Research Aims

This research attempted to identify the most eco-friendly way of redeveloping both Lamella and Airey properties.

The research questions are therefore:

What are the current defects within both property types?

What are the possible ways of rectifying these defects?

The ultimate question of this research asks:

Is it more practicable to redevelop or to demolish and rebuild in terms of impact on the carbon footprint?

11.4 Research Results

11.4.1 Questionnaire Survey

One of the main data inputs of this research paper is a questionnaire that was sent out to industry and to fellow academics within Leeds Beckett University, School of the Built Environment, Engineering and Computing. The questionnaire acquired 16 responses, a response rate of 25%. The question format and responses were as follows:

Question 1—Have you ever heard of post-war non-traditional buildings within the UK?

Results showed 25% of respondents haven't heard about non-traditional buildings.

Question 2—If your previous answer was yes, how did you come across buildings of this sort?

Some of the interesting answers for this question are:

My current employer.[1]

News story relating to housing estate in Oulton.[2]

Question 3—Which of these post-war non-traditional dwellings do you know of? (Fig. 11.1).

---

[1] Quantity Surveyor practicioner from a professional social house developer company.
[2] Architectural Technologist student undergraduate degree programme Leeds Beckett University (2016–2019).

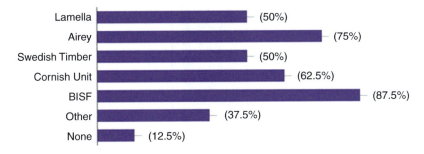

Fig. 11.1 This figure showing percent of participants that know each non-traditional property type

Question 4—Do you know if any of these non-traditional dwellings are deemed to have major defects by the Building Research Establishment?

The results were similar to that of question one with 25% of respondents not knowing about the Building Research Establishment's stance on non-traditional dwellings.

Question 5—Are you aware of the current structural and ecological issues with non-traditional buildings? If yes, please explain.

This question raises many different points due to the open nature of the question. The most useful answers were:

> A number of these dwelling were built as temporary housing due to a housing shortage following the war. The components used to build these properties, such as concrete, can be defective and deteriorate over time. These properties are generally hard to heat for residents and cost a substantial amount of money for housing providers to maintain.[3]

> Structurally, it depends on which non-trad building you are dealing with and if the issue is an inherent defect due to the construction method of the building or if the defect has occurred later in its life… some can have no/little defects aside from the wear and tear caused by 50 years of use … whereas others are more prone to structural issues (Airey/BISF's)…[4].

Question 6—What is your current understanding of fire protection in non-traditional buildings within the UK? (Fig. 11.2).

Question 7—Do you think that UK building regulations impede the efficient development of non-traditional structures under the current legislation? (Fig. 11.3).

Question 8—Do you believe that the UK government should be encouraging the future development of these buildings? (Fig. 11.4).

Question 9—In addition to the previous implications, are you aware of the advantages and disadvantages of updating these non-traditional buildings? If so, please explain.

The two main themes to the answers were, e.g.:

---

[3] Senior Architectural Technologist practicioner from a professional social house developer company.
[4] QProject Surveyor practicioner from a professional social house developer company.

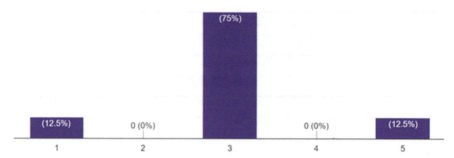

Fig. 11.2 This figure showing participants' understanding of fire protection in non-traditional dwellings, with 0 being no understanding and 5 being professional in the field

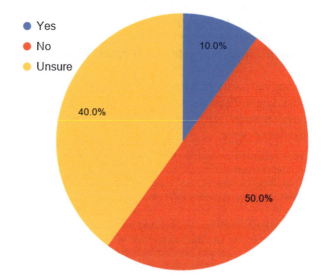

Fig. 11.3 This figure showing participant stance on whether the UK building regulations impeded the development of non-traditional dwellings

Yes… In simplistic terms I would state the following 4 advantages: Reuse of existing buildings will have a lower carbon footprint than knocking them down and starting again. Social Aspects, People enjoy living in them and don't want to move out. Local authorities don't have the facilities to decant residents and move them elsewhere while their home is knocked down and rebuilt. Upgrading of existing buildings will have a lower lifecycle cost than rebuilding, especially if the repair work can extend the lifespan up to 30 years.[5]

I am a firm believer that it is more economically advantageous to demolish the majority of non-trad housing stock and replace with new housing. Non-trads were not built with longevity in mind, they were a quick fix and yes you can externally refurbish (EWI)[6] and structural strengthen (SEWI,[7] concrete repairs, bracing, etc) non-trads but the cost to do this is still

---

[5] QProject Surveyor practicioner from a professional social house developer company.

[6] External wall Insulation.

[7] Structural external wall insulation.

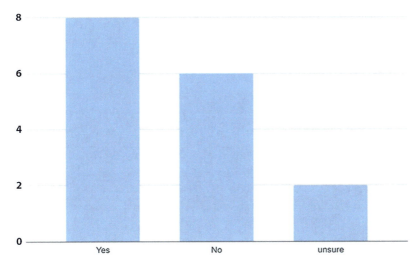

Fig. 11.4 This figure showing questionnaire feedback on whether the UK government should be encouraging the development of these buildings

significant and could be spent on demolishing and rebuilding houses. There are other considerations to this such as there is already a housing shortage, if non-trads were demolished and rebuilt where would the people currently residing in non-trads be temporarily housed? But overall I believe it is more economically advantageous to demolish and rebuild.[8]

Question 10—Depending on your profession, would you be interested in the future development of these non-traditional properties (where possible), or are there any restrictions and/or concerns that would prevent you from doing so?

The majority of the participants agreed to develop non-traditional properties in the future with two participants adding additional useful information:

> I think that there are restrictions on the re use of non-trads but this is often based on a poor understanding of what are the solutions that can be used to resolve these issues (Senior Architectural Technologist, 2019).

> I would be interested to see if there are more financially viable repairs that can be implemented to increase the lifespan of non-trads but at the moment, I do not believe the technologies are out there to do this and as mentioned earlier on, externally refurbishing properties with EWI is an expensive endeavour (Project Surveyor, 2019).

11.4.2 Existing Literature Review

In relation to alternative non-traditional dwellings, there are a high volume of Airey properties with "26,000" (Harrison, 2012) across the UK. Airey properties (as shown in Fig. 11.5) were constructed between "1945–55…" by "…W Airey & Sons

---

[8] Project Surveyor practitioner from a professional social house developer company.

Fig. 11.5 Example of Airey property (Google, 2020)

Ltd R Costain Ltd" (Harrison, 2012). These dwellings have a characteristic external wall of exposed aggregate PRC panels throughout with upper panels oversailing lower panels, in a hanging tile fashion. The corners of an Airey property have splayed PRC corner panels, in order to cover the storey height tapered PRC columns, which support the hanging panels.

Airey dwellings are illustrated in the BRE book on page 135, Non-traditional Houses: Identifying Non-traditional Houses in the UK 1918–1975. The BRE lists their three major structural defects: "Cracking of PRC columns; Water penetration through PRC panels; High chloride content in PRC panels" (Harrison, 2012). An in-depth study was carried out by BRE, "In the early 1980s, investigation of fire damage to an Airey house revealed cracking to the structural PRC columns caused by an inadequate cover to the embedded steel reinforcement and chemical changes to the surrounding concrete" (Harrison, 2012). For these reasons the BRE have labelled the building type to be designated defective.

After the questionnaire research was completed and following the responses from the project surveyor used in the examples above, research was undertaken into SEWI and concrete repairs. This review highlighted two major design possibilities for Airey property development. The first option for Airey properties is to use a SEWI option with minimal works done as possible. The steps of this are:

Inspecting the existing concrete frame system for structural faults and repair using Mapei or similar concrete strengthening methods, thus repairing PRC columns and increasing the protection from fire damage.

A structural external insulation to be installed—"The panels are joined with a rigid mesh which is mechanically clipped together to form a continuous monolithic structural system which stops movement in the walls and ties the property together" (Structherm, 2019), thus increasing thermal performance and the structural integrity of the property.

The second development option for Airey properties is to completely demolish and rebuild anew.

Lamella properties (as shown in Fig. 11.6) were built between "1946–1948" by "F Hills & Sons Ltd" (Harrison, 2012). Lamella properties have a "Distinctively shaped Lamella gable roof covered with tiles or metal sheeting, with projecting flat topped dormer windows" (Harrison, 2012). The vaulted roof is made up of criss-crossing parallel timber short battens. These are hinged together and bolted to form an interlocking diamond structure. This is then covered with sarking felt, battens and plain tiles. From the ground to the underside of the projecting eaves on the first floor, the construction is relatively commonplace, i.e. a no-fines concrete slab with 300-mm-thick masonry wall construction.

Fig. 11.6 Example of Lamella property (Google, 2020)

This building type is illustrated in the BRE book, *Non-traditional Houses: Identifying Non-traditional Houses in the UK 1918–75*, on page 579. Due to the construction, the major issues are deficiencies in fire stopping at eaves height and above, little or no thermal insulation above the ground floor masonry cavity wall structure and the current structural instability of the supporting timber framework due to moisture ingress from poor water tightness.

11.4.3 Professional Experience

During the first author's career, he worked alongside a senior architectural technologist, mentioned above in the questionnaire, in a practice near Huddersfield. During his time at this company, notes were taken in his reflective practice work diary, which was under review by supervisors in the company. This section will extrapolate data from this work diary, focusing on three methods of development for Lamella properties.

The first option for Lamella properties is a redevelopment option with minimal work done as possible. The steps for this are:

Removal of the existing roof finish.

Galvanised straps fixed back into the existing Lamella roof structure. Straps to be built back into the gable wall.

Repair of the existing roof timber structure with like-for-like members.

Fire stopping to be installed at the party wall to provide a form of fire compartmentalisation between the two properties.

Insulation to be installed between the Lamella grid systems providing 100-mm-thick insulation across the roof structure.

Cover the Lamella timber frame with a breather membrane roofing felt; then counter batten using treated roof battens over the felt; and finally reinstall the tiled roof finish.

The second option requires more demolition. The steps for this option are:

Existing timber Lamella roof structure to be removed to wall plate level located at the first floor level.

Steel purlins to be installed to span from gable wall to party wall to support the roof structure.

Timber stud frame walls to be constructed to the first floor level to construct the lower mansard roof section.

Roof to be formed from loose rafters. Roof to be strapped back to the gable and fire stopping to be fitted at the party wall line similar to option one.

Completed wall and roof construction to be insulated and clad in a breather membrane and then to receive a roof tile finish similar to the existing construction on treated timber battens.

The third and final option is to completely demolish and rebuild anew.

11.5 Discussion

11.5.1 Questionnaire

In the review of the questionnaire, one of the key factors that stood out is how little is known about these property types amongst the non-specialists, with 25% stating that they never knew they existed. This shows that there is a research and educational gap between the specialist and non-specialist parties. This can be backed up by the senior architectural technologist's quote, "I think that there are restrictions on the re-use of non-trads but this is often based on a poor understanding of what the solutions are that can be used to resolve these issues". This research paper highlights this gap and aims to increase the knowledge base of the industry.

The second key difference spotted in the questionnaire is the current clash of mind-set between the specialists. On the one hand, the majority of the parties involved with developing non-traditional dwellings regard cost as a main factor, and because of this they believe "it is more economically advantageous to demolish the majority of non-trad housing stock and replace it with new housing"[1]. On the other hand, the other party, normally the principal designer, thinks more about the environmental factors and other key factors of the development and has the standpoint of "Reuse of existing buildings will have a lower carbon footprint than knocking them down and starting again. Also, in terms of the social aspects, people enjoy living in them and don't want to move out. Local authorities don't have the facilities to decant residents and move them elsewhere while their home is knocked down and rebuilt. Upgrading of existing buildings will have a lower life cycle cost than rebuilding, especially if the repair work can extend the lifespan up to 30 years"[3].

11.5.2 Airey

There are both positives and negatives to both options of development. The major ones are listed below:

Option 1—Redevelop the existing building envelope

Advantages	Disadvantages
Existing Airey concrete frame is retained, which will reduce the carbon footprint of sourcing a new structure. It also makes the project shorter and more cost-effective	Proposed works may be impossible due to existing PRC columns being too eroded to be able to structurally strengthen them
Proposed works will rectify the structural issues by the use of the concrete structural strengthening	No upgrade of internal elements or party wall
Fire issues will be resolved by the coating of a concrete column with strengthening mortars and meshes	
Thermal efficiency will be increased due to the new thermal element wrap	

Advantages	Disadvantages
The aesthetic upgrade of the Airey property is transformational	
A more cost-efficient method of development	
No need to decant tenants due to the majority of the works being external	

Option 2—Demolish and rebuild

Advantages	Disadvantages
New building could be built to a Passivhaus standard	Greatly increased carbon footprint on construction compared to option 1
Could make use of renewable energy sources	Increased construction cost and time for development when compared to option 1
New buildings' lifespan could greatly outlive the 30-year lifetime warranty on SEWI systems	

11.5.3 Lamella

Looking over the possible options for developing Lamella properties, the key advantages and disadvantages of each method are:

Option 1—Redevelop the existing building envelope

Advantages	Disadvantages
Existing lamella frame is retained, which will reduce the carbon footprint of sourcing a new structure. It also makes the project shorter and more cost-effective	Proposed works may raise more structural issues due to unknown characteristics of the timber lamella frame
Existing key aesthetical bell-shaped gable retained	The proposed repairs to the roof space by reusing the existing structure may not enable all cold bridging issues to be resolved
Proposed works will rectify the structural issues by the use of the galvanised straps	All thermal improvement works are based on the amount of insulation that can be fitted within the existing timber roof structure, so they could struggle to achieve required U-values
Fire stopping at party wall lines will be resolved	
Thermal efficiency will be increased within the lamella timber structure	

Option 2—Partial demolish with redevelopment

Advantages	Disadvantages
The existing ground floor element is left untouched, thus reducing the carbon footprint compared to full demolition	Increased carbon footprint from sourcing the new roof structure
Existing non-traditional element of the roof structure is removed, furthering the lifespan of the property	Increased construction cost and time for development when compared to option 1
Structural issues and fire stopping at party wall lines will be resolved. This work would also be easier to undertake due to the ease of access when the existing roof structure is removed	Proposed works lose the characteristic bell-shaped gable of the house type
Thermal efficiency of the existing roof will be increased and could achieve greater thermal efficiency compared to option 1 as insulation thicknesses could be increased within the proposed roof structure	
Cold bridging could be eliminated as the works will allow a holistic approach to improving the properties	

Option 3—Complete demolition and rebuild

Advantages	Disadvantages
New building could be built to a Passivhaus standard	Greatly increased carbon footprint in construction compared to options 1 and 2
Could make use of renewable energy sources	Increased construction cost and time for development when compared to options 1 and 2
	Proposed works lose the characteristic bell-shaped gable of the house type

11.6 Conclusion

This paper considers relevant academic, professional and trade literature and interviews concerning post-war non-traditional dwellings. These dwellings account for roughly 2% of the total UK domestic housing stock [using "25 million homes in the UK" (News.bbc.co.uk, 2019)]. The majority of these dwellings have poor thermal properties due to the method of construction; therefore, it is of utmost importance to rectify and redevelop these outdated properties if we are to hit our target of "emissions to net zero by 2050" (GOV.UK, 2019). The paper highlights this disparity and helps to identify the most eco-friendly way of redeveloping both Lamella and Airey properties as a way of promoting this issue of research and development for the industry. The benefits and disadvantages of updating both these types of properties have been compared in order to propose a tentative solution for the most ecologically viable way for their development.

11.6.1 Airey

The solution is tentative because of the uncertainty of ascertaining PRC's chloride content, which leads to structural damage. However, in an ideal situation if the concrete columns are in a sound condition to repair and reinforce, then redeveloping the property using SEWI is the best option for an eco-friendly method of development.

11.6.2 Lamella

When developing the Lamella properties, longevity does need to be considered as option 1 (redevelop existing building envelope) is the most eco-friendly option. However, this does not increase the longevity of the dwelling due to the fact that the existing timber structure is retained, for which little or no structural test may have been completed. This draws us to the conclusion that the best option for Lamella properties is to re-roof with a mansard roof construction method, thus ensuring the longevity of the property with no deficiencies to the insulation provided at roof level. However, it loses the characteristic bell-shaped gable of the house type.

11.6.3 Research Review

The initial aims of the research paper have been met but not with sufficient evidence to definitely say what is the most eco-friendly way of redeveloping both Lamella and Airey properties, due to the limitations of time and resources. These limitations render the research as a first pass on possibilities for redeveloping and solving the existing UK housing crisis for non-traditional dwellings. This paper does, however, point out the current constraints and limitations for a more thorough feasibility study in the future.

References

Creswell, J., & Creswell, J. (2017). *Research design qualitative quantitative and mixed methods 5th edition.* [online] Slideshare.net. https://www.slideshare.net/dangledat12/research-design-qualitative-quantitative-and-mixed-methods-5th-edition

Google. (2020). *Google maps.* [online] https://www.google.com/maps/@53.6013988,-1.7942904,3a,75y,50.33h,86.45t/data=!3m7!1e1!3m5!1sBhnO9ct6d1ImWBqrfN84uw!2e0!6s%2F%2Fgeo2.ggpht.com%2Fcbk%3Fpanoid%3DBhnO9ct6d1ImWBqrfN84uw%26output%3Dthumbnail%26cb_client%3Dmaps_sv.tactile.gps%26thumb%3D2%26w%3D203%26h%3D100%26yaw%3D292.8007%26pitch%3D0%26thumbfov%3D100!7i13312!8i6656

GOV.UK.(2019).*UKbecomesfirstmajoreconomytopassnetzeroemissionslaw.*[online]https://www.gov.uk/government/news/uk-becomes-first-major-economy-to-pass-net-zero-emissions-law

Harrison, H. (2012). *Non-traditional houses*. IHS BRE Press.
McPherson, G. (2019). 54 inspirational and shocking quotes on climate change. [online] *Green Coast*. https://greencoast.org/quotes-on-climate-change/
News.bbc.co.uk. (2019). BBC news. In *Depth*. [online] http://news.bbc.co.uk/1/shared/spl/hi/guides/456900/456991/html/default.stm
Structherm. (2019). *Wrexham aireys*. [ebook] Structherm Ltd. https://www.structherm.co.uk/wp-content/uploads/2019/07/Wrexham-Westdale-10513-2019.pdf
Williams, P. (n.d.). Non-traditional houses; identifying non-traditional houses in the UK 1918-1975. *Housing Studies, 21*(3), 441–442.

Chapter 12
Implications of a Natural Ventilation Retrofit of an Office Building

Ashvin Manga and Christopher Allen

12.1 Introduction

> Learn to look after your staff first and the rest will follow. Richard Branson

The relationship between the physical office environment and employee satisfaction has received global attention as a critical aspect contributing to business success (Budie et al., 2018). Raising satisfaction with the physical work environment has become a major corporate real estate strategy (Budie et al., 2018). Work environment satisfaction is the extent to which the physical work environment meets the needs of office users (Kim & de Dear, 2013). However, business parks are often poorly designed in terms of energy efficiency, indoor air quality and occupant thermal comfort (Khatami, 2014). In addition, the affordability of mechanical ventilation systems resulted in a decline of vernacular and passive design strategies used in office buildings; this has resulted in the increase of worldwide installed capacity of air-conditioning systems from below 4000 GW in 1990 to over 11,000 GW in 2016 (Asfour, 2017). Energy consumption for indoor space cooling is anticipated to triple until 2050 (Braungardt et al., 2019), whilst the built environment sector contributes up to 30% of global annual greenhouse gas emissions and consumes up to 40% of all energy (Zhai & Helman, 2019; Geng et al., 2019).

Natural ventilation has been recognised as one of the most promising sustainable strategies to reduce building energy consumption, improve thermal comfort and maintain a healthy indoor environment (Tong et al., 2016; Wang, 2017). However, many local modern office buildings were constructed in the absence of passive design principles when research shows that most people favour natural ventilation and openable windows (Carrilho & Linden, 2016). In addition, they tend to show

---

A. Manga (✉) · C. Allen
Department of Construction Management, Nelson Mandela University, School of the Built Environment, Port Elizabeth, South Africa

increased thermal tolerance when occupying naturally ventilated buildings. Natural ventilation retrofits provide a possible solution. However, the perceived negative aspects to natural ventilation, namely, external ambient air pollution and increased internal particulate matter counts, need to be carefully considered in the decision-making process (Tong et al., 2016). However, data to inform this decision-making is limited, and therefore the aim of this research is to explore the impact a natural ventilation retrofit has on indoor air quality, occupant thermal comfort and ventilation in a commercial office building located in Port Elizabeth, South Africa.

12.2 Indoor Air Quality

'Indoor air quality or IAQ' is what we experience as the temperature, humidity, ventilation and chemical or biological contaminants of the air inside our buildings (Brown, 2019). Many people spend 90% of their day indoors where air pollution can be two to five times more polluted than levels outside (Marta et al., 2017). Furthermore, building environments are exposed to all kinds of textiles, equipment, paper, cleaning products and maintenance activities producing contaminants that leave indoor air very different from 'fresh outside air' (Brown, 2019). The health risks from exposure to indoor air pollution may therefore be higher than those related to outdoor pollution (Cincinelli & Martellini, 2017). In particular, poor indoor air quality can be dangerous to vulnerable groups such as children, young adults, the elderly or those suffering chronic respiratory and/or cardiovascular diseases (Cincinelli & Martellini, 2017), the same groups that have been most impacted by the COVID-19 pandemic (NCIRD, 2020). IAQ has thus received increasing attention from the international scientific community, political institutions and environmental governance as a means for improving the comfort, health and well-being of building occupants (Abouleish, 2020).

12.3 Indoor Carbon Dioxide

Carbon dioxide (CO_2) is a colourless, tasteless, odourless and non-flammable gas that is heavier than air and may accumulate at lower spaces, causing a deficiency of oxygen (Azuma et al., 2018). The normal outdoor CO_2 concentrations are approximately 380 ppm, although in urban areas these have been reported to be as high as 500 ppm because of the increased human activity (Azuma et al., 2018). The main source of CO_2 in the non-industrial indoor environment is human metabolism (Park et al., 2019). The indoor CO_2 concentration is an indicator of indoor air quality acceptability, air flow exchange suitability and whether there is enough fresh air within indoor spaces (Park et al., 2019). The average indoor CO_2 concentration ranges from 800 to 1000 ppm, with an upper limit of 1000 ppm for CO_2 concentrations in commercial buildings (Park et al., 2019). Ventilation with ambient air is often used for reducing indoor CO_2 concentration. Research has also indicated that for every 100 ppm decrease in the differential between indoor and outdoor carbon dioxide concentration (dCO_2), office workers experienced fewer SBS symptoms, including 60% fewer reports of sore throat and 70% fewer reports of symptoms of wheezing (Park et al., 2019). They further report that CO_2 affects decision-making at thresholds of 600 ppm, which is below the normally accepted comfort range of 1000 ppm (Park et al., 2019). A study in human subjects stated that inhalation exposure to 1000 ppm CO_2 for a short term caused marked variations in respiratory movement amplitude, peripheral blood flow increases and the cerebral cortex functional state (Azuma et al., 2018).

12.4 Sick Building Syndrome (SBS)

Sick building syndrome presents symptomatically as the complex spectrum of ill health complaints, such as mucous membrane irritation (rhinorrhoea, nasal congestion, sore throat, eye irritation), asthma symptoms (chest tightness, wheezing), neurotoxic effects (a headache, fatigue, irritability), gastrointestinal disturbance, skin dryness and sensitivity to odours (Nag, 2019). These symptoms may present among occupants in office buildings, schools, public buildings, hospitals and recreational facilities (Nag, 2019). Building-related sicknesses have been observed as being pervasive in modern high-rise buildings (Nag, 2019). These buildings are designed to be airtight for energy-saving. Subsequently, the windows remain sealed, depriving the building of natural ventilation and daylighting, and HVAC systems recirculate the air in the building, with minimal replacement of fresh air (Nag, 2019). Concerns about human health due to deteriorating indoor environmental quality are gradually increasing with a public outcry among the building occupants as well as challenging lawsuits (Nag, 2019). Typically, maintaining allowable IAQ in office buildings depends on effective ventilation systems in operation. Ineffective or inadequate ventilation systems result in inefficient removal of pollutants from indoor air and display of SBS signs among the occupants (Nag, 2019).

12.5 Natural Ventilation

Natural ventilation essentially describes the air movement created by naturally occurring pressure differences; this pressure difference can be created either by wind or by temperature differences (Alatala, 2016). Proper utilisation of natural ventilation can provide large ventilation rates without the consumption of energy (Wang & Zhai, 2018). As natural ventilation is driven by wind or buoyancy, the naturally ventilated building form includes wind-driven ventilation form, buoyancy-driven ventilation form and the combination of these two. When wind hits a building, positive pressure is created on the windward side and negative pressure on the leeward side of the building (Alatala, 2016), forcing air into the building from the windward side and out from the leeward side, creating cross-ventilation through the building. Natural ventilation is an increasingly popular green-building technology that has proven to be an effective solution to lower building cooling energy and to improve indoor air quality in various climates and types of buildings (Chen et al., 2019).

12.6 Hybrid Ventilation Systems

Ventilation has been shown to play a critical role in improving indoor environments (Meng et al., 2019). Hybrid ventilation is an effective means of minimising ventilation energy and improving indoor climate. It has been found that air-conditioners are often operated for 6–12 h per day, consuming electricity at a rate of 40–45 kWh/m^2 (Weerasuriya et al., 2019). In buildings with hybrid ventilation systems, transition spaces such as corridors, with motorised inlets, have more flexible thermal comfort limits than office spaces and can thus be used to bring in cooler outdoor air increasing cooling energy savings (Yuan et al., 2018). In addition to providing thermal comfort, natural ventilation has the advantage of improving indoor air quality, reducing energy consumption in buildings and eliminating what is known as the 'sick building syndrome'. From an economics perspective, natural ventilation can reduce a building's capital cost by about 10–50%, compared to an air-conditioned building of similar dimensions (Weerasuriya et al., 2019). A building's potential for natural ventilation not only depends on its dimensions and architectural features but also relates to climatic (e.g. wind speed, direction, outdoor temperature, solar radiation) characteristics of the neighbourhood (e.g. geometry and orientation of buildings) and the behaviours of residents (Weerasuriya et al., 2019).

12.7 Research Methodology

Two quantitative research designs were used in this research project. First, an experimental design was used to determine the change in indoor air quality before and after the building's existing non-openable windows were retrofitted into openable windows. Indoor air quality was measured using a medical-grade indoor air quality device, the AirVisual Node by IQAir. Second, a survey design was used to conduct a pre- and post-survey of sick building syndrome symptoms experienced by employees. The survey was conducted before and after the natural ventilation retrofit. The experimental and survey design allowed this research to correlate the impact openable windows had on indoor air quality and subsequently the impact the change in indoor air quality had on occupant satisfaction and well-being.

12.8 Building Type and Location

The building selected for this research represents typical modern open-planned office buildings in Port Elizabeth, South Africa. The office building selected is currently and has been occupied by an engineering consulting firm since 2009. The building is located along a main road (Fig. 12.1) and had 22 office users during the research period. The office space before the natural ventilation retrofit was mechanically ventilated only. The maintenance of the mechanical ventilation was subcontracted to a specialist firm and to the knowledge of the researcher was maintained regularly. The indoor levels of particulate matter reflect a maintained mechanical ventilation system as office particulate concentrations during the mechanically ventilated office period reached a maximum of 13 µ/m^3.

Figure 12.2 illustrates a street view of the office building. The windows depicted are non-openable. Figures 12.3 and 12.4 illustrate the before and after the existing windows were retrofitted. The existing non-openable aluminium casement window was retrofitted into an openable window. The retrofit took 1 day; ten windows were retrofitted at a cost of R20,000.00 (£950.00). The air quality device was located at a height of 1.6 m and was placed centrally within the office. The manufacture of the air quality monitor states valid measurements for an indoor space of up to 100 m^2, and the research area was 80 m^2 (Chloe, 2020).

Fig. 12.1 Selected office building

Fig. 12.2 Street view

Fig. 12.3 Before retrofit

12.9 Research Method

The indoor air quality device was first placed in the office on the 18th of June 2019. From 18 June to 14 August 2019 (8 weeks), the device captured air quality data for the period prior to the natural ventilation retrofit. During this time, the pre-SBS survey was circulated among office users. The existing windows were retrofitted on the 15th of August, and indoor air quality data for the post-natural ventilation retrofit period commenced. The indoor air quality device captured data for 5 weeks after

Fig. 12.4 After retrofit

Table 12.1 Overview of indoor air quality metrics

Indoor air quality results				
Mechanical ventilation		Hybrid ventilation		
IAQ metric	Mean	IAQ metric	Mean	Change %
PM2.5 (µg/m³)	2.89	PM2.5 (µg/m³)	2.23	−23
PM10 (µg/m³)	4.03	PM10 (µg/m³)	3.31	−18
Carbon dioxide (ppm)	685.31	Carbon dioxide (ppm)	643.51	−6
Temperature (Celsius)	22.0	Temperature (Celsius)	22.2	+1
Humidity (RH%)	51.22	Humidity (RH%)	51.16	−0.1

the retrofit, during which the employees had complete control over the openable windows. At the end of the 5-week post-retrofit period, the post-SBS survey was circulated among the office users.

12.10 Research Results: Indoor Air Quality

Table 12.1 is a summary of the indoor air quality metrics recorded by the IAQ device. The mean score for each IAQ parameter was calculated for both mechanical and hybrid ventilation. The improvement in IAQ metrics is displayed by the change percentage column. The most notable improvements are reflected by the 23% reduction in PM2.5, the 18% reduction in PM10 and the 6% reduction in carbon dioxide concentrations.

Table 12.2 Carbon dioxide

Office hours in which CO_2 level is >1000 pm	
Before operable windows	36 office hours
After operable windows	0 office hours
Comment	CO_2 levels exceeded 1000 ppm for 8% of total measured office hours

12.11 Research Results: Carbon Dioxide

The analysis of the 14-week real-time data showed in Table 12.2 revealed that the pre-operable window period had 12 incidents in which office carbon dioxide levels exceeded 1000 ppm. The office carbon dioxide level exceeded 1000 ppm for a total of 36 h. On average, office carbon dioxide levels would exceed 1000 ppm twice to three times a week.

Once the operable windows were in use, the office carbon dioxide level did not exceed 1000 ppm for the remainder of the experiment. Operable windows, controlled by employees, effectively reduced >1000 ppm carbon dioxide incidents to 0.

12.12 Research Results: SBS Survey

The sick building syndrome (SBS) survey was conducted using a 5-point Likert scale design. A mean score was calculated for each symptom. The symptoms were then ranked, and the pre- and post-survey results were tabled and compared.

The pre-openable window sick building syndrome questionnaire indicated that the top three symptoms experienced by employees are tiredness, difficulty/poor concentration and nasal congestion. The pre-openable window SBS survey results formed the benchmark to which the post-openable SBS survey was compared against. The impact openable windows had on SBS symptoms experienced by employees is highlighted by the following reduction in the mean scores of the top three SBS symptoms identified by the pre-openable window SBS survey:

Tiredness—18.25% improvement

Difficulty/poor concentration—20.80% improvement

Blocked or stuffy nose—25.82% improvement

Table 12.3 provides an overview of the mean score for SBS symptoms experienced by employees before and after the natural ventilation retrofit. There was a reduction in mean score across almost all the SBS symptoms experience by employees. Notable findings include a reduction in 'dry throat' by 35%, 'runny nose' by 28% and watery eyes by 29%.

Table 12.3 SBS survey results

		Mean		
No.	SBS symptoms	Pre	Post	Change %
1	Tiredness	3.89	3.18	−18
2	Difficulty/poor concentration	3.22	2.55	−21
3	Blocked or stuffy nose	3.06	2.27	−26
4	Watery eyes	3.05	2.18	−29
5	Sensitivity to odours	2.95	2.18	−26
6	Sneezing	2.78	2.64	−5
7	Runny nose	2.78	2	−28
8	Dry throat	2.78	1.82	−35
9	Headache	2.68	2.36	−12
10	Coughing	2.67	2.18	−18
11	Dryness and irritation of the skin	2.65	2	−25
12	Dizziness	1.84	1.9	3
13	A sensation of difficulty in breathing	1.78	1.45	−19
14	Tightness of the chest	1.78	1.55	−13

12.13 Discussion

The US Environmental Protection Agency stated that indoor PM10 concentrations should not exceed 150 (µg/m³) and PM2.5 should not exceed 35 (µg/m³). The office PM2.5 and PM10 concentrations are significantly lower than the regulations provided by the EPA. However, the natural ventilation retrofit notably contributed to a further reduction of office PM2.5 and PM10 concentrations. The increased ventilation provided by the natural ventilation retrofit contributed significantly to improved indoor office air quality by improving ventilation and introducing clean outdoor air. According to Zhai et al. (2019), it has been found that sick building syndrome (SBS) symptoms are substantially reduced in office buildings with natural ventilation compared to buildings with air-conditioning. This research concurred with those results and indicated that indoor PM2.5 concentrations improved by 22.8% and PM10 concentrations by 17.9% after retrofitting with openable windows. Nasal congestion ranked as the third most experienced SBS symptom. The post-SBS survey indicated a 25.8% improvement reducing the mean score response for 'blocked or stuffy nose' from a mean score of 3.06 to 2.27.

At first glance the office carbon dioxide concentration appears to have been insignificantly improved by the natural ventilation retrofit. However, the significant research findings are indicated by analysis of the real-time carbon dioxide levels. Prior to the installation and use of openable windows, the office space experienced 12 incidents in which office carbon dioxide levels exceeded 1000 ppm. The office space carbon dioxide level exceeded 1000 ppm for a total of 36 h (8% of total measured office use). During the 5-week post-openable window installation period, the office carbon dioxide level did not once exceed 1000 ppm. The reduction of

>1000 ppm incidents was notable in that the carbon dioxide level was measured at 3-s intervals. The reduction in >1000 ppm carbon dioxide levels reflects in the survey findings with the top two pre-openable window SBS complaints being tiredness and difficulty/poor concentration. The openable windows contributed to 18.3% reduction in tiredness experienced by employees and 20.1% in difficulty/poor concentration. Both these symptoms have a major impact on productivity in offices, so nearly one in five improvements would have massive return on investment as employees make up the majority of business costs.

The review of the literature and the research findings correlate strongly indicating that natural ventilation can significantly improve office environments and employee satisfaction. Natural ventilation is a significant sustainable solution for decreasing the energy usage in buildings, improving thermal comfort and maintaining a healthy indoor environment (Tong et al., 2016).

12.14 Conclusion

This research quantified the impact a natural ventilation retrofit had on improving the indoor air quality of a previously mechanically only ventilated office building. Furthermore, the IAQ findings were correlated with a sick building syndrome survey that was circulated before and after the office building was retrofitted. The relationship between improved indoor air quality and employee well-being is clear with quantified reduction in SBS symptoms experienced by employees. The effectiveness of the natural ventilation solution was also demonstrated with accurate IAQ data indicating significant improvements to IAQ metrics.

A notable observation when analysing the real-time carbon dioxide data, after the openable windows were in use by employees, is that the carbon dioxide levels would reach 850–950 ppm and steadily lower to 650–750 ppm. The data indicates that office users could sense when office carbon dioxide levels where above 800 ppm and instinctively opened the windows. Prior to the openable windows, the carbon dioxide level would continue to rise until after office hours.

The importance of measurement is made clear by this research in that until indoor air quality is measured by organisations the quality of the indoor air breathed by employees is unknown. Furthermore, the possible health implications and medical costs remain unknown. The cost of the retrofit was a once-off capital expenditure of R20 000 with savings immediately in terms of reduced air-conditioning running time. Since the openable windows are controlled by the employees, there is no operational expenditure for the benefits derived from the natural ventilation retrofit, whilst the improvement in tiredness and concentration levels would markedly improve productivity, providing a major return on investment.

In climates such as in Port Elizabeth, South Africa, hybrid ventilation systems provide a true solution to sustainable ventilation within buildings. The sealed office building design with mechanical ventilation is an inappropriate design for warm and temperate climates. Warm and temperate climates can exploit passive design principles for a large portion of the year relying on mechanical ventilation occasionally. In conclusion, natural ventilation should be considered during the design phase of any new construction. However, natural ventilation retrofits can be effective in improving indoor air quality of existing buildings without increasing operational costs over the long term.

References

Abouleish, M. Y. Z. (2020). Indoor air quality and coronavirus disease (COVID-19). *Public Health*.
Alatala, E. (2016) Enhancing the Building Performance of Low-Cost Schools in Pakistan – A Study of Natural Ventilation, Thermal Comfort and Moisture Safety [Online]. Chalmers University of Technology. Available from: https://publications.lib.chalmers.se/records/fulltext/245948/245948.pdf
Asfour, O. S. (2017). Natural ventilation in buildings: An overview. *Communications, 2*(4), 14–23.
Azuma, K., Kagi, N., Yanagi, U., & Osawa, H. (2018). Effects of low-level inhalation exposure to carbon dioxide in indoor environments: A short review on human health and psychomotor performance. *Environment International, 121*, 51–56.
Braungardt, S., Bürger, V., Zieger, J., & Bosselaar, L. (2019). How to include cooling in the EU renewable energy directive? Strategies and policy implications. *Energy Policy, 129*, 260–267.
Brown, N. J. (2019). *Indoor air quality* [Electronic version]. Ithaca, NY: Cornell University, Workplace Health and Safety Program.
Budie, B., Appel-Meulenbroek, R., Kemperman, A., & Weijs-Perree, M. (2018). Employee satisfaction with the physical work environment: The importance of a need based approach. *International Journal of Strategic Property Management, 23*(1), 36–49.
Carrilho, G., & Linden, P. (2016). Ten questions about natural ventilation of non-domestic buildings. *Building and Environment, 107*, 263–273.
Chen, Y., Tong, Z., Samuelson, H., Wu, W., & Malkawi, A. (2019). Realizing natural ventilation potential through window control: The impact of occupant behavior. *Energy Procedia, 158*, 3215–3221.
Chloe. (2020). *AirVisual Pro basic information*. [Online] https://support.airvisual.com/en/articles/3029442-what-area-coverage-are-the-airvisual-pro-s-measurements-valid-for
Cincinelli, A., & Martellini, T. (2017). Indoor air quality and health. *International Journal of Environmental Research and Public Health, 14*(11), 1286.
Geng, Y., Ji, W., Wang, Z., Lin, B., & Zhu, Y. (2019). A review of operating performance in green buildings: Energy use, indoor environmental quality and occupant satisfaction. *Energy and Buildings, 183*, 500–514.
Khatami, N. (2014). Retrofitted natural ventilation systems for a lightweight office building. Loughborough University. Thesis. https://hdl.handle.net/2134/17820
Kim, J., & de Dear, R. (2013). Workspace satisfaction: The privacy-communication trade-off inopen-plan offices. *Journal of Environmental Psychology, 36*, 18–26.
Marta, S., Canha, N., Lage, J., & Candeias, S. (2017). Indoor air quality during sleep under different ventilation patterns. *Atmospheric Pollution Research, 8*, 1132–1142.
Meng, X., Wang, Y., Xing, X., & Xu, Y. (2019). Experimental study on the performance of hybrid buoyancy-driven natural ventilation with a mechanical exhaust system in an industrial building. *Energy & Buildings, 208*, 109674.
Nag, P. K. (2019). *Office buildings: Health, safety and environment*. Springer.
National Center for Immunization and Respiratory Diseases (NCIRD). (2020). *Groups at higher risk for severe illness*. [Online] https://www.cdc.gov/coronavirus/2019-ncov/need-extra-precautions/groups-at-higher-risk.html

Park, J., Loftness, V., Aziz, A., & Wang, T. H. (2019). Critical factors and thresholds for user satisfaction on air quality in office environments. *Building and Environment, 164*, 106310.

Tong, Z., Chen, Y., Malkawi, A., Liu, Z., & Freeman, R. B. (2016). Energy saving potential of natural ventilation in China: The impact of ambient air pollution. *Applied Energy, 179*, 660–668.

Wang, B. (2017). Design-based natural ventilation evaluation in early stage for high performance buildings. *Sustainable Cities and Society. 45*. https://doi.org/10.1016/j.scs.2018.11.024

Wang, R., Zhai, X. (2018) Handbook of energy systems in green buildings (R Wang and X Zhaieds). Springer-Verlag Berlin Heidelberg.

Weerasuriya, A. U., Zhang, X., Gan, V. J. L., & Tan, Y. (2019). A holistic framework to utilize natural ventilation to optimize energy performance of residential high-rise buildings. *Building and Environment, 153*, 218–232.

Yuan, S., Vallianos, C., Athienitis, A., & Rao, J. (2018). A study of hybrid ventilation in an institutional building for predictive control. *Building and Environment, 128*, 1–11.

Zhai, Z. J., & Helman, J. M. (2019). Implications of climate changes to building energy and design. *Sustainable Cities and Society, 44*, 511–519.

Zhai, Y., Honnekeri, A., Pigman, M., Fountain, M., Zhang, H., Zhou, X., & Arens, E. (2019). Energy and buildings use of adaptive control and its effects on human comfort in a naturally ventilated office in Alameda, California. *Energy & Buildings, 203*, 109435.

Part IV
Education and Sustainability

Chapter 13
Developing a Sustainable Urban Environment Through Teaching Asset Management at a Postgraduate Level

David Thorpe and Nasim Aghili

13.1 Introduction

The Population Division of the Department of Economic and Social Affairs of the United Nations has estimated that globally, more people live in urban areas than in rural areas, with 55 per cent of the world's population residing in urban areas in 2018. This proportion is projected to be 68% by 2050 (United Nations, 2019). It has been estimated that cities are responsible for 75 percent of global carbon dioxide (CO_2) and other greenhouse gas (GHG) emissions. Key aspects of urban life, such as transportation and buildings, are considered to be major sources of these emissions (United Nations Environment Programme, 2020).

While producing significant percentages of GHG emissions, cities and other urban areas also have the potential to be affected by climate change and its consequences like global warming, which in turn is associated with rising sea levels and extreme weather events, and the potential to facilitate the incidence of tropical diseases, all of which have the potential to significantly impact on the urban living environment, including services, housing, infrastructure, work conditions, living conditions, safety, security and health. Thus it is important that cities, including the key building and construction sector, undertake initiatives to minimise the impact of climate change and improve their resilience to the effects of this impact. It is been stated that such success is likely to be possible only through a coordinated approach to fight climate change, an approach that is being taken by many cities through initiatives like using renewable energy sources, cleaner production and limiting industrial emissions (United Nations Environment Programme, 2020).

D. Thorpe (✉) · N. Aghili
School of Civil Engineering and Surveying, University of Southern Queensland, Toowoomba, QLD, Australia
e-mail: David.Thorpe@usq.edu.au

Fig. 13.1 Conceptual framework of paper

One process for achieving a sustainable urban environment is to use a strategically focused life cycle asset management approach, based on theoretical and practical principles, in conjunction with good project and risk management to the ongoing planning, development and management of the city's transportation, building, infrastructure and other assets that are significant sources of the operational and embodied energy that produce GHG emissions in the city. The objective of the discussion in this paper is to discuss an approach to achieving this goal, using the following process, which is further developed and illustrated in Fig. 13.1:

Review the issues in urban sustainability management.

Discuss the role of asset management in facilitating urban sustainability.

Using the example of existing study courses, discuss how this role can be facilitated by postgraduate education in sustainable asset management, supported by other courses.

Discuss the effectiveness of these courses in facilitating urban sustainability and develop conclusions.

13.2 Issues in Urban Sustainability Management

The issues in urban sustainability are underscored by a study of 79 cities, undertaken by the C40 Cities Climate Leadership Group in conjunction with the University of Leeds, the University of New South Wales (Australia) and Arup, which found that over 70% of greenhouse gas (GHG) emissions come from utilities and housing, capital, transportation, food supply and government services. These emissions included the supply of goods and services outside city boundaries (such as power, externally manufactured goods and water supplies), which while originating outside cities form a significant component of their requirements (C40 Cities, 2018).

These issues occur throughout the asset life cycle, which can be divided into the phases of planning, design, construction, operation (including maintenance) and retirement. Planning and design are important, as decisions made in them impact on the subsequent phases of the life cycle. Managing the carbon footprint of the city will also require addressing the 17 sustainable development goals of the United Nations, which from an urban point of view include social aspects (e.g., poverty, hunger, health, education, safe and secure cities), resources (e.g., water, energy, sustainable consumption and production), resilience, climate change and sustainable ecosystems (United Nations, 2015). It will also require coordinated activities by professional engineers, asset managers and other built environment professionals. For example, while professional engineers are responsible for planning, designing, delivering and maintaining significant public and private works, they are also expected to undertake sustainable engineering practices and adhere to sustainable management principles (Engineers Australia, 2014), a requirement that is reinforced in the Engineers Australia Code of Ethics and Guidelines on Professional Conduct (2019). Such requirements would be expected to guide the actions of engineers and other professionals in minimising the impact of urban development and management on the world's carbon footprint.

The asset life cycle components in which emissions occur are primarily construction and operation, including asset retirement. As an example, construction is a significant contributor to the carbon footprint of cities. It has been estimated that building construction and operations (including the manufacture of materials and products for building construction, such as steel, cement and glass) accounted for 36% of global final energy use and nearly 40% of energy-related carbon dioxide (CO_2) emissions in 2017 (International Energy Agency and the United Nations Environment Programme, 2018). The World Green Building Council has similarly stated that buildings account for 39% of energy-related global CO_2 emissions, of which 28% comes from operational carbon and 11% arises from the energy used to produce building and construction materials, which is usually referred to as embodied carbon (World Green Building Council, 2019). A further perspective on the energy produced by the construction sector is an Australian study that used a two-region globally closed model of Australia and the Rest of the World which found that from 2009 to 2013 the embodied energy component of the construction sector was a total of 20% of Australia's carbon footprint, consisting of 1.9% of direct GHG emissions and 18.1% of embodied energy, which was mainly attributed to electricity, gas and water; materials; and construction services (Yu et al., 2017).

These examples show that it is important to achieve a sustainable urban environment that minimises the percentage of greenhouse gas emissions and its impacts on the carbon footprint. At the moment, this percentage is high, both in cities as a whole and in the development and management of their buildings and infrastructure, which are activities in which engineers and other built environment professionals have a significant role. While the process of achieving a sustainable urban environment has commenced, there is much to be achieved. Doing so requires sustainability focused, life cycle-oriented built environment management that spans all life cycle activities, such as planning, design, construction, operation, maintenance and retirement. This process will require sustainable management of the design, construction and operation of the built environment and include activities like recycling materials, strategic use of information systems and the minimisation of waste in the construction and operation of assets. Through good planning and ongoing management, this approach assists city government and other parties to maximise sustainable management of the urban built environment, such as power generation, roads and transportation, water and sewerage management and greening of the urban environment itself. As engineering and other built environment professionals would be expected to have a leading role in developing a sustainable urban environment, their education in its development and implementation is expected to be a significant component of this process.

Therefore, there are two questions required to be addressed. They are:

RQ1: Can a strategic approach to asset management facilitate the development and management of urban sustainability?

RQ2: How can postgraduate education in sustainable asset management and related study courses support the achievement of a sustainable urban environment?

13.3 The Role of Asset Management in Facilitating Urban Sustainability

13.3.1 The Sustainable Asset Management Process

In order to address the first question (RQ1), the first step is to consider the management of environmental factors in the urban environment. Addressing this step requires an understanding of the terms "asset" and "asset management," followed by consideration of the life cycle of urban assets and whether using a sustainable asset management approach is a suitable process for managing these factors.

While there are a number of definitions of assets, a good definition is that from the International Asset management—Overview, Principles and Terminology Standard (ISO5500:2014), in which an asset can be defined as an "item, thing or entity that has potential or actual value to an organisation." In turn, the term "asset management" can be defined as a "coordinated activity of an organisation to realise value from assets" (International Organization for Standardization, 2014a). These definitions indicate that assets and their management have an economic value. They also demonstrate that such management of assets requires to be undertaken in a way that maximises their life cycle performance.

From a strategic point of view, effective asset management requires the recognition of a range of stakeholders – owners, managers, users and external stakeholders—who may be impacted by the asset although they may not use it. From a management point of view, assets should achieve a number of outcome-oriented goals in areas like provision of adequate level of service at the required level of demand, meeting serviceability and life cycle performance goals, having optimum service lives and achieving maximum life cycle benefit and lowest life cycle cost subject to other requirements (University of Southern Queensland, 2020a).

Managing the emissions produced by the construction phase, for example, requires the use of sustainable construction processes. One approach is to consider the use of alternative materials in the construction process that are energy efficient to develop and install and that where possible are recyclable. For example, advanced materials like fibre composites, green cement and concrete, use of recycled materials in road surfacing and other innovations are improving the sustainability of the construction process from the material aspect. Other issues in managing construction phase emissions include managing energy usage and the negative impact on environmental pollution from the component activities of construction and refurbishment. Low et al. (cited in Medineckiene et al., 2010) identified five aspects (energy efficiency, water efficiency, environmental protection, indoor environmental quality and other green features that focus on the adoption of green practices and new technologies with potential environmental benefit) as areas of environmental impact of the construction process. Managing these issues requires assessing the combination of variables affecting matters like building (or infrastructure) production impact on the environment, and financial and social conditions, through using approaches like multicriteria analysis (Medineckiene et al., 2010). Sustainable management of the other phases of the asset like cycle will similarly require consideration of their component activities.

13.3.2 Systems Approach to Sustainable Asset Management

To facilitate the delivery of sustainable asset management, assets are best managed by an asset management system, which can be considered a set of interrelated and interacting elements of an organisation, with the function of establishing the organisational asset management policy and objectives, and the processes needed to achieve them (International Organization for Standardization, 2014a), accompanied by a strategic asset management plan. Such an asset management system would normally have the components of leadership, planning, support (of the system), operation, performance evaluation and improvement (International Organization for Standardization, 2014b).

An asset management system would be expected to be underpinned by an asset management information system, supported by an asset register (or inventory), and contain a number of sub-systems, such as financial management, maintenance management, operations management and decision support. Information technology tools like building information modelling (BIM), geographic information systems (GIS), and tools like the use of radio frequency identification (RFID) for identification and tracking, along with managing big data, can further aid sustainable life cycle management through facilitating construction and demolition waste management (Kabirifar et al., 2020).

As an example of these technologies, BIM initially commenced as a three-dimensional (3D) model, which can express visually and in other ways the three primary spatial dimensions of width, height and depth. It has subsequently added additional dimensions of 4D (planning) and 5D (costing) and has further developed additional dimensions, including sustainability and facilities management (Charef et al., 2018). These last two dimensions make BIM of particular interest with respect to sustainable asset management, as it is not only able to record detailed information in digital form about buildings and infrastructure components and their functional parts (Xu et al., 2014) but has the potential to be a useful tool to optimize the environmental performance of building elements and buildings (Habibi, 2017). Chong et al. (2017) noted that the research trend of BIM for sustainability was apparent in the design, construction, use of products and materials and energy efficiency. Zhang et al. (2016) suggested that the use of BIM in the architectural, engineering and construction industry is helpful for environmental sustainability monitoring and management over a building's full life cycle and could be used at the conceptual design stage to build sustainability into the design solution and allow sustainability to become a key component of the design, construction and delivery of a building.

At the same time, the use of BIM in its current form has limitations. Charef et al. (2018) undertook a systematic review and online survey of architectural, engineering and construction stakeholders across Europe. They found that while there was agreement about the first five dimensions (up to cost) of the BIM model, there was some confusion among the practitioners using the sixth and seventh dimensions (6D and 7D) of BIM (sustainability and facilities management). Lu et al. (2017) reviewed the use of BIM for green buildings. However, the use of BIM for facility management (FM) during the operation phase is still limited and found that there was weak interoperability among various green BIM applications, limited capability of BIM applications supporting the construction and operation phases of projects, lack of industry standards, low industrial acceptance of green BIM applications, low accuracy of BIM-based prediction models and a lack of appropriate project delivery methods. Thus, while technologies like BIM have considerable potential for application to the sustainable development and management of urban buildings and infrastructure, there are a number of issues that need to be resolved before they can be fully used by practitioners.

In summary, a strategic approach to asset management, using a well-structured and managed life cycle approach supported by good systems, is able to strongly facilitate and contribute to the development and management of urban sustainability. This contribution is achieved through planning and design that consider sustainability, a sustainable approach to construction, use of sustainable and recyclable materials, use of good processes underpinned by an asset management system aligned with corporate objectives and supported by good information systems, a focus on minimising the use of energy and an understanding and consideration of sustainable management goals.

13.4 Postgraduate Education in Sustainable Asset Management

A strong understanding of sustainability-focused asset management principles by engineers and other professionals practising in this field is essential for effective sustainable outcomes to the asset management principles. Therefore, to address the second question (RQ2) on how postgraduate education in sustainable asset management and related study courses can support the achievement of a sustainable urban environment, it is necessary to consider how such education develops the knowledge and skills required by professionals in engineering and the built environment in asset management.

One course offered in this field is the postgraduate course Asset Management in an Engineering Environment (University of Southern Queensland, 2020a), which is part of the Master of Engineering Science and other engineering and built environment postgraduate coursework programs offered by the University of Southern Queensland. This course, which utilises the principles of good teaching principles like student-centred learning (Biggs, 1999) and authentic assessment (Gulikers et al., 2004), adopts a strategic life cycle approach to managing engineering assets and has a strong sustainability emphasis. Following an introduction to asset management, this course initially discusses the strategic asset management principles of the strategic asset management framework, the asset life cycle and asset management economics, before focusing on the application of these principles to asset management operations and maintenance, integrated asset management and asset management systems. It concludes with a discussion of current and emerging issues in asset management. The course has a strong stakeholder-oriented sustainable asset management focus, which is illustrated in Fig. 13.2.

Figure 13.2 emphasises the interaction between physical infrastructure assets, which are founded on the natural environment and are linked with the economic and social environments, and the various communities of stakeholders in the assets (owners and managers, the community of asset users and the external community, which are people or organisations served indirectly by the asset or are affected by the asset). Balancing these requirements results in the asset being required to meet the three environmental, economic and social components of sustainability. This sustainable asset management theme is maintained throughout the course.

This course, which was first offered in 2004 as an online course only, is taught in a blended learning mode and is studied both at the Toowoomba (Queensland, Australia) campus of the University and online. It is normally assessed by a single assignment worth 50% of marks and an examination worth 50% of marks. In 2020, 67 learners were enrolled in this course, of which 44 studied online and the remaining 23 studied on campus. The main question in the assignment asks learners to assume the role of an asset manager who has been tasked to manage a middle-level engineering asset that has several issues and is required to develop a plan to rectify this position. The other questions are normally in asset depreciation and the use of engineering economics principles. The examination asks learners to apply

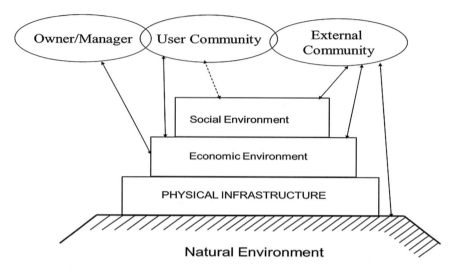

Fig. 13.2 Interaction of communities and their environments – asset management in an engineering environment. Source: University of Southern Queensland (2020a)

principles learnt in the whole course to example engineering applications nominated by them.

While generally learner feedback at the end of the course has been quite positive, there have also been some less positive comments, including about relevance of the course to particular engineering and built environment disciplines such as surveying, and concerns from time to time about the currency of some of the material taught in the course. A possible weakness of the course is that, being strategic in nature, it does not deal with asset operations and maintenance in the detail required by practitioners. To address these concerns, the course is regularly reviewed and updated to include current professional issues and advances in asset management, including information systems. Recent initiatives have included strengthening the use of asset management standards and incorporating sections on BIM and GIS. These sections are expected to be expanded over time. In particular, it has been stated that there is a growing need for universities to provide their graduates with appropriate BIM-related skills in architecture, civil engineering, building construction and construction project management programs, including both teaching and practising the collaboration and integration required for successful implementation (Forsythe et al., 2013). Future expected developments of this course include a stronger emphasis on sustainable and recyclable materials and on energy management. Other initiatives have included making the course relevant to a wide range of engineering and built environment initiatives, such as including an example relating to an electricity power distribution network.

The focus of this course on sustainable engineering and built environment has been strengthened by the development and delivery of companion courses. One of these courses is Management of Technological Risk (University of Southern

Queensland, 2020b), which teaches risk management in accordance with the ISO 31000:2018 Risk Management Guidelines (International Organization for Standardization, 2018) and then applies the principles taught to the management of project risk and process risk. This course assists asset managers and other engineering and built environment practitioners to understand and manage the risks associated with good asset management principles. A further course, Advanced Engineering Project Management (University of Southern Queensland, 2020c), which includes a specific section dealing with sustainability in project management, teaches the management of engineering projects at a project and program level, which is a valuable skill for implementing sustainable asset management principles, including the sustainable management of construction activities.

The above discussion has answered the second question with respect to how postgraduate education in sustainable asset management and related study courses can support a sustainable urban environment. Such support can be provided through a strategically focused postgraduate asset management course with a strong sustainability orientation that is supported by companion courses in related areas like risk management, project management and other related areas, all of which are regularly updated.

13.5 Discussion and Conclusion

Urban facilities consume significant natural resources, are significant emitters of greenhouse gases and have other impacts on the natural environment. The built environment, and in particular its construction and operational life cycle phases, is a major source of these emissions as well as contributing to other negative impacts on the environment such as waste and pollution. Addressing these issues from a sustainable urban environment point of view will require the achievement of relevant global sustainable development goals for the assets that comprise the buildings and infrastructure of cities, using a strategic asset management approach. It would also require coordinated activities by engineers, asset managers and other built environment professionals responsible for their management. These professionals, many of whom are required by their professional organisations to practice sustainability, would be expected to have the knowledge and skills to effectively achieve the objectives of sustainable urban development and management. Two questions, one on whether a strategic approach to asset management can facilitate the development and management of urban sustainability (RQ1) and the other on how postgraduate education in sustainable asset management and related courses can support the achievement of a sustainable urban environment (RQ2), have been accordingly proposed to provide a framework for addressing the development of the necessary skills to achieve this goal.

The first question (RQ1) has been addressed through developing an understanding of the strategic asset management process, the standards that support it, stakeholder management and its requirements, including construction and the use of advanced asset management systems. While it is recognised that further development of the strategic asset management process is required, the role of strategic and forward thinking asset management in developing a sustainable approach to asset management has been confirmed. This process is contingent on ensuring that this process is current and addresses key issues. For example, the uncertainty about what the sixth and seventh dimensions of BIM (sustainability and asset management) represent requires resolution, as both dimensions are significant for sustainable asset management.

To address the second question (RQ2), a postgraduate course in asset management in an engineering environment, along with supporting courses in risk management and project management, has been reviewed from the point of view of how well it deals with sustainable whole of life asset management. While enhancements to this course in areas like application to a wider range of engineering disciplines and the further development of course material on information systems like BIM and GIS are required, it is concluded that a strategically focused postgraduate engineering asset management course with a strong sustainability focus can, with the support of suitable companion courses, provide the necessary theoretical underpinning to enable engineers and other built environment professionals to effectively develop and manage a safe and sustainable urban environment. Complementing this course and its companion courses in the future with a broader sustainable engineering and management course that addresses global sustainable development goals would further enhance its contribution to a sustainable and liveable urban environment.

In conclusion, the development, maintenance and ongoing management of a sustainable urban environment, with a view to reducing its carbon footprint, minimising its waste and increasing its liveability, are essential. Strategic, whole of life asset management of this environment, along with development of engineers and other built environment professionals through postgraduate education in sustainable asset management, complemented by appropriate companion courses, are considered key components of this process. Ongoing enhancement of these courses and the future addition of a broader sustainable engineering and management course would be expected to further enhance the education and development of professionals charged with the development and management of a sustainable urban environment.

References

Biggs, J. (1999). What the student does: Teaching for enhanced learning. *Higher Education Research and Development, 18*(1), 57–75.

C40 Cities. (2018). *Consumption-based GHG emissions of C40 cities*. C40Cities [Online]. https://www.c40.org/researches/consumption-based-emissions

Charef, R., Alakaa, H., & Emmitt, S. (2018). Beyond the third dimension of BIM: A systematic review of literature and assessment of professional views. *Journal of Building Engineering, 18*, 242–257.

Chong, H., Lee, C., & Wang, X. (2017). A mixed review of the adoption of Building Information Modelling (BIM) for sustainability. *Journal of Cleaner Production, 142*, 4114–4126.

Engineers Australia. (2014). *Sustainability policy*. Engineers Australia.

Engineers Australia. (2019). *Code of ethics and guidelines on professional conduct*. Engineers Australia.

Forsythe, P., Jupp, J., & Sawhney, A. (2013). Building information modelling in tertiary construction project management education: A programme-wide implementation strategy. *Journal for Education in the Built Environment, 8*(1), 16–34.

Gulikers, J. T. M., Bastiaens, T. J., & Kirschner, P. A. (2004). Perceptions of authentic assessment: Five dimensions of authenticity. In *Proceedings, Second biannual Northumbria/EARLI SIG assessment conference*. Bergen.

Habibi, S. (2017). The promise of BIM for improving building performance. *Energy and Buildings, 153*, 525–548.

International Energy Agency and the United Nations Environment Programme. (2018). *Global Status Report: Towards a zero-emission, efficient and resilient buildings and construction sector*. United Nations.

International Organization for Standardization. (2014a). *ISO 55000:2014 asset management—Overview, principles and terminology*. ISO.

International Organization for Standardization. (2014b). *ISO 55001:2014 asset management—Asset management—Management systems—Requirements*. ISO.

International Organization for Standardization. (2018). *ISO 31000 Risk management—Guidelines*. ISO.

Kabirifar, K., Mojtahedi, M., Wang, C., & Tam, V. (2020). Construction and demolition waste management contributing factors couple with reduce, reuse, and recycle strategies for effective waste management: A review. *Journal of Cleaner Production, 263*, 121265.

Lu, Y., Wu, Z., Chang, R., & Li, Y. (2017). Building Information Modelling (BIM) for green buildings: A critical review and future dimensions. *Automation in Construction, 83*, 134–148.

Medineckiene, M., Turskis, Z., & Zavadskas, E. K. (2010). Sustainable construction taking into account the building impact on the environment. *Journal of Environmental Engineering and Landscape Management, 18*(2), 118–127.

United Nations. (2015). *Transforming our world: The 2030 agenda for sustainable development*. United Nations.

United Nations. (2019). *World urbanization prospects: The 2018 revision*. United Nations.

United Nations Environment Programme. (2020). *Cities and climate change*. United Nations [Online]. https://www.unenvironment.org/explore-topics/resource-efficiency/what-we-do/cities/cities-and-climate-change

University of Southern Queensland. (2020a). *Asset management in an engineering environment*. University of Southern Queensland.

University of Southern Queensland. (2020b). *Management of technological risk*. University of Southern Queensland.

University of Southern Queensland. (2020c). *Advanced engineering project management*. University of Southern Queensland.

World Green Building Council. (2019). *Bringing embodied carbon upfront—Coordinated action for the building and construction sector to tackle embodied carbon*. World Building Council.

Xu, X., Ding, L., Luo, H., & Ma, L. (2014). From building information modeling to city information modelling. *Journal of Information Technology in Construction, 19*, 292–307.

Yu, M., Wiedmann, T., Crawford, R., & Tait, C. (2017). The carbon footprint of Australia's construction sector. *Procedia Engineering, 180*(2017), 211–220.

Zhang, J., Schmidt, K., & Li, H. (2016). BIM and sustainability education: Incorporating instructional needs into curriculum planning in CEM programs accredited by ACCE. *Sustainability, 8*(6), 525.

Chapter 14
The Impact of Department of Construction Management Facebook Environment-Related Posts

John Smallwood

14.1 Introduction

The construction industry has become one of the major contributors to adverse impacts on the natural environment as projects are often associated with negative environmental impacts, which result in the degradation of the natural environment, which is a product of soil pollution, water pollution, air pollution, deforestation, destruction of wildlife, and land degradation. Furthermore, construction activities generate waste, noise, ground vibration, dust, hazardous emissions, disturbance of natural features, and the destruction of archaeological artefacts during the construction process and its activities, in which cumulative impacts result in environmental degradation (Chen & Li, 2006). This has resulted in increased awareness of the natural environment and the desire for a more sustainable approach by all stakeholders. Hence, it is imperative for all contractors to consider implementing appropriate environmental management practices to reduce the impact of the construction process and its activities on the natural environment and society as a whole, instead of only focusing on time, cost, and quality within a project as traditionally practiced.

Interviews conducted by Carpenter et al. (2016) among sustainability officers and student leaders at 21 leading sustainable universities in the USA revealed that the top American sustainable universities rely on social media to reach large audiences. Generally, their target audiences are students, staff, faculty members, local community residents, campus administration leaders, news media professionals, and alumni. Furthermore, sustainability officers primarily rely on Facebook to encourage interaction with their posted content.

J. Smallwood (✉)
Department of Construction Management, Nelson Mandela University, Port Elizabeth, South Africa
e-mail: John.Smallwood@mandela.ac.za

© The Author(s), under exclusive license to Springer Nature Switzerland AG 2022
C. Gorse et al. (eds.), *Climate Emergency – Managing, Building, and Delivering the Sustainable Development Goals*, https://doi.org/10.1007/978-3-030-79450-7_14

Narula et al. (2019) maintain the media is playing a role in shaping attitudes of responsibility in the general population. Social media is a mode of communication, sharing feelings and sharing information with each other; however, it can also enhance the knowledge of the general population. They emphasise that there are a range of topical environmental issues, which include, inter alia, worsening environmental conditions, climate change, the problem of waste disposal, and increasing pollution. Therefore, development of environmental awareness and the promotion of sustainable development have become goals or policies of not only governments, and non-government organisations (NGOs), but other stakeholders.

Black (1993) defines public relations practice as "The art and science of achieving harmony with the environment through mutual understanding based on truth and full information." He states that there are two distinct branches of practice, namely, the reactive and proactive sectors. Reactive includes reacting to problems, dealing with crises, and managing change, and the proactive includes planned programmes that serve both the organisation and the public's interest.

Wright and Hinson (2009) state that social media deliver web-based information created by people with the intention of facilitating communication, which now represents one of the world's major sources of social interaction as people share stories and experiences with each other. However, their fourth study conducted among public relations practitioners globally indicates that public relations practitioners should measure the amount of communication that is being disseminated with respect to their organisations, or client organisations, through blogs and other social media (4.2/5.0), and/or analyse the content of what is being communicated with respect to their organisations, or their clients, in blogs and other social media (4.4/5.0). The mean scores presented in parentheses range between 1.00 and 5.00, 1.00 representing the lowest point and 5.00 the highest point.

Given the state of the planet, the impact of the built environment in terms of sustainability, the importance of public relations, and that the department undertakes environment-related posts, a study was conducted to determine the impact of 'environment-related' posts on the Department of Construction Management Facebook page during a calendar year to determine which attracted the most interest.

14.2 Review of the Literature

14.2.1 Public Relations

Black (1993) describes the role of public relations by presenting the hexagon model, the six sides representing the several factors which influence the role and scope of public relations: the publics of concern; issues of concern; media; the nature of the organisation; situational timing factors, and resources. However, this paper focuses on one specific issue of concern, namely, the environment, and in terms of the media 'side', social media in the form of Facebook. The nature of the 'organisation', namely, a Department of Construction Management within a university, implies that the built environment should be the focus of such environment-related posts.

Situational timing factors in turn imply that 'current' issues at the time of the post inform, to an extent, the topic of the post.

14.2.2 The Impact of Social Media

The study conducted by Wright and Hinson (2009), cited in the 'Introduction', determined the following: social media have enhanced the practice of public relations (4.0/5.0); blogs have enhanced the practice of public relations (3.8/5.0), and social media offer organisations a low-cost way to develop relationships with members of various strategic publics (4.0/5.0). These findings are notable, as they are 2009 vintage – 11 years old. Narula et al. (2019) maintain social media is playing an important role in sharing a range of environmental issues, suitable solutions thereto, updating the audience regarding the different types of disaster, and precautionary measures.

Can social media make a difference? Porter et al. (2007 in Carpenter et al., 2016) state that communicators who use social media perceive that they have more power in creating change than those who do not use social media to reach publics.

14.2.3 Current Reality

Before the advent of the Internet, audiences had to endure listening to or seeing advertisements repeatedly, and the advertisements were in control of the media that people listened to or watched (Evans, 2011). However, the Internet evolved in the 1990s, and audiences no longer had to endure commercials, as they could fast forward through commercials and ignore the 'noise' of marketing. Social media provides the forum for people to acquire information in a natural and more honest way than being subjected to marketing messages. Target audiences can be reached through user-generated content sites and social media communities. The reality is that the users and community members are now in 'control' and do not want to be 'drilled' with marketing messages.

14.3 Research

14.3.1 Research Method and Sample Stratum

The research method can best be described as experimental in that posts pertaining to different issues were undertaken periodically by the author, the page administrator. The impact of environment-related posts was compared to the impact of non-environment related posts, and then the impact of four categories of environment-related posts, namely, URL referral, photo, video, and media release, was compared.

Posts were constrained to the built environment, and construction, COVID-19 upon the onset thereof, particularly construction health and safety (CH&S), and in addition, tertiary construction management education and related issues such as seminars. In terms of the review of the degree of interest, the review spanned a period of a year from 15 June 2019 to 14 June 2020 and addressed issues such as 'reach', 'clicks', and 'reactions, comments, and shares'. Reach is the number of people who saw any of the page posts. Reach can be broken down into people who saw the posts with or without advertising (paid or organic posts). All posts were without advertising.

14.3.2 Research Results

Table 14.1 indicates that a total of 210 posts were made during the period, 50 (23.8%) of which were environment-related. The 50 environment-related posts resulted in a reach of 4902, 248 clicks and 93 reactions, comments, and shares. The totals equate to a mean reach of 98.0, 4.96 clicks and 1.86 reactions, comments, and shares. In terms of comparisons, non-environmental posts realised a mean reach of 124.4 per post, compared to environmental posts' 98.0. In terms of clicks, non-environmental posts realised a mean reach of 13.52 per post, compared to environmental posts' 4.96. In terms of reactions, comments, and shares, non-environmental posts realised a mean reach of 2.41 per post, compared to environmental posts' 1.86.

Table 14.2 provides an analysis of the Department of Construction Management's environment-related Facebook posts in terms of type of post. The single (2.0%) media release realised a mean reach of 239.0, followed by video (145.0), photo (141.4), and URL referral (87.29), the mean for all four types being 98.04. However, in terms of clicks, photo realised a mean of 15.20, followed by media release (13.00), video (11.00), and URL referral (3.26), the mean for all four being 4.96. In terms of reactions, comments, and shares, photo realised a mean of 5.20, followed by media release (4.00), video (1.50), and URL referral (1.43), the mean for all four being 1.86.

Table 14.1 Summary of the Department of Construction Management's Facebook posts for the period 15 June 2019 to 14 June 2020

Category	Total No.	%	Reach No.	%	No./post	Click No.	%	No./post	R/C/S No.	%	no./post
Environment	50	23.8	4902	19.8	98.0	248	10.3	4.96	93	19.5	1.86
Non-environment	160	76.2	19900	80.2	124.4	2164	89.7	13.52	385	80.5	2.41
Total	210	100.0	24802	100.0	118.0	2412	100.0	11.49	478	100.0	2.28

Table 14.2 Analysis of the Department of Construction Management's environment-related Facebook posts for the period 15 June 2019 to 14 June 2020 in terms of type of post

Category	Total No.	Total %	Reach No.	Reach %	Reach No./post	Click No.	Click %	Click No./post	R/C/S No.	R/C/S %	R/C/S No./post
URL referral	42	84.0	3666	74.8	87.29	137	55.2	3.26	60	64.5	1.43
Photo	5	10.0	707	14.4	141.4	76	30.7	15.20	26	28.0	5.20
Video	2	4.0	290	5.9	145.0	22	8.9	11.00	3	3.2	1.50
Media release	1	2.0	239	4.9	239.0	13	5.2	13.00	4	4.3	4.00
Total	50	100.0	4902	100.0	98.04	248	100.0	4.96	93	100.0	1.86

Table 14.3 presents the Department of Construction Management environment-related Facebook posts for the period 15 June 2019 to 14 June 2020. The first column presents the date of the post, followed to the right by the post description, the type of post, reach, clicks, and then reactions, comments, and shares (R/C/S) in the extreme right-hand column.

A total of 210 posts were made during the period, 50 (23.8%) of which were environment related. The 50 posts resulted in a reach of 4902, 248 clicks and 93 reactions, comments, and shares. The totals equate to an average reach of 98.04, 4.96 clicks, and 1.86 reactions, comments, and shares. Furthermore, the top ten posts (20.0%) accounted for 35.1% of the reach, 44.4% of the clicks, and 46.2% of reactions, comments, and shares.

The top ten environment-related posts in terms of reach are 'ACHASM Contribution to "Covid-19" "Construction Return to Work" Documentation' media release on 28/05/20 (239); 'SEEDS 2019 Conference Dinner, Ipswich, United Kingdom on 11 September - we received two awards' photo on 03/10/19 (220); 'Mitigating the spread of the Coronavirus' video on 03/03/20 (179); 'WATER MOST IN Belgium!!!' (Elevated canal) photo on 28/09/19 (170); 'Morocco to commence construction of world largest sea water desalination plant in 2021' URL referral on 22/06/19 (169); 'We are delighted to highlight X's second place in the Greenovate Property Awards 2019 Competition for his submission titled "Implications of a natural ventilation retrofit of an office building" supervised by Y' photo on 11/12/19 (161); 'A bridge made of grass' URL referral on 18/06/19 (157); 'Asbestos-containing materials exist in many of our buildings and structures!' URL referral on (145); 'R1.2 billion solar farm approved for the Garden Route' on 07/02/20 (143); and 'Former Bricklayer Turns Stones Into Works of Art' URL referral on 10/01/20 (139).

The 'ACHASM Contribution to "Covid-19" "Construction Return to Work" Documentation' media release on 28/05/20 (239) addressed the contribution of a group of Candidate and Professional Construction Health and Safety Agents (CHSAs) to the development of the construction industry's COVID-19 regulations and guidelines under the leadership of the Executive Director, ACHASM, who is also a research associate in the Department of Construction Management. This constituted a major intervention on the part of construction H&S practitioners.

Table 14.3 Analysis of the Department of Construction Management's environment-related Facebook posts for the period 15 June 2019 to 14 June 2020

Date	Post description	Type	Reach No.	Reach %	Clicks	R/C/S
28/05/20	ACHASM Contribution to 'Covid-19' 'Construction Return to Work' Documentation	Media release	239	4.88	13	4
03/10/19	SEEDS 2019 Conference Dinner, Ipswich, UK, 11 September - we received two awards	Photo	220	4.49	33	18
03/03/20	Mitigating the spread of the Coronavirus	Video	179	3.65	9	1
28/09/19	WATER MOST IN Belgium!!! (Elevated canal)	Photo	170	3.47	14	4
22/06/19	Morocco to commence construction of world largest sea water desalination plant in 2021	URL referral	169	3.45	7	2
11/12/19	X's second place in the Greenovate Property Awards 2019 Competition for his submission titled 'Implications of a natural ventilation retrofit of an office building' supervised by Y	Photos	161	3.28	15	3
18/06/19	A bridge made of grass	URL referral	157	3.20	3	0
15/07/19	Asbestos-containing materials exist in many of our buildings and structures!	URL referral	145	2.96	10	3
07/02/20	R1.2 billion solar farm approved for the Garden Route	URL referral	143	2.92	2	2
10/01/20	Former bricklayer turns stones into works of art	URL referral	139	2.84	4	6
29/12/19	An Eye for Renewal ǀ Carte Blanche ǀ M-Net	URL referral	136	2.77	4	4
10/09/19	Johannesburg unveils the completed S-bend wall mural project:	URL referral	122	2.49	0	2
22/10/19	'There is ingenuity in Africa': The architect who builds with trash	URL referral	113	2.31	5	7
06/04/20	The National Institute for Occupational Health (NIOH) has arranged a PPE USE and COVID-19 Training Session	URL referral	111	2.26	9	2
05/10/19	Automated 'decking' swimming pool cover	Video	111	2.26	13	2
20/12/19	Prefab homes on stilts include solar panels, water collection systems and organic gardens	URL referral	109	2.22	2	1
30/04/20	Covid-19 occupational health and safety measures in workplaces Covid-19 (C19 OHS)	URL referral	108	2.20	18	1
14/05/20	South Africa's COVID-19 strategy needs updating: here's why and how	URL referral	104	2.12	3	0
17/04/20	Site operating procedures protecting your workforce during coronavirus (Covid-19)	URL referral	103	2.10	6	2
08/01/20	Transparent solar panels will turn windows into green energy collectors	URL referral	103	2.10	4	8
24/04/20	Housebuilders begin 'controlled' restart of sites	URL referral	99	2.02	9	1

(continued)

Table 14.3 (continued)

Date	Post description	Type	Reach No.	Reach %	Clicks	R/C/S
26/09/19	Astronauts make concrete in space for the first time	URL referral	98	2.00	0	1
30/04/20	Presentations to the 'virtual' first Asia Pacific SDEWES conference on sustainable development of energy, water and environment systems	Photo	92	1.88	11	1
11/10/19	Plastic brick invention to launch in Cape Town	URL referral	91	1.86	6	0
09/03/20	Hemp wood, the 'new' oak. 20% stronger and grows 100 times faster!	URL referral	86	1.75	4	2
27/03/20	Australian university develops 'pandemic drone' to spot Covid-19 in crowds	URL referral	84	1.71	1	0
03/01/20	New York City passes law for bird-friendly exterior requirement for buildings	URL referral	83	1.69	2	0
21/04/20	Call for the designation of construction sector as essential services	URL referral	82	1.67	2	1
13/04/20	Commercial real estate must do more than merely adapt to coronavirus	URL referral	82	1.67	1	1
07/03/20	How Taiwan used big data, transparency and a central command to protect its people from coronavirus	URL referral	81	1.65	2	0
09/03/20	Ten coronavirus questions for construction firms	URL referral	80	1.63	2	2
09/01/20	House inside a rock	URL referral	75	1.53	3	3
28/04/20	Learning during COVID-19	URL referral	73	1.49	5	3
20/12/19	Tesla's new solar roof cheaper than A regular roof with solar panels	URL referral	73	1.49	2	0
29/02/20	Water theft: Thirty years of looting	URL referral	71	1.45	2	0
27/02/20	Sevierville company first in the world to produce hemp hardwood flooring	URL referral	68	1.39	0	2
14/05/20	Crushing coronavirus uncertainty: The big 'unlock' for our economies	URL referral	65	1.33	3	0
12/12/19	December 1998: White Paper on the Energy Policy of the Republic of South Africa	Photos	64	1.31	3	0
08/06/20	'What are the responsibilities of employers during COVID-19'	URL referral	62	1.26	4	0
11/03/20	Construction Health & Safety New Zealand (CHASNZ) has released a succinct resource that provides an overview of COVID-19 toolbox and outlines some easy steps to take to mitigate the risk of catching the virus	URL referral	62	1.26	0	0

(continued)

Table 14.3 (continued)

Date	Post description	Type	Reach No.	%	Clicks	R/C/S
20/12/19	UK renewables generate more electricity than fossil fuels for first time	URL referral	62	1.26	0	0
08/06/20	Covid-19: The folly of correcting mistakes when heading the wrong way	URL referral	61	1.24	0	0
18/01/20	This new building material has cement-like strength—and it's alive	URL referral	57	1.16	1	0
11/06/20	Research group predicts severe recession due to 'flawed, self-inflicted' hard lockdown	URL referral	53	1.08	2	0
08/03/20	Coronavirus: Get over it!	URL referral	49	1.00	5	0
19/03/20	Crane working at remote Antarctic site	URL referral	48	0.98	1	0
05/06/20	Free content for World Environment Day	URL referral	46	0.94	0	0
06/04/20	The power of good design	URL referral	40	0.82	3	4
28/01/20	Foresight Africa: Top priorities for the continent 2020–2030	URL referral	37	0.75	0	0
11/06/20	Spain: Taking sustainable energy to the next level	URL referral	36	0.73	0	0
Total	50		4902		248	93
Mean			98.04		4.96	1.86

The 'SEEDS 2019 Conference Dinner, Ipswich, United Kingdom on 11 September - we received two awards' photo on 03/10/19 (220) featured Professor X (Sustainability Retrofit Award) and Professor Y (Health and Safety Award), a research associate and a professor in the Department of Construction Management. Previous 'Impact of Facebook posts' research determined that posts relating to staff and students realised high reaches.

'Mitigating the spread of the Coronavirus' video on 03/03/20 (179) featured the Singaporean Minister for Health addressing their parliament with respect to the nature of COVID-19.

The 'WATER MOST IN Belgium!!!' (Elevated canal) photo on 28/09/19 (170) presented a notable photo of innovative engineering in terms of a canal negotiating local topography.

The 'Morocco to commence construction of world largest sea water desalination plant in 2021' URL referral on 22/06/19 (169) is relevant to the metropole in which the Department of Construction Management's university is located in that the metropole has had water restrictions in place for years, and the City of Cape Town has three operational desalinisation plants, i.e. the issue is 'close to home'.

'X's second place in the Greenovate Property Awards 2019 Competition for his submission titled "Implications of a natural ventilation retrofit of an office building" supervised by Y' photo on 11/12/19 (161) featured a member of staff and a student in the Department of Construction Management. A stated above, previous 'Impact of Facebook posts' research determined that posts relating to staff and students realised high reaches (Smallwood, 2018).

'A bridge made of grass' URL referral on 18/06/19 (157) featured a bridge courtesy of BBC News. Every year the last remaining Inca rope bridge still in use is cast down, and a new one erected across the Apurimac river in the Cusco region of Peru. The Q'eswachaka bridge is woven by hand and has been in place for at least 600 years. Once part of the network that linked the most important cities and towns of the Inca empire, it was declared a World Heritage Site by UNESCO in 2013. No modern materials, tools or machines are used in the whole process of building the bridge – only grass and human power (Busque, 2019). Figure 14.1 provides a view of the bridge under re-construction.

'Asbestos-containing materials exist in many of our buildings and structures!' URL referral on 15/07/19 (145) is a post by SGS, Geneva, Switzerland, announcing the launch of its all-new Asbestos e-learning training course. Asbestos-containing materials are still an issue in South Africa, and the South African Asbestos Regulations 2001 were gazetted on 10 February 2002. These resulted in, inter alia, the substitution of asbestos fibre with synthetic fibre.

Fig. 14.1 Q'eswachaka bridge spanning the Apurimac river in the Cusco region of Peru (Busque, 2019)

Fig. 14.2 Wall section (Televičiūtė, 2016)

The 'R1.2 billion solar farm approved for the Garden Route' on 07/02/20 (143) featured a topical issue in South Africa, namely, electricity supply. The reason being the challenges courtesy of coal-fired power stations in the form of blackouts/power outages and the emissions from the power stations. Solar energy being a form of sustainable energy, in addition to the value of such construction projects, means that such posts are likely to attract interest.

'Former Bricklayer Turns Stones Into Works of Art' URL referral on 10/01/20 (139) features Johnny Clasper from Yorkshire, UK, who evolved from a bricklayer to a stonemason. He is cited as turning stones and rocks into anything ranging from patios to innovative sculptures and captivating mosaics (Televičiūtė, 2016) (Fig. 14.2).

The bottom ten environment-related posts in terms of reach are as follows: 11/06/20 'Spain: taking sustainable energy to the next level' URL referral on 11/06/20 (36); 'Foresight Africa: Top priorities for the continent 2020–2030' URL referral on 28/01/20 (37); 'The power of good design' URL referral on 06/04/20 (40); 'Free content for World Environment Day' URL referral on 05/06/20 (46); 'Crane working at remote Antarctic site' URL referral on 19/03/20 (48); 'CoronaVirus: Get over it!' URL referral on 08/03/20 (49); 'Research group predicts severe recession due to 'flawed, self-inflicted' hard lockdown' URL referral on 11/06/20 (53); 'This new building material has cement-like strength—and it's alive' URL referral on 18/01/20 (57); 'Covid-19: The folly of correcting mistakes when heading the wrong way' URL referral on 08/06/20 (61), and 'UK renewables generate more electricity than fossil fuels for first time' URL referral on 20/12/19 (62).

14.4 Conclusions

Given the reach, clicks, and reactions, comments, and shares, overall, the Department of Construction Management's Facebook page is relevant.

Given the respective reaches, clicks, and reactions, comments, and shares, non-environmental posts can be deemed to have had a greater impact than environment-related posts. However, the reaction to environment-related posts indicates that such posts are of interest and relevant.

Although the non-environment-related posts were not presented and analysed in the paper, it should be noted that five of the top ten posts overall in terms of reach were student/staff/industry personalia-related, which prior research has identified as having the greatest impact in terms of the Department of Construction Management's Facebook posts (Smallwood, 2018).

Based upon the findings, it can be concluded that certain environment-related posts are of greater interest than others. Those that are of greater interest include topical issues, personalia linked to the Department, and unique issues. Topical issues include COVID-19, access to water due to the local drought, solar energy, and asbestos-containing materials. Personalia linked to the Department include the receipt of awards by staff and students. Unique issues include 'WATER MOST IN Belgium!!! (Elevated canal)', 'A bridge made of grass', and 'Former Bricklayer Turns Stones Into Works of Art'.

The two personalia linked to the Department posts constituted 4.0% of the posts, and although they accounted for 7.7% of the reach, they accounted for 19.4% of the clicks and 22.6% of the reactions, comments, and shares. This further underscores the finding of prior research, namely, that personalia-related posts were identified as having the greatest impact in terms of the Department of Construction Management's Facebook posts (Smallwood, 2018). Furthermore, it leads to the conclusion that clicks and reactions, comments, and shares are important indicators and are ultimately the 'measure of impact'.

Furthermore, it can be concluded that Facebook plays an important role in sharing a range of environmental issues and suitable solutions thereto as contended by Narula et al. (2019).

14.5 Recommendations

It is recommended that Facebook page statistics be regularly reviewed to determine the impact and relevance of posts. Attention should be focused on clicks and reactions, comments, and shares.

Facebook page administrators should evolve a 'cocktail' of posts in terms of an ideal mix to optimise interest in, to inform, and to raise the level of environmental awareness and the impact of their pages. However, personalia linked to the department or organisation should feature prominently in posts. Doing so is likely to promote interest in a page, the issues addressed, and the 'owner'.

References

Black, S. (1993). *The essentials of public relations*. Kogan Page Ltd.
Busque, J. (2019). *A bridge made of grass*. https://www.bbc.com/news/in-pictures-48628325?fbcli d=IwAR1_6UZFtHjtneFeBngcrGVnIn9aKNog8ir6qrGVXDwYVIigAEsipzHCvOA
Carpenter, S., Takahashi, B., Cunningham, C., & Lertpratchya, A. P. (2016). The roles of social media in promoting sustainability in higher education. *International Journal of Communication, 10*(2016), 4863–4881.
Chen, Z., & Li, H. (2006). *Environmental management in construction: A quantitative approach*. Taylor & Francis.
Evans, L. (2011). *Social media marketing*. Que Publishing.
Narula, S., Rai, S., & Sharma, A. (2019). *Environmental awareness and the role of social media*. IGI Global.
Smallwood, J. J. (2018). The impact of posts on interest in a Department of Construction Management Facebook page. In *Proceedings of the Second European and Mediterranean structural engineering and construction conference*, 23–28 July 2018, EPE-03-1—EPE-03-5.
Televičiūtė, J. (2016). *Former bricklayer turns stones into works of art*. https://www.bored-panda.com/stone-sculptures-mosaic-walls-johnny-clasper/?fbclid=IwAR3aVvtZHf2Ne 3jtyMiFRLhgTvxlkHULTZSCQ2EqWyLYukWKjW7WNZu4aM8&utm_source=web. facebook&utm_medium=referral&utm_campaign=organic
Wright, D. K., & Hinson, M. D. (2009). An updated look at the impact of social media on public relations practice. *Public Relations Journal, 3*(2) Spring 2009.

Chapter 15
'The Breakfast Room Game': A Case of an Innovative Construction Project Management Simulation for Year 6 Children

Ian C. Stewart

15.1 Introduction

The following is the case of an academic/practitioner partnership that created an innovative Lego-based construction project management game/simulation for use in primary schools with year 6 pupils. The paper will show how the professionals involved (academic, construction project manager, headteacher) engaged with each other, the pedagogic design and operation of the game with some reflections on what was learned about the conditions for professional outreach of this type. The author runs a unit on the MSc Management of Projects at the University of Manchester which involves the use of live case studies of a range of different project types in the North West, donated by local companies. There is a certain degree of annual turnover in these cases so there is always a search for new ones. A project manager from local construction firm Beaumont Morgan offered his apartment complex construction as a case, in return for help with Corporate Social Responsibility (CSR) obligations attached to his project. He was charged with creating 'social value' for the community around his project, particularly for the primary school. This situation is becoming very common in construction projects being implemented in areas of social need, especially when the local councils are involved in the site acquisition.

In discussions with the headteacher of the primary school, the ideas of somehow bringing in construction practices, STEM, sustainability and life skills were drawn together. The headteacher asked whether actual construction materials could be brought to the school for the children to use, which of course would be a health and

---

I. C. Stewart (✉)
Department of Mechanical, Aerospace and Civil Engineering, The University of Manchester, Manchester, UK
e-mail: I.C.Stewart@manchester.ac.uk

safety problem and so an obvious 'no'. However, this led to thinking of the possibilities of other construction media and from this came the idea of using Lego to simulate some kind of construction activity. From this came the question of the kinds of learning outcome that the headteacher wanted.

The discussion on how such an activity could be performed and controlled led naturally to the need for project management and the opportunity for teaching the children construction project management techniques simultaneously with actually using them. Findings of studies reveal that using simulation in project management education can improve students' knowledge level and develop their soft skills that are required in managing projects (Geithner and Menzel, 2016; Savelsbergh et al., 2016; Konstantinou, 2015). Starting with Posner (1987), studies of project management (PM) skills and competencies required by employers reveal these to be leadership, communication, dealing with ambiguity, cultural awareness, relationship management, team working, time management, self-efficacy, planning, conflict resolution, problem-solving and dealing with complexity. These are also the kinds of things considered to be 'life skills'. Therefore, PM could be an ideal means for children to experience and develop these life skills. If a construction project were to be simulated, what kind of structure would be relevant to the children? The idea of a purpose-built free-standing 'Breakfast Room' was proposed, as these are becoming a common part of primary schools in deprived areas. They are also fairly simple structures, so well within a child's imagination.

One typical trait of a profession is the transmission of practices and esoteric knowledge from experienced to neophyte. This is common in graduate employment, where graduates are taken beyond their university courses and introduced to the esoteric practices of the cadre. However, as will be shown, there is potential for such transmission regarding construction project work to begin far earlier. It is possible that outreach can achieve more than simply telling children about construction project work, but actually creating a realistic experience of the work, that is accessible at their scale. This is more likely to have memorable impact on them. This kind of activity would be important for developing young people who are familiar with project work and see it as a first career. Though there is a little academic research in STEM outreach generally (i.e. Landry et al., 2019; Warner, 2019), there is virtually nothing on the experience of practitioners as they get involved in this. This paper will show that through partnership with an academic with interests in PM pedagogy, it is possible for construction project management practitioners to convert their knowledge and experience into compelling teaching and learning experiences that will be desirable to schools. Thus, they could satisfy CSR obligations with greater ease, creating benefit for the profession, for schools and for children, and even enjoy doing it. For construction PMs with CSR requirements, this paper will answer two questions of interest:

What are the interactions around the game between the principal stakeholders; academic practitioner; and teacher and child?

Which construction project management practices can actually be conveyed to children and used in one lesson?

15.2 A Pedagogy of Lego-Based Simulation and the 'Intermediate Impossible'

When the serious use of Lego in teaching and learning is discussed, often the concept of 'serious play' is thought of. However, this work must be distanced from the 'Lego serious play' concept. Lego serious play is a branded facilitation methodology using Lego to represent business or organisational matters, to encourage creativity, participation and ownership of problems (see Hayes & Graham, 2019; Hadida, 2013; Burgi et al., 2004). The work described here is different to that. The model constructed is not a symbol or a metaphor; although it might be a simplification of something larger and more complex, it is actually the thing to be built and success in construction of the model that represent success in the learning activity.

To learn project management, a person has to actually manage a project. Learning by experience is essential, especially if the learner has no actual prior experience to reflect on. The Lego-based simulations created by the author are building with specific intent, to create model experiences with some degree of connection to the real world while participating in the activity of building model objects. In both playing with and using Lego 'seriously' in such a simulation, many interesting problems (designed-in and emergent) are met and solved. Progress is visible and easily measured, feedback is instant and errors are forgivable and easily corrected. The end, completion of the model, the material instantiation of the learning, is satisfying but also so should be the process.

Although not 'serious play', the sensation of 'play' is something to be aspired to in that the activity should feel effortless: the participant should feel in control and maybe even have a smile on their face as they do it. According to Race (2015), learners need to invest their emotions into a learning activity in order to feel 'belonging'. Emotional engagement is fundamental to effective learning, especially when designing simulations, to develop a sense of 'commitment to the exercise…a sense of personal accomplishment or failure for results obtained… conflict, time- pressure or whatever in order to place the participant in as close to a real life situation as possible' (Denholm et al., 2012).

Race (2005) in his book *Making Learning Happen* asserted the importance of knowing what factors make learning effective. His theory begins with knowing why individuals want to learn. At the 'front' end when designing a Lego-based learning experience, the intended learning outcomes have to be known. The most powerful question is 'What should the learner be able to do by the end of this?' As Race (2015) says '…it's not knowledge till we do things with it…'. These outcomes then have to have valence to the 'Why?'. Keys and Wolfe (1990) define a simulation as '…a simplified and contrived situation that contains enough verisimilitude or illusion of reality (or sufficient correspondence to the phenomena it purports to represent)'. Such illusions try to create or introduce some semblance of reality or authenticity into a learning environment to deliver higher-order learning outcomes involving synthesis, creation and action while simultaneously delivering soft-skill development. The rules or process for whatever activity is being simulated also have to be known as these are what will structure the use of the knowledge or skill being taught through the simulation.

The following learning outcomes for the game were developed:

Discuss realistic construction project choices—structure, site, functionality, aesthetics, cost, budget and fitness for purpose (quality).

Identify the consequence of decisions in the difference between what the client wanted and what the team delivered.

Practice life skills such as knowing how to/when to ask for help, interacting with unknown adults, following instructions, improvising, negotiation, information handling, teamwork and leadership.

Complete a piece of teamwork under realistic constraints of cost and time.

After the intended learning outcomes for the simulation were set, what followed with building a Lego-based simulation are the same considerations for producing any kind of simulation for learning. The next important question is, 'how real' should the simulation be? As Thavikulwat and Pillutla (2010) identify, decisions have to be made as to how much of the 'real' is to be represented: '…extraneous details, hazards, cost and inconveniences must be stripped away…producing an accelerated frame of action so that they can be more efficient than real world operating environments'. How much of the project management could actually be taught and then usefully experienced in a couple of hours? How much experience and life skill opportunity could be packed into the activity? Of course, it must be remembered that the intended participants are children, so a softening of the illusory real is needed. Would they have resilience to cope when things went wrong as they might? Failure should not be so emotionally crushing as to block off learning, but the possibility has to be real enough to spur vigorous performance. Of course, when designing learning activities for children, the play element can be more explicit; however there is little need to design it in—the children are quite capable of finding it by themselves.

The problem driving the learning has to be real and the scenario created has to have sufficient realism for learners to enter into it and see a connection between what they are doing and actual project work. Like building with Lego, PM is about putting something new into the world, solving problems and battling constraints. The learners are likely to experience some very real emotions during the activity and hopefully obtain experiences that are directly related to project work. At the 'back' end when it is finished, with careful debriefing, it must be ensured that they make sense of that experience and its relation to reality of project work. The participant's learning is real.

In between the front and back end 'realities' is the 'intermediate impossible', to slightly warp the concept of de Bono (de Bono, 1986). Intermediate impossibles are described by Roffe (1999) as '…impractical, or even ridiculous, ideas [which] trigger the imagination of a group and stimulate them'. Here is where the Lego comes in. Following the rules of the activity leads to the creation of objects that might be impossible or even ridiculous in the real world, but via completion of those objects comes completion of the learning; the impossible is an intermediate stage on the way to the learning outcomes. The need for learning construction PM techniques is real; the need for a breakfast room and the constraints it is being procured under is real. However, the resulting room cannot actually be used to eat breakfast in. The rooftop garden available as an environmental extra has no stairs going up to it, and it features a hibachi grill which would never be allowed in a school. It is fun, but impossible. The completion of the impossible structure is an intermediate stage between the motivating problem and the satisfaction of the learning outcomes. The Lego model is a representation of their learning experience, without which the learning would be more difficult, perhaps impossible. It is a physical indicator of their progress in managing their project.

The main design issues and choices are presented in Table 15.1.

In parallel with the simulation design decisions are the considerations of how the work will function as a learning activity. For this, the Race model (Race, 2015) is

Table 15.1 Key simulation design considerations

Simulation design consideration	Response
Verisimilitude—The question of how much reality is 'enough' to function in a realistic way but still allow for learning to play as well as playing to learn	The construction has to be realistic—a free-standing 'breakfast room' building with a slab floor, four walls, a roof, seating and tables for 30 children, glazing, doors
	Children will be split into three teams—a PM team to design, cost and schedule and control the works, a construction team to transport and build and a supplier team to fabricate, hold and release construction components
	Time pressure will be created by a sped-up clock running at 1 min = 1 day, 30 days for construction. Cost and quality pressure will be created by a set budget but also by some conflicting client demands in the brief
	PM tools—MoSCoW analysis of client brief, scheduling with visual tools such as Gantt chart, and project costing. Children would be familiarised with tools/theories before actually using them, via an interactive teaching session before the simulation
	Realistic prices and nomenclature for the construction elements
Extraneous details—The question of where to draw the boundary between the real and the simulated experience	Decision was made not to include any risk management or risk events
Acceleration/timing—The question of how to sequence the actions in the simulation, how forgiving the simulation should be of children falling behind	There are 2 h available. Sense of urgency is created through the use of a sped-up clock
	Expected sequence of actions—PM team does MoSCoW analysis of brief, design sketching and costing of design, check against budget, develop schedule and generate purchase orders. Supplier team puts their stock in order, receives purchase orders and releases stock at the correct time to the construction team. Construction team checks for quality, receives products, transports to site and assembles
	The adults should be able to monitor the game clock and give help where needed if children as seen to fall behind or stall on a particular task

(continued)

Table 15.1 (continued)

Simulation design consideration	Response
Efficiency/actions—The question of what actions students are to take in coordination with the events in the simulation, how forgiving the simulated reality should be of the wrong actions	The project management team make setup decisions based on analysis of client brief and costing of design ideas (the supplier team holds the cost data, PM team will have to collaborate with them over designs). This will involve some calculation. These drive the scheduling; thereafter the schedule triggers procurement and construction activity. The remaining actions are transferring purchase orders and monitoring progress against the schedule All relevant decisions will be made before construction; it is unlikely once the purchase orders have been placed, for anything to go wrong except perhaps delay in physical construction of the model
Material realisation/operating environment—Resources needed to operate or create the work of the simulation and its outcomes	A sufficient amount of Lego is required to create the prefabricated wall panels to speed up construction and make sorting, storage and handling easier A set of PM documentation is required to control the project; Gantt chart with pre-made duration bars featuring lead time and fitting time to speed up assembly of the schedule for placing procurement orders. Price list for the supplier team, list of available construction elements for PM team design work, purchase orders and delivery notes. Instruction lists for the three teams should not go over one page Three tables will be needed, one for each team with enough area to do the work as a team and a room with sufficient circulation space and a PC with projector for the PM lecture
Personal commitment—The question of how to create commitment to the illusion that we create such that the penalties and rewards created through it are felt as motivating	It is a 'game' but there is no internal competition between the teams. The teams collaborate to compete against the clock to deliver on time, cost and quality There is a certificate of achievement for each child on successful completion There should be positive peer pressure effects from being in teams and excitement from the novelty of this form of learning activity
Application of learning—The question of what an individual is supposed to know beforehand, what has to be learned in the process of the simulation	It is unlikely that the children will have any PM or construction knowledge before the simulation Teaching around construction project management is provided at the start. An interactive lecture is presented such that as a theory is introduced, the students participate in examples of it in application. The teaching is as short as possible to allow them to apply the learning as soon as possible Good performance has to be achieved quickly in order to keep up with the movement of time
Personal and group consequences—The question of how far the performance of an individual should affect the group and vice versa	PM team has to make decisions, whereas the construction and supplier teams are more reactive, driven by schedule and purchase orders. PM team will affect the performance of the other two teams; therefore they will need the most help The teams are small enough that one not pulling their weight would affect the performance of the group in the simulation. Careful selection of the participants will prevent this

15 'The Breakfast Room Game': A Case of an Innovative Construction Project... 181

Table 15.2 Race model categories and matching design points

Race model category	Design point
Want/need	Children need to have fun and have curiosity stimulated in order to learn, they want something to tell other children about or their parents, they want to display competence, they want to be able to work with friends in a team, they want to be able to win, achievement has to seem possible from the start
Doing	Production of model room; application of PM to control project; completion of project management documentation; making decisions which are realistic in terms of PM; collaboration between PM, construction and supplier teams; reasonably fast transition from planning to building with Lego, achievable within 2 h
Making sense	Seeing how documentation allows for control of project, applying PM theories to decisions, seeing Lego construction take shape and being completed, realising how to make a contribution to the team, lessons learned/debrief, questions for the practitioner or academic
Feedback	Identifying effect of contribution to team, successful completion of tasks, identifying effect of project control techniques, visual confirmation through the Lego structure, responses from team mates and other project groups, coaching and feedback from adults in process and at end, certificate of achievement
Verbalising	Use of speech in communicating within team and between teams, describing problems to adults, giving instructions, giving encouragement and directions, providing lessons learned, hearing each other's lessons learned
Assessing	Comparing finished artefact to the initial client brief, self-judgement of competence based on contribution to the team

used. Unlike the perpetual cycle of Kolb, the Race model for experiential learning has a definite start point—the want/need that sets the rest of the process in motion. Table 15.2 shows the Race categories and the matching pedagogic design points.

Verbalising and assessing stages are essential to drive the learning home, to ensure that the children are aware that they were not simply having fun with Lego; the Lego was just an intermediate stage on the way to completing the learning outcomes. Children will be directed to compare the final model to the original brief, budget and schedule—to determine if the work has been brought in on time, cost and quality—the very soul of PM. Central to this stage is the 'lessons learned' activity, allowing the children to make judgements of the competences that they have gained or demonstrated and to discuss how the game could be improved.

15.3 Observations from Operation of the Game

Table 15.3 shows how the teaching session was organised.

A short lecture introducing PM theory was given, with fun examples to engage the children, Barnes triangle, then Gantt charts, with the idea of having problems getting to school on time. Suggested innovations like getting dressed before

Table 15.3 How the simulation sits in the teaching session

30 mins	30 mins	30 mins	30 mins
Teaching session	Project preparation	Construction	Debrief
The children receive the business case and the brief The children are introduced to some basic PM concepts in scheduling/costs The children are put into teams and get familiar with their information and ask questions	The clock begins—1 min = 1 day—deadline is day 60 The teams collectively interpret the brief, collaborate over a possible design and schedule, checking decisions against brief and impact on schedule/budget Procurement orders are created	PM team places procurement orders and monitors progress against schedule Supplier team receives orders and dispatches components Construction teams obtain components from suppliers, assemble the chosen design	What went well, what did not—what did you learn about project management? What advice would you give future project managers? This can also be project over-run time if needed

washing or eating breakfast before we wake up kept laughs coming. Bottom-up vs. top-down costing was introduced via planning a pizza party and MoSCoW analysis using the wishlist of the client. Interestingly, most of the children already knew what a project and project management were, through their parent's work. After this, the game was introduced. The children (a group of nine, five girls, four boys), were divided into three teams on three separate tables: PM, supplier and construction/logistics.

The PM team dealt with incomplete information to produce a design based on the client scope and budget. Then they developed a fully costed schedule and wrote purchase orders for prefabricated components (mostly panels of blank wall, glazing or doors to be fixed together) to suit their interpretation of the brief. There was a concern that the paperwork would be too boring for them to deal with, but they enjoyed playing with the ready-made blocks on the Gantt chart (composed of lead time and fitting time) and debating the order of tasks and deliveries. One boy was thrilled to do all cost calculations by hand and got annoyed at offers of a calculator. They naturally de-prioritised fancy fittings and extensive glazing through MoSCoW analysis, the client wanted a lot of natural light—a 'should have'—but glass panels were the most expensive type of wall panel. They did this to make room for environmental features like the rooftop garden and aerogenerators, but staying within budget.

The supplier team had a price list based on realistic construction costs, which were set by a quantity surveyor from Beaumont Morgan. Only the supplier team knew these costs, so they had to be consulted on implications of the design choices. They prefabricated the main structural components and then 'sold' them to the PM team as they became available. With a pile of Lego in the room, the principal challenge of the supplier team was to keep everyone's hands off it until it was time to build, but the girls fiercely asserted that nothing left their premises without a signed

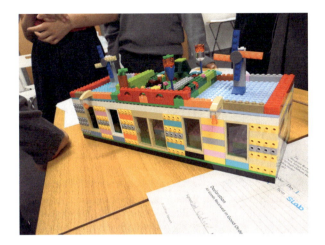

Fig. 15.1 The completed breakfast room

purchase order. They had the component lead time information, so they calculated when these would be completed and so made them available on time.

A construction and logistics team placed orders according to the PM team schedule, checked the goods ex-works, rejected anything that did not pass their quality tests and then had responsibility to transport them from the supplier to the site and assemble them. They probably had the most fun as only they were allowed to build. The PM team drifted over to stand behind them, helpfully shouting encouragement, brandishing the Gantt chart and reminding them what day it was on the schedule. There was no competition as such in the game, but the teams were competing against the sped-up clock. The PM team felt this very keenly and made sure construction team knew about it.

They successfully completed construction of their breakfast room 5 days ahead of schedule and under budget, as in Fig. 15.1. The 'lessons learned'/debrief was based on two questions. In response to 'What did you learn about project management?', the children said that PM was stressful, there was a lot of shouting and 'lots of paper and information which gets out of control easily'; suppliers seem to be able to 'lie back and wait for the money to flow in.' In response to 'What advice would you give project managers following your experience?', the advice of the children was that PMs should try to stay calm, do things ahead of schedule and should 'provide snacks for project workers when there are delays'. They exited the classroom very proud of their achievement certificate, holding it high.

15.4 How Skills Are Developed

It is impossible to claim that entirely new skills are developed after just 2 h of exposure to experiential construction project management education. However, certain behaviours could be observed, and it can be argued that the practice of these led to

Table 15.4 Skills or competencies acquired in the game

Skills and competencies experienced	How
Science	Observing the effects of project management
Technology	Use of PM techniques, i.e. Gantt charts, to control the project, use of appropriate nomenclature, awareness of prefabrication
Engineering	Making design choices from scope to realisation, organisation of production, organisation of supply chain, prefabrication
Mathematics	Calculation of costs, quantities and procurement, cost modelling of different designs, budgetary control, impact of lead times
Sustainability	Decision making around environmental construction options
Soft skills	Teamwork, communication, information management, decision-making, dealing with incomplete information, collaboration, responsibility, giving/receiving instructions, interacting with unknown adults, working under time pressure, resilience

successful completion of the task. To that extent then, certain skills were evidently practiced. Also, some activities, such as use of Gantt charts, were entirely new to the children. Completion was dependent on scheduling, so successful completion meant that this technical skill was successfully acquired to some extent. Table 15.4 shows the skills and competencies experienced and what activities in the game supported this learning.

15.5 The Interactions and Exchanges in Production of the Game

The practitioner, headteacher and academic together designed the learning outcomes.

- The headteacher
- Set the requirements for value of the activity
- Chose the children to participate—these were clearly hand-chosen children, bright, inquisitive, continually asking questions
- Provided a suitable room, furniture and a teacher in attendance

- The practitioner
- Reached out to the school
- Requested the help of the academic
- Advised on construction processes
- Provided the construction terminologies and costings
- Paid for the required Lego pieces
- Wrote and presented a brief presentation on what it was like to be in charge of making a building
- Gave encouragement to the children, especially when designing the wall panels

- The academic
- Created the game architecture
- Made decisions on the operation and the project management theories/documentation required
- Wrote and delivered the theory lecture and activities
- Purchased the Lego
- Provided some MSc student labour to help
- Gave support and advice to the practitioner in writing their presentation and ensuring its suitability for the audience

The role of the academic in the room was as per any other teaching context, to create a teaching experience and then to ensure effective participation in learning. There are differences between working with children rather than adults; there were more enthusiasm and spontaneous questioning. It was notable how the children encouraged each other and took pride in their role in their team.

The role of the practitioner in this kind of situation is under-researched. From direct observation in the class, it could be seen that the practitioner acted as a model of construction professional behaviour. There was a rapt audience for their anecdotes. The practitioner's story of controlling his apartment complex project was of interest; the children were aware of constructions in their neighbourhoods. They seemed to enjoy facts about size, scale complexity, decisions and difficult situations and talking about what they would do in those situations. There are positive ego and esteem factors for practitioners through sharing accounts of their successes and triumphs over adversity. The children responded to encouragement from practitioner, especially when the PM team was struggling to make design decisions/compromises. Through engagement in this activity, the practitioner also obtained evidence for CSR, outreach and claims in support of professional chartership. The practitioner now wants to be able to run the game themselves, to use it in subsequent outreach activities and so will need training in it from the academic. Interaction with the academic also gave the practitioner opportunity to reflect on their own practice and to observe interaction between education professionals from two different types of institution, getting a better sense of the concerns of headteachers which will enable him to speak their 'language' when making offers to other schools.

15.6 Future Developments

The game was made quite forgiving, due to concerns over resilience of the year 6 pupils. However, there are some options for making it more challenging:

- Inclusion of risk events—through a deck of cards, roll of the dice, or spin of a wheel
- Play with competing suppliers
- Play with competing projects, limited supply
- Lower budgets requiring more value engineering and client negotiation

The expanding role of 'social value' in construction projects means that major contractors are looking for ways to deliver £millions of benefit to the communities in which their construction projects are situated. Manchester City Council Construction Skills Network invited the author to present the game to the major construction companies in the area. From this meeting, Liang O'Rourke approached the author to build a version of this game for their corporate social responsibility department.

15.7 Conclusion

Throughout the design work the author was concerned, as the children had grown up digitally, would they still know how Lego 'worked'? There was no need for concern. Between the lines of the children's comments in the lessons learned/debrief, it was clear they had an experience which reflected actual practice. PM was a valid means of teaching and practicing life skills for children as per the APM strategy to have 'project management as a life skill for all'. It was notable that children appeared to have an awareness of PM through parental work and an intrinsic appreciation of deadlines and costs. They naturally prioritised sustainability considerations, finding ways to integrate aerogenerators and rooftop garden into the design and ensuring the budget would cover it. It was clearly possible to communicate the basic theory and techniques necessary to control a project. These included Barnes triangle, Gantt charts, top-down and bottom-up costing and MoSCoW analysis. Theories can be accepted if they are quickly demonstrated through practice.

Developing construction games and simulations is an excellent means to bring together practitioner and academic. Construction project managers charged with creating social value and academics looking for outreach activities have much to offer each other and together can create enriching innovative teaching and learning activities to engage future project managers and satisfy the CSR requirements placed on project organisations. The interactions identified in this paper show that each contributes unique value, the synergistic effect of which gets children excited about construction project management and hopefully creates lasting positive memories.

In terms of academic research in this area, although this paper is a case rather than a piece of research, it does raise questions worthy of future research. As regards the game, there are the classic questions that are asked of all teaching innovations—Does it actually lead to learning gain? Do the children retain the learning and PM skills after a week, a year? A further question is how this might be fair beyond bright hand-picked children—reaching out to groups that may not have considered a managerial role in the construction industry. To answer these would require post-test work and more lengthy interactions with primary schools. As regards the stakeholders, particularly the practitioners, what are their real motivations for participating in these kinds of activity, what is their subjective experience of working with an academic and what do they feel they gain by participating in this kind of outreach? The outreach experience from the point of view of the construction professional is under-researched. As regards this game/simulation as a type of outreach, has it changed any attitude towards the construction industry in the children? Answers to this question are the ultimate indicator of whether these kinds of activity succeed in stimulating a new generation of capable construction professionals.

References

Burgi, P., Bart, V., & Lentz, J. (2004). Case study: Modeling how their business really works prepares managers for sudden change. *Strategy & Leadership, 32*(2), 28–35.

de Bono, E. (1986). *Six thinking hats*. Viking.

Denholm, J. A., Protopsaltis, A., & de Freitas, S. (2012). Team-based mixed-reality (TBMR) games in higher education. *International Journal of Game-Based Learning, 3*(1), 18–33.

Geithner, S., & Menzel, D. (2016). Effectiveness of learning through experience and reflection in a project management simulation. *Simulation & Gaming, 47*(2), 228–256.

Hadida, A. (2013). Let your hands do the thinking! *Strategic Direction, 29*(2), 3–5.

Hayes, C., & Graham, Y. (2019). Understanding the building of professional identities with the LEGO® SERIOUS PLAY® method using situational mapping and analysis. *Higher Education, Skills and Work - Based Learning, 9*(1), 99–112.

Keys, B., & Wolfe, J. (1990). The role of management games and simulations in education and research. *Journal of Management, 16*(2), 307–336.

Konstantinou, E. (2015). Professionalism in project management: Redefining the role of the project practitioner. *Project Management Journal, 46*(2), 21–35.

Landry, J., Barnett, H., Chapman, D., & McCullough, R. (2019). Four strategies for driving a university pre-college computing outreach program. *Journal of Information Systems Education, 30*(3), 191–201.

Posner, B. (1987). What it takes to be a good project manager. *Project Management Journal, 18*(1), 51–54.

Race, P. (2015). *The lecturer's toolkit*. Routledge.

Race, P. (2005). *Making learning happen*. Sage.

Roffe, I. (1999). Innovation and creativity in organisations: A review of the implications for training and development. *Journal of European Industrial Training, 23*(4/5), 224–237.

Savelsbergh, C., Havermans, L., & Storm, P. (2016). Development paths of project managers: What and how do project managers learn from their experiences? *International Journal of Project Management Pergamon, 34*(4), 559–569.

Thavikulwat, P., & Pillutla, S. (2010). A constructivist approach to designing business simulations for strategic management. *Simulation & Gaming, 41*(2), 208–230.

Warner, J. (2019). K–12 STEM outreach: Sharing our profession with the next generation. *Institute of Transportation Engineers, 89*(8), 22–23.

Part V
Health Safety and Wellbeing

Chapter 16
Underground Utility Services on Irish Construction Projects: Current Work Practices and the Effectiveness of the Health and Safety Authority (HSA) Code of Practice

Shane Carmody, Michael Curran, and John P. Spillane

16.1 Introduction

Underground utility services are an inherent feature in the formation of any infrastructure project, and nearly all urban utilities are buried underground (Canto-Perello & Curiel-Esparza, 2013). Some basic functioning utilities include gas, electricity and telecommunications (Healey, 1995), and their demand is increasing dramatically in response to rapid urban growth and development of new communication technologies such as broadband (Jaw & Hashim, 2013). These services are crucial to the surrounding environment and everyday life, for example, providing the removal of waste through sewage ducts, communication through fibre-optic cables, provision of drinking water and supplying gas to enable heating and cooking in homes. Jaw et al. (2018) assert that these essential services have fostered the need for holistic, data-driven planning of underground spaces for the sustainable development of a city along with rapid urbanisation. Furthermore, Lester and Leonard (2007) note that this growing demand for utility services has resulted in much construction of new utility installations, maintenance and repair systems. However, Metje et al. (2015) argue that services can be extremely dangerous if the wrong methods and tools are incorrectly used, coupled with operative inexperience. Talmaki (2012) agrees that excavator operators undertaking tasks in the presence of underground utilities have the ever-present risk of striking buried utilities. A utility

---

S. Carmody · J. P. Spillane
School of Engineering, University of Limerick, Limerick, Ireland

M. Curran (✉)
School of Natural and Built Environment, Queens University Belfast, Belfast, UK
e-mail: mcurran23@qub.ac.uk

strike occurs when any element of the utility network infrastructure is hit, leading to damage during excavation (USAG, 2016).

In Ireland, the dangers of working with underground utility services have been acknowledged by its governmental statutory organisations. The Electricity Supply Board (ESB) (2013) has published guidance on 'Avoidance of Electrical Hazards when Digging/Drilling' and states that the possibility of serious injury or death is a very realistic factor when working near or with services. Also, the Health and Safety Authority (HSA) (2016) has issued a 'Code of Practice for Avoiding Danger from Underground Services'. However, on review, previous research acknowledging and highlighting the effectiveness of these guidelines is scant. Therefore, it is necessary to identify and evaluate why utility strikes are still occurring and causing disruption, if these guidelines are being utilised effectively. To address these issues and to fulfil a relatively new topic in the research area, it is paramount to acknowledge and generate results based on actual events that emerge, when considering these inherently complex environments. Thus, concentrating on this relevant facet of research, the aim of this study is to evaluate current work practices within the Irish underground utility service industry. Specifically, it aims to analyse services that are regulated and implemented by the HSA's Code of Practice and to determine if these practices impact safety on utility projects from a positive or negative perspective. This is achieved by undertaking a sequential mixed-method approach, incorporating a combination of qualitative techniques for analysis, including a literature review and semi-structured interviews and manually assessing the resultant data using coding techniques. The literature primarily focuses on the HSA's Code of Practice in Ireland whilst comparing its effectiveness in conjunction with similar guidelines in the United Kingdom (UK) and Australia. In addressing this aim, it is anticipated that this study will assist and aid on-site management and operative teams in identifying the risks when working with underground utility services and improving the safety of all stakeholders concerned on such projects in Ireland.

16.2 Underground Utility Services in Ireland, UK and Australia

Underground utilities are difficult to assess from the ground surface (Hao et al., 2012), and reliable and accurate information of underground utilities is crucial for planners and developers for underground spaces development (Jaw et al., 2018). Talmaki and Kamat (2014) argue that the problem of underground utility lines being struck by excavating equipment is long-standing and significant. As underground utilities cannot be physically seen, they are much more dangerous and susceptible to being struck or damaged compared to overhead services. However, there are a range of systems and methods to preventing these strikes from occurring. The following provides an extensive review of the HSA's Code of Practice in Ireland while comparing with similar guidelines in the UK and Australia.

16.2.1 Client Duties

In Ireland, the HSA's Code of Practice (2016) outlines strict guidelines based on construction regulations for the client to abide by when working with underground services. The Code of Practice states that the client has a legal duty to only employ competent designers and contractors and highlights that the client plays a very important role when it comes to safety and health on construction projects. However, in the UK, the Health and Safety Executive (HSE) (2014) place no such requirement on the client in their equivalent guidelines. Furthermore, the Australian Constructor's Association (ACA, 2014) lacks detail regarding the client. Thus, compared to these other jurisdictions, the Irish Code of Practice ensures much more involvement from the client and recognises their role and duty of care. Moreover, Metje et al. (2015) support that it is the duty of the client to record the occurrence of utility strike incidents in a variety of different ways as part of their health and safety reporting procedures.

16.2.2 Designer Duties

Comparing the Irish, UK, and Australian guidelines, the duties of the designer are very similar, where they are requested to always mitigate the dangers of underground services by methods of designing the risk out of the project or task. The HSA (2016) enforces strict guidelines requesting designers to formulate a preliminary health and safety plan if a particular risk is highlighted within a project (i.e. underground service), as defined in the Safety, Health and Welfare at Work (Construction) Regulations 2013 (HSA, 2013). The HSA also requires a designer to ensure that the most up-to-date as-built drawings are obtained and passed onto the contractor to notify them of any risks that may remain. Both the HSE (2014) and ACA (2014) require that the designer should only identify service risks and highlight the potential safety and constructability impact that will arise from these services. However, Metje et al. (2015) argue that a lack of care around services and lack of information are reasons why utility strikes continue to occur. Nevertheless, Whittaker (2019) believes that it is very easy to adjust the constructability of a project around dangers when they are highlighted.

16.2.3 Contractor Duties

The requirements of the contractor in Ireland are vague in comparison to that in the UK and Australia. The HSA (2016) guidelines request the contractor to ensure that risk assessments are completed and that the team members receive 'adequate training'. However, in the UK the contractor must ensure a safe system of work is in place, including clear information on the type of services that are in the area and the status of the services (HSE, 2014). The contractor must also state the method of work to be used to mitigate the risk of damaging the utility lines. A publication for working underground in New South Wales (2007) indicates that the principal contractor must prepare a health and safety management plan that is site specific whilst ensuring that site-specific risk assessments are completed on a continuous basis. Furthermore, the ACA (2014) requires the contractor to liaise with designers to complete any risk assessments that highlight potential work practices that may cause a strike.

16.2.4 Using as-Built Drawings

The implementation of relevant as-built drawings is advocated for in all codes of practice and safety guidelines across Ireland, the UK and Australia. Cazzanigaa et al. (2013) notes that the ability to gain knowledge on the whereabouts and the status of a service is extremely useful; however, Talmaki et al. (2013) argue that utility strikes are caused by incomplete, out-of- date and inaccurate data. For example, Shevlin (2011) supports that in the USA, many cities have underground utility lines that are over half a century old. HSE (2014) guidelines are extremely useful as they visually illustrate what a utility map may look like and how to interpret it. Conversely, the ACA (2014) guidelines lack any basic information regarding the interpretation of maps and drawings. Similarly, the HSA's (2016) Code of Practice includes no such diagrams of explanation of maps; however, it clearly highlights all the limitations of maps, which is an extremely important aspect to have included in safety guidelines. Talmaki and Kamat (2014) argue that the accuracy of as-built drawings for the location and depth of utility lines cannot be considered reliable, causing marking techniques to be approximate and inaccurate in nature (Bernold, 2005).

16.2.5 Excavation Techniques

Many failed excavation cases are reported worldwide every year (Jaw & Hashim, 2013), causing many serious accidents (Rogers et al., 2012). McMahon et al. (2005) state that more than four million holes are excavated in the UK every year, with approximately 60,000 utility strikes occurring (ICE, 2017). Furthermore, 500,000 utility strikes occur annually in the USA (Talmaki & Kamat, 2014). In Ireland and the UK, excavators are one of the most common methods of striking services (HSE, 2014; HSA, 2016), and Metje et al. (2015) claim that hand tools are also a very common cause of accidents if they are not used correctly. The ACA (2014) advises the best method to use is air lancing; however, this method is rarely used in Ireland or the UK. As a result, most utility strikes in Ireland happen as a result of plant items such as excavators and drilling machines, due to the lack of fine control and visibility that the operator has (Barhale, 2006). Makana et al. (2018) further argue that excavation techniques including no trial holes or slip trenches dug to confirm utilities, and the inappropriate and inadequate tools made available are also causal factors of utility strikes.

16.3 Research Method

This study is part of a primary investigation which aims to contribute to both industry and academia. Taking into consideration the theoretical stance and reasoning that this research is founded on, a critical realism approach is adopted. Also, as the nature of the study mainly concerns the opinions of human participants, the ontological approach is that of a subjectivist. On completion of an informative literature review, the research method adopted consists of six exploratory semi-structured interviews with six construction professionals, based within three Irish companies who specialise in extensive underground utility service works. All three companies operate mainly with earthworks and underground excavating for the installation of utility services. Guest et al. (2006) argue that data saturation can be achieved with six interviews, and these participants are selected based on criterion and convenience sampling strategies: firstly, by identifying their credentials and experiences with underground utility services and, secondly, by arranging interviews depending on the participant availability at a suitable time convenient to them. A semi-structured interview approach is chosen as it provides a conversational nature which allows a more flexible response from the interviewees (Lingard et al., 2008). Also, this method allows questions to lead from one to another, enabling the interviewee to provide as much information as possible (Curran et al., 2018).

The identities of the interviewees involved remain anonymous, and confidential information (such as company names, addresses, client details, etc.) is not be disclosed. To ensure non- bias within the study, the job title of each interviewee is uniform throughout the three companies (i.e. groundsmen, engineer or director) whenever possible. Consideration for ethical issues is also acknowledged, and each participant is informed of the nature of the research, its overall purpose and what the resultant data is used for. Experience within the underground utility service industry is a key factor when choosing potential participants and also a knowledge and awareness of the HSA's Code of Practice. The literature provides a range of key themes which in turn provides the basis of the questions for the interviews. All six interviews are recorded in handwritten note format, and the interviews took place on each site, where health and safety guidelines are adhered to, including the use of appropriate personal protective equipment (PPE).

16.4 Results and Analysis

The interviews commenced by gaining general background information about each participant and their respective project, followed by extensively discussing service industry issues and how safety can be maximised on work areas surrounding services. The semi-structured interview process allowed the operatives to voice their opinions about the industry and the problems that surround safe working practices. It should also be noted that all the participants (named Operatives 1 to 6) were highly involved in dealing with utility services and had witnessed a strike on a service during their industry careers. A range of operatives were interviewed, including groundsmen, engineers and directors. Table 16.1 illustrates the various factors that influence strikes.

The most common issues identified are 'No scanning', 'No as-built drawings', 'Key indicators not studied', 'Poor plant operator' and 'Poor communication'. Nevertheless, Table 16.2 illustrates a concise list of strategies on how to improve safety around services. The most common strategies include 'scan the full works area', 'accurate and on time as-builts' and 'communication between staff before work takes place'.

The final part of the interviews centred around their evaluation of the current HSA's Code of Practice and whether they felt that it is an effective guideline to prevent strikes from occurring on-site. All interviewees stated that they were familiar with the Code of Practice, and five of the six operatives believed that the Code is effective. It is worth documenting that the findings from the individual interviews are specific to this research, thus not a generalised view. Nevertheless, this study provides a foundation to advance and expand further, supporting continuous research into the HSA guidelines and safe working practices in the underground utility services industry in Ireland.

Table 16.1 Various factors that influence service strikes

	Case A		Case B		Case C	
(O = Operative)	O1	O2	O3	O4	O5	O6
Poor plant operator	x				x	x
Lack of warning tape	x	x				
No blinding material	x					
Cable not in location prescribed		x			x	
No scanning		x	x	x	x	x
No as-built drawings			x	x	x	x
Key indicators not studied			x	x	x	
Services not installed as per spec	x					
Poor communication			x	x		x
Lack of relevant experience				x	x	

Table 16.2 How to improve safety around services

(O = Operative)	Case A O1	O2	Case B O3	O4	Case C O5	O6
Hand dig around service	x			x		
More civil inspector's present on site	x		x			
Liability lies with installer	x					
Scan the full works area	x	x	x	x	x	x
Communication between staff before work takes place		x		x		
Service to be installed as per spec		x				x
Accurate and on time as-builts			x	x		x
Key indicators to be studied before work takes place			x			
Spotter to be in place at all times					x	
All services to be surveyed after changes take place						x

16.5 Discussion

16.5.1 Various Factors that Influence Service Strikes

Collectively, ten fundamental issues and problems were identified and discussed throughout the six interviews. The analysis undertaken identifies 'No scanning', 'No as-built drawings' and 'Key indicators not studied' as the most common factors that influence strikes. These were recurring themes across all six interviews, and many of the respondents shared almost identical views, even though they were based on different projects and employed by different companies. The HSA (2016) clearly states that it is the duty of the designer to provide up-to-date as-built drawings before any project commences. Thus, it is interesting to note why these are not being implemented by companies prior to a project commencing. Sărăcin (2017) supports that the precise knowledge of the positioning and routes of services has become a necessity in the service industry, which leaves no excuse for companies not to have access to as-built drawings.

Furthermore, the HSA (2016) advises the use of utility scanners on projects before any work commences. The Code provides further detail on each scanner and the limitations and benefits of each, as do the HSE (2014) and ACA (2014) guidelines. It is difficult to comprehend why issues such as no scanning and as-built drawings are huge contributors towards strikes on projects, particularly when a statutory code exists. The Code clearly states why and how scanners and maps should be used, yet this information appears to be ignored. Nevertheless, Makana et al. (2018) support that a common cause of utility strikes is not using appropriate geophysical methods to scan an underground area.

Another recurring theme was 'Key indicators not being studied'. Three of the operatives felt that this area needs to be explored further. The HSA guidelines only have an extremely brief section on what these key indicators might be and do not provide any significant information to the reader. However, the client's duty is clearly outlined, and it is stated that competent and experienced designers and contractors should be employed where applicable (HSA, 2016). One of the respondents stated that most issues can be minimised and eradicated by hiring experienced and competent contractors and designers; however, there is always a risk of human error. Nevertheless, another respondent remarked that if the Code is observed properly by operatives, then there should be minimal strikes occurring on-site.

16.5.2 How to Improve Safety around Services

Ten improvements and strategies can be implemented to increase safety around underground utility services. All six operatives declared that 'Scanning the full works area' is paramount to improve safety. The HSE (2014) and HSA (2016) actively encourage the use of scanners to locate services within the work area. 'Accurate and on time as-built drawings' was another important strategy identified by the respondents, as Costello et al. (2007) note that the accuracy of as-builts can be very misleading at times. Also, the time it takes to access certain types of as-built drawings was a concern for some of the interviewees, as this has the potential to stop work from commencing. Fibre-optic cable as-built drawings can take nearly a whole month before the operative receives them on-site. HSA (2016) guidelines state that it is up to the service providers' discretion to do everything that is reasonably practical to ensure such information is made available to whoever requires it.

The operatives argued that this is a grey area for utility companies as the Code of Practice suggests that it is the designer's duty to ensure that the most up-to-date as-built drawings are passed onto the contractor. While this duty should be fulfilled, it still does not address the issue of the length of time to access as-built drawings. It is interesting to note that in an emergency situation, companies are advised to undertake work assuming services are in the area. Furthermore, if as-built drawings cannot be obtained, the HSA (2016) advises that the utility company requests a representative on-site to help locate services. Two operatives believed that 'More civil inspectors present on site' would maximise the safety around services as projects commence. The role of the inspector helps to identify the service when excavation takes place and its approximate location before excavating. The civil inspector would also sign off on any work done to the service, which in turn increases the chances of the 'Service being installed as per specification'. These factors should dramatically increase the safety around services and thus reduce strike rates. It is not a requirement to have civil inspectors on site when work is undertaken; however, one of the directors interviewed argued that it would be in the best interests of the company to request that an inspector is present on-site when work is taking place. Makana et al. (2016) discuss the development of PAS 128 (BSI, 2014), a new standard specification for underground utility detection, verification and location, which aims to address some of these shortcomings identified.

16.5.3 The HSA Code of Practice

The interviews concluded with a discussion on the HSA's Code of Practice for Avoiding Danger from Underground Services. All six interviewees were familiar with the Code, and five of them understood its purpose to aid potential users within the industry. Two of the respondents believed it could be improved in certain ways. One of the main improvements suggested is the inclusion of a section based on key indicators and how they can be clearly identified. Experienced and competent contractors should understand what key indicators are and how they can be utilised to prevent strikes from occurring; however, some guidance is still necessary. One respondent suggested that it would take an inexperienced operative far less time to understand what key indicators are from using a guidance document, rather than weeks or months of on-site experience. Nevertheless, there is no doubt that on-site experience and guidance from more experienced operatives are vital to young operatives, but it still takes considerable time to provide the education and training.

Another theme conveyed during the interviews was that the Code of Practice is struggling to keep up with new trenchless technologies that now currently exist. Based on the literature research, the HSA's Code of Practice (2016) has no guidance on trenchless technologies which creates a potentially serious issue for utility companies. Although these practices are uncommon in Ireland at present, it is more than likely that companies will implement new and improved methods in the not too distant future. The main excavation techniques currently being utilised are via excavators or hand tools, but one respondent acknowledged that if trenchless technologies are to advance in Ireland, then the current HSA guidelines should be updated and streamlined to include information on these technological advancements.

16.6 Conclusion and Recommendations

Essentially, this exploratory study focuses on the impact that the HSA's Code of Practice for Avoiding Danger from Underground Services in Ireland has on projects which are undertaken within the underground utility service industry in Ireland. Underground utility services are an inherent feature in the formation of any infrastructure project, and these services are crucial to the surrounding environment and everyday life. Therefore, site operatives and management teams are tasked with ensuring that groundworks are conducted safely, and the threat of underground strikes is minimised through the early identification of issues and adoption of resultant strategies to counteract the issues identified. Considering the results captured from the six interviews and data analysis, various issues within the underground utility service industry which contribute to strikes occurring in Ireland include 'No scanning', 'No as-built drawings', 'Key indicators not studied' and 'Poor communication between operators'. Some of the various strategies identified to counteract these issues and to improve safety around services include 'Scanning the full works area', 'Accurate and on time as-built drawings' and 'More civil inspectors on site'. Moreover, the results indicate that the HSA's Code of Practice is effective in preventing strikes from occurring, as five out of the six operatives agreed that it was a helpful aid to them while working on underground utility services.

However, the findings established from the interviews and data analysis are specific to this research, and only a concise, subjective view of the topic is produced, not a generalised one. Nevertheless, this study provides a foundation to advance and expand further, supporting continuous research into HSA guidelines for underground utility services in Ireland. Many of the issues that exist within the underground utility service industry in Ireland are very clearly articulated and highlighted in the HSA guidelines; however, it is evident that they are not being followed properly, or at all in some extreme cases, by operatives on-site. Based on the research, it is evident that the Code does assist in the prevention of strikes, but it is not being utilised to its full potential. The findings in this paper can be developed further, and it is anticipated that a broader analytical context can be addressed in a subsequent journal publication, where additional theoretical points of departure and areas of discussion can be articulated. To gain a richer understanding of current work practices and the effectiveness of the guidelines in these inherently risky environments, alternative qualitative research methods can be implemented in further research such as action research and ethnography.

It is recommended that more individual interviews and focus group seminars are considered for qualitative analysis, and a sequential selection strategy is incorporated using criterion selection, such as quota and random sampling. From a quantitative perspective, a questionnaire survey should be composed and distributed to a larger sample to further strengthen the research. There is also an opportunity to develop and update the current HSA's Code of Practice and introduce compulsory elements such as training and educational programmes for new, inexperienced site operatives. Furthermore, underground utility service operatives should hold a specific safety card prior to undertaking any groundworks, similar to the Safe Pass card held by general construction operatives. Nevertheless, this study provides a foundation for informing and confirming the validity and necessity of the research and ensuing investigation going forward. Overall, the key contribution of this research reinforces that the HSA's Code of Practice is effective in promoting safety in the underground utility services industry in Ireland. However, the research also highlights that the Code of Practice is not being followed and utilised effectively by operatives in the industry and provides recommendations for appropriate implementation and use.

References

Australian Constructors Association (ACA). (2014). *Construction and building industry safety guideline underground services* (pp. 1–17). Australian Constructors Association.

Barhale. (2006). *Tool box talk—Underground services*. http://intranet.scottishwatersolutions.co.uk/portal/page/portal/SWS_PUB_HEALTH_AND_SAFETY/SWSE_PG_HEALTH_AND_SAFETY/BP%20Information/Site%20Mobiliztion/Barhale%20-%20TBT%20Avoiding%20Underground%20Services.pdf

Bernold, L. E. (2005). Accident prevention through equipment mounted buried utility detection. In *Trends and current best practices in construction safety and health*.

BSI (British Standards Institution). (2014). *PAS 128:2014 specification for underground utility detection, verification and location*. BSI.

Canto-Perello, J., & Curiel-Esparza, J. (2013). Assessing governance issues of urban utility tunnels. *Tunnelling and Underground Space Technology, 33*, 82–87.

Cazzanigaa, N. E., Carriona, D., Migliaccioa, F., & Barzaghia, R. (2013). A shared database of underground utility lines for 3D mapping and GIS applications. *ISPRS-International Archives of the Photogrammetry, Remote Sensing and Spatial Information Sciences, 1*, 105–108.

Costello, S., Chapman, D., Rogers, C., & Metje, N. (2007). Underground asset location and condition assessment technologies. *Tunnelling and Underground Space Technology, 22*(5–6), 524–542.

Curran, M., Spillane, J., & Clarke-Hagan, D. (2018). External stakeholders in urban construction development projects: Who are they and how are they engaged? In C. Gorse & C. J. Neilson (Eds.), *Proceeding of the 34th annual ARCOM conference, 3–5 September 2018* (pp. 139–148). Association of Researchers in Construction Management.

Electricity Supply Board (ESB). (2013). *Avoidance of electrical hazards when digging/drilling* (pp. 1–9). The ESB Networks.

Guest, G., Bunce, A., & Johnson, L. (2006). How many interviews are enough? An experiment with data saturation and variability. *Field Methods, 10*(18), 59–82.

Hao, T., Rogers, C. D. F., Metje, N., Chapman, D. N., Muggleton, J. M., Foo, K. Y., Wang, P., Pennock, S. R., Atkins, P. R., Swingler, S. G., & Parker, J. (2012). Condition assessment of the buried utility service infrastructure. *Tunnelling and Underground Space Technology, 28*, 331–344.

Healey, P. (1995). *Managing cities: The new urban context* (1st ed., pp. 374–375). Chichester: Wiley.

Health and Safety Authority (HSA). (2013). *Safety, health and welfare at work (construction) regulations 2013*. https://www.hsa.ie/eng/Legislation/Regulations_and_Orders/Construction_Regulati ons_2013/

Health and Safety Authority (HSA). (2016). *Code of practice for avoiding dangers from underground services* (pp. 1–51). The Health and Safety Authority.

Health and Safety Executive (HSE). (2014). *Avoiding danger from underground services* (pp. 1–40). The Health and Safety Executive.

ICE (Institute of Civil Engineers). (2017). *ICE – Mock trial, service strike*. https://www.aps.org.uk/events/ice-mock-trial-service-strike

Jaw, S. W., & Hashim, M. (2013). Locational accuracy of underground utility mapping using ground penetrating radar. *Tunnelling and Underground Space Technology, 35*, 20–29.

Jaw, S. W., Van Son, R., Soon, V. K. H., Schrotter, G., Kiah, R. L. W., Ni, S. T. S., & Yan, J. (2018). The need for a reliable map of utility networks for planning underground spaces. In *Proceeding of the 17th international conference on ground penetrating radar (GPR)* (pp. 1–6). IEEE.

Lester, J., & Leonard, B. E. (2007). Innovative process to characterize buried utilities using ground penetrating radar. *Automation in Construction, 16*(4), 546–555.

Lingard, H., Townsend, K., Bradley, L., & Brown, K. (2008). Alternative work schedule interventions in the Australian construction industry: A comparative case study analysis. *Construction Management and Economics, 26*(10), 1101–1112.

Makana, L., Metje, N., Jefferson, I., & Rogers, C. (2016). *What do utility strikes really cost?* A Report by the University of Birmingham – School of Civil Engineering.

Makana, L. O., Metje, N., Jefferson, I., Sackey, M., & Rogers, C. D. (2018). Cost estimation of utility strikes: Towards proactive management of street works. *Infrastructure Asset Management, 7*(2), 64–76.

McMahon, W., Burtwell, M. H., & Evans, M. (2005). *Minimising street works disruption: The real costs of street works to the utility industry and society*. UK Water Industry Research.

Metje, N., Ahmad, B., & Crossland, S. M. (2015). Causes, impacts and costs of strikes on buried utility assets. In *Proceedings of the institution of civil engineers-municipal engineer* (Vol. 168, No. 3, pp. 165–174), Thomas Telford Ltd.

New South Wales Government. (2007). *Work near underground assets – Guide* (pp. 1–55). New South Wales Government.

Rogers, C. D. F., Hao, T., Costello, S. B., Burrow, M. P. N., Metje, N., Chapman, D. N., Parker, J., Armitage, R. J., Anspach, J. H., Muggleton, J. M., & Foo, K. Y. (2012). Condition assessment of the surface and buried infrastructure – A proposal for integration. *Tunnelling and Underground Space Technology, 28*, 202–211.

Sărăcin, A. (2017). Using geo radar systems for mapping underground utility networks. *Procedia Engineering, 209*, 216–223.

Shevlin, T. (2011). *What a water main break tells us about the city's aging infrastructure*. https://www.newport-now.com/2011/03/25/what-a-water-main-break-tells-us-about-the-city's-aging-infrastructure/

Talmaki, S. A. (2012). *Real-time visualization for prevention of excavation related utility strikes*. Doctoral dissertation, The University of Michigan.

Talmaki, S., & Kamat, V. R. (2014). Real-time hybrid virtuality for prevention of excavation related utility strikes. *Journal of Computing in Civil Engineering, 28*(3), 04014001.

Talmaki, S., Kamat, V. R., & Cai, H. (2013). Geometric modeling of geospatial data for visualization-assisted excavation. *Advanced Engineering Informatics, 27*(2), 283–298.

Utility Strike Avoidance Group (USAG). (2016). *2014 utility strike damages report*. USAG.

Whittaker, J. (2019). *Importance of underground utility detection and mapping* [online]. Medium, https://medium.com/@aimlocatingva/importance-of-underground-utility-detectionand-mapping-3d842f88ff51

Chapter 17
Health and Safety Practices and Performance on Public Sector Projects: Site Managers' Perceptions

Nomakhwezi Mafuya and John Smallwood

17.1 Introduction

The Construction Industry Development Board (CIDB) (2009) highlights the considerable number of accidents, fatalities, and other injuries that occur in the South African construction industry in their report 'Construction Health & Safety Status & Recommendations'. The report also cited the high level of non-compliance with H&S legislative requirements, which the CIDB contends is indicative of a deficiency of effective management and supervision of H&S on construction sites as well as planning from the inception/conception of projects within the context of project management. Furthermore, the report also cited a lack of sufficiently skilled, experienced, and knowledgeable persons to manage H&S on construction sites.

The Construction Regulations, which constitute the primary regulations in terms of managing H&S in the South African construction industry, allocate a range of H&S responsibilities to clients, designers, quantity surveyors, principal contractors, and contractors (Republic of South Africa, 2014).

Given the status of H&S in South African construction, the requirements of H&S legislation, the functions of SMs, and the status of H&S on ECDRPW's projects, a study was conducted among ECDRPW's projects' contractors' SMs, the aim being to assess H&S-related experiences and practices on ECDRPW's projects. The objectives were to determine the:

- Frequency at which 16 H&S actions are undertaken;
- Frequency of H&S training;
- Frequency at which four ergonomics problems are encountered;

N. Mafuya · J. Smallwood (✉)
Department of Construction Management, Nelson Mandela University, Port Elizabeth, South Africa
e-mail: John.Smallwood@mandela.ac.za

- Extent to which ten factors contribute to exposure to hazards on construction sites, and
- Frequency at which 24 factors are the cause of construction accidents.

17.2 Review of the Literature

17.2.1 Health and Safety Legislation

In terms of the South African Construction Regulations (Republic of South Africa (RSA) (2014), clients are required to, inter alia, prepare an H&S specification based on their baseline risk assessment (BRA), which is then provided to designers. Designers in turn are required to, inter alia: consider the H&S specification; submit a report to the client before tender stage that includes all the relevant H&S information with respect to the design that may affect the pricing of the work, the geotechnical-science aspects, and the loading that the structure is designed to withstand; inform the client of any known or anticipated dangers or hazards relating to the construction work and make available all relevant information required for the safe execution of the work upon being designed or when the design is changed; modify the design or make use of substitute materials where the design necessitates the use of dangerous procedures or materials hazardous to H&S; and consider hazards relating to subsequent maintenance of the structure and make provision in the design for that work to be performed to minimise the risk. Hazard identification and risk assessment (HIRA) and appropriate responses in the form of amendments to design, details, and specification are necessary to mitigate design originated hazards, which process should be structured and documented.

The H&S specification, which in theory should have been revised to include any relevant H&S information included in the designer report, must then be included in the tender documentation by clients. Thereafter, clients must, inter alia: ensure that potential principal contractors (PCs) have made provision for the cost of H&S in their tenders; ensure that the PC to be appointed has the necessary competencies and resources; ensure that every PC is registered for workers' compensation insurance cover and in good standing; discuss and negotiate with the PC the contents of the PC's H&S plan and thereafter approve it; take reasonable steps to ensure that each contractor's H&S plan is implemented and maintained; ensure that periodic H&S audits and documentation verification are conducted at agreed intervals, but at least once every 30 days; ensure that the H&S file is kept and maintained by the PC, and appoint a competent person in writing as an agent when a construction work permit is required.

Key contractor-related interventions in terms of the Occupational Health and Safety Act (RSA, 1993) and the Construction Regulations (RSA, 2014) include HIRA, which has implications in terms of awareness and competencies in the form of knowledge and skills. Consequently, H&S information and training are critical for managers, supervisors, and workers. Toolbox talks can inform and raise the level of awareness, H&S induction informs with respect to key H&S issues on sites such as hazards, daily H&S task instructions inform and caution workers with respect to hazards related to activities, and HIRA training empowers supervisors and workers to conduct HIRA. Safe work procedures (SWPs) guide supervisors and workers in terms of undertaking activities in a healthy and safe manner and should follow HIRA. Therefore, adequate supervision is necessary to ensure that SWPs are followed and that any personal protective equipment (PPE) required is in fact worn.

17.2.2 Ergonomics Problems

Based upon previous studies conducted in South Africa, the following were identified as the top ten ergonomics problems: repetitive movements; climbing and descending; handling heavy materials; use of body force; exposure to noise; bending or twisting the back; reaching overhead; reaching away from the body; working in awkward positions; and handling heavy equipment (Smallwood, 1997, 2002; Smallwood et al., 2000).

17.2.3 Contributing Factors in Accidents

A hundred accidents were reviewed during a study conducted by Haslam et al. (2005).

The first category includes 'immediate accident circumstances and shaping factors'. Problems arising from workers, which include all site-based personnel, or the work team, especially worker actions or behaviour and worker capabilities, were judged to have been involved in over 70% of the accidents. Workplace factors, most notably poor housekeeping and problems with the site layout and space availability, were considered to have contributed 49% of the accident studies. Local hazards on-site were a feature in many of the 100 accidents reviewed. Shortcomings with equipment, including personal protective equipment (PPE), were identified in 56% of the incidents.

The second category are 'originating influences', in which case inadequacies with risk management were considered to have been present in 94% of the accidents. In terms of 'construction design and processes', Haslam et al. state that the elimination or reduction of risks through design or alternative methods of construction is highly desirable. In terms of 'project management', a clear influence from problems with project management was identified in only a quarter of the accident studies, although this is likely to have been because the precise effects are difficult to corroborate. In terms of 'risk management', deficiencies related thereto were identified in most of the 100 accidents studied. Accidents invariably involve an inadequately controlled risk, which is indicative of a failing of management. In terms of 'client and economic influences', there was limited direct evidence of the influence of client requirements or the economic climate on the accidents studied, although they undoubtedly do affect construction H&S. The findings relative to 'H&S education and training' were three-fold: H&S being overlooked in the context of heavy workloads and other priorities; taking shortcuts to save effort and time, and inaccurate perception of risk. Underlying each of these are inadequate H&S knowledge, which indicates deficiencies with H&S education and training.

17.3 Research

17.3.1 Research Method and Sample Stratum

The study adopted a quantitative approach and focused on the seven school construction projects underway in the Sarah Baartman District of the Eastern Cape province administered by the ECDRPW and included, inter alia, 40 contractors' SMs that were involved with the projects. The SMs were surveyed using a self-administered questionnaire, which was deemed appropriate due to the geographical location of the school projects and the fact that respondents were SMs. The questionnaire consisted of 22 questions and 6 sub-questions, 21 of which were closed-ended and 7 open-ended. Six of the close-ended questions were demographic, and ten were Likert scale-type questions. Thirty-three responses were received and included in the analysis of the data, which equates to a response rate of 82.5%. The analysis of the data entailed the computation of descriptive statistics in the form of frequencies and a measure of central tendency in the form of a mean score (MS).

17.3.2 Research Results

57.6% of respondents have worked for their current employer ≤ 1 year, 33.3% for $> 1 \leq 5$ years, and 9.1% for > 5 years. The mean length of time respondents had worked for their current employer is 2.3 years. 39.4% of respondents have worked in construction $> 1 \leq 5$ years, followed by 24.2% for > 20 years, 15.2% for ≤ 1 year, 12.1% for $> 10 \leq 20$ years, and 9.1% for $> 5 \leq 10$ years. The mean length of time respondents had worked in construction is 11.9 years. In summary, 84.8% of respondents have worked in construction for > 1 year, 45.4% for > 5 years, and 36.3% > 10 years. Therefore, the respondents can be deemed experienced, which contributes to the reliability of the findings.

36.4% of respondents were ≤ 30 years of age, followed by 27.3% $> 30 \leq 40$, followed by 15.2% relative for each of $> 40 \leq 50$ and $> 50 \leq 60$, and only 6.1% > 60 years. In summary, 63.7% were ≤ 40 years of age, and 36.5% were > 40 years. Per definition relative to workers, people > 40 years of age are 'older workers'. The mean age was 37 years. 39.4% of respondents were female, and

60.6% were male. In terms of qualifications, 53.1% of the respondents have a National Diploma or higher qualification.

Table 17.1 presents the frequency at which 16 H&S actions are undertaken on ECDRPW's projects in terms of percentage responses to a frequency range never to always, and MSs ranging between 1.00 and 5.00.

The relatively high level of response relative to the 'unsure' option in many cases is attributable to the respondents being SMs, and not necessarily in the position to respond with certainty. However, in many cases they could have deduced the status quo. For example, the frequency that first ranked 'ensure that the designer takes the H&S specification into consideration during design stage' is the case, which attracted 28.1% unsure response, could have been deduced by perusing the PC's version of the H&S specification, and deliberation of the realities reflected in the design and details and on-site-related realities. However, it would have been more difficult to deduce the frequency that 'assemble H&S expertise at design development stage' is the case, which attracted 40.6% unsure response, although an inquiry directed to the CHSA, if one was appointed, would provide an indication.

It is notable that all the actions have MSs >3.00, which indicates that in general, the actions can be deemed to be taken frequently, as opposed to infrequently.

It is also notable that 15/16 (93.8%) MSs are >4.20 ≤ 5.00 – between often and always/always. The top 5/15 (33.3%) MSs are in the upper half of the range, namely >4.60 ≤ 5.00 – 'ensure that the designer takes the H&S specification into consideration during design stage', 'assess whether the PC to be appointed has the H&S competencies and resources to carry out the construction work safely', 'provide the H&S specification to contractors at bidding stage', 'monitor the development of the H&S file', and 'ensure that the H&S specification is specific to the activities to be undertaken'. These are notable findings as they are requirements of the Construction Regulations, and indicate that the intended integration of design, procurement, construction, and use phase are being realised. The first five actions in this range are followed by 'BoQ include H&S items', 'ensure adequate provision for H&S is reviewed prior to awarding the project', 'approve the H&S plan relative to the H&S specification and Construction Regulations', 'assemble H&S expertise at design development stage', 'receive the H&S plan from the contractor prior to commencing work on site', 'ensure the baseline risk assessment (BRA) is provided to the designer(s) prior to commencement of the bidding process', 'monitor construction activities relative to design HIRAs', 'monitor construction activities relative to the H&S plan', 'discuss and negotiate with the PC the contents of the plan at the end of the review process', and 'ensure the design HIRA process is documented prior to commencement of the bidding process'. Most of these are requirements of the Construction Regulations and underscore that the intended integration of design, procurement, and construction are being realised. However, the facilitation of adequate financial provision for H&S has been a contentious issue.

The last ranked action 'revisit the process if the design changes at any point' has a MS > 3.40 ≤ 4.20, which indicates it can be deemed to be taken between sometimes and often/often. However, the 4.12 MS is very close to the upper limit of the range.

Table 17.1 Frequency at which 16 H&S actions are undertaken on ECDRPW's projects

Action	Response (%) Unsure	Never	Rarely	Sometimes	Often	Always	MS	Rank
Ensure that the designer takes the H&S specification into consideration during design stage	28.1	0.0	0.0	0.0	6.3	65.6	4.91	1
Assess whether the PC to be appointed has the H&S competencies & resources to carry out the construction work safely	15.6	0.0	0.0	3.1	9.4	71.9	4.81	2
Provide the H&S specification to contractors at bidding stage	18.2	3.0	0.0	0.0	18.2	60.6	4.63	3
Monitor the development of the H&S file	15.6	3.1	3.1	0.0	9.4	68.8	4.63	4
Ensure that the H&S specification is specific to the activities to be undertaken	9.4	0.0	3.1	6.3	12.5	68.8	4.62	5
BoQ include H&S items	6.5	0.0	6.5	6.5	9.7	71.0	4.55	6
Ensure adequate provision for H&S is reviewed prior to awarding the project	12.5	0.0	3.1	9.4	12.5	62.5	4.54	7
Approve the H&S plan relative to the H&S specification and construction regulations	18.2	0.0	6.1	0.0	21.2	54.5	4.52	8
Assemble H&S expertise at design development stage	40.6	0.0	3.1	3.1	15.6	37.5	4.47	9
Receive the H&S plan from the contractor prior to commencing work on site	9.1	6.1	3.0	3.0	9.1	69.7	4.47	10
Ensure the baseline risk assessment (BRA) is provided to the designer(s) prior to commencement of the bidding process	25.0	3.1	3.1	3.1	12.5	53.1	4.46	11
Monitor construction activities relative to design HIRAs	21.2	0.0	3.0	3.0	30.3	42.4	4.42	12
Monitor construction activities relative to the H&S plan	12.1	0.0	6.1	6.1	24.2	51.5	4.38	13
Discuss and negotiate with the PC the contents of the plan at the end of the review process	18.2	3.0	6.1	0.0	24.2	48.5	4.33	14
Ensure the design HIRA process is documented prior to commencement of the bidding process	21.2	3.0	6.1	6.1	15.2	48.5	4.27	15
Revisit the process if the design changes at any point	24.2	0.0	6.1	9.1	30.3	30.3	4.12	16

Table 17.2 presents the frequency of H&S training in terms of percentage responses to a range annually to daily, and MSs ranging between 0.00 and 7.00, the midpoint of the range being 3.50. It is notable that all the MSs are above the midpoint of 3.50, which indicates that in general, the respondents' organisations can be deemed to present the training frequently, as opposed to infrequently.

It is notable that the MSs of Daily Safe Task Instructions (DSTIs), general H&S training, and 'toolbox talks' are > 6.13 ≤ 7.00 – between weekly and daily/daily.

The MS for hazard identification and risk assessment (HIRA) is > 4.39 ≤ 5.26, which indicates that it can be deemed to be conducted between monthly and fortnightly/fortnightly.

The MS for H&S induction is > 3.51 ≤ 4.39, which indicates it can be deemed to be presented between quarterly and monthly/monthly. This is not unexpected, as H&S induction is presented to persons previously unexposed to sites.

Table 17.3 presents the frequency at which four ergonomics problems are encountered on projects in terms of percentage responses to a frequency range never to daily, and MSs ranging between 1.00 and 5.00. It is notable that all the MSs are > 3.00, which indicates that in general the ergonomics problems can be deemed to be encountered frequently, as opposed to infrequently.

It is notable that the MS of repetitive movements is > 4.20 ≤ 5.00, which indicates it can be deemed to be encountered between 'daily to weekly' and daily/daily.

3/4 (75.0%) of the problems' MSs are > 3.40 ≤ 4.20, which indicates use of body force, handling heavy material/equipment, and climbing and descending can be deemed to be encountered between 'weekly to fortnightly' and 'daily to weekly'/'daily to weekly'.

Table 17.4 indicates the extent to which ten factors contribute to exposure to hazards on construction sites in terms of percentage responses to a scale of 1 (minor) to 5 (major), and a MS ranging between 1.00 and 5.00.

It is notable that all the MSs are above the midpoint score of 3.00, which indicates that in general the respondents can be deemed to perceive all factors contribute to exposure to hazards on construction sites.

It is notable that no MSs are > 4.20 ≤ 5.00 – factors contribute between a near major to major and major extent to exposure to hazards on construction sites.

However, 6/10 (60.0%) MSs are > 3.40 ≤ 4.20, which indicates the factors contribute between some extent to a near major/near major extent – inadequate H&S management, inadequate H&S supervision, inadequate H&S experience, inadequate H&S knowledge, inadequate awareness, and inadequate H&S expertise.

The remaining 4/10 (40.0%) MSs are > 2.60 ≤ 3.40, which indicates the factors contribute between a near minor extent and some extent/some extent – inadequate health promotion, inadequate H&S culture, inadequate PPE programmes, and inadequate training.

Table 17.5 presents the frequency at which 24 factors are the cause of construction accidents in terms of percentage responses to a frequency range never to always, and MSs ranging between 1.00 and 5.00.

Table 17.2 Frequency of H&S training on ECDRPW's projects

Type	Response (%) Unsure	Never	Annually	Bi-annually	Quarterly	Monthly	Fortnightly	Weekly	Daily	MS	Rank
DSTIs	10.0	0.0	3.3	0.0	0.0	3.3	3.3	6.7	73.3	6.52	1
General H&S	21.9	0.0	3.1	3.1	0.0	0.0	3.1	12.5	56.3	6.32	2
Toolbox talks	6.3	0.0	0.0	0.0	0.0	0.0	0.0	68.8	25.0	6.27	3
HIRA	32.1	0.0	3.6	0.0	7.1	32.1	3.6	3.6	17.9	4.68	4
H&S induction	9.4	0.0	21.9	6.3	9.4	12.5	6.3	15.6	18.8	4.07	5

Table 17.3 Frequency at which four ergonomics problems are encountered

Problem	Response (%) Unsure	Never	Fortnightly to monthly	Weekly to Fortnightly	Daily to weekly	Daily	MS	Rank
Repetitive movements	12.5	0.0	3.1	3.1	28.1	53.1	4.50	1
Use of body force	12.1	6.1	9.1	3.0	45.5	24.2	3.83	2
Handling heavy material/equipment	6.1	6.1	12.1	21.2	36.4	18.2	3.52	3
Climbing and descending	6.1	9.1	15.2	12.1	33.3	24.2	3.52	4

Table 17.4 Extent to which ten factors contribute to exposure to hazards on construction sites

Factor	Response (%) Unsure	Minor 1	2	3	4	Major 5	MS	Rank
Inadequate H&S management	0.0	6.5	22.6	9.7	22.6	38.7	3.65	1
Inadequate H&S supervision	0.0	9.1	18.2	9.1	27.3	36.4	3.64	2
Inadequate H&S experience	0.0	12.1	12.1	18.2	21.2	36.4	3.58	3
Inadequate H&S knowledge	0.0	15.2	6.1	24.2	15.2	39.4	3.58	3
Inadequate awareness	0.0	13.3	6.7	30.0	16.7	33.3	3.50	5
Inadequate H&S expertise	0.0	9.1	21.2	15.2	21.2	33.3	3.48	6
Inadequate health promotion	0.0	15.6	12.5	15.6	31.3	25.0	3.38	7
Inadequate H&S culture	3.3	10.0	23.3	16.7	20.0	26.7	3.31	8
Inadequate PPE programmes	0.0	21.2	9.1	24.2	15.2	30.3	3.24	9
Inadequate training	0.0	12.5	28.1	18.8	25.0	15.6	3.03	10

It is notable that only 2/24 (8.3%) factors have MSs > 3.00, which indicates that in general, employee negligence and inadequate worker participation can be deemed to be frequently the cause of construction accidents, as opposed to infrequently.

It is notable that none of the factors' MSs are > 4.20 ≤ 5.00 – between often and always/always.

8/24 (33.3%) MSs are > 2.60 ≤ 3.40, which indicates the factors are rarely to sometimes/sometimes the cause of construction accidents – employee negligence, inadequate worker participation, unsafe acts, non-compliance with H&S regulations, non-compliance to SWPs, unskilled workers, poor housekeeping, and non-compliance to SOPs.

The remaining 16/24 (66.7%) MSs are > 1.80 ≤ 2.60, which indicates the factors are between never and rarely/rarely the cause of construction accidents. 5/16 (31.3%) of the MSs are in the upper half of the range, namely >2.20 ≤ 2.60 – unsafe conditions, inadequate H&S training, inadequate H&S supervision, inadequate monitoring of construction activities relative to the H&S plan, and work factors. The remaining 11/16 (68.7%) MSs are in the lower half of the range, namely > 1.80 ≤ 2.20 – inadequate construction HIRAs, inadequate risk management, poor

Table 17.5 Frequency at which factors are the cause of construction accidents

Factor	Unsure	Never	Rarely	Sometimes	Often	Always	MS	Rank
Employee negligence	9.1	6.1	12.1	42.4	18.2	12.1	3.20	1
Inadequate worker participation	3.1	9.4	18.8	31.3	34.4	3.1	3.03	2
Unsafe acts	3.2	9.7	32.3	19.4	25.8	9.7	2.93	3
Non-compliance with H&S regulations	9.1	12.1	33.3	12.1	21.1	12.1	2.87	4
Non-compliance to SWPs	18.2	12.1	15.2	36.4	12.1	6.1	2.81	5
Unskilled workers	6.1	18.2	21.2	24.2	24.2	6.1	2.77	6
Poor housekeeping	6.3	12.5	28.1	34.4	9.4	9.4	2.73	7
Non-compliance to SOPs	12.5	9.4	37.5	25.0	9.4	6.3	2.61	8
Unsafe conditions	3.1	18.8	37.5	21.9	12.5	6.3	2.48	9
Inadequate H&S training	9.4	25.0	21.9	25.0	15.6	3.1	2.45	10
Inadequate H&S supervision	3.0	39.4	15.2	21.2	6.1	15.2	2.41	11
Inadequate monitoring of construction activities relative to the H&S plan	10.0	30.0	26.7	16.7	10.0	6.7	2.30	12
Work factors	16.1	32.3	9.7	32.3	6.5	3.2	2.27	13
Inadequate construction HIRAs	21.9	34.4	15.6	15.6	3.1	9.4	2.20	14
Inadequate risk management (in general)	12.9	38.7	19.4	12.9	6.5	9.7	2.19	15
Inadequate poor working platforms	15.2	36.4	15.2	18.2	12.1	3.0	2.18	16
Inadequate baseline risk assessment	16.1	38.7	16.1	12.9	9.7	6.5	2.15	17
Extreme temperatures	13.8	27.6	27.6	24.1	6.9	0.0	2.12	18
Inadequate H&S management	9.4	43.8	15.6	18.8	6.3	6.3	2.07	19
Inadequate monitoring of construction activities relative to design HIRAs	12.1	39.4	21.2	18.2	6.1	3.0	2.00	20
Inadequate maintenance of plant and equipment	3.1	46.9	21.9	15.6	6.3	6.3	2.00	20
Poor site layout	12.1	42.4	18.2	21.2	3.0	3.0	1.93	22
Inadequate PPE	3.1	59.4	9.4	15.6	6.3	6.3	1.87	23
Inadequate design HIRAs	21.9	37.5	25.0	9.4	3.1	3.1	1.84	24

working platforms, inadequate baseline risk assessment, extreme temperatures, inadequate H&S management, inadequate monitoring of construction activities relative to design HIRAs, inadequate maintenance of plant and equipment, poor site layout, inadequate PPE, and inadequate design HIRAs.

Respondents were requested to provide comments with respect to the meaning of H&S. 6.1% did not, however, 93.9% made a comment or more: 1 No. (69.7%); 2 No. (24.2%), and 3 No. (0.0%). Selected comments are as follows:

- 'Brother's keeper.'
- 'Must always be healthy and safe.'
- 'H&S is No. 1.'
- 'Life depends on it.'
- 'H&S of persons.'
- 'Protective and alert.'

Thereafter, regarding their views with respect to the Construction Regulations, 15.2% did not provide a comment; however, 84.8% made a comment or more: 1 No. (75.8%); 2 No. (9.1%), and 3 No. (0.0%). Selected comments are as follows:

- 'Employer/employees benefit.'
- 'Minimise/control injuries.'
- 'Proactive. Reduce problems. Contractors good base to work from.'
- 'Clearer, they are in place.'
- 'Important.'

Finally, for comments in general with respect to H&S, 42.4% did not provide a comment; however, 57.6% made a comment or more: 1 No. (45.5%); 2 No. (9.1%), and 3 No. (3.0%). Selected comments are as follows:

- 'Lost time injury = affects production.'
- 'Work smartly.'
- 'Appointed supervisors have inadequate H&S knowledge.'
- 'Challenging to follow.'

17.4 Discussion

The findings relative to the frequency at which H&S actions are undertaken by primarily clients and designers on projects are corroborated to a degree by previous South African research (CIDB, 2009). Furthermore, the findings are courtesy of a different stakeholder group in the form of contractors' SMs, which underscore their reliability.

The findings relative to the frequency at which ergonomics problems are encountered corroborates with prior South African research findings (Smallwood, 1997, 2002; Smallwood et al., 2000). Furthermore, these findings indicate that ergonomics problems persist, underscoring the physically demanding nature of construction.

The findings in terms of the extent to which factors contribute to exposure to hazards and the extent to which factors are the causes of accidents on construction sites corroborates with the findings of Haslam et al. (2005). These findings highlight the need for a multi-stakeholder approach to H&S on public sector projects, to ensure an enabling project environment exists, which promotes optimum attention to H&S.

Limited research has been conducted relative to H&S practices on public sector projects in South Africa, and therefore the findings have contributed to the related body of knowledge. Further such research will enable the collective findings to be interrogated in the future.

The qualitative findings in the form of comments with respect to the meaning of H&S, the Construction Regulations, and H&S in general are notable, as they indicate an understanding and appreciation of the reality of H&S endeavours, namely that they are concerned with the health and wellbeing of people, the preservation of people's lives, and overall performance, as opposed to compliance. Such a culture constitutes an opportunity to consolidate thereon, by addressing shortcomings and intensifying and developing further H&S interventions.

17.5 Conclusions

Given the frequency at which 16 H&S actions are undertaken on ECDRPW's projects, it can be concluded that clients and designers are, to a degree, complying with the requirements of the Construction Regulations and contributing to project H&S endeavours, and a multi-stakeholder approach relative to H&S is realised on ECDRPW's projects.

Given the frequency of H&S training on projects, it can be concluded that training is undertaken frequently as opposed to infrequently on ECDRPW's projects and that contractors are to a degree complying with the requirements of the Construction Regulations.

Given the frequency at which four ergonomics problems are encountered, it can be concluded that construction is physically demanding and daily activities entail exposure to ergonomics hazards.

Given the extent to which ten factors contribute to exposure to hazards on construction sites, it can be concluded that hazards are primarily attributable to contractors and a range of contractor-related inadequacies contribute thereto.

Given the frequency at which 24 factors are the cause of construction accidents, it can be concluded that contractors are primarily the origin of the causes and that the management of contractors has a predominating role to play, especially in terms of H&S training and worker participation in H&S.

Given the respondents' comments, it can be concluded that they have a positive disposition with respect to H&S-related regulations, and an understanding and appreciation of the intentions thereof, and the issues with respect to construction H&S.

Given that seven schools in a district administered by the ECDRPW and that only 33 respondents contributed to the study, it can be concluded that the findings are indicative, as opposed to representative. However, given the 'research gap' in terms of an assessment of H&S-related experiences and practices on public sector projects and, specifically, ECDRPW's projects, the findings provide an indication of aspects that require attention.

17.6 Recommendations

H&S awareness should be maintained at industry, industry sector, organisation, and project level among all stakeholders.

Given the frequency of ergonomics problems, the construction process and its activities need to be re-engineered. However, designers have a key role to play to facilitate such re-engineering in terms of design, details, and specifications.

Given the frequency at which employee-related factors are the cause of construction accidents, worker H&S training must be intensified, worker participation must be promoted, and constant H&S feedback must be provided.

Furthermore, the ECDRPW should ensure that H&S is a value on all projects, which implies that it is addressed from stage 1, included in the project charter, addressed throughout the other five stages, and addressed during all project-related meetings. The management of all stakeholders should be committed to H&S, be involved therein, and support H&S endeavours.

Planning by all stakeholders in general is important to ensure that H&S is integrated into the design, procurement, and construction processes.

References

Construction Industry Development Board (CIDB). (2009). *Construction health & safety in South Africa status & recommendations.* CIDB.

Haslam, R. A., Hide, S. A., Gibb, A. G. F., Gyi, D. E., Pavitt, T., Atkinson, S., & Duff, A. R. (2005). Contributing factors in construction accidents. *Applied Ergonomics, 36*, 401–415.

Republic of South Africa. (1993). *Government gazette no. 14918. Occupational health & safety act: no. 85 of 1993.*

Republic of South Africa. (2014). *No. R. 84 occupational health and safety act, 1993 construction regulations 2014. Government gazette no. 37305.*

Smallwood, J. J. (1997). Ergonomics in construction. *ergonomicsSA, 9*(1), 6–23.

Smallwood, J. J., Deacon, C. H., & Venter, D. J. L. (2000). Ergonomics in construction: Workers' perceptions of strain. *ergonomicsSA, 12*(1), 2–12.

Smallwood, J. J. (2002). Construction ergonomics: General contractor (GC) perceptions. *ergonomicsSA, 14*(1), 8–18.

Chapter 18
Understanding the Concept of Resilience in Construction Safety Management Systems

Isaac Aidoo, Frank Fugar, Emmanuel Adinyira, and Nana Benyi Ansah

18.1 Introduction

Health and safety (H&S) is an important consideration in all organisations, and it is of outmost importance in the construction industry because the industry is ranked among the topmost contributors to occupational accidents. Although a lot of success has been chocked in health and safety performance in other fields, the construction industry continues to lag. H&S management in the construction industry is relatively more complicated because of the industry's uniqueness in relation to its complex environmental, organisational, and technical characteristics (Bosch Rekveldt et al., 2011). Traditional approaches to finding solutions to these problems tend to be institutionalised through policies, plans, procedures, and processes for H&S management. These approaches are not easily adaptable to the natural and inevitable changes in the work that is conducted and the H&S risks that are encountered (Wachter & Yorio, 2014). Traditional H&S management does not always improve the results of H&S because they are centred exclusively on the technical requirements and on obtaining short-term results (Weinstein, 1996). Organisations only act when accidents or injuries happen. Traditional H&S management programmes are isolated and many times not integrated with the rest of the functions of an organisation. There are no predictive or anticipatory capabilities where organisations can anticipate future states of the system and foresee the effects of corrective actions (Kontogiannis et al., 2016). It is against this background that Pęciłło (2016) suggests a resilience engineering approach as a prospective solution to the gaps of the traditional H&S management and H&S culture approaches. Resilience H&S management responds to the changing and unforeseen H&S-related issues associated

I. Aidoo (✉) · F. Fugar · E. Adinyira · N. B. Ansah
Department of Construction Technology and Management, Kwame Nkrumah University of Science and Technology, Kumasi, Ghana

with the increasingly complex nature of sociotechnical systems. Hollnagel et al. (2013) also support the view that resilience engineering which is a multifaceted, theoretical approach to designing and managing complex, dynamic-adaptive sociotechnical organisations is an appropriate option to the traditional methods of safety management.

This study aims to bring to the fore the understanding of the concept of resilience in construction health and safety management systems (H&SMS) in the construction industry. It sought to show the various concepts of resilience through definitions in the various field of studies and to particularly explore the position of literature on the concept of resilient H&SMS. This study therefore collated a wide range of studies on the concept of resilience in various fields of studies to achieve the aim of the study. Publications from 2009 to 2019 relevant to this study were reviewed to support existing literature.

18.2 Background of Resilience

The term "resilience" has been used in different fields of studies over the past decades. In the field of engineering, it was applied to illustrate the property of timber to clarify why some types of woods are able to withstand abrupt and intense pressure without breaking. This was termed modulus of resilience for measuring the ability of material to endure severe conditions (Hollnagel, 2014). Holling (1973), in the study of ecology, referred to the resilience of an ecosystem as its ability to undergo changes and still survive. He further juxtaposed resilience with stability, defined as the ability of a system to return to its equilibrium state after a temporary disturbance, but further claimed that resilience and stability were two important properties of ecological systems. Alexander (2013) suggests that the resilience concept was first applied by Rankine who used it to illustrate the mechanical strength and deformation of steel beams. However, most scholars attribute the first adoption of resilience to Holling (1973, 1996) who examined the stability of ecological systems to establish characteristics of rebounding by absorbing stresses (Folke, 2006). Resilience has assumed an important dimension and now dominates thinking of almost every field of studies that has to be undertaken by humankind. The concept is now talked about at almost every international discussions and is a major key concept in international, national, and local policy and development dialogues.

18.3 Resilience from Different Perspectives

The resilience concept has varied origins. Resilience is an emerging concept being used in several disciplines to manage safety (Thoren, 2014). The concept was first used by scientist in the field of engineering, ecology, and natural and physical sciences; however it has soon been taken over by social scientists. The concept of resilience is adopted in research from different fields such as mechanics, psychology, social sciences, aviation, medicine, ecology, urban, education, information system, transportation, organised crime, leadership, and economics (VanBreda, 2011; Hollnagel, 2013; Parker, 2020). Bhamra et al. (2011) reveal the fact that the concept of resilience is multidisciplinary and multifaceted. The notion of resilience is firmly based on ecology, and the definitions used in various studies have charted the suggestions of Holling (1973), whose studies were focused on stability of the ecosystem. The concept of resilience has been studied at the organisational phase. The resilience concept was used originally in organisations to identify the need to respond to variations in the business environment (McAslan, 2010). In the organisational field, resilience is analysed based on different perspectives, for example using an individual, sectorial, organisational, social, as well as a supply chain focus (Lengnick-Hall et al. 2011). Ruiz-Martin et al. (2018) suggest that there are three main lines in the conceptualisation of organisational resilience thus (1) resilience as a feature of an organisation (i.e. something that an organisation has), (2) resilience as an outcome of the organisation's activities (i.e. something that an organisation does); and (3) resilience as a measure of the disturbances that an organisation can tolerate. Chewning et al. (2012) and Alvarenga et al. (2017) investigated resilience in the context of information systems. In economics, Caldera Sánchez et al. (2016) referred to it as adoption of proactive measures to avoid crises and identify relevant early-warning indicators. Social sciences have adopted the concept of resilience and has moved from the individual to the societal level (Godschalk, 2003). Resilience is a common concept in the context of disaster risk reduction and is obviously more important to be debated in the field of adaptation. The concept of resilience has been embraced in the policies of numerous administrations, including those of the USA and Canada; United Nations; European Commission; etc. (Pecillo, 2016). A recent study (Gowan et al., 2014) suggests that a resilient community is well-placed to manage hazards, to reduce their effects, and/or to recover quickly from any negative impacts, resulting in a similar or improved state as compared to before the hazard happened.

18.4 The Concept of Resilience in Safety Management

Suddaby (2010) posits that for any construct to have a clear concept, four elements are key. It must have a good definition, scope conditions or contextual circumstances, semantic relations with other concepts, and lastly coherence and logical consistency. 'Concept', when it's used as a noun, refers to a thought or a notion, conceived in the mind. When used as an adjective, it refers to the organisation of an idea around a central theme or idea.

Conceptually, resilience H&S management is hinged on the fundamental theme of resilience engineering (RE) as delineated by the RE scholarly community and its quest to better realise and manage disruptions that may contribute to incidence and accidents in systems and organisations. Conceptually, the authors interpret these thoughts in the context of H&SMS in construction industry based on the literature and other authors in the construction industry. It provides a platform upon which future research works on resilient H&SMS in the construction industry could be developed.

18.4.1 Concept Building Towards Resilience in Safety Management Systems

The focus on technical requirements in the practice of safety management occasionally improves H&S performance with a short-term result. Often times, organisations only react when accidents or fatalities happened making the system a reactive one. The adduced positions make traditional H&S management programs isolated and disintegrated with the core functions of the organisations. Going forward there seem to be no predictive or anticipatory capabilities where organisations try to anticipate (proactive) future states of the system and foresee the effects of corrective actions (Kontogiannis et al., 2016).

The concept of resilience shows a new way of dealing with H&S and accidents in complex sociotechnical systems (Akselsson et al., 2009; Shirali et al., 2016). Drawing from the work of Dekker et al. (2008) and Hollnagel (2011a), the concept of resilience H&S management is better understood. In their work, they state that for a system to achieve resilience, it must have four capabilities for its functionalities and operations, i.e. ability to respond, ability to monitor, ability to learn, and ability to anticipate. This is also supported by Pecillo (2016) and Shirali et al. (2015) all cited by Trinh et al. (2018), stating that the basic concept underpinning resilience engineering is that in a domain where there are scarce resources, complex systems, and several conflicting goals, an organisation or a system must frequently manage its safety issues and build a H&S system through a resilient process that can anticipate (knowing what to expect), monitor (knowing what to look for), respond (knowing what to do), and learn (knowing what can happen). The theory of resilience engineering augments the concept of H&S management by offering the four resilience processes for managing H&S in a system or organisation.

18.5 Research Methodology

One important knowledge resource area and a significant way to developing new areas of research is publishing in scientific journals (Al-Sharif & Kaka, 2004). A systematic review of related literature on the concept of resilience in H&S management systems published from 2009 to 2020 was conducted adopting an approach used by previous researchers (Torraco, 2005; Whittemore, 2005; York, 2008; Pillay, 2017). The study was conducted using Google Scholar, Semantic Scholar, and

Microsoft Academic utilising resilience concept, resilience safety, and H&S management as keywords. The Preferred Reporting Items for Systematic Reviews and Meta-Analyses (PRISMA) flow chart for search and selection of articles with guidelines adopted from Liberati et al. (2009) was used. This is made up of four key stages, thus identification of the relevant literature, records screened at abstract stage, full-text reading to check for eligibility and inclusion, and excluding criteria as illustrated in Fig. 18.1. A total of 343 papers from 2009 to 2020 were loaded onto Mendeley. Duplicates were eliminated by employing Mendeley "Check for duplicates" feature, and titles and abstract were also screened for required content bringing the number of articles down to 307. To qualify for inclusion, articles must be published in the English language, peer-reviewed, and comprised of some of the following: area of field, concept, definitions, and classification. Two hundred records were excluded due to their irrelevance to the main topic. Eighty-three articles were excluded with reason that despite having the keyword 'resilience', they were not related to the topic, leaving 24 papers systematically reviewed for

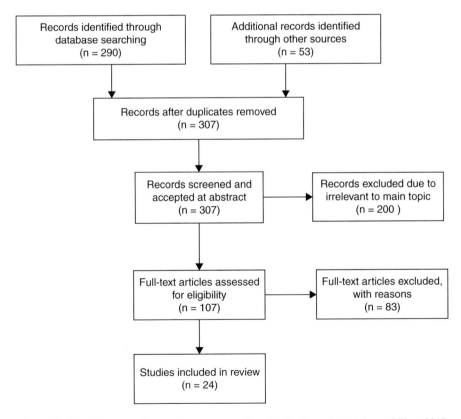

Fig. 18.1 PRISMA flow diagram for paper search and selection adapted from (Pillay, 2017; Liberati et al., 2009)

eligibility and inclusion. Much as this does not cover all articles on resilience engineering, they provide an illustrative view of what has been published on the topic.

18.6 Results and Discussions

This section examines the definitions and area of dimensions used to explain the concept of resilience. Being a novel in the area of H&S, it can be said that is a communal understanding of resilience. A close examination of the literature reviewed suggests different ways researchers have understood resilience with many using the terms interchangeably as can be seen in Table 18.1.

18.6.1 Concept of Resilience

Throughout the review, resilience was generally seen as the ability of a system to continue to perform and meet its objectives in the face of challenges. There is a general agreement around the concept that resilience is achieved by an amalgamation of absorbing the challenges encountered and mutating by adapting and changing to continue to succeed in the face of challenges. This is collaborated by several authors of the concept of resilience (Hollnagel, 2009; Shirali et al. 2012, b; Pillay et al., 2010; Blay, 2017). The resilience concept was generally described by adaptive capacity, flexibility, coping ability, persistence, adaptability abortive capability, and recovery capability.

In the area of resilience engineering, countless efforts in research are aimed at connecting resilience to some describable features. One of the most dispersed methods led to disintegrating resilience into four capabilities, recognising that a resilient system needs to be suitably balanced among them (Hollnagel, 2011a). A resilient system must be able to respond to any type of happenings, to monitor ongoing developments, to anticipate future threats and opportunities, and to learn from past event's failures and successes. The validity of these four capabilities has been extensively accepted for representing successfully how systems and people feel comfortable with unexpected and unforeseen events in everyday work (Rankin et al. 2014, b) and to encourage proactive strategies for managing daily operations and functions (Praetorius & Hollnagel, 2014). The same capabilities have been used to define a devoted framework, which ruminates legal, institutional, organisational, and procurement aspects of societal resilience (Becker et al., 2014). Thus, these capabilities are further looked at below:

Table 18.1 Resilience concept definitions and dimension (author's construct)

	Research area	Reference	Definitions of resilience	Emphasis/dimension
1	Resilience engineering	Shirali et al. (2012a, b)	'a paradigm for safety management that concentrates on how to help people to create foresight, and to anticipate the different forms of risk in order to cope with complexities under pressure and move towards success'	Conducted a survey of RE using safety culture and management factor
2	Resilience engineering	Costella et al. (2009)	'a paradigm of safety management that focuses on how to help people cope with complexity under pressure to achieve success'	Proposed a new method for assessing safety management systems from RE perspective and conducted a RE audit at a Brazilian manufacturing company
3	Resilience engineering	Pillay et al. (2010)	'Developing an organisation's behavioural and cognitive capability such that it is able to effectively adjust and continue performing optimally near its safe operating envelop in the presence of everyday threats and environmental stressors at all levels of the organisation'	Proposed a six-factor questionnaire based on managerial resilience and LIOH which could be used for assessing RE in gold mining; introduced realty gradient of safety (RGS) as a quantitative measure of RE
4	Resilience engineering	Han et al. (2010)	'a proactive management approach that allows future risk to be anticipated and the safety level in an organisation to be maintained, based on perceptions of current and changing safety levels and recognition of acceptable limits'	Investigated how organisations perceived change in safety culture and restored it to an acceptable state using a RE perspective, proposed a model of safety management based on interactions between workers, managers, and organisation in construction
5	Resilience engineering	Shirali et al. (2012a, b)	'Inherent capacity of a plant to cope with complex and unexpected events'	Assessed the challenges of developing RE and adaptive capacity in a chemical process plant
6	Resilience engineering	Madni and Jackson (2009)	The ability to anticipate, adapt, and recover from disruptions	Adaptation, anticipate

(continued)

Table 18.1 (continued)

	Research area	Reference	Definitions of resilience	Emphasis/dimension
7	Resilience engineering	Woods (2015)	Resilience refers to the ability manage/regulate adaptive Capacities of systems that are layered networks, and are also a part of larger layered networks, so as to produce sustained adaptability over longer scales	Assesses resilience and how to engineer it in complex adaptive systems—Rebound, robustness, brittleness, adoptability
8	Organisational	Akgün and Keskin (2014)	The ability of firms to develop specific responses to disruptions and engage in transformative activities	
9	Organisational resilience	Stephenson et al. (2010)	'The ability of an organisation to survive, and potentially thrive in an environment of change and uncertainty'	
10	Organisational resilience	Haimes (2009)	'The ability of a system to withstand a major disruption within an acceptable degradation parameters and to recover within an acceptable time and composite costs and risks'	
11	Organisational resilience	Hollnagel (2009)	'(a resilient system) is able effectively to adjust it functioning prior to, during, or following changes and disturbances, so that it can continue to perform as required after a disruption or a major mishap, and in the presence of continuous stresses'	Response, monitor, anticipate, and learn
12	Social-ecological resilience	Carpenter et al. (2001)	(1) Amount of disturbance a system can absorb resilience and remain within a domain of attraction; (2) capacity for learning and adaptation (3) degree to which the system is capable of self-organising	Adaptive capacity, learning, innovation

(continued)

Table 18.1 (continued)

	Research area	Reference	Definitions of resilience	Emphasis/dimension
13	Development resilience	Pasteur (2011); Barrett and Constas (2014)	Capacity of a person, household, or other aggregate unit to avoid poverty in the face of various stressors and in the wake of myriad shocks over time	Vulnerability, robustness
14	Socioeconomic resilience	Mancini et al. (2012)	Socioeconomic resilience refers to the policy-induced ability of an economy to recover from or adjust to the negative impacts of adverse exogenous shocks and to benefit from positive shocks	Economic response capacity
15	Ecological resilience		Measure of the amount of change needed to change an ecosystem from one set of processes and structures to a different set of processes and structures	Ecology
16	Supply chain	Ponomarov and Holcomb (2009)	The ability of the supply chain to prepare for unexpected events and adapt to and recover from disruptions	
17	Supply chain	Carvalho et al. (2012)	The ability of a system either to return to its original state or to shift to a superior state desirable following disturbance	
18	Project management	Geambasu (2011)	The ability to restore capacity and continuously adapt to changes and to achieve its objectives in the face of disruptive events	
19	Project management	Schroeder and Hatton (2012)	The capacity to evolve in response to risks emerging after the project planning stage	Risk management. The article describes the nature of resilience and outlines a number of practical strategies to build resilience into projects to respond to risk more effectively
20	Project management	Hillson (2014)	The capacity to maintain purpose and integrity under external or internal shocks	Explores the other types of risk that are usually missed from the typical risk process
21	Project management	Turner and Kutsch (2015)	The art of noticing, interpreting, containing, preparing for, and recovering from disruption	

(continued)

Table 18.1 (continued)

	Research area	Reference	Definitions of resilience	Emphasis/dimension
22	Project management	Giezen et al. (2015)	The capacity to overcome unexpected events	The concept of strategic capacity for analysing the decision-making process on mega project. It consists of three elements: Strategic ambiguity, redundancy, and resilience
23	Project management	Zhu (2016)	The ability to cope with uncertainty	
24	Project management	Blay (2017)	The capability to respond to, prepare for, and reduce the impact of disruptions caused by changes in the project environment	Identifies challenges in applying the concept of organisational resilience to project organisations

18.6.2 Ability to Respond Response: Dealing with the Actual

Responding to external and internal disorders or, more in general, to any feedback or indicator is a crucial need for every organisation. A system must be able to differentiate between what is urgent and what is important and provide effective and on-time responses to maintain productivity and ensure safety. This differentiation is necessary to ensure that the real-time competencies that are required to manage unimagined or extreme events at the sharp end are not eroded in the continuous attempt to predetermine corresponding responses to all the possible events (Hollnagel, 2011b, c). Responding is linked to sharp-end decision-making. Defining the cognitive threshold over which additional resources are required is generally a difficult task (Cuvelier & Falzon, 2011). Various examples of prompt responding to system disruption can be discovered in the healthcare literature, particularly in the emergency departments, which call for strong responding capabilities due to the inevitable variability of daily practices (Wears et al., 2006; Nemeth et al., 2008; Wachs et al., 2016). In summary, the analysis of this capability should aim at providing material for understanding what capacities are required to properly respond to everyday situations.

18.6.3 Ability to Monitor: Dealing with the Critical

Hollnagel (2015) opines that ability to monitor is about knowing what to look for, i.e. how to *monitor* that which is or can become a threat in the near term. The monitoring must cover both that which happens in the environment and that which happens in the system itself, i.e. its performance. This is the ability to address the *critical*. The ability to respond is connected to the capability of understanding actual threats timely and precisely. An effective response is thus linked to an effective system's status monitoring and its operational environment, i.e. acquiring signals related to both positive and negative situations (Valdez Banda et al., 2015; Di Gravio et al., 2015a, b; Wreathall, 2011). Consequently, the ability to monitor should aim at defining relevant indicators to gain an understanding of present-day working conditions.

18.6.4 Ability to Anticipate: Dealing with the Potential

Traditionally, the ability to anticipate is associated with conjecturing future happenings based on historic data (Hollnagel, 2014). Granted that these signs offer an overall understanding of H&S levels, they may fail. The difference between anticipating and monitoring is generally the different time scale of observations and the related point of view. The ability to anticipate normally extends the focus of monitoring's leading indicators to cope with long-term changes, threats, opportunities, and environmental potential status. Monitoring and anticipating at different organisational levels might become complementary. For sharp-end operators, coping with near-term issues is an operational everyday activity (monitoring); at the blunt end, more attention is paid generally to strategic decision-making, relying on a long-term analysis (anticipating). The distinction is not a dichotomy because strategic decisions affect and are affected by operational behaviours and vice versa (Tjørhom & Aase, 2011). Typical anticipating features are related to understand if, and how, future events like threats and opportunities are modeled (e.g. qualitatively, quantitatively, etc.), understanding which efforts in terms of expertise and funding are employed for this purpose (Wilson et al., 2009). The analysis of the ability to anticipate should look at detecting upcoming threats and opportunities timely and efficiently and increase system's preparedness.

18.6.5 Ability to Learn: Dealing with the Facts

The traditional H&S management systems lay emphasis only on learning on adverse events, i.e. accidents, incidents, and near misses. A resilient system must be able to learn from previous happenings. It is essential to understand what has occurred and be able to learn the right lessons from it. The ability to learn addresses the factual (Hollnagel, 2011b, c). Denning (2006) and Lundberg et al. (2014) express the same view that learning is, and must be, and forms the basis for any viable system. It is also fundamental to resilience. Any system or organisation that refuses to learn from its past events either positively or negatively will end up spending too many resources, any time it faces a similar problem. Most importantly, learning helps the organisation to improve its response to an issue. By gathering and reflecting upon the incidents, crises, and accidents, the system may improve its barriers and processes for coping with an effect and re-adjust its structures to better endure known trepidations. From a resilience perspective, learning is perfectly a constant function; however, in many real situations, learning arises because of major disorders, i.e. ad hoc. Learning can, therefore, be based on feedback as well as feed-forward (Lundberg et al., 2014). Thus, in effect, learning should lead to selecting worthy indicators and give means to anticipate potential future threats and opportunities.

The review suggests that resilience covers a wide range of concepts, ranging from abstract to existing facets, with each reflecting different solutions in terms of being reactive, proactive, and adaptive, not only to prevent negative outcomes but also to support and strengthen outcomes of processes and systems. According to Manyena (2006), cited by Pillay (2017), a major challenge confronting researchers is the variances in definition of resilience (Hollnagel, 2009; Pillay, 2017). Pillay (2017) further states that resilience in itself is context specific and that some organisations may display different levels of resilience in their H&SMS.

18.7 Conclusion

This study aimed at understanding the concept of resilience in construction H&S management systems in the construction industry through a systematic literature review of previous works and studies. Traditional H&S management does not always improve the results of safety because they are centred exclusively on the technical requirements and on obtaining short-term results. Organisations only act when accidents or injuries happen. Resilience is generally seen as the ability of a system to continue to perform and meet its objectives in the face of challenges. There is a general agreement around the concept that resilience is achieved by an amalgamation of absorbing the challenges encountered and mutating by adapting and changing so as to continue to succeed in the face of challenges. It emerged from the review that there are four capabilities that underpin the concept of resilience in H&SMS, i.e. ability to respond, ability to monitor, ability to learn, and ability to anticipate. That the development of a comprehensive framework of resilience in H&SMS hinged on these capabilities is worth considering implementing if construction H&S management maturity is a goal. Further findings revealed in this study have provided a solid foundation for further studies and analysis on resilience H&SMS and other safety-related concept in the construction industry. Construction organisations would be informed of the need to critically review H&S systems to make resilience an important part of their operational processes.

References

Akgün, A. E., & Keskin, H. (2014). Organisational resilience capacity and firm product innovativeness and performance. *International Journal of Production Research, 52*(23), 6918–6937.

Akselsson, R., Ek, Å., Koornneef, F., Stewart, S., & Ward, M. (2009). Resilience safety culture in aviation organisations. In *17th world congress on ergonomics*. International Ergonomics Association.

Alvarenga, M. Z., Santos, W. R., & Pelissari, A. S. (2017). Orelacionamento collaborative e os sistemas e tecnologias de informação impactam a resiliência das cadeias de suprimentos? *Espacios (Caracas), 38*(1), 3–21.

Al-Sharif, F., & Kaka, A. (2004). PFI/PPP topic coverage in construction journals. In: Khosrowshahi, F. (Ed.), 20th Annual ARCOM Conference, 1-3 September 2004, Heriot Watt University. Association of Researchers in Construction Management, Vol. 1, 711–719

Alhajeri, M. (2014). Health and safety in the construction industry: challenges and solutions in the UAE

Alexander, D. E. (2013). Resilience and disaster risk reduction: an etymological journey. *Natural Hazards and Earth System Sciences, 13*(11), 2707–2716.

Barrett, C. B., & Constas, M. A. (2014). Toward a theory of resilience for international development applications. *Proceedings of the National Academy of Sciences of the United States of America, 111*(40), 14625–14630. https://doi.org/10.1073/pnas.1320880111

Becker, P., Abrahamsson, M., & Tehler, H. (2014). An emergent means to assurgent ends: Societal resilience for safety and sustainability. In C. P. Nemeth & E. Hollnagel (Eds.), *Become. Resilient* (p. 1e12). Ashgate Publishing Ltd.

Bhamra, R., Dani, S., & Burnard, K. (2011). Resilience: The concept, a literature review and future directions. *International Journal of Production Research, 49*(18), 5375–5393.

Blay, K. B. (2017). *Resilience in projects: definition, dimensions, antecedents and consequences*. Ph.D., Loughborough University.

Bosch-Rekveldt, M. et al. (2011). Grasping project complexity in large engineering projects: The TOE (Technical, Organizational and Environmental) framework. *International Journal of Project Management. International Project Management Association, 29*(6), 728–739. https://doi.org/10.1016/j.ijproman.2010.07.008

Caldera Sánchez, A., et al. (2016). Strengthening economic resilience: Insights from the post-1970 record of severe recessions and financial crises. *OECD Economic Policy*.

Carpenter, S., Walker, B., Anderies, J. et al. (2001). From metaphor to measurement: Resilience of what to what? *Ecosystems 4*, 765–781. https://doi.org/10.1007/s10021-001-0045-9

Carvalho, H., Barroso, A. P., Machado, V. H., Azevedo, S., & Cruz-Machado, V. (2012). Supply chain redesign for resilience using simulation. *Computers and Industrial Engineering, 62*(1), 329–341.

Chewning, L. V., Lai, C. H., & Doerfel, M. L. (2012). Organizational resilience and using information and communication technologies to rebuild communication structures. *Management Communication Quarterly, 27*(2), 237–263.

Costella, M. F., Saurin, T. A., & de Macedo Guimarães, L. B. (2009). A method for assessing health and safety management systems from the resilience engineering perspective. *Safety Science, 47*, 1056–1067. https://doi.org/10.1016/j.ssci.2008.11.006.

Cuvelier, L., & Falzon, P. (2011). *Coping with uncertainty*. Resilient Decis Anaesth.

Dekker, S. W. A., Hollnagel, E., Woods, D. D., & Cook, R. (2008). *Resilience engineering: New directions for maintaining safety in complex systems*. Lund University School of Aviation, Sweden.

Denning, P. (2006). Hastily formed networks. *CACM, 49*, 15–20.

Di Gravio, G., Mancini, M., Patriarca, R., & Costantino, F. (2015a). Overall safety performance of air traffic management system: Forecasting and monitoring. *Safety Science, 72*, 351–362. https://doi.org/10.1016/j.ssci.2014.10.003.

Di Gravio, G., Mancini, M., Patriarca, R., & Costantino, F. (2015b). Overall safety performance of the air traffic management system: Indicators and analysis. *Journal of Air Transport Management, 44*, 65–69. https://doi.org/10.1016/j.jairtraman.2015.02.005.

Folke, C. (2006). Resilience: The emergence of a perspective for social–ecological systems analyses. *Global Environmental Change, 16*, 253–267.

Geambasu, G. (2011). *Expect the unexpected: An exploratory study on the conditions and factors driving the resilience of infrastructure projects*. Ph.D., École Polytechnique Fédérale de Lausanne.

Giezen, M., Salet, W., & Bertolini, L. (2015). Adding value to the decision-making process of mega projects: Fostering strategic ambiguity, redundancy, and resilience. *Transport Policy, 44*, 169–178.

Godschalk, D. R. (2003). Urban hazards mitigation: Creating resilient cities. *Natural Hazards Review, 4*, 136–143. https://doi.org/10.1061/(ASCE)1527-6988(2003)4:3(136)

Gowan, M. E., Kirk, R. C. & Sloan, J. A. (2014). Building resiliency: a cross-sectional study examining relationships among health-related quality of life, well-being, and disaster preparedness. *Health Qual Life Outcomes 12*, 85. https://doi.org/10.1186/1477-7525-12-85

Gürbilek, N. (2013). A conceptual framework for resilience engineering in construction safety. *Journal of Chemical Information and Modeling*. https://doi.org/10.1017/CBO9781107415324.004

Haimes, Y. Y. (2009). On the definition of resilience in systems. *Risk Analysis, 29*, 498–501. https://doi.org/10.1111/j.1539-6924.2009.01216.x.

Han, U., Lee, S. H., & Pena-Mora, F. (2010). System dynamics modelling of a safety culture based on resilience engineering. In *Construction research congress 2010* (pp. 389–397). ASCE. https://doi.org/10.1061/41109(373)39

Henrik, E., Sander, E., van der Leeuw., Charles, L., Redman, D. J., Meffert, G. D., Christine, A., & Thomas, E. (2010). Urban transitions: On urban resilience and human-dominated ecosystems. *AMBIO, 39*(8), 531–545.

Herrera, I. A., Nordskag, A. O., Myhre, G., & Halvorsen, K. (2009). Aviation safety and maintenance under major organizational changes, investigating non-existing accidents. *Accident Analysis and Prevention 41*, 1155e63. https://doi.org/10.1016/j.aap.2008.06.007

Hillson, D. (2014). *How to manage the risks you didn't know you were taking*. Presented at the PMI® Global Congress 2014, Newtown Square.

Holling, C. S. (1973). Resilience and stability of ecological systems. *Annual Review of Ecology and Systematics, 4*, 1–23. https://doi.org/10.1146/annurev.es.04.110173.000245.

Holling, C. S. (1996). Engineering resilience versus ecological resilience. *Engineering within Ecological Constraints, 31*, 32.

Hollnagel, E. (2009). The four cornerstones of resilience engineering. In C. P. Nemeth, E. Hollnagel, & S. W. A. Dekker (Eds.), *Resilience engineering perspectives, Volume 2: Preparation and restoration* (pp. 117–133). Ashgate.

Hollnagel, E. (2011a). Epilogue: RAG – the resilience analysis grid. In: Hollnagel, E., Pariès, J., Woods, D. D., et al. (Eds.), *Resilience engineering perspectives volume 3: resilience engineering in practice*. Farnham: Ashgate, 275–296.

Hollnagel, E. (2011b). Prologue: The scope of resilience engineering. In E. Hollnagel, J. Pariès, D. D. Woods, & J. Wreathall (Eds.), *Resilience engineering in practice: A guidebook, MINES ParisTech*. Ashgate Publishing, Ltd.

Hollnagel, E. (2011c). To learn or not to learn, that is the question. In E. Hollnagel, J. Paries, D. D. Woods, & J. Wreathall (Eds.), *Resilience engineering in practice: A guidebook*. Ashgate Publishing, Ltd.

Hollnagel, E. (2013). Building research & information: Resilience engineering and the built environment. *Building Research & Information*. https://doi.org/10.1080/09613218.2014.862607

Hollnagel, E. (2014). *Safety–I and Safety–II. The past and the future of safety management*. Farnham: Ashgate.

Hollnagel, E., Braithwaite, J., & Wears, R. L. (2013). *Resilient health care*. Ashgate. ICAO (2014), ICAO Safety Report. 2014 ed.

Hollnagel, E. (2015). Safety-I and Safety-II, the past and future of safety management. *Ashgate Safety Hygiene News, 45*(10), 50–52. http://dx.doi.org/10.1007/s10111-015-0345-z

Kontogiannis, T., et al. (2016). Total safety management: Principles, processes and methods. *Safety Science*. https://doi.org/10.1016/j.ssci.2016.09.015

Lengnick-Hall, C. A., Beck, T. E., & Lengnick-Hall, M. L. (2011). Developing a capacity for organizational resilience through strategic human resource management. *Human Resource Management Review. Elsevier Inc., 21*(3), 243–255. https://doi.org/10.1016/j.hrmr.2010.07.001

Liberati, A., et al. (2009). The PRISMA statement for reporting systematic reviews and meta-analyses of studies that evaluate health care interventions: Explanation and elaboration. *Annals of Internal Medicine, 151*, W1–W30. https://doi.org/10.7326/0003-4819-151-4-200908180-00136

Lundberg, J., Törnqvist, E. K., & Nadjm-Tehrani, S. (2014). Establishing conversation spaces in hastily formed networks: The worst fire in modern Swedish history. *Disasters, 38*, 790–807.

Lundberg, J., & Johansson, B. J. (2015). Systemic resilience model, *Reliability Engineering and System Safety. Elsevier, 141*, 22–32. https://doi.org/10.1016/j.ress.2015.03.013

Madni, A. M., & Jackson, S. (2009). Towards a conceptual framework for resilience engineering. *Systems Journal, IEEE, 3*, 181–191. https://doi.org/10.1109/JSYST.2009.2017397.

Mancini, A., Salvati, L., Sateriano, A., Mancino, G., & Ferrara, A. (2012). Conceptualizing and measuring the 'economy' dimension in the evaluation of socio-ecological resilience: A brief commentary. *International Journal of Latest Trends in Finance and Economic Sciences, 2*, 190–196.

Manyena, S. B. (2006). Rural local authorities and disaster resilience in Zimbabwe. *Disaster Prevention and Management: An International Journal, 15*(5), 810–820. https://doi.org/10.1108/09653560610712757

McAslan, A. (2010). The concept of resilience. Understanding its origins, meaning and utility. Available from: http://torrensresilience.org/images/pdfs/resilience%20origins%20and%20utility.pdf

Nemeth, C., Wears, R., Woods, D., Hollnagel, E., & Cook, R. (2008) Minding the gaps: Creating resilience in health care. In K. Henriksen, J. B. Battles, M.A. Keyes, & M. L. Grady (Eds.), *Advances in patient safety: New directions and alternative approaches* (Vol. 3: Performance and Tools). Agency for Healthcare Research and Quality (US).

Parker, D. J. (2020). Disaster resilience—A challenged science. *Environmental Hazards, 19*(1), 1–9. https://doi.org/10.1080/17477891.2019.1694857.

Pasteur, K. (2011). *From vulnerability to resilience: A framework for analysis and action to build community resilience*. Practical Action Publishing.

Pęciłło, M. (2016). The concept of resilience in OSH management: A review of approaches. *International Journal of Occupational Safety and Ergonomics (JOSE), 22*(2), 291–300. https://doi.org/10.1080/10803548.2015.1126142.

Pecillo, M. (2016). The concept of resilience in OSH management: A review of approaches. *Central Institute for Labour Protection – National Research Institute (CIOP-PIB), Poland International Journal of Occupational Safety and Ergonomics (JOSE), 22*(2), 291–300. https://doi.org/10.1080/10803548.2015.1126142.

Pillay, M. (2017). Resilience engineering: An integrative review of fundamental concepts and directions for future research in safety management. *Open Journal of Safety Science and Technology, 7*, 129–160. https://doi.org/10.4236/ojsst.2017.74012.

Pillay, M., Borys, D., Else, D., & Tuck, M. (2010). Safety culture and resilience engineering: Theory and application in improving gold mining safety. In *Gravity Gold 2010* (pp. 129–140). AusIMM.

Ponomarov, S. Y., & Holcomb, M. C. (2009). Understanding the concept of supply chain resilience. *The International Journal of Logistics Management, 20*, 124–143.

Praetorius, G., & Hollnagel, E. (2014). Control and resilience within the maritime traffic management domain. *J Cogn Eng Decis Mak, 8*, 303e17. https://doi.org/10.1177/1555343414560022.

Rankin, A., Lundberg, J., Woltjer, R., Rollenhagen, C., & Hollnagel, E. (2014a). Resilience in everyday operations: A framework for analyzing adaptations in high-risk work. *Journal of Cognitive Engineering and Decision Making, 8*, 78–97.

Rankin, A., Lundberg, J., & Woltjer, R. (2014b). A framework for learning from the adaptive performance. In C. P. Nemeth & E. Hollnagel (Eds.), *Become. Resilient* (pp. 79–96). Ashgate Publishing, Ltd.

Rosa, L. V. et al. (2017). A resilience engineering approach for sustainable safety in green construction. *Journal of Sustainable Development of Energy, Water and Environment Systems, 5*(4), 480–495. https://doi.org/10.13044/j.sdewes.d5.0174

Ruiz-Martin, C., López-Paredes, A., & Wainer, G. (2018). What we know and do not know about organizational resilience. *International Journal of Production Management and Engineering, 6*(1), 11–28.

Schroeder, K., & Hatton, M. (2012). Rethinking risk in development projects: From management to resilience. *Development in Practice, 22*(3), 409–416.

Shirali, G. A., Motamedzade, M., Mohammadfam, I., Ebrahimipour, V., & Moghimbeigi, A. (2015). Assessment of resilience engineering factors based on system properties in a process industry. *Cognition, Technology & Work, 18*(1), 19–31. https://doi.org/10.1007/s10111-015-0343-1.

Shirali, G. A., Shekari, M., & Angali, K. A. (2016). Quantitative assessment of resilience safety culture using principal components analysis and numerical taxonomy: A case study in a petrochemical plant. *Journal of Loss Prevention in the Process Industries, 40*, 277–284. https://doi.org/10.1016/j.jlp.2016.01.007.

Shirali, G. H. A., Mohammadfam, I., Motamedzade, M., Ebrahimipour, V., & Moghimbeigi, A. (2012b). Assessing resilience engineering based on safety culture and managerial factors. *Process Safety Progress, 31*, 17–18. https://doi.org/10.1002/prs.10485.

Shirali, G. H. A., Motamedzade, M., Mohammadfam, I., Ebrahimipour, V., & Moghimbeigi, A. (2012a). Challenges in building resilience engineering (RE) and adaptive capacity: A field study in a chemical plant. *Process Safety and Environmental Protection, 90*, 83–90. https://doi.org/10.1016/j.psep.2011.08.003.

Stephenson, A., Seville, E., Vargo, J., & Roger, D. (2010). *Benchmark resilience: A study of the resilience of organisations in the Auckland region*. Resilient Organisations Research Report 2010/03b, Resilient Organisations Research.

Suddaby, R. (2010). Editor's comments: Construct clarity in theories of management and organization. *Academy of Management Review, 35*(3), 346–357. https://doi.org/10.5465/AMR.2010.51141319.

Thoren, H. (2014). Resilience as a unifying concept. *International Studies in the Philosophy of Science, 28*(3), 303–324. https://doi.org/10.1080/02698595.2014.953343.

Tjørhom, B., & Aase, K. (2011). The art of balance: Using upward resilience traits to deal with conflicting goals. In E. Hollnagel, J. Pariès, D. D. Woods, & J. Wreathall (Eds.), *Resilience engineering in practice: A guidebook* (pp. 157–170). Ashgate Publishing, Ltd.

Torraco, R. J. (2005). Writing integrative literature reviews: Guidelines and examples. *Human Resource Development Review, 4*, 356–367. https://doi.org/10.1177/1534484305278283.

Trinh, M. T., Feng, Y., & Jin, X. (2018). *Conceptual model for developing resilient safety culture in the construction environment*. American Society of Civil Engineers. https://doi.org/10.1061/(ASCE)CO.1943-7862.0001522.

Turner, N., & Kutsch, E. (2015). Project resilience: Moving beyond traditional risk management. *PM World J, 4*(11).

Valdez Banda, O. A., Goerlandt, F., Montewka, J., & Kujala, P. (2015). A risk analysis of winter navigation in Finnish sea areas. *Accident; Analysis and Prevention, 79*, 100–116. https://doi.org/10.1016/j.aap.2015.03.024.

VanBreda A. (2011). Resilience theory: a literature review with special chapters on deployment resilience in military families & resilience theory in social work. Available from: http://vanbreda.org/adrian/resilience/resilience_theory_review.pdf

Wachs, P., Saurin, T. A., Righi, A. W., & Wears, R. L. (2016). Resilience skills as emergent phenomena: A study of emergency departments in Brazil and the United States. *Applied Ergonomics*. https://doi.org/10.1016/j.apergo.2016.02.012.

Wachter, J. K., & Yorio, P. L. (2014). A system of safety management practices and worker engagement for reducing and preventing accidents: An empirical and theoretical investigation. *Accident Analysis and Prevention, 68*, 117–130.

Wears, R. L., Perry, S., & McFauls, A. (2006). "Free fall" e a case study of resilience, its degradation, and recovery in an emergency department. In: *Proceedings of the resilience engineering symposium* (pp. 8–10). Sophia-Antip [Online].

Weinstein, M. H. (1996). Improving safety programs through total quality. *Occupational Hazards, 58*(8), 27.

Whittemore, R. (2005). The integrative review: Updated methodology. *Journal of Advanced Nursing, 52*, 546–553. https://doi.org/10.1111/j.1365-2648.2005.03621.x.

Wilson, J. R., Ryan, B., Schock, A., Ferreira, P., Smith, S., & Pitsopoulos, J. (2009). Understanding safety and production risks in rail engineering planning and protection. *Ergonomics, 52*, 774–790. https://doi.org/10.1080/00140130802642211.

Woods, D. (2015). Four concepts for resilience and the implications for the future of resilience engineering. *Reliability Engineering and System Safety, 141*, 5–9. https://doi.org/10.1016/j.ress.2015.03.018.

Wreathall, J. (2011). Monitoring e a critical ability in resilience engineering. In: *Resilience engineering in practice: A guidebook, safety, and security in work and its environment* (pp. 61–68).

York, L. (2008). Editorial: What we know, what we don't know, what we need to know: Integrative literature reviews are research. *Human Resource Development Review, 7*, 139–141. https://doi.org/10.1177/1534484308316395.

Zhu, J. (2016). *A system-of-systems framework for assessment of resilience in complex construction projects*. Ph.D., Florida International University.

Chapter 19
Optimisation of the Process for Generation, Delivery and Impact Assessment of Toolbox Talks on a Construction Site with Multiple Cultures

Máire Feely, James G. Bradley, and John Spillane

19.1 Introduction

Toolbox talks (TBT) are a traditional and potentially impactful form of safety communication in construction. TBT materials are in demand and provided by notable construction-oriented organisations (CPWR, 2019). Adult learners have a unique way of learning, many in the industry being kinaesthetic learners. Current approaches emphasize teaching, rather than learning strategies, and appear to contradict recent competency-based developments in education science (Laberge et al., 2014). If operatives are not taught with the consideration of the different learning styles, learning by doing in this case (Garavan et al., 2012), then are they getting the optimum outcome from a TBT? Are they gaining enough knowledge from these TBTs to combat concerning risks?

The researcher while on a construction site observed the different methods of composition and delivery of a TBT, delivered by both internal staff and external consultants, and the likelihood of the impact it had on the workers. Evidence does indicate that TBTs are perceived to be important and may be a component of effective safety programs in construction (Olson et al., 2016). It was also observed that a focus on site safety compliance often results in the completion of TBTs as a necessary evil or tick-box exercise so they are frequently delivered to show compliance, but little attention is given to ways in which the TBT should be executed for maximum impact.

The researcher observations on site showed that if TBTs are not interpreted or absorbed properly, then safety risks can emerge. Issues observed included

---

M. Feely · J. G. Bradley (✉) · J. Spillane
Construction Management and Engineering, School of Engineering, University of Limerick, Limerick, Ireland
e-mail: Jim.Bradley@ul.ie; John.Spillane@ul.ie

miscommunication of the TBT message and misunderstanding of the message. These occurred due to a combination of factors such as lack of information provided, language and learning barriers, poor generation of TBTs and authority of personnel delivering the TBTs.

Literature shows that while TBTs are valuable, there are opportunities for improving the frequency and quality of them. Literature explains the different learning processes, both pedagogy and andragogy, which should be considered when tackling the process of a TBT (Ellis & Loewen, 2007). Site safety, cleanliness and attitudes of workers could have huge negative implications as a result of substandard delivery of safety messages. By not focusing on the best way to teach safety that caters for operator learner types, efficacy is already lost. If the capability of a TBT is limited by this, workers will unplug themselves from being alert to the risks addressed in the safety bulletin and put the excuse down to language or learning barriers. This has the potential to skew a worker's perception of a TBT and leads them to view it simply as a 'tick-box' exercise with no thought given to the significance it could have on their safety.

The literature also shows that the inclusion of safety anecdotes about close calls, injuries or fatalities is currently a rare occurrence but has the potential to have huge impacts on workers as they can relate to it (Varley & Boldt, 2002; Kaskutas et al., 2016). Views from the literature have contributed to this research on determining how to optimise the generation, delivery and ease of implementation of a TBT and what methods could be introduced to gain this optimisation. The use of anecdotes, large text, pictures and bullet points and the manner and authority of delivery along with including demonstrations when designing and transmitting a TBT will be examined when producing a favourable outcome. The suggested solutions also focus on learning styles of workers, the delivery formats and impact assessment and tweak these to form a process for optimum impact.

19.2 Literature Review

19.2.1 Employers' Onus

The management of health and safety at work regulations states that employers must make adequate information and training available for their employees. This regulation is very broad and does not give in-depth details to support management in enforcing this within their company. TBTs have the potential to have a positive, lasting impact on a worker's safety. Kaskutas et al. (2016) agree that these talks improve communication, empower workers, reduce injuries and improve safety. Many articles are indicated for further study to be done in assessing the impact of a TBT, but no research papers were found on this, which posed the question of how one would go about this. Bahn and Barratt-Pugh (2012) stated the evaluation of training will provide organisations with informative evidence to develop the most appropriate mechanism for their specific context.

Toolbox talks should be viewed as opportunities for managers to have a conversation with their employees about risks on their site with special mention of past experiences as well as any current near misses or incidents that have occurred recently. Eggerth et al. (2018) agreed with this, stating the relevance of the TBT can be increased by contextualising it in terms that the workers can identify with. However, this only deals with the delivery and not necessarily the impact on workers when conducting these talks. When defining a TBT, the American Society of Safety Professionals (ASSP.Org, 2018) defines it as a 10–15-min instructional session on hazards and steps to take to prevent these from occurring. Similarly, Kaskutas et al. (2016) say TBTs are informal work-site training sessions, designed to improve safety and prevent work-related incidents that require minimal effort and resources to deliver. The Health and Safety Authority (HSA) in Ireland (2020) simply defines it as a short safety talk. No definition was presented with regard to detailing what a TBT should include, how to deliver, what impact TBTs should have on workers and how the impact could be monitored.

With the fact that TBTs are delivered to adults, then the way adults learn needs to be considered. Ellis and Loewen (2007) define a learning style as the characteristic way in which people orientate to solving problems. There are three main types of learners: visual, learning through visual presentation; auditory, through an auditory manner; and kinaesthetic, learning by being '"hands on'. Construction employees' work is primarily hands on, falling into the kinaesthetic category of learners. Little is said in the literature linking effectiveness, in other words the impact of a TBT, with the teaching styles tailored for kinaesthetic labourers. A lecture approach to a TBT seems to be time effective and effortless for employers and most popular format for getting safety messages complete. However, Burke et al. (2006) declared that these lecture-type safety talks are one of the least engaging methods of safety training.

The way adults learn is often overlooked in the construction industry. To learn is viewed by Garavan et al. (2012) as to gain knowledge or skill in a particular area. They look at the learning approaches, pedagogy and andragogy. The pedagogy learning process lies with the teacher, where the learner is inert. Examples of this approach include lectures, assigned reading and exams, whereas the andragogy process builds on the learner's experiences, group discussions, problem-solving activities and self-directed learning. Holmes and Abington-Cooper (2000) state that the mission of educators is to assist adults in developing their full potential and note that andragogy is the teaching method that is used to achieve this. Building from this, and the fact that construction workers are adult learners, an androgogy approach should be used where TBT delivery methods involve more participative methods using stories of past near misses and experiences of the learners, group discussions and learning from each other - all of which are absent from a pedagogy-based lecture format.

Learners learn from what they actively do (Garavan et al., 2012) especially if they can relate to it. In addition, life requires us to do, more that it requires us to know, in order to function as stated by Holt (2001), and this directly correlates to a construction site. The power of knowledge is fine, but in reality, if there is no practical competency, then construction projects would never succeed. If a worker's day typically involves all practical activities, then why should a TBT be anything other than another practical activity?

Jannadi (1996) instructed that in order to reduce injuries on site, top management must be accountable and committed to safety policies. When reviewing the literature, most agreed and stated that the authority of the person delivering any safety talk was paramount to it having an optimum effect on workers. Jeschke et al. (2017) challenged this and stated that construction site supervisors are more likely to have a significant impact on safety, compared to top managers and safety managers. This suggests that experienced foremen in their given field should be giving these TBTs to workers of the same trade as them. Traits of a trainer should include credibility, perspective and the ability to connect. Boendermaker et al. (2003) identified 37 characteristics for a trainer, the most important including competency, being critical of the learning process, good at communicating and having respect for the trainee. While authority is important, knowledge and experience are what will grab a worker's attention which could require a different person each time depending on the TBT topic. A TBT shouldn't be viewed as a lecture but more as a discussion that requires the participants' attention and involvement if the transfer of knowledge is to succeed.

Burke et al. (2006) note that health and safety training is globally recognised as a means of reducing the costs associated with injuries and illnesses. Costs can be seen in many forms, time, capital and quality being the main three. Herbert W. Heinrich's law (Heinrich, 1941) states for every accident causing major injury in the workplace, there are 29 accidents that cause minor injury and 300 accidents causing no injury. A construction company should be looking in-depth at reducing the near misses and minor injuries, which will in turn aid in the reduction of major accidents. Costs must be controlled and managed on any building site, but what the literature fails to look at is the time and cost expense of running TBTs for a construction company. Mustard and Orchard's (Iwh.on.ca, 2020) research states that 5% of the total cost of a project's budget goes into health and safety. Take the example of a London site, where salaries averaged £18 per hour and TBTs lasted 20 min. The HSA (2020) recommends having five to ten people present; however on bigger sites, this number can be significantly higher. If TBTs are delivered 5 days a week, it will cost a company £300 per week, £15,000 a year to deliver TBTs. This figure however does not account for preparation, facilities or resources. The total cost of workplace self-reported injuries and ill health in 2016/2017 in the UK was £15 billion. (HSA.ie, 2020). By increasing capital put into prevention, through using TBTs as a means of communicating health and safety issues to the workforce on site, then this may lead to a reduction in negative, costly accidents on site.

19.3 Research Method

To gain information on how to optimise the structure and delivery of a TBT, qualitative action research has been undertaken. Themes from the literature reviewed formed the baseline for interviewing construction managers and workers. The valuable information gathered from their experience contributed to gaining further knowledge into the types of learners they were and their perspectives on current TBT design and delivery. A focus on adult learning types was fundamental to this methodology. A new design of TBT was part of this study and was critiqued through asking participants for their impressions on possible barriers to implementation, the advantages, disadvantages and suggested improvements. The general operators were then able to give their first-hand experience of their views on a TBT and the effects it had on them.

Two TBTs were carried out back to back on a London construction site. The first one (TBT-A) simulated the current methods used on the site, that is, a lecture-style approach. The second one (TBT-B) focused on facilitating kinaesthetic learners with aspects such as participation and inclusion of real-life experiences, both regarded as important in the literature reviewed. A comparison of the two TBTs (TBT-A and TBT-B) was then completed. Constant factors applied to the study included the duration of 10 to 15 min, as indicated by the American Society of Safety Professionals (ASSP.Org, 2018); the context, which was on the safety topic of general fire safety; and finally, the audience, consisting of 1 contract manager, 1 health and safety manager, 2 project managers, 1 site manager, 2 site engineers, 4 foremen and 9 labourers. TBT-A and TBT-B were delivered back to back, and a feedback session took place immediately after TBT-B was delivered, to discuss and compare both TBTs. An easily completed form was developed and used, which facilitated this comparison. Immediately before the feedback session and interviews, the researcher delivered a short talk explaining the different learner types so that the participants could understand what type of learner they were.

The feedback session focused on gathering information for comparison of the two talks regarding the delivery methods, structure and impact TBT-A and TBT-B had on each participant. The delivery methods and structure were an important comparison to distinguish the efficacy of both TBTs. The impact each TBT had on the construction workers allowed for the barriers of a TBT's effectiveness and the opportunities for improvement to be distinguished. The discussion at the end of the talks was paramount to evaluating which of the talks suited each individual better.

Post-completion of both TBTs and subsequent discussion session, interviews were conducted with 40 people. This was to delve deeper into the topic of learner types and the identification of elements of TBTs that were felt would accommodate for their different learning styles.

Survey questions were developed using themes uncovered in the literature review such as managers and workers' perspective of the TBT, their current method of delivery and the efficacy of TBTs, barriers they encountered in relation to the effectiveness of a TBT and what opportunities there is for improvements. The semi-structured interviews were conducted, and the interviewees' responses filled in on a survey form while completing the interview. The data collection involved using Likert scales, option and ranking lists and yes/no questions. Using these types of scoring forced the interviewee to think about their responses while still being uncomplicated to complete.

The results gathered on the construction site were then analysed with the literature review to design suggested changes to be made to current TBTs.

19.4 Research Results

Forty people, whose roles included project managers, supervisors, health and safety officers, engineers and general operatives, were interviewed on a construction site in London. Themes formulating the interviews included their perception of TBTs, their current and desired participation in the talks and their opinion on the delivery of the current format against what they saw as beneficial. The semi-structured interviews were carried out after two separate TBTs, TBT-A and TBT-B, were delivered to the participants.

Of the 40 participants interviewed, 82% either strongly agreed or agreed that there are huge benefits to having a TBT. Eighteen people said that one of the main benefits they get from a TBT is that it highlighted current topics on-site that are a concern. Allowing workers to ask questions on the given topic was a lower benefit to them. A further benefit was noted that TBTs could assess the competence of both the operatives, to see if they understood the material and the trainer and to check if the trainers were giving the correct information in the right way. While some people said that there were enough resources and money put into TBTs, the vast majority said that more resources and money was needed – no one said less were needed. The creation of booklets, including simple diagrams in different languages for each worker, was a suggestion made, should additional resources be made available to allow the booklets to be generated. Other suggestions involved putting more time into the design and method of delivery rather than rushing through the TBTs.

Despite 88% people saying that they see a TBT as being very important, the answers to whether people remembered the last TBT they had were very mixed. Nine said they did not remember any, 11 said they remembered some, and 15 said they remembered the content in great detail. People's participation level and willingness to partake in these talks were rated very high with 65% of respondents saying that they frequently or always participate. On discussion, some felt they had no choice whether they participated or not. However, these answers did not correspond to comments from managers who gave the TBTs, who said that many workers would not pay attention or partake in any discussion. This may indicate that operators were giving self-protective answers and they did not want to be seen as not participating.

There was a mixed response around how long a TBT should last and how often they should be given due to inconsistency in terms of duration and frequency of

TBTs on site. There is inconsistency because different foremen were giving the talks on different topics with each TBT having different duration, and some foremen delivered TBTs more frequently than others. Nine believed that TBTs should be done once per week with 13 suggesting daily TBTs. Only three believed TBTs should last for 5 min or less and 14 said 10–15 min. Additional notes made were that TBTs should be done when new risks were identified, new issues arose from safety audits or as foremen deemed necessary. The general consensus was that each week might be different depending on what was happening on site. Management stated that the company usually completes two TBTs per week.

A discussion alone or a discussion paired with a demonstration was deemed the most beneficial delivery method for workers. Comments included that as many methods as possible should be used to gain optimum impact. Two foremen said that incorporating language translation systems would also aid understanding. One project manager suggested that after discussion and demonstration were used, the operatives should then be monitored doing the task (where applicable) and any questions and issues could then be addressed and further improvements made.

With regard to who should deliver TBTs, foremen, site managers and head of safety were regarded as being the most appropriate. Comments were made that perhaps the workers themselves should be giving the talks, as it directly affects their work and work environment. Others stated that foremen could be seen as too friendly to have an optimum impact and that managers should supervise TBTs being delivered, ensuring all points are addressed and received. External bodies and top managers were the least desired people to be delivering TBTs, despite them having the greatest authority.

Following on from this, the conversations around external TBTs were very split. Fifteen people thought that TBTs worked well or very well, explaining that TBTs can shed new light on specific issues and give an alternative perspective on safety and thus reduce accidents. An external body may have general experience giving talks but without specificity for different sites. The operatives who agreed that TBTs delivered by externals worked well said that they were more likely to pay attention to external bodies and that it increased their self-awareness. Although 35% of people declared external TBTs worked very well, over 50% of people's opinion were neutral or below on the matter. They explained that the effectiveness would depend on the topic and whether or not it was relevant to the site itself. Others argued that it undermines management and supervisors and that site should be able to carry out their own TBTs. Some management claimed that operatives are more likely to listen to someone they know and trust and that external bodies are not aware of the issues faced on the specific site. They also said that there can be language difficulties and if operatives were given a TBT from their own supervisor, they may be able to ask questions in their native language or find someone to translate during the day.

A total of 50% of operatives said they would change the teaching style of how TBTs are being delivered as this was their main concern. Allowing workers to ask questions on the given topic was ranked low, perhaps because their current TBT format does not involve a discussion. They also stated that TBTs should be more topical and relevant to their own work and not just the site as a whole. Language and learning barriers were the ranked highest in terms of barriers blocking TBTs from having significant impact. When discussing this, they stated that an easy, concise format was a much-preferred method of delivery.

Two TBTs, TBT-A and TBT-B, were delivered to the group of people on site. TBT-A had all text and was delivered in a lecture style, suiting auditory and visual learners, while TBT-B included a discussion and demonstration and therefore required more involvement from workers, better suiting kinaesthetic learners. Two different TBTs were given to 20 people, operatives, head of safety, foremen, site managers and project managers on the topic of fire safety. They were then asked if they knew what learner type they were, visual, auditory or kinaesthetic, and which TBT they preferred with an explanation of why. Despite 6 operatives being unsure of their preferred method of learning, 60% said they were kinaesthetic learners.

Of the 20 people present, 18 preferred TBT-B. When having a post-talk discussion with them, they said that they found the TBT-B easier to understand, more engaging and made more sense. They noted that the demonstrations and use of easy language made it easier to listen, engage with and learn from the TBT. The issues they found with TBT-A was that they did not understand every word spoken, due to English not being everyone's native language. They had no visual to help with what was being said nor any language translation tools to aid them.

19.5 Discussion

TBTs are an opportunity for managers to have informative discussions on safety with their workers. However, the findings of this study indicated that the overall process of TBTs requires optimisation. The optimised solution needs to be implementable, provide a learning experience that is 'sticky' and accepted among workers. From the researcher's conversations with site workers, important themes that impact on the quality of the TBT were identified and include delivery style and trainer, consistency in terms of frequency and duration, language barriers and learner type issues. Changes that could be made and implemented were based on findings in the literature review, the researcher's observations and from analysis of the interview data.

The authority of the person delivering any safety talk is paramount for optimum effect on workers, and Boendermaker et al. (2003) identify the main characteristics of a trainer as being competent, having the respect of the learners and having a critical view of the learning process. While the literature states that top managers should be accountable and committed to safety policies, operators said that foremen would be the best for delivering and the role of the managers should be to supervise these to ensure the delivery was up to company standard. This would then enforce a standard that could be implemented throughout a company and facilitate impact assessment allowing the TBT to be tweaked and changed to get the optimum outcome. It also coincides with Bahn and Barratt-Pugh (2012) noting that the evaluation of training will provide organisations with the evidence to develop the most appropriate mechanism for their specific context. Another proposal on top of managing the talks would be to offer some education and training to the people giving TBTs on how to deliver them, i.e. train the trainer.

While participation levels were ranked high among workers, managers said from their observations it was actually quite low. To enhance concentration and participation, workers need to want to be involved in TBTs. The teaching style in which TBTs are being given was the main element that participants would change, so if the current format was adjusted to suit operatives, their willingness to participate would increase. The inclusion of relatable material based on experience and real-life examples or stories would also increase engagement.

The teaching style was ranked highest in terms of what needed to be improved. Many resources are available for the different types of TBTs that can be given, along with sample TBTs. However, none of them tell trainers how to adapt these to the learner types of the operatives. TBTs need to change to suit the predominately kinaesthetic learner's style on site with an andragogy approach taken. For the method of delivery, an option including a demonstration and discussion had the highest preference among workers, while a lecture style had the lowest. This corresponds with Burke et al. (2006) who states that these lecture-type safety talks are one of the least engaging methods of safety training. By including a more engaging conversational style of TBT, people will begin to relate to the material, learn from each other and teach each other. Sharing of anecdotes among workers would leave a lasting effect and ensure that the safety talks are more memorable.

Consistency issues of duration and frequency of TBTs arose from site. Despite the surveyed managers stating two TBTs per week are completed, foremen explained that they decided themselves if more TBTs needed to be done depending on the work being completed during any given week. Furthermore, a main contractor and or safety officer may require an extra talk to be done on a specific topic during the week. This drives inconsistency in how TBTs are delivered and their impact on operatives and managers. An external person coming in giving a TBT would also add to disruption. Opinions on external TBTs sparked strong views from managers and supervisors, finding their authority undermined by external trainers. They also said that an external person would not know the specific issues faced on site and thus workers would not find their training insightful. Workers should know and trust who is going to deliver these safety talks to maximise their impact.

The American Society of Safety Professionals (ASSP.Org, 2018) recommends a TBT should last 10 to 15 min. The mixed response of how long the current TBTs last versus how long it should last was highlighted by workers stating that they did not know what the optimum time would be for them because of the inconsistency they encountered. Linking back to the literature review stating a TBT should last 15 min, this is what should be used for a standard. If a company has a strict number of TBTs to be done and the duration is to be the same, this consistency facilitates a routine for everyone. Consistency can be achieved by having a set time allocation and place and same number of TBTs given per week, and this would be accommodating for operatives who move around different sites.

One of the barriers encountered was that TBTs are not being delivered to fully suit each learner type. This can then limit the 'stickiness' of TBTs on workers, i.e. retention of information. When asking what learner type each person was, it was found that a great deal of people did not understand what was being asked, even after the researcher had explained the differences. After further explanation, six of the participants were still unsure which learner type they were. During some of the interviews, the researcher found that some of the operatives looked for the 'right answer' and they were found to be looking into each other's answers. It was then explained to them that it was not a quiz and that it was just their opinion the researcher needed, that there was no right or wrong. From the literature, it is clear that the way adults learn requires that an andragogy approach to teaching should be taken. 60% of workers who participated in this study said their learner type was kinaesthetic with just 4 workers falling into the visual and auditory categories.

A major issue seen on the site was language barriers reducing the impact of TBTs. The company that was involved in the pilot study was based in London; however the operatives are predominantly non-native, and English is not their first language. All the improvements already outlined could be implemented, but the language barrier has the potential to negate all of the positive changes. In other words, any new process for TBTs will fail if the language barrier is not minimised. As Cocerhan and Bradley (2019) stated, a Google translate application could help foreign workers be more effective and productive while enhancing site safety. Suggestions by those interviewed also agreed that assistive technology would help them to fully understand the TBTs, with no important detail slipping through due to translation errors.

A TBT directly affects a worker's performance and safety and is a chance for them to gain knowledge on tasks they will be completing on site. Increasing the efficacy of TBTs will have a positive impact on workers. In order to do this, supervision of the talks from managers is required to fix any issues with the current process. Workers should be allowed and encouraged to give feedback of the receiving talks. The TBT needs to be discussion, demonstration and anecdote-based, and workers should be encouraged to drive these discussions. Consistency is also required and can be achieved through allotted times for TBTs, and a standard for their duration and frequency agreed and implemented. Assistive technology should be used to combat any language barriers that arise, and the workers learner types need considering for optimal impact.

19.6 Conclusion

The basic hypothesis that this study sets out to examine was that inconsistencies with content, design and delivery of TBTs had a negative effect on their impact. Observations and interviews with experienced workers completed on a London construction site provided data on the inconsistencies blocking effective TBTs that proved that this hypothesis was correct.

The focal points of the interviews were people's perception of TBTs and their participation levels and thoughts on the delivery method of TBTs. The results complemented what was found in the literature review, and both sources were used to form suggested changes that would lead to improvements to the current approach.

Consideration of current barriers as described in the literature such as the delivery methods, consistency and lack of assessment, coupled with the findings from this research, allowed a method for optimising the overall process of a TBT to be proposed. Changes to bring improvements include the following: (a) More resources and money should be going into TBTs in terms of supporting the design and delivery so as to improve the suitability of the format to allow maximum benefit. (b) The style of training needs to be suitable for the adult learners in the TBT group, making the material accessible and, more importantly, relatable for the workers. For example, construction workers who have the kinaesthetic learning style require training delivered in a way that suits them—in other words formatting the TBT to include demonstration and discussion to make the training more interactive. (c) Train the trainers and ensure that they understand the importance of matching the training delivery method with the needs and learning styles of the people in front of them. Andragogy-style teaching needs to be considered by people giving the TBT. (d) More resources need to go into supervising the people delivering TBTs, so that workers and managers get the most out of TBTs and continuous improvement can be facilitated. These suggestions if implemented would deliver a consistent and impactful TBT experience for all workers on site irrespective of level, language or learning style.

Possible future research in this topic would be assessing the effectiveness of the implementation of the changes suggested above by focusing on work habits, accidents, workers' behaviour and perceptions before and after the TBT. Monitoring the changes being made and the impact such changes have on workers would allow for further adjustments to be made if necessary. As in any improvement, there is a need for resources to facilitate the TBT system changes, but the return on investment could be significant.

References

Assp.org. (2018). *Measuring the effectiveness of toolbox safety training.* [online] https://www.assp.org/news-and-articles/2018/11/08/measuring-the-effectiveness-of-toolbox-safety-training

Bahn, S., & Barratt-Pugh, L. (2012). Emerging issues of health and safety training delivery in Australia: Quality and transferability. *Procedia-Social and Behavioral Sciences, 62*, 213–222.

Boendermaker, P., Conradi, M., Schuling, J., Meyboom-de Jong, B., Zwierstra, R., & Metz, J. (2003). Core characteristics of the competent general practice trainer, a Delphi study. *Advances in Health Sciences Education, 8*(2), 111–116.

Burke, M., Sarpy, S., Smith-Crowe, K., Chan-Serafin, S., Salvador, R., & Islam, G. (2006). Relative effectiveness of worker safety and health training methods. *American Journal of Public Health, 96*(2), 315–324. https://www.ncbi.nlm.nih.gov/pmc/articles/PMC1470479/.

Cocerhan, I., & Bradley, J. (2019). *Investigation into how technology can overcome language barriers experienced by construction workers from Eastern Europe on Sites in London.*

Cpwr.com. (2019). *Handouts & toolbox talks*. CPWR. https://www.cpwr.com/publications/handouts-and-toolbox-talks

Eggerth, D., Keller, B., Cunningham, T., & Flynn, M. (2018). Evaluation of toolbox safety training in construction: The impact of narratives. *American Journal of Industrial Medicine, 61*(12), 997–1004.

Ellis, R., & Loewen, S. (2007). Confirming the operational definitions of explicit and implicit knowledge in Ellis (2005): Responding to isemonger. *Studies in Second Language Acquisition, 29*(01).

Garavan, T., Hogan, C., & Cahir-O'Donnell, A. (2012). *Making training & development work*. Oak Tree Press.

Health and Safety Authority. (2020). *Home*. [online] https://www.hsa.ie/eng/

Heinrich, H. (1941). *Industrial accident prevention*. McGraw-Hill.

Holmes, G., & Abington-Cooper, M. (2000). Pedagogy vs. andragogy: A false dichotomy? *The Journal of Technology Studies, 26*(2), 50–55.

Holt, A. (2001). *Principles of construction safety*. Blackwell Science.

Iwh.on.ca. (2020). [online]. https://www.iwh.on.ca/sites/iwh/files/iwh/reports/iwh_accomplishments_2017.pdf

Jannadi, M. (1996). Factors affecting the safety of the construction industry. *Building Research & Information, 24*(2), 108–112.

Jeschke, K., Kines, P., Rasmussen, L., Andersen, L., Dyreborg, J., Ajslev, J., Kabel, A., Jensen, E., & Andersen, L. (2017). Process evaluation of a toolbox-training program for construction foremen in Denmark. *Safety Science, 94*, 152–160.

Kaskutas, V., Jaegers, L., Dale, A. M., & Evanoff, B. (2016). Toolbox talks: Insights for improvement. *Professional Safety, 61*, 33–37. https://www.onepetro.org/journal-paper/ASSE-16-01-33.

Laberge, M., MacEachen, E., & Calvet, B. (2014). Why are occupational health and safety training approaches not effective? Understanding young worker learning processes using an ergonomic lens. *Safety Science, 68*, 250–257.

Olson, R., Varga, A., Cannon, A., Jones, J., Gilbert-Jones, I., & Zoller, E. (2016). Toolbox talks to prevent construction fatalities: Empirical development and evaluation. *Safety Science, 86*, 122–131.

Varley, F. D., & Boldt, C. M. (2002). Mining Publication: Developing toolbox training materials for mining. DHHS (NIOSH) Publication No. 2002-156. *Info Circular, 9463*, 39–44.

Part VI
Programme Management, Project Management and Supply Chains

Chapter 20
Bagging a Bargain Begets Amnesia: Insights of Integrating Responsible Sourcing into Building Information Modelling

Sophie Ball and Colin A. Booth

20.1 Introduction

The construction sector currently accounts for 7% of UK GDP, and collective concerns from the public due to increased awareness of the planet's environmental limits mean architecture, engineering and construction (AEC) organisations have amplified their accountability to stakeholders (Upstill-Goddard et al., 2013). The sectors have obvious impacts on the environment—including pollution, alteration of landscapes and high levels of manufacturing and waste (Zhou & Lowe, 2003). The industry has, therefore, become expressive when considering transparency, accountability and legitimacy; however, research suggests some AEC organisations trail behind in these traits (Glass, 2012).

The evolution of corporate social responsibility (CSR) means organisations need to consider smart decisions within internal processing. Amplified consciousness leads organisations to become positive influencers of socio-environmental activities, which address changing social orders (Finster & Hernke, 2014). One emerging aspect is responsible sourcing (RS). Through investment in supply chains, organisations can align with ethically sound materials and products. This addresses sustainability issues, as well as having a positive impact on purchasing decisions, which can create stakeholder confidence (Glass, 2012).

Building Information Modelling (BIM) has established itself as a key tool for many AEC organisations to aid project management, allowing for collective

S. Ball (✉)
Kier Construction, The Old Mill, Chapel Lane, Warmley, Bristol, UK
e-mail: sophie.ball@kier.co.uk

C. A. Booth
Centre for Architecture and Built Environment Research (CABER), Faculty of Environment and Technology, University of the West of England (UWE), Bristol, UK

collaboration by stakeholders from pre-construction through to operation and maintenance. Also, BES6001 provides a framework for RS construction products. The standard provides support for both manufacturing and purchasing to effectively manage supply chains. Therefore, this study aims to explore the benefits and barriers of implementing and integrating RS into the BIM systems of AEC organisations.

20.2 Literature Review

Consumers are main drivers of sustainable product manufacturing by demanding producer responsibility, radical reduction of resource use and renewable energy for production and buildings that have strong environmental certifications (Finster & Hernke, 2014). There is growing awareness that current practices are incompatible with increasing demands. This is due to high resource use and rising scarcity in materials, meaning improvement in production and waste management is required. Industry must respond by delivering effective and efficient outputs.

The term 'supply chain' continues to change depending on perspective, but generally, it is described as the management of manufacturing or logistics of transport and distribution. The manufacturing of a supply chain is self-explanatory—the extraction of natural resources and production of items. Logistics is the functional area of a supply chain which controls the efficient and effective forward and reverse flows of goods from the point of origin to the point of consumption (Christopher & Peck, 2004). Supply chains deplete natural resources, and though it can be surmised that some natural resources are renewable (water, timber, soil), the rate of extraction interrupts healthy replenishment rates (Agrawal & Lee, 2016).

Frameworks for assessing buildings (such as BREEAM, LEED) do not necessarily consider the materials used throughout a project. It is apparent that the AEC sector, therefore, requires an overarching method to regard the quality, environment and health and safety of developments (Ghumra et al., 2009). Resilient supply chains are highly coveted, and organisations are continually looking to achieve robust processing. Traditionally, supply chains are viewed in a linear sense—from cradle to grave. As organisations look to improve their processes, responsible sourcing has emerged as a new procurement method (Presutti Jr, 2003). RS is achieved by investigating supply chains and devising actions, which ensure the use of ethically sound products. Acquiring materials with certified status means organisations can make positive purchasing decisions (Glass, 2012). By choosing responsible materials, the impacts of construction can be reduced before reaching site.

As with most change, there is a value-action gap. This means that whilst initiatives may appear valuable, they still have low uptake (Guo et al., 2015). Bowen et al. (2001) explored the value aspect and found that 60% of managers strongly agreed that their company should take responsibility for the impacts of their suppliers. There are various reasons why organisations do not investigate their supply chains to make more sustainable choices—including lack of awareness (Glass, 2012), outsourcing, risk, time (Khoo et al., 2001), quality management

(Upstill-Goddard et al., 2012) and cost (Agrawal & Lee, 2016). These barriers mean certain areas of the sector are wary to begin sustainable processing. Users' influence over sustainable building is blighted by the lack of established platforms between designers and consumers, along with lack of cohesion between sector stakeholders (Rohracher, 2001). To move forward, key stakeholders must be consulted throughout development to reduce clashes of interest (Meehan & Bryde, 2011).

BIM seeks to integrate processes and enable collaborative working environments, including all stages of building life cycles. It uses software programmes to model both object-oriented and spatial semantics to unlock more efficient methods of designing and creating assets (Mahamadu et al., 2013). A range of stakeholders can utilise the framework dependent on interest and layer information into one central file, which can be used throughout the building life cycle (Gruen et al., 2009). By creating a coordinated model, BIM has emerged as a solution to facilitate more sustainable construction. The use of BIM throughout projects creates feedback loops, which simulate stakeholder interests. This can be reused in future projects to improve material efficiency and reduce waste, timescales and errors (Whyte & Hartmann, 2017).

The underlying principle of successful supply chains is the collaborative exchange of information (Christopher & Peck, 2004). BIM allows users to electronically store and share information allowing for quick and remote access. Organisations may continually analyse and update information on products to ensure value. BIM, therefore, promotes responsible choices producing sustainable construction, as designs can be continuously analysed to maximise benefits.

Local sourcing has been linked to minimising waste (Jiao et al., 2013). By utilising BIM to assess materials through a four-phase framework, users are provided with a method to select the most sustainable component. Pairing this with effective scheduling and material calculations reduces a development of carbon footprint through streamlining of design and construction programmes, which can reflect positively on an organisations' reputation (Chen et al., 2010). However, whilst the use of BIM has obvious advantages, there is still a limited amount of uptake. BIM is still in its early stages of development in terms of clashes with language, intellectual property rights and data security (Mahamadu et al., 2013). One gap in BIM's software is within product information—using libraries of materials is useful for modelling purposes, but there is no indication of sourcing. This creates issues with quality and procurement (Eastman et al., 2011). Consequently, there remains a gap in the framework which could be closed with the right influx of product information.

Organisations readily recognise that making quality a stakeholder priority allows businesses to attain total quality management (Christopher & Peck, 2004). One solution to rising demand for construction procedures is accurate information on materials, which can be utilised by designers and contractors (Mustow, 2006). BES6001 is a UK framework standard for responsibly sourced construction materials through a points-based system. The BRE Global (2016) describes BES6001 as 'Responsible Sourcing of Construction Products is demonstrated through an ethos of supply chain management and product stewardship and encompasses social, economic and environmental dimensions.' This covers the following areas: organisational management, supply chain management requirements and environmental and social requirements. Three aspects of supply chain management are assessed under separate areas (material traceability, environmental management systems and health and safety management systems) which links to other standards.

Presently, the application of BES6001 is varied. For example, the UK concrete industry already has more than 91% of sector organisations carrying the BES6001 certification with 85% rated very good or excellent. Though BES6001 provides a point of reference and allows developments to access valuable credits into schemes such as BREEAM, there remains an issue with reliable summarising of overall performance. This lack of 360-degree perspective is partly due to the complicated and costly nature of life cycle analysis (LCA). There is also a lack of investigation as to the positive application of this within organisations, but previous research does imply links between certification and application (Upstill-Goddard et al., 2012).

From a review of the literature, it may be ascertained that most studies review RS and BIM as separate, and the potential for integration remains unexplored. Therefore, this is the first known study that explores the overlap of RS and BIM into one sustainable construction process.

20.3 Research Methodology

20.3.1 Literature Review

A review of existing responsible sourcing and building information modelling literature followed the PRISMA evidence-based transparent and complete reporting process, whereby articles were identified, screened and checked for eligibility before inclusion in a systematic or structured review (Moher et al., 2009). A host of databases (including Scopus, Web of Science, Emerald, Wiley-Blackwell, Taylor and Francis, amongst others) were searched for any accessible articles.

20.3.2 Data Collection

An objectivity-based methodology was utilised for this study. A quantitative cross-sectional questionnaire survey, aligned to the study's aim, was adopted as the method of inquiry. This enabled the capacity to collect a large amount of data in a controlled setting with minimal influence by the researcher. The choice of this strategy was borne out of the need to capture both the benefits and the barriers of integrating and implementing responsible sourcing in BIM systems. The instrument was developed through an iterative process of literature reviews and consulting experts to refine the measurement items, before piloting with industry and academic professionals. Feedback from the pilot exercise was used to amend and address possible issues before the final version of the questionnaire was distributed.

The main purpose of the questionnaire was to solicit construction professional's perceptions of the benefits and the barriers of integrating and implementing responsible sourcing in building information modelling. Section 1 was designed to capture the participant's personal details; Section 2 listed nine factors used to determine apparent advantages of integrating and implementing responsible sourcing in building information modelling; and Sect. 3 listed eight factors used to gauge perceived challenges against integrating and implementing responsible sourcing in building information modelling. For Sects. 2 and 3, participants were asked to record their ratings for each factor on a 5-point Likert-type scale (*1, strongly disagree, 2, disagree, 3, neutral, 4, agree, and 5, strongly agree*) in a horizontal grid system stored as ordinal data. A small number of open-ended questions (dual approach) were also included to elicit rich qualitative data, alongside the quantitative approach.

Ethical approval was sought before the final questionnaire was shared. Approval meant all participants were informed in a participant information cover letter that their involvement was entirely voluntary and their decision to complete and return their completed questionnaire was their consent to take part in the study. As their responses would be anonymous, participants were also informed that there would be no opportunity to withdraw once the questionnaire had been returned.

20.3.3 Data Analysis

The primary data was entered into Excel (2016 version) and analysed by using a descriptive statistical tool—the central tendency measure of weighted average. Demographical information was analysed by means of frequency analysis to provide a snapshot of the respondents' characteristics.

The following weighted average formula was used to calculate the average score for each factor, where WAS_i denotes the weighted average score for each factor i, α_j denotes the numerical value for each ranking level in which 1 is allocated to the lowest rank and 5 is allocated to the highest rank, n_{ij} denotes the number of respondents for factor i with ranking level αj and N denotes the total number of respondents for the question.

Formula for the Weighted Average Score

$$\text{WAS}_i = \frac{\sum_{j=1}^{5}\left(\alpha_j n_{ij}\right)}{N} \qquad (20.1)$$

An additional formulation was required to address the weakness of the weight average score, which did not account for the degree of variation between the responses. Hence, a coefficient of variation was added to each of the weighted average scores to compute the beneficial index value (BIV), which determined the final rankings as shown in the formula below.

Formula for the Benefit/Barrier Index Value

$$\text{BIV}_i = \text{WAS}_i + \frac{\text{WAS}_i}{\delta_i} \qquad (20.2)$$

where BIV_i denotes the beneficial index value for each factor i and δ_i denotes the standard deviation for each factor i.

20.4 Research Results and Discussion

Findings from the analysis of these responses are presented and discussed beneath under the main sections: (1) Participant and Organisational Demographics; (2) Benefits to Implementing and Integrating Responsible Sourcing Within BIM Systems; and (3) Barriers to Implementing and Integrating Responsible Sourcing Within BIM Systems.

20.4.1 Participant and Organisational Demographics

The respondents were from organisations within procurement, management, design or construction—no one considered their role outside this scope. Most respondents (75%) were from 'medium-sized organisations' (>151 employees). All participants were professionally certified or degree-level educated, and all organisations already had a sustainability policy (60%) or were in the process of creating one.

Resilient supply chains are crucial to resource heavy sectors, and it is encouraging that most respondents have policies dedicated to RS (75%) and CSR (67%). Acquiring materials which are responsible allows organisations to make positive purchasing choices (Glass, 2012). Of the organisations surveyed, 100% of respondents value RS within their supply chain.

Over 80% of respondents believe in mandatory responsible sourcing policies. A robust supply chain is key for any strong performing organisation, and the implementation of mandatory standards seems a simple modification to influence change within the sector. The implementation of RS initiatives within organisations does produce various positive outcomes including gaining competitive advantage, customer loyalty and increased brand value (Cohen & Robbins, 2011).

20.4.2 Benefits to Implementing and Integrating Responsible Sourcing within BIM Systems

The questionnaire listed nine factors considered to be advantages to organisations in the construction industry (Table 20.1). Analysis of the questionnaire responses (Table 20.2) has been used to list the beneficial factors of implementing and integrating responsible sourcing within BIM systems in a ranked order of importance (Table 20.3).

20 Bagging a Bargain Begets Amnesia: Insights of Integrating Responsible Sourcing...

Table 20.1 List of beneficial factors

1	RSa	RS and BIM are key drivers for the progression of our organisation
2	RSb	RS and BIM provide a competitive advantage for our organisation
3	RSc	RS and BIM promote collaboration
4	RSd	RS and BIM can be implemented without substantial changes to our existing systems
5	RSe	RS and BIM can be implemented without extensive staff upskilling
6	RSf	RS and BIM allow our organisation to exceed minimum compliance/regulation
7	RSg	RS and BIM help our organisation to be accountable for its sustainability impacts
8	RSh	RS and BIM encourage ethical related decisions
9	RSi	RS and BIM add value for our stakeholders

Table 20.2 Calculation of parameter values for the beneficial factors

	ASS	Variance	SD	BIV	BIV rank
RSa	3.58	0.74	0.86	7.74	6
RSb	4.33	0.22	0.47	13.53	1
RSc	4.08	0.91	0.95	8.36	4
RSd	2.00	0.83	0.91	4.19	9
RSe	2.42	1.24	1.11	4.58	8
RSf	3.58	0.24	0.49	10.85	3
RSg	3.92	1.24	1.11	7.43	7
RSh	3.92	0.91	0.95	8.02	5
RSi	3.67	0.22	0.47	11.44	2

Table 20.3 List of beneficial factors in ranked order

1	RSb	RS and BIM provide a competitive advantage for our organisation
2	RSi	RS and BIM add value for our stakeholders
3	RSf	RS and BIM allow our organisation to exceed minimum compliance/regulation
4	RSc	RS and BIM promote collaboration
5	RSh	RS and BIM encourage ethical related decisions
6	RSa	RS and BIM are key drivers for the progression of our organisation
7	RSg	RS and BIM help our organisation to be accountable for its sustainability impacts
8	Rse	RS and BIM can be implemented without extensive staff upskilling
9	RSd	RS and BIM can be implemented without substantial changes to our existing systems

Based on the final ranking, the three top benefits were that RS and BIM provide a competitive advantage (RSb), add value for stakeholders (RSi) and enable minimum compliance (RSf), whilst the lowest three ranking benefits were accountability (RSg), training (RSe) and changes to existing systems (RSd). The following section attempts to explain and discuss these findings.

The AEC sector remains a multi-faceted and complex market, facing a paradigm shift to increase productivity, efficiency, value, quality and sustainability. This requires resilient and collaborative business models to maintain dominance. The pressures as a result of activism, legislation and awareness mean that sustainability must be embraced and companies must strive to incorporate it as part of their supply chains to ensure competitive advantage (Bubicz et al., 2019). Arayici et al. (2011) undertook a systematic review of BIM implementation and found that AEC SMEs experience higher efficiency and competitive advantage. This barrier also remains at the forefront in more recent studies (Hong et al., 2017; Charef et al., 2019; Oraee et al., 2019; Chan et al., 2019). Kaner et al. (2008) also supports this notion through a study which analysed case studies within the AEC industry. This was as a result of reduced design lead times and reduction of clashes and errors. This study reinforces the importance of competitive advantage as it was found to be the highest ranked barrier.

Organisations strive to be viewed as positive societal influencers due to the development of CSR; therefore it can be surmised that the perceived increase in access to clients encourages organisations to maintain positive public relations (Upstill-Goddard et al., 2013). A firm's reputation is improved through tangible action, rather than corporate communication. By effectively managing supply chains, it can be evidenced that an organisation is meeting stakeholder expectations. Ofori (2000) describes supply chain management as essential to the performance and competitiveness of construction enterprises. This is because of the range of influencers across each project—a variety of materials, components, stakeholders' consultants and products are used. Additionally, Hoejmose et al. (2013) recognised that an organisation's desire to achieve competitive advantage drives responsible decision-making. The implementation of responsible initiatives within organisations has been found to produce various positive outcomes including gaining competitive advantage, customer loyalty and increased brand value (Cohen & Robbins, 2011).

The research has identified stakeholder value as another key benefit. In addition to this, it can be inferred that an organisation's ability to assure quality delivery has worth. This is mirrored by a study by Ageron et al. (2012) which noted that management and stakeholder's opinions were a key influencer over sustainable supply chain management. Strategies, which empathise with stakeholder's opinions and are pro-active, consequently produce strong stakeholder relationships (Epstein, 2008). The benefit of responding to public concern to drive attitudes and promotion of their organisation with self-regulation are echoed by Finster and Hernke (2014). Environmental consciousness has increased in recent years, which in turn enables organisations to be positive influencers. This combined with enthusiasm of employees where organisational processes are concerned is vital when introducing responsible practice (Upstill-Goddard et al., 2013). From the participants' responses, it can be deduced that public relations are important drivers for environmental responses within their organisations. Stakeholders are at the centre of ecological modernisation and benefit from developing a bespoke form of governance. This research therefore supports Cohen and Robbins (2011) findings that the implementation of initiatives within organisations does produce positive outcomes. Chan et al. (2019) reinforce the theory that growing stakeholder demand for CSR and benevolent activities are a determinant in the maintenance of a successful supply chain.

A previous barrier to RS and BIM uptake has been client reluctancy. This is based primarily on return on investment and the appreciation and definition of value. Some clients consider only tangible values without considering intangible factors (Becerik-Gerber & Rice, 2010). This study emphasises the changing attitudes of AEC stakeholders in terms of perceived value of RS and BIM. However, whilst stakeholder value was the highest-ranking barrier within this study, it should also be noted that depending on a project size, funding stream and affiliations, it can be difficult to manage stakeholder expectations (Bubicz et al., 2019).

When considering the AEC sector, the external pressures from legislative bodies must be recognised. Both consumers and legislation have prompted organisations to become more environmentally friendly (Frota Neto et al., 2008). However, in recent years compliance has become a minimum, and stakeholders have begun to expect more. Companies seem to agree that RS enables exceeding legislative standards. This implies companies comply because of regulation and risk of penalties. Seuring and Muller (2008) explored the impact of compliance with legislation and mandatory changes in the adoption of new technologies. Opinions seem to be confirmed by results in terms of current practice. It appears that whilst many organisations accept the benefit of accountability and transparency, some of the changes already in their policies may be as a result of mandatory legislation. Whilst some academics disagree with legislative governance, participants in this research acknowledge the advantage of mandatory legislative changes (Boyd et al., 2007). The positive for regulation is that due to its compulsory nature, implementation does not require a positive employee attitude (Cohen & Robbins, 2011).

20.4.3 Barriers to Implementing and Integrating Responsible Sourcing Within BIM Systems

The questionnaire listed eight factors considered to be challenges to organisations in the construction industry (Table 20.4). Analysis of the questionnaire responses (Table 20.5) has been used to list the barrier factors of implementing and integrating responsible sourcing within BIM systems in a ranked order of importance (Table 20.6).

Table 20.4 List of barrier factors

1	BBa	We have not evaluated the implementation of RS and BIM
2	BBb	Our employees would require significant upskilling to successfully implement BIM
3	BBc	We use existing in-house collaboration software to deliver projects
4	BBd	We do not feel BIM protects our organisations' data and ideas
5	BBe	Other businesses we work with do not use BIM
6	BBf	Introducing RS and BIM will not help our organisation gain more clients
7	BBg	RS and BIM are too expensive
8	BBh	Responsibly sourced materials are not widely available

Table 20.5 Calculation of parameter values for the barrier factors

	ASS	Variance	SE	BIV	BIV rank
Bba	3.33	2.22	1.49	5.57	3
BBb	2.92	1.58	1.29	5.24	5
BBc	2.67	1.22	1.11	5.08	7
BBd	3.42	1.22	1.10	6.52	2
Bbe	2.33	1.56	1.25	4.20	8
BBf	2.83	1.47	1.21	5.17	6
BBg	4.00	1.00	1.00	8.00	1
BBh	2.50	0.75	0.87	5.39	4

Table 20.6 List of barrier factors in ranked order

1	BBg	RS and BIM are too expensive
2	BBd	We do not feel BIM protects our organisations' data and ideas
3	BBa	We have not evaluated the implementation of RS and BIM
4	BBh	Responsibly sourced materials are not widely available
5	BBb	Our employees would require significant upskilling to successfully implement BIM
6	BBf	Introducing RS and BIM will not help our organisation gain more clients
7	BBc	We use existing in-house collaboration software to deliver projects
8	Bbe	Other businesses we work with do not use BIM

Based on the results presented above, the three key barriers included cost (BBg), intellectual property challenges (BBd) and lack of evaluation into implementation (BBa), whilst the lowest three ranking barriers included upskilling (BBb), competitive advantage (BBf) and in-house software (BBc). The following section attempts to explain and discuss these findings.

The persistent reduction of cost is paramount to successful profit margins. Typically, the main barrier of sustainability investments has been cost—this is maintained within this study as the highest-ranking barrier for RS and BIM implementation. Giunipero et al. (2019) emphasised that cost and economic uncertainty were key hindrances to responsible supply chain management. Harwood et al. (2011) noted the importance of considering RS cost-risk. By managing risk and reducing conflicts between social responsibility and traditional procurement, AEC SMEs can maintain profitability whilst cutting capital cost. Undoubtedly, profits are an area which every organisation aims to improve whilst maintaining the delivery of quality outputs. Bowen et al. (2001) also identified a barrier to RS as cost due to larger capital cost and longer lead times.

An organisation's business case for RS products depends on both upstream suppliers and the consumer; a robust and transparent relationship between all units must be maintained (Meckenstock et al., 2015). Budget often dictates a project and costs of RS materials often fall on the client. As RS materials do not necessarily impact the use of a product, the value can be difficult to envisage. Carter and Jennings (2002) recognised three main barriers to RS when considering associated costs which included uncertainty over what 'responsible' means. This study reinforces that the perception of RS products is that they have longer lead times and higher initial costs, which mirrors the findings of Agrawal and Lee (2016). This perception means organisations are dissuaded from including them in their bids for tender, despite this being the earliest opportunity for setting out sustainability aspirations. The aforementioned issues create trade-offs for AEC organisations. Guo et al. (2015) studied this conflict and assessed whether it is important to buy from a responsible organisation or a risky cheaper organisation. They concluded that as the consumer begins to demand a more sustainable market, it will become more beneficial to choose RS.

In terms of BIM application, several studies have highlighted key barriers in relation to cost, including the upfront cost of applications and software and the implementation and upskilling of staff, due to a lack of industry professionals (Thomson & Miner, 2010; Ganah & John, 2014; Liu et al., 2015). When considering perceived costs of education to successfully apply BIM, it can be split into two elements—the cost of upskilling or hiring competent individuals and behavioural costs to ensure support and organisational change to integrate BIM effectively.

Compared to traditional two-dimensional drawings and specifications, BIM produces a coordinated extractable electronic design which can be shared quickly and efficiently. At the end of the design phase, a BIM model can have significant value for owners (Porwal & Hewage, 2013). Respondents in this study also highlighted protection of creative data and ideas as an important barrier. BIM is still in its early stages of development in terms of clashes with language, intellectual property rights and data security (Jiang & Chen, 2014). In line with this, Ghaffarianhoseini et al. (2016) suggested that widespread uptake is linked to issues with cyber-security, unclear legal jargon and intellectual property rights. Kivits and Furneaux (2013) also indicated a need for resolving the numerous financial and legal issues which currently limit the AEC sector's adoption of BIM.

Whilst BIM enables the collaborative and constant exchange of information and the full design accessible to team members, some organisations feel that there is the opportunity for unauthorised access and copyright infringements (Chien et al., 2014). BIM implementation also raises contractual issues relating to responsibilities, risk, information and asset sharing and data ownership. Additionally, there is the issue of wording within contract clauses, for instance, BIM protocol requires additional comprehensive intellectual property licences which are not provided and understood by all members of a design team (Udom, 2012).

Whilst sustainability and responsible sourcing are key to AEC organisations—with 92% of respondents operating an RS policy within their supply chains—there remains an issue of complexity. As sustainability is sub-divided into economic, social and environmental dimensions, it can be difficult to interpret how to incorporate sustainable behaviours. Despite respondents' concerns over the costs of RS materials, it would seem many have not completed active research into RS and BIM implementation. Though there are growing concerns over resource use and accountability, it remains difficult to achieve a responsible supply chain, partially due to a lack of awareness and fake views of sustainable products (Glass, 2012).

To achieve sustainable construction, the choice of product is crucial, and every decision affects supply chain management (Glass, 2012). Häkkinen and Belloni (2011) highlighted one technical barrier for sustainable construction—a lack of common frameworks. BES6001 has been highlighted as a solution to provide accurate information on materials which echoes the ethos and determination to reduce negative environmental impacts (Mustow, 2006). Furthermore, a barrier to sustainable construction is the depth in which products must be analysed, and the complications of this when outsourcing, time scales and cost are considered. All can inhibit the effective greening of an organisation's supply chain (Bowen et al., 2001; Boyd et al., 2007). This means that the use of an integrated framework has the potential to alleviate barriers to both RS and BIM.

20.5 Conclusion

This study makes a distinct contribution to knowledge by responding to a research gap at the interface of implementing and integrating responsible sourcing within BIM systems for the architecture, engineering and construction (AEC) sectors. The main findings of the study reveal the top perceived benefits of RS and BIM implementation from the perspective of AEC SME organisation as competitive advantage and value for stakeholders. There is also an appreciation that the use of RS and BIM enables compliance standards to be exceeded. The key barriers for uptake are cost, intellectual property rights and creativity and a lack of understanding and awareness of the practical application of both RS and BIM.

Over the last decade, increasing environmental awareness has led to a growing appreciation of sustainable construction, responsible sourcing and the implementation of all certification frameworks and how BIM can ensure and enable these processes. Whilst respondents within this study are aware of potential and tangible benefits to RS and their integration into BIM, actual uptake within the industry remains limited. Any change from traditional technology involves behaviour change to create a social norm. Tools need to be used to alleviate the adoption gap. Rogers (1995) divides organisations into five groups: innovators, early adopters, early majority, late majority and the laggards. The AEC sector is considered one of the more innovating markets attributable to attempts to reduce time and fierce competition.

Whilst many respondents did not operate dedicated sustainability teams, the opinions gauged through the research are considered valuable. Through this study, there are several recommendations for action:

- Firstly, one gap in BIM's software is within product information—using libraries of materials is useful for modelling purposes, but there is no indication of sourcing. This creates issues with quality and procurement. Consequently, there remains a hole in the framework which could be removed with the right influx of product information. By creating environments where there is easy access to sources which provide information on RS, it may be possible to increase the amount of RS products currently utilised by the AEC sector.
- Secondly, the study revealed that stakeholders are the primary driver for the implementation of responsible sourcing and BIM. Consequently, the recommendation is that rather than relying on the morals, budgets and aspirations of clientele, responsible sourcing should be made mandatory on all construction projects which require the use of BIM.
- Thirdly, an increase or continuation of investigation into the organisational barriers to uptake of both RS and BIM would allow a framework to be generated to pinpoint key areas for improvement.

These discussions are preliminary and require wider research to ascertain general attitudes to responsible sourcing within the AEC sector. Despite this, the survey investigated a new method to improving the AEC's uptake of RS by using BIM, and therefore results provide a starting point for research into the practical integration of frameworks into BIM. The aim was to gauge stakeholder opinion—key benefits and barriers have been successfully identified, and further research should build on validating these responses. It would be useful to conduct interviews with industry professionals who are aware of practical application—which should guide further research into the topic.

References

Ageron, B., Gunasekeran, A., & Spalanzani, A. (2012). Sustainable supply management: An empirical study. *International Journal of Production Economics, 1*(16), 168–182.

Agrawal, V. V., & Lee, D. (2016). Responsible sourcing. In A. Atasu (Ed.), *Environmentally responsible supply chains* (Springer series in supply chain management) (Vol. 3). Springer.

Arayici, Y., Coates, P., Koskela, L., Kagioglou, M., Usher, C., & O'Reilly, K. (2011). Technology adoption in the BIM implementation for lean architectural practice. *Automation in Construction, 20*(2).

Becerik-Gerber, B., & Rice, S. (2010). The perceived value of building information modeling in the US building industry. *Electronic Journal of Information Technology in Construction, 15*(2), 185–201.

Bowen, F. E., Cousins, P. D., Lamming, R. C., & Faruk, A. C. (2001). Horses for courses explaining the gap between the theory and practice of green supply. *GMI, 35*(1), 41–60.

Boyd, D. E., Spekman, R. E., Kamauff, J. W., & Werhane, P. (2007). Corporate social responsibility in global supply chains: A procedural justice perspective. *Long Range Planning, 40*(1), 347–356.

Bubicz, M. E., Barbosa-Pavoa, A., & Carvalho, A. (2019). Incorporating social aspects in sustainable supply chains: Trends and future directions. *Journal of Cleaner Production, 237*(1).

BRE Global. (2016). PN327 BES 6001v3.1 Guidance. Available from: http://www.greenbooklive.com

Carter, C. R., & Jennings, M. M. (2002). Social responsibility and supply chain relationships. *Transportation Research, 38*(1), 37–52.

Chan, D., Olawumi, T., & Ho, A. (2019). Critical success factors for building information modelling (BIM) implementation in Hong Kong. *Engineering, Construction and Architectural Management, 1*(1).

Charef, R., Emmitt, S., Alaka, H., & Fouchal, F. (2019). Building information modelling adoption in the european union: An overview. *Journal of Building Engineering, 25*(3).

Chen, D., Mocker, M., Preston, D., & Teubner, A. (2010). Information systems strategy: Reconceptualization, measurement, and implications. *MIS Quarterly, 32*(4), 233–259.

Chien, K.-F., Wu, Z.-H., & Huang, S.-C. (2014). Identifying and assessing critical risk factors for BIM projects: Empirical study. *Automation in Construction, 45*(1), 1–15.

Christopher, M., & Peck, H. (2004). Building the resilient supply chain. *The International Journal of Logistics Management, 15*(2), 1–14.

Cohen, N., & Robbins, P. (2011). *Green business: An A-Z guide*. Sage.

Eastman, C., Teicholz, P., Sacks, R., & Liston, K. (2011). *BIM handbook*. Wiley.

Eltantawy, R. A., Fox, G. L., & Giunipero, L. (2009). Supply management ethical responsibility: Reputation and performance impacts. *Supply Chain Management: An International Journal, 14*(2), 99–108.

Epstein, M. (2008). *Making sustainability work: Best practices in managing, measuring corporate social, environmental and economic impacts*. Greenleaf Publishing.

Finster, M., & Hernke, M. T. (2014). Benefits organisations pursue when seeking competitive advantage by improving environmental performance. *Industrial Ecology as a Source of Competitive Advantage, 8*(5), 652–662.

Frota Neto, J. Q., Bloemhof-Ruwaard, J. M., van Nunen, J. A. E. E., & van Heck, E. (2008). Designing and evaluating sustainable logistics networks. *International Journal of Production Economics, 111*(2), 22.

Ganah, A. A., & John, G. A. (2014). Achieving level 2 BIM by 2016 in the UK. *Computing in Civil and Building Engineering, 11*(1), 143–150.

Ghaffarianhoseini, T., Tookey, J., Ghaffarianhoseini, A., Naismith, N., Azhar, S., Efimova, O., & Raahemifar, K. (2016). Building information modelling (BIM) uptake: Clear benefits, understanding its implementation, risks and challenges. *Renewable and Sustainable Energy Reviews, 75*(1), 1046–1053.

Ghumra, S., Miles Watkins, M., Phillips, P., Glass, J., Frost, M. W., & Anderson, J. (2009). Developing an LCA-based tool for infrastructure projects. In *Proceedings of 25th annual conference of association of researchers in construction management* (Vol. 1, pp. 1003–1010).

Giunipero, L. C., Bittner, S., Shanks, I., & Cho, M. H. (2019). Analyzing the sourcing literature: Over two decades of research. *Journal of Purchasing and Supply Management, 25*(5), 100–221.

Glass, J. (2012). The state of sustainability reporting in the construction sector. *Smart and Sustainable Built Environment, 1*(1), 87–104.

Gruen, A., Behnisch, M., & Kohler, N. (2009). Perspectives in the reality-based generation, nD modelling, and operation of buildings and building stocks. *Building Research & Information, 37*(5–6), 503–519.

Guo, R., Lee, H. L., & Swinney, R. (2015). Responsible sourcing in supply chains. *Management Science, 62*(9), 2722–2744.

Häkkinen, T., & Belloni, K. (2011). Barriers and drivers for sustainable building. *Building Research & Information, 39*(3), 239–255.

Harwood, I., Humby, S., & Harwood, A. (2011). On the resilience of corporate social responsibility. *European Management Journal, 29*(4), 283–290.

Hoejmose, S., Brammer, S., & Millington, A. (2013). An empirical examination of the relationship between business strategy and socially responsible supply chain management. *International Journal of Operations & Production Management, 33*(5), 589–621.

Hong, Y., Hammad, A. W. A., Sepasgozar, S., & Akbarnezhad, A. (2017). BIM adoption model for small and medium construction organisations in Australia. *Engineering. Construction and Architectural Management, 11*(1), 20.

Jiang, J., & Chen, J. (2014). Contractual collaboration systems of mega complex construction project organisations. *Bridges, 10*.

Jiao, Y., Wang, Y., Zhang, S., Li, Y., Yang, B., & Yuan, L. (2013). A cloud approach to unified life-cycle data management in architecture, engineering, construction and facilities management: Integrating BIMs and SNS. *Advanced Engineering Informatics, 27*(1), 173–188.

Kaner, I., Sacks, R., Kassian, W., & Quitt, T. (2008). Case studies of BIM adoption for precast concrete design by mid-sized structural engineering firms. *Journal of Information Technology in Construction, 13*.

Khoo, H. H., Spedding, A., Bainbridge, I., & Taplin, D. M. R. (2001). Creating a green supply chain. In *Voluntary environmental agreements: Process, practice and future use* (pp. 71–88). Autumn.

Kivits, R. A., & Furneaux, C. (2013). BIM: Enabling sustainability and asset management through knowledge management. *The Scientific World Journal, 1*.

Liu, S., Xie, B., Tivendal, L., & Liu, C. (2015). Critical barriers to BIM implementation in the AEC industry. *Int. Journal of Marketing Studies, 7*(6), 100–123.

Mahamadu, A., Mahdjoubi, L., & Booth C. (2013). Challenges to BIM-cloud integration: Implication of security issues on secure collaboration. In *IEEE International conference on cloud computing technology and science*. IEEE.

Meckenstock, J., Barbosa-Povoa, A., & Carvalho, A. (2015). The wicked character of sustainable supply chain management: Evidence from sustainability reports. *Business Strategy and the Environment, 25*(7), 449–477.

Meehan, J., & Bryde, D. (2011). Sustainable procurement practice. *Business Strategy and the Environment, 20*(2), 94–106.

Moher, D., Liberati, A., Tetzlaff, J., & Altman, D. G. (2009). The prisma group preferred reporting items for systematic reviews and meta-analyses: The PRISMA statement. *PLoS Medicine, 6*(7), e1000097.

Mustow, S. (2006). Procurement of ethical construction products. *Proceedings of the Institution of Civil Engineers Engineering Sustainability, 159*(ES1), 11–21.

Ofori, G. (2000). Greening the construction supply chain in Singapore. *European Journal of Purchasing and Supply Management, 6*(304), 195–206.

Oraee, M., Hosseini, M. R., Edwards, D. J., Li, H., Papadonikolaki, E., & Cao, D. (2019). Collaboration barriers in BIM-based construction networks: A conceptual model. *International Journal of Project Management, 37*(1), 839–854.

Porwal, K., & Hewage, N. (2013). Building information Modeling (BIM) partnering framework for public construction projects. *Automation in Construction, 31*, 204–214.

Presutti, W. D., Jr. (2003). Supply management and e-procurement: Creating value added in the supply chain. *Industrial Market Management, 32*(5), 219–226.

Rogers, E. M. (1995). *Diffusion of innovations* (4th ed.). The Free Press.

Rohracher, H. (2001). Managing the technological transition to sustainable construction of buildings: A socio-technical perspective. *Technology Analysis and Strategic Management, 13*, 137–150.

Seuring, S., & Muller, M. (2008). Core issues in sustainable supply chain management – A Delphi study. *Business, Strategy and the Environment, 17*(1), 455–466.

Thomson, D. B., & Miner, R. G. (2010). Building information modeling – BIM: Contractual risks are changing with technology. *Guest Essays.*

Udom, K. (2012). *BIM: Mapping out the legal issues.* NBS (online).

Upstill-Goddard, J., Glass J., Dainty, A., & Nicholson, I. (2012). Integrating responsible sourcing in the construction supply chain In S. D. Smith (Ed.), *Proceedings 28th annual ARCOM conference* (pp. 1311–1319), 3–5 September 2012, Association of Researchers in Construction Management.

Upstill-Goddard, J., Glass, J., Dainty, A., & Nicholson, I. (2013). Characterising the relationship between responsible sourcing and organisational reputation in construction firms. In *Sustainable building conference,* Coventry University.

Whyte, J. K., & Hartmann, T. (2017). How digitizing building information transforms the built environment. *Building Research and Information, 45*(6), 591–595.

Zhou, L., & Lowe, D. (2003). Economic challenges of sustainable construction. engineering. In *Proceedings of the RICS foundation construction and building research conference* (pp. 1–2).

Chapter 21
An Analysis of Adversarial/Cooperative Attitudes in Construction Contracting: How Approaches to Adversarial Procurement Might Have a Lasting Effect on Project Culture

Emily Harrison and John Heathcote

21.1 Introduction

The research aim of this study was to investigate whether the issue of resistance to partnering in construction lies with individuals' behavioural tendencies created by an adversarial culture. The study focuses on construction professionals, whose attitude might have been most affected by the construction industry's approach to procurement, the quantity surveyor. The study tested, using the prisoner's dilemma experiment (an explanation for the game is made in Fig. 21.1), how construction professionals, both from a main contractor and from a partnering framework background, compared to a control group act in a test of *cooperation* vs *defection*. This would allow the study to identify any group bias towards cooperation or competition and demonstrate the potential of an adversarial competitive environment to influence behaviours in what has been shown to be a negative outcome for project value.

Problem Statement: Construction professionals act in an adversarial manner towards construction contracting, despite research that proves this problematic for project success and greater sustainable outcomes.

Research Question: Are construction professionals conditioned by their environment to act against what is mutually beneficial for project success?

Research Aim: The overall aim of this research was to test how construction professionals operate in the prisoner's dilemma experiment.

Research Design: An experiment is proposed, which asks participants to 'play' the prisoner's dilemma game.

E. Harrison (✉) · J. Heathcote
Leeds Beckett University, School of the Built Environment, Engineering and Computing, Leeds, UK
e-mail: E.Harrison2598@student.leedsbeckett.ac.uk

> Investopedia defines the Prisoner's Dilemma as *"a paradox in decision analysis, [where] two individuals acting in their own self-interests do not produce the optimal outcome. [It] is set up in such a way that both parties choose to protect themselves at the expense of the other participant. As a result, both participants find themselves in a worse state than if they had cooperated with each other in the decision-making process"* (Chappelhow, 2019).
>
> Utilising 'game theory' this basic game allows for two possible outcomes, players either 'cooperate' or 'defect'. Depending on what their adversary 'plays' a score is arrived at. If both players cooperate, both receive 3 points each. If both 'defect' they both receive '1'point each. If one 'defects' and the other 'cooperates' then the defector receives 5 points and the 'co-operator' loses scoring '0'. This makes the game challenge about tricking the other into cooperating, so you can take advantage of them by defecting and scoring 5 points; or communicating, through play/action, that you intend to cooperate and inducing a series of cooperative plays that gains each player 3 points each time. Collectively a cooperative strategy beats a defecting strategy, but individually a 'sneaky defector' might out score their opponent. Game strategies will reveal the strategies of a player; their tendency to favour a defect or a cooperative strategy.

Fig. 21.1 An explanation of the game; 'prisoner's dilemma' which forms the basis of the experiment

Despite the considerable and comprehensive research, experiments and interviews carried out by the academics in the project management field, stating the proven benefits of partnering, cooperation and collaboration, the resistance to change within the construction sector is still robust, and attributed this to the *competitive tendering* and the structure of the industry around this competitive bidding premise. The industry itself may have entrenched values which trickle down into taught approaches and management styles, ultimately affecting a project outcome in one way or another. The construction professionals, closest to the construction contract, are quantity surveyors; this group often takes up the role as project managers and are, by nature of their profession, close to the importance of the cost management of the project. Existing literature has made the case for UK construction to take a more holistic approach to projects in order to achieve a more favourable and valuable outcome for all parties.

It can be argued the 1994 *Constructing the Team* Report, written by Sir John Latham, was the catalyst which sought to drive the UK construction industry to modernise, an industry which beforehand was argued as primitive and unrecognisable (Gardiner, 2014). Open tendering resulting in cutthroat bidding, which Reeves argued, resulted in 'contractor tender price[s] invariably inadequate to do the job… the contractor starved the supply chain of cash …[and led to] a glut of litigation' (Gardiner, 2014). Adamson and Pollington also note that at this time the 'construction industry spent more on litigation, than on training, research or development' (2006). As such Latham proposed that at the heart of any project, there should be a strong sense of teamwork and partnering in order to collaborate for the common good of the project (1998), in turn strengthening the role of the project manager. Although others had tried previously, Latham had the first compelling argument for moving towards collaboration and based much of the report on emphasising the importance and approach to implementing these radical ideas to constructing a team, as the industry could not 'wait 30 years for another Banwell or 50 years for another Simon', his predecessors who had both tried and failed to make progressive change to the UK construction industry.

Latham anticipated that implementation of his recommendations would happen when the client adopted the report's suggested approaches. These changes cannot be met without client agreement and suggest that clients need to be *better clients/ employers* (including the UK government) (Latham, 1994). Furthermore, the client still needed to ensure the full development of a design, before the construction team are procured, resulting in the requirement for a full design team to scrutinise and problem-solve ahead of any tenders being sent out, which would then result in less risk, often in less cost and more certainty of outcome, for all involved in the project (Latham, 1994).

One prominent outcome of the Latham report was the step towards and ultimate entrenchment of adjudication in UK law, which provided a cost-effective alternative to litigation. This can be shown by the vast reduction of disputes ending up in court over the past two decades. However, Morrell, the UK's first Government Construction Advisor, argued that the industry is still adversarial and that disputes are merely managed more effectively. Of the key outcomes of the Latham report, the NEC suite of contracts is also argued, a contract which obliged any party to alert the other as soon as any problems arose or otherwise lose the right to later base a claim on the project (Gardiner, 2014). However according to a 2010 RICS survey, just 7% of contracts in use were NEC, with JCT still being the preference (Langdon, 2010). The NEC did cover 26% of the value of the work, with NEC projects often tending to be much larger; however just 1% of all contracts were formal partnering agreements (Langdon, 2010). Suggesting that 16 years on, the Latham report may have not made as much of an impact as many think.

Following on from Egan and advancing Latham's ideas with a further report *Re-Thinking Construction* in 1998, Egan looked at the supply chain in more detail, alongside the notion of continuous improvement. Egan also highlighted the need for a change to the culture and structure of the construction industry, who at the time, and some argue still, is unsupportive of any improvement (1998). His ideas at the time of his writing of the report were not however unfamiliar. Supply chain analysis and supply chain transparency are mainstream to the manufacturing industry, where partnering and production partners and shared learning are the common approach taken. There is respect and understanding in manufacturing that without each other they will not succeed, an idea which is yet to permeate the construction industry. The manufacturing industry employ longer-term partnerships with their tier 2 contractors, also known as the 'doers' on any construction project; this has proven to lower cost through the efficiency of working over more than one project (Heathcote et al., 2018). Egan however raised this idea of shared gains and argues for a 'framework solution' that utilised the 'increased use of partnerships and long-term framework agreements [which] will help ... deliver greater process efficiency', ultimately benefiting all parties (1998). Not only were these preferred contractual arrangements imagined to promote teamwork, but the facilitative leadership of the project manager would aim to create collaborative relationships with the various actors to the wider project team that made up any project's supply chain. This was also an attempt at avoiding teams having to learn on the job at the client's expense (1998). In theory an idea which although many do attempt to follow is difficult to implement due to the varying nature of projects and at times the desires or intent of contractors.

Dawkins analysed the works of Robert Axelrod, applying his research to the complexities of human nature to argue that 'nice guys can finish first', an idea which historically has not had much support (2006). Axelrod, fascinated by this simple gambling game, used the *prisoner's dilemma* to look at various strategies applied by players of the game and how 'successive rounds of the game give us the opportunity to build up trust, or mistrust, to reciprocate or placate, forgive or avenge' (Dawkins, 2006). Axelrod ran an open tournament inviting programmers to propose differing strategies towards the game. In Axelrod's test, the most points achieved by a player was 600, where both parties cooperated consistently. Axelrod categorised cooperative strategies as 'nice' strategies, which was defined as 'one that is never the first to defect' (Dawkins, 2006). The most prominent of these strategies was 'Tit for Tat' and was created by a professor of game theory, Anatol Rapoport. It is capable of defecting, but only in the form of retaliation. This is also known as a *forgiving* strategy, one that is capable of retaliation, *but* had a short-term memory and 'is swift to overlook old misdeeds [and] lets bygones be bygones' (Dawkins, 2006). These approaches (nice and forgiving) scored much higher than other 'nasty' strategies that did hold grudges, and this was because 'unforgiving strategies [didn't] do very well [as cooperative and/or forgiving strategies] they can't break out of runs of mutual recrimination' (Dawkins, 2006). When Axelrod invited suggestions for strategies, out of the 15 that entered, 8 were nice strategies, and all eight were the top scoring ones. Following the outcome of this experiment, two schools of thought developed; *nasty* vs *nice*. As such a second competition was arranged, the *nasty* school worked to try and exploit the obvious *nice* strategies, with new *nasty* strategies that could now exploit *nice* strategies that had been revealed by the earlier tournament. In many cases nasty strategies were able to exploit *nice* approaches; however the *forgiving* strategies again came out on top as they overcame the retaliating nature of *nastier* others. When the competition ran for a third time, some strategies died out, others came more numerous, and after 1000 generations, stability was reached. Although these approaches were all ran through a computer programme, they can be applied to everyday life. For example, '*always cooperate* is not stable against invasion by nasty strategies such as *always defect*', there needs to be some threat or action of recrimination; otherwise, just like in everyday life, your working environment, and very specifically the construction industry, you will be taken advantage of. Dawkins looked at the *Tit for Tat* approach and argued that 'tit for tat individuals, cooperating with one another in cosy little enclaves, may prosper so well that they grow from small local clusters, into larges [ones]' (2006). In applying this to the construction industry, it could be argued that principal contractors, working collaboratively with their supply chain, may work to benefit and result in prosperous outcomes for all parties. Dawkins also argues that 'individuals tend to resemble their immediate neighbours' (2006). Where there is a perpetuating culture in specifically the UK construction industry, actions and approaches end up being copied and repeated as individuals move from company to company or onto different projects, which also filters down to the whole design team, which, history can show, stagnates the development of ideas and progressive development. The *non-envious* approaches as highlighted by Axelrod and Dawkins are quite happy to win just as much as the other party, which is the notion and thinking of both Latham and Egan which asserts that partnering and collaboration are mutually beneficial for all parties.

Heathcote et al. highlight the common issue of game playing on construction contracts and argue that 'far from removing cost uncertainty for the client organisation, and transfer risk to the contractor, this "game" increases it, and necessarily means that misinformation on progress is presented for much of the project' (2018). The grounds for their research are based on the competitive nature of bidding, and the emphases based on the need to win. Contractors as such have developed an adversarial approach to 'asymmetric negotiation', where, essentially, they bluff their way onto construction contracts (Heathcote et al., 2018). The lack of structure within the tender process of the construction industry allows this to happen as the lack of barriers to new entrants, and in many cases old entrants, results in attractive yet unachievable bids being submitted. Once contracted, and the client then committed, the contractor then has the opportunity to build their profit margin through generating claims on the project, effectively exploiting the client (Heathcote et al., 2018). There has been attempts to 'blacklist' contractors where this has happened, but due to this bid process and the nature of single projects not being moderated, this has been difficult to implement as for each tender the contractor is effectively starting from a clean slate (Heathcote et al., 2018). As the industry is susceptible to swings in the availability of work, plus these low barriers to new applicants, low bids are perpetually reinforced by these structural factors. McMeeken argued that despite the extensive research of many parties, most notably Latham and Egan, 'its translation to in-situ construction can be said to have had a limited effect' (2008). Cheung furthers this point by arguing that '"partnering" has been corrupted to represent the traditional adversarial approach with a different rhetoric' (2003). Chan et al. go a step forward to better understand what partnering entails and what barriers the construction industry has in place affecting its implementation. The Construction Industry Institute defines partnering as 'a long-term commitment between two of more organisations for the purpose of achieving specific business objectives by maximising the effectiveness of each participants resources. This requires changing traditional relationships to a shared culture without regard to organisational boundaries. The relationship is based on trust, dedication to common goals, and an understanding of each other's individual expectations and values' (1991). Chan et al. also argued that 'partnering can provide the basis for project stakeholders to reorient themselves towards a win-win environment to problem solving and foster synergistic teamwork amongst them' (2003). Although there is no singular definition or suggested approach to partnering, the general consensus is that partnering lowers the risk of cost overruns and delays as a result of better time and cost control over the project (Chan et al., 2003). Not only does partnering provide all of the above, if done correctly, but it also provides a framework for conflict resolution (Moore et al., 1992).

Partnering may not solve all the problems of the construction industry and is totally dependent on those who drive it; however, it is argued by some to have driven an increased opportunity for innovation (Chan et al., 2003). These may not be drastic industry changing innovations although there is no limit, but merely value engineering proposals born from the open communication and trust of partnering teams. Uher proposes that 'partnering has nothing to do with old adversarial management practices... [and] embraces a new approach by way of trust, open communication, teamwork and shared goals' (1999). As noted previously, the ideas raised by Uher and Chan of shared goals and win-win projects are not new and are well utilised throughout manufacturing. Outsiders may see these as common-sense approaches; however, the historical nature of the construction industry and culture construction professionals are raised in have bred hostility and mistrust. 'Many parties do not trust the other party due to past experience and fear of the unknown' (Larson, 1995). As such, this results in an ability or an unwillingness to change the myopic way of thinking (Chan et al. 2003). Trust may be affected by an individual's past personal experience of litigation and dispute and operating in an adversarial relationship with the contract team (Albanese, 1994), whereas Cowan et al. take the pessimistic view that developing trust for a contractor, or a client, might be a *risk* in itself (1992). Furthermore, Cheung argues that partnering is paid lip service and is in fact a change in rhetoric only, not in behaviour (2003). If this is the case and a company or an industries' commitment towards partnering is insincere, then it is bound to fail (Hellard, 1996). However, the fault may not be solely in the hands of top management; Lazar argues the issue of a lack of understanding in mid-level staff and how to enforce partnering could be a factor as top management's ideas and approaches have not filtered down (1997). Conversely, the commercial pressures on all levels of a company may affect the approach to partnering; even if top management is fully bought into partnering contracts, the demand of budget limitations and programming constraints can result in a compromise to these values to get a project over the line for executive sign-off. As such Chan et al. focus on the requirement for 'project stakeholders ... to reach a balance between commercial interests and partnering attitudes' (2003). As the research shows that adversarial behaviour negatively affects not only the project cost and delivery but may cause compromises down the line due to this, affecting the benefits realisation initially perceived by the stakeholder. Chan looks at various notable research papers and key thinkers in the partnering management approach in order to whittle down the barriers to partnering to nine key areas. Realised through an analysis of this research to look into the frequency of reference, these key areas are listed below:

1. Misunderstanding of the partnering concept
2. Relationship problems
3. Cultural barriers
4. Uneven commitment
5. Communication problems
6. Lack of continued improvement
7. Inefficient problem-solving
8. Insufficient efforts to keep partnering going
9. Discreditable relationships

Heathcote and Brayston argue that in order to overcome obstacles within a project, such as those above, there is a significant advantage of the use of a facilitator to navigate the issues any project throws up (2017). The facilitator not only manages stakeholder expectation and drives the team towards an innovative approach that is valuable but also maintains objectivity throughout the process (Heathcote & Brayston, 2017). This approach, similar to that of supply chain transparency, instead looks at the individuals involved with the process and how to better utilise the team to better define a problem and thus accurately problem-solve, bringing all levels to work together instead of working against and even blaming each other. This produces a more valuable outcome and works to move forward instead of dwelling on past issues. It could be argued the rise of the prominence of project managers and their importance within a project is down to the valuable work they do as facilitators.

Kahneman and Tversky take a different perspective to that of Heathcote and Chan, as they argue the phenomenon of 'The Planning Fallacy' which find humans inherently optimistic and overconfident in their approach to planning (1979). Research has traced over optimisation in many areas of human cognition in everyday life, coining the term 'rose tinted glasses', as we often exaggerate our own talents (Lovallo & Kahneman, 2003). It can be argued that these traits affect bid production as there will always be a lack of foresight and underestimation of the risks a project may throw up. Even the most experienced construction professionals may be susceptible to the planning fallacy, taking credit for any positive outcomes whilst attributing any negatives to external factors and bodies. This incites a reaction from the other parties involved and could be a factor for adversarial behaviour. Supported by a large body of empirical research, Hofstader reiterates these findings stating, 'it always takes longer than you expect, even when you take into account Hoftsader's Law' (1979). The recursive nature of the role and of projects does not necessarily result in learned behaviour as Kahnemen and Tversky argue that errors made are instead due to a disregard of risk (1979). The complexity of estimating a programme or cost for a project can also become overwhelming due to the melting pot of trades, materials, clients and consultants involved. What initially presents as over-optimism quickly turns to procrastination due to this complexity, ultimately affecting the bid submission (Brunnermeier et al., 2008). Flyvbjerg notes that these findings show that 'project cost overruns and benefit underruns are significantly more common than cost underruns and benefit overruns' (2018). In order to attempt to avoid this, Lovallo and Kahnmen look at the notion of reference class forecasting, or an 'outside view' is a way we're then able to frame the project and break down the complexities, providing checks and balances against our own 'inside view' (2003). In practice this may be easier for contracted to a framework agreement, as the similarity of the projects allows the project team to pre-empt any known risks and effectively price and plan for their resolution. This supports Egan's view of the importance of frameworks to project success.

21.2 Research Review and Methodology

The following information provides justification for the method of research chosen for this study; in order to gain actual actions of construction professionals instead of their beliefs, an experimental strategy was used in order to gain clarity between their espoused opinions and true actions. Although they may perceive their approach to be collaborative, in action they may in fact behave adversarially dependent on the other parties' reactions and responses. The adaptability of this approach lends itself to the majority of contexts.

The generalisability of this research however creates a limitation for the approach taken. The assumption that the results of this research will apply to other settings has been considered and is taken into account due to the complexity of society. This research applies to the UK, namely the Leeds construction scene, but however, due to the nature of the experiment, can readily be reconducted in various locations throughout the UK if not the world.

Research Method To test the inherent attitudes in the sample group towards cooperative and adversarial strategies, an experiment needed to be devised that allowed for the target group sample of construction professionals to be compared to a random non-construction-based group; this dictated the use of an 'unpaired' design (Polit & Beck, 2012). But also, it required an experiment which would lead the subjects into revealing their inherent biases. The game, the 'prisoner's dilemma', seemed to offer the perfect choice. The 'prisoner's dilemma' has formed the base for investigations into 'game theory' for some time (Myerson, 1991), but in this study we sought to utilise it to detect adversarial tendencies that the subjects themselves might be unaware of.

Population and Sample The research is about testing a cooperative or competitive tendency of a particular professional group. The research had access to $n = 3$ distinct group samples; one was *main contractors*, the group most likely to be enveloped in a competitive culture. The second group was formed by a quantity surveying professional practice, whose practice worked only with clients that were in a *partnering framework*. Meaning that they did not have to engage in much tendering of competitive bids, this group came from a competitive culture but had experienced more cooperative partnering, which may have had a moderating effect on their outlook. And thirdly, a control group is made up of people who were not involved in this industry or were involved in contracting, who were expected to be the most cooperative players of the three sample groups.

21.3 Findings

The control group after $n = 8$ rounds of the 'prisoner's dilemma' game were found to be the most *cooperative*, with a percentage of 63% *cooperative* strategies compared to 37% *defecting* strategies.

The main contractor group after $n = 8$ rounds of the prisoner's dilemma game were found to have a tendency to be more likely to *defect* than the control group, with a percentage of 36% *cooperative* to 64% *defecting*.

The project management framework partnership professional group after $n = 10$ rounds of the 'prisoner's dilemma' game were found to have a tendency to be more likely to *defect* than the control group, but *less* than the main contractor group, with a percentage of 45% *cooperative* to 55% *defecting*.

Both main contractor and project management framework groups included people who were heavy *defectors* playing *defect* for either all or most plays. These strategies were explained by those players as 'making sense to them, because it is the only way to avoid losing'. This is an interesting perspective, implying that not beating the other side was regarded as a loss. Only the *control* group included a game that was completely *cooperating*, which results in the best combined score. The project management framework group included a mix of strategies, but included games that were *mostly cooperative*, but not entirely cooperative. It would seem that maintaining a *cooperative* approach against a determined *defector* is a difficult task for even a player with a tendency to *cooperate* (Fig. 21.2).

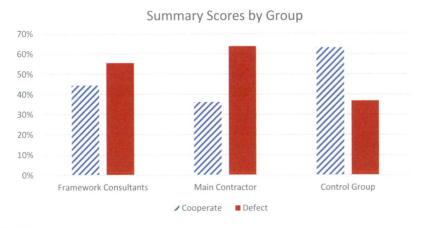

Fig. 21.2 Summary of game scores by group

An ongoing lockdown in the UK due to COVID-19, during the period assigned for data collection, and the nature of the collection of this data, meant that the volume of results was limited. This was because meeting in groups and having access to offices was prohibited.

21.4 Discussion and Conclusions

Whilst the Quantity Surveyors working in a partnering framework employed more cooperative and less competitive strategies, than the Quantity Surveyors in the main contactor group, they were still notably more competitive when compared to the control group. Both professional groups displayed measurably greater tendencies towards competitive/adversarial strategies than the control group. The differential is notable (though might represent a type II error, due to being statistically underpowered), sufficiently to warrant pursuing the research to achieve statistical significance.

Peter Drucker, the management guru, has been credited with suggesting that *a company's culture would trump any attempt to create a strategy that was incompatible with its culture*. These findings support the idea that the culture towards competitive strategies is entrenched in UK construction industry's professionals. Observably this has had far-reaching effects not readily overturned by nationally led studies which strongly purported a change to more cooperative 'partnering'. This research's support for an apparent cultural bias towards competitive tendencies offers an explanation for why numerous attempts to introduce a more partnership and collaborative approach to project contracting are resisted and fail in implementation in the UK construction industry. Notable within the findings are the extreme competitive strategies pursued by some of the intervention groups, whether there are in the main contractor or the project management framework groups. Although some nod towards more collaborative strategies were evident in the framework partnering group, that sample still included extreme examples of competitive strategies. Those approaches were not represented at all in the control group.

Moving towards a more collaborative and partnered approach that serves sustainable innovative outcomes for UK construction will have to include an industry change that also addresses the cultural bias found in this research.

Furthering the research to increase the volume of experimental games to a volume that permits the removal of the possibility of a type II error is recommended.

References

Adamson, D., & Pollington, A. (2006). *Change in the construction industry: An account of the UK construction industry reform movement* (pp. 1993–2003). Routledge.
Albanese, R. (1994). Team building process: Key to better project results. *Journal of Engineering Management., 10*(6), 36–44.

Brunnermeier, M., Parker, J., & Papakonstantinou, F. (2008). *An economic model of the planning fallacy*. National Bureau of Economic Research. Working Paper 14228.

Chan, A. P. C., Chan, D. W. M., & Ho, K. S. K. (2003). Partnering in construction: Critical study of problems for implementation. *Journal of Management in Engineering, 19*(3), 126–135.

Chappelhow, J. (2019). *Prisoners dilemma*. Investopedia. [online] https://www.investopedia.com/terms/p/prisoners-dilemma.asp#:~:text=The%20prisoner's%20dilemma%20is%20a,expense%20of%20the%20other%20participant

Cheung, S.-O., Ng, T. S. T., Wong, S.-P., & Suen, H. C. H. (2003). Behavioural aspects in construction partnering. *International Journal of Project Management, 21*, 333–343.

Construction Industry Institute (CII). (1991). *In search of partnering excellence*.

Cowan, C., Gray, C., & Larson, E. (1992). Project partnering. *Project Management Journal., 22*(4), 5–12.

Dawkins, R. (2006). *The selfish gene*. Oxford University Press.

Egan, J. (1998). *Re-thinking construction*. The Report of the construction task force. Department of Trade and Industry.

Flyvbjerg, B. (2018). Planning fallacy or hiding hand: Which is the better explanation. *World Development, 103*, 383–386.

Gardiner, J. (2014). *Latham's Report: Did it change us?* https://www.building.co.uk/focus/lathams-report-did-it-change-us/5069333.article

Heathcote, J., & Brayston, M. (2017). 'Value Management': The importance of whole team representation, Stakeholder positioning and facilitation, to successful Value workshop outcomes. In *Leeds Sustainability Institute. SEEDS conference* 2017.

Heathcote, J., Kazemi, H., & Wilson, M. (2018). Illustrating game playing on construction contracts: The negative impact of procurement strategies. A proposal for research. In *ARCOM 34th annual conference*.

Hellard, R. B. (1996). The Partnering Philosophy—A procurement strategy for satisfaction through a teamwork solution to project equality. *Journal of Construction Procurement., 2*(1), 41–55.

Hofstadter, D. (1979). *Godel, escher, bach: An eternal golden braid*. Basic Books.

Kahneman, D., & Tversky, A. (1979). Intuitive prediction: Biases and corrective procedures. *TIMS Studies in Management Science, 12*, 313–327.

Lovallo, D., & Kahneman, D. (2003). *Delusions of success: How optimism undermines executives decisions*. Harvard Business Review: Decision Making.

Langdon, D. (2010). *Contracts in use: A survey of building contracts in use in 2007*. RICS.

Larson, E. (1995). Project partnering: Results of study of 280 construction projects. *Journal of Management in Engineering., 11*(2), 30–35.

Lazar, F. D. (1997). Partnering–new benefits from peering inside the black box. *Journal of Management in Engineering., 13*(6), 75–78.

Latham, M. (1994). *Constructing the team: Joint review of procurement and contractual arrangements in the United Kingdom construction industry*. Crown.

McMeeken, R. (2008). *Egan 10 years on*. https://www.building.co.uk/news/egan-10-years-on/3113047.article

Moore, C., Mosley, D., & Slagle, M. (1992). Partnering guidelines for win-win project management. *Project Management Journal., 22*(1), 18–21.

Myerson, R. B. (1991). *Game theory: Analysis of conflict* (p. 1). Harvard University Press. Chapter-preview links, pp. vii–xi.

Polit, D. F., & Beck, C. T. (2012). *Nursing research: Generating and assessing evidence for nursing practice* (9th ed.). Wolters Kluwer. Lippincott Williams & Wilkins.

Uher, T. E. (1999). Partnering performance in Australia. *Journal of Construction Performance., 15*(2), 163–176.

Chapter 22
"Megaprojects to Mega-Uncertainty" Is About Risk Management to Perform

Charlene Chatelier, Adekunle S. Oyegoke, Saheed Ajayi, and John Heathcote

22.1 Introduction

Traditional project management methodologies revolved around sound technical and procedural factors, like scheduling, scoping, budgeting, risk management and quality assurance, amongst others. However, a high percentage of projects are still being reported as failed even with the use of these well-established methodologies and frameworks. In the case of megaprojects, only 35% succeed (Merrow, 2011), characterised as economical, technological, aesthetic or political (Flyvbjerg, 2013). Evidence confirms, based on the most common project success indicator, that only approximately one in a thousand projects manages to deliver on all three targets, illustrating that the success rate is dismissive (Flyvbjerg, 2017). This paper objects this destructive and dictated view because it builds a negative perspective on the performance of megaprojects.

These projects deserve better evaluation models, considering they normally attract investments well above $1 billion worth and include most likely intangible benefits, appealing long-term outcomes and very uncertain conditions ((Müller et al., 2012). The most common causes for these failures have directly been linked to cost and schedule overrun (Kaming et al., 1997), business case and scope creep (PMI, 2016), uncertainty (Lock, 2003), strategic risk (Forbes, 2017), risk management (Oyegoke, 2019), criticality of people (Dvir et al., 2006) and perceptions and project complexity (Liu et al., 2016), amongst others.

Megaprojects are regarded as a multi-trillion-dollar global delivery model for large investments for most industries (mining, mineral processing plants, oil and

---

C. Chatelier (✉) · A. S. Oyegoke · S. Ajayi · J. Heathcote
Leeds Sustainability Institute, Leeds Beckett University, School of the Built Environment and Engineering, Leeds, UK
e-mail: C.Chatelier@leedsbeckett.ac.uk

gas, IT, supply chains, aerospace, defence, mega-events like Rugby World Cup). It affects the way we shop as consumers, our energy bills, ways of travel and how we use everyday technology like the Internet. The current conservative valuation for the global megaproject market is USD 6–9 trillion per year, (Flyvbjerg, 2014). Frey (2017) continues to predict that megaproject spending will escalate within the next decade to around USD 22 trillion per year accounting for 24% of global gross domestic product (GDP). In this perspective that is larger than any nation's GDP including China and America. Recently the largest project (China Belt) costs well over USD 1 trillion. That is more than the total market capitalisation of Apple.

The current global pandemic has further intensified the issue and put the performance of megaprojects to its thorniest lifetime test yet, with megaprojects announced exponentially at all levels for social reconstruction and economic reform after the devastating impact of COVID-19. The outcomes (benefits and value) will be expected to be preserved for many years to come, considering the huge sacrifices made in terms of time, cost and scope. The audiences and stakeholders of these immediate projects will critically evaluate these projects' successes based on outcomes delivered, not on the time and not on the initial investment required; change is demanded and expected to be delivered at the sacrifice of time and cost. Roles of politicians changed overnight where they have become the steering project managers with a global audience and the fate of many lives depending on their navigations. The risk management employed throughout the pandemic will be crucial for the better value of life itself, with the outcomes impacting millions of lives not just today, this year or when a vaccine is available but for many generations to come. Worries are growing about GDP depletions, economic retractions, social value exploitations and serious health risks. Now with megaprojects identified as a delivery tool for major change and developments, the pressure to perform whilst preserving value has never been so significant.

Academic knowledge in megaprojects has therefore become more crucial considering the substantial impact big projects have socially, economically and environmentally (Flyvbjerg et al., 2003). Immediate, research attention is therefore required to better understand why the current theory in particular 'the iron triangle' is not enhancing performance but instead risks the deliverance of better outcomes for the staggering 90% failing megaprojects (Merrow, 2011) by enabling uncertainty and subsequently risk based on the findings of this paper.

According to Bryde (2008), are not only the most cited but also the most used Iron Triangle criteria measures of project success. In their study relating to the project manager's experience of projects, White and Fortune (2002) found that the iron triangle was used as a primary way of defining project success by most project managers. Research by Müller and Turner (2007) stresses how the iron triangle is valued by both inexperienced and experienced project managers. This persistent popularity may be as a result of its simplicity. Subsequently explained by Jugdev and Müller (2005) that when projects are delivered to these criteria, the declaration of success seems relatively simple or perhaps too simple for complex models like megaprojects to serve an effective purpose.

This paper is linked to a broader risk management research stream initiated and supported by Leeds Beckett University for PhD studies. The paper firstly recognises the failure or success indicators for megaprojects and then explores how these factors are incorporated or dealt with when evaluating performance analysing nine UK-based case studies:

(1) Firstly by the iron triplets. (2) Secondly by considering the increased complexity and uncertainty of megaprojects in contemporary global project environments. The paper finally concludes that the data represented although not scientifically significant offer hope for a broader perspective of risk management to better influence the performance of megaprojects. Several hypotheses are drawn upon that call for more research and academic collaboration.

22.2 Literature Review

The so-called megaproject pathologies, i.e. the chronic budget overruns, and failure of such projects have highjacked the focus of literature, and relatively diminutive devotion has been given to the specific needs of evaluating large projects. This paper suggests that conceptualising megaprojects as both evolving and dynamic systems would provide a useful basis for performance evaluation. Literature is therefore drawn from two strands, success and failure criteria of megaprojects and how evaluation for megaprojects might improve.

22.2.1 Project Performance Evaluation

Many government reports demonstrate that projects have been judged against time, cost and scope criteria (see IPA, 2017 report), and the results have not been promising, with nearly 65% of megaprojects failing to point out ICT projects have been particularly inclined to failure, with high percentages (around 80%) (Savolainen et al., 2012). Lastly, performance evaluation and transport infrastructure projects can be controversial according to Ika (2009), whilst Turner et al. (2012) reckon it is related to the time frame and stakeholder's perspective.

Turning a blank eye on the ongoing ambiguity that exists regarding the determining factors for project success will continue to have significant consequences for how megaproject 'success' and 'failure' are defined, subsequently risking future stakeholder value or existing value depletion.

	Project Criteria for Mega Projects							
Variables	Iron Triangle 1969	Kliem&Ludin (1992)	Atkinson (1999)	Nelson (2005)	Prince 2 (2009)	Pinto (2010)	PMI PMBOK (2013)	Williams et.al. (2015)
	Time	Time	Time	Time	Time	Time	Time	Time
	Cost	Cost	Cost	Cost	Cost	Cost	Budget	Budget
	Scope	Scope	Scope	Product	Scope	Quality	Scope	Quality
		Soft Aspects	Quality	Value	Benefits	Client acceptance	Requirements	Client satisfaction
				Use	Quality		Quality	Client relationship issues
				Learning	Risk		Risk	

Fig. 22.1 Chatelier (2020)—Project criteria suggested over time adopted

22.2.2 A Review of the Iron Triangle

The 'triangle' has been criticised for putting too much emphasis on cost, time and scope. Not only does this limit the evaluation to other important factors like risk, benefits, value, stakeholders, etc. but narrows perspectives of how the project manager or practitioner is being appraised or rewarded for their capabilities to meet these criteria (Wateridge, 1998). Several authors had suggested some improvements; see a summary below in order of publication.

These adaptions, therefore, emphasise the inadequacies of this widely accepted method (Fig. 22.1).

> The weight assigned to the Iron Triangle as the primary determinant for project success has been criticised by Atkinson (1999). He proclaimed that is a 'phenomenon' and Cost and Time 'only guesses' suggesting that the protects require a new criteria for success evaluation. This debate was supported by Gardiner and Stewart (2000), who estimated that about 50–70% of projects have a severe schedule and budget overruns, stressing that initial estimates are insufficient for weighing success, especially when used to benchmark management processes.

22.2.3 Subjectivity in Project Evaluation

It is well established that subjectivity is a factor that affects performance evaluation practices. Over the years, different methods have been projected for project evaluation complexity whilst segregating complexity away from risk (He et al., 2015). However, the recorded successes of these methods are not known especially in the context of these staggering failure rates and therefore excluded.

22.2.4 Differentiating Characteristics of Megaprojects

Accepting Characteristics and Differentiating Megaprojects

It is not news that megaprojects are different from projects due to its multidimensional characteristics which differentiate itself from standard projects. Additionally, Merrow (2011) sets the expectation for megaprojects to be more visual due to these characteristics. Often megaprojects are characterised by long schedules stretching over decades worth over hundreds of millions or billions often affected by an enormous amount of uncertainties and risks (Bruzelius & Flyvbjerg, 2002). These arguments should be acknowledged as they set precedent why the simplicity of the iron triangle does not compensate for the complexity of megaprojects.

Substantial Stakeholder Involvement

Megaprojects attract a lot of public attention; balancing stakeholder interest can become very political especially when there are negative implications. Numerous authors in literature have expressed the criticality of project stakeholder management for project success (Boonstra et al., 2008). Findings from Mulholland (2019) study stresses the significance of following a processual and pluralistic approach about stakeholders demonstrating the need for a more focus. As both the amalgamation and dismantle of stakeholder influences and interests as they evolve over the project life cycle, the evaluation of project success should, therefore, mirror this evolution of stakeholders.

Organisational Structure

Firstly, in megaproject sometimes referred to as meta-organisations, stakeholders/core members or investment partners join at different time intervals, and these parties have diverse priorities, beliefs, preferences and planning techniques, causing major problems in collective action (Gil & Tether, 2011). Variance is often caused by structural changes, prone to megaprojects and not supported by the iron triplets. An automatic dissmal is triggered that might sent out a very negative view to society and stakeholder relying on the outputs or anticipated benefits.

Eccentric Cost

If delayed or altered, the cost implications could have huge significance for these projects. Flyvbjerg et al. (2003) advocate that extensive escalation of cost seems to be the norm for megaprojects, instead of the exception. Berechman and Chen (2011) accentuate risks associated with a cost overrun should be included in the decision-making process and project evaluation.

Extreme Time Delays

Influencing factors like government policies can cause major delays for these projects. Additionally, a study by Oyegoke and Al Kiyumi (2017) found that extra costs and project time overruns are the most significant effects causing delays for megaprojects in Oman, therefore supporting the argument that forecasting both estimates of cost and schedules remains difficult for megaprojects. However, such difficulty ought to be considered for performance evaluation for a more accurate and engaging reflection.

Complexity and Uncertainty

High-level complexity – Due to the uniqueness of these projects, their complexities are often not duplicated. There is a growing assumption that uncertainty is instigating complexity derived from project environments. According to the European Cooperation in Science and Technology (COST), megaprojects are characterised by 'extreme complexity (both in technical and human terms)' Cost 2018. Chatelier (2020) suggested a model additionally assumes that uncertainty is simultaneously instigating risk management.

Unique First-Time Environments

Siggelkow and Rivkin (2009) stress performance is directed referring to interactions between low- and high-level choices and environmental factors (Bingham et al., 2009). Because of the first-time initiation factor quite often, there is no prior knowledge of the political, social and technical conditions or legal and financial structures.

Ability to Derive Exponential Benefits (Author, 2020)

In his book *Megaprojects and Risk*, Bent Flyvbjerg expresses megaprojects as qualitatively significant for both economic and social development stages (Flyvbjerg et al., 2003). The definition proposed by authors of *Oxford Handbook of Megaproject Management* best applies to the context of this paper 'Megaprojects are large-scale, complex ventures that typically cost $1 billion or more, take many years to develop and build, involve multiple public and private stakeholders, are transformational, and impact millions of people' (Flyvbjerg, 2017, p. 2).

Common consensus exists that the above dynamics have one denominator in common – direct impact on uncertainty; this general acceptance should therefore allow for these dimensions to be incorporated when evaluating project success.

22.2.5 Reasons for Megaproject Failure in Context of the Iron Triangle

Although the literature has highlighted several reasons why projects fail and even suggested solutions, not enough has been done to investigate and address the root cause of these failures as highlighted in the context of this paper with specific reference to megaprojects.

Project Life Cycle

Considering the longer delivery period (5–10 years) for megaprojects, PMI (2016) indicates that project and benefits should be tracked throughout their life cycle from the project initiation stage up to execution and post-execution. There is no evidence that once the business case has been approved, projects are tracking and monitoring intended outcomes throughout the entire project. However, such reporting might be regarded as redundant if the project was misrepresented for funding purposes. Delays cause extension of the project life cycle and raise uncertainty based on influencing variables.

Business Case

It was argued by Pollack (2007) that projects are failing due to the mismatch between the project management literature which tends to assume that success is based on 'the existence of a pre-existing business plan, with clearly defined goals and constraints, with goals that can be decomposed with clear customer requirements' (Pollack, 2007, p. 217). However, his observation was contradictory in a sense that 'highly detailed or rigid plans have been identified as limiting freedom to make decisions' (Bohle et al., 2016).

Scope Creep

Very dynamic and multi-interpretable scope may change dramatically in the course of the development process, causing simple projects to turn into a manifold of ambition and complexity with a lot of complications.

Limitations of the Iron Triangle

It was proposed by Van der Hoorn and Whitty (2015) that the iron triangle forms part of the artefacts identified in project management due to its lack of validity in relation to how projects are managed in practice.

In the construction industry, it was suggested by Collins and Baccarini (2004) that success in projects should surpass expectations for quality, cost and time objectives. Toor and Ogunlana (2010) additionally noted that other pointers such as resource efficiency, safety, effectiveness, conflict and dispute reduction and stakeholder satisfaction are progressively important for performance in construction.

De Wit's (1988) highlighted that the iron triangle will only be classified as a traditional task-related criterion, based on early research work that distinguishes between psychosocial and tasks success criteria, thus excluding criteria such as stakeholder and customer satisfaction and team relationships.

The simple use of the iron triangle overshadows its adequate contribution to practice (Chatelier et al. 2020).

22.3 Research Methodology

The methodology used forms part of the proposed research currently under development, but preliminary analysis has already yielded very interesting findings regarding the future of megaproject performance.

An extensive literature review was initially performed. Followed by an inductive cross-case analysis, the technique adopted followed a structured process by using constructed cases to arrive at 'cross-case' trends. Theoretical propositions are then derived from these 'patterns'. The approach is mainly inspired by Eisenhardt (1989); he consequently formulates a theoretical process where findings could be generalised following the review of cases of a specific domain Eisenhardt (1989, p. 545). The multiple case study approach focused on one core question (Eisenhardt, 1989): Can the performance of megaprojects be successfully measured using traditional project management theory—the iron triangle? To address the previously mentioned, the paper sought to answer a set of secondary questions including what challenges and factors had constituted the successes or failures of megaprojects? What form of organisation is a megaproject? How does it differentiate from a standard project? Which actors influence performances for megaprojects? The qualitative data of nine case studies were collected from a series of different sources (published reports, project reports, journals and news articles) and collated into an Excel spreadsheet for easy reference.

22.3.1 Originality/Value

However, Eisenhardt (1989) advised against adding cases when there is incremental improvement upon reaching theoretical saturation. Small convenience sampling was applied for this study, and therefore no scientific significance obtained; however, a greater sample would be recommended to increase validity.

The study has gone beyond the focus of previous literature highlighting endless listings of causes and cures of MP failures; instead, this paper critically analysed the performance criteria currently used and encouraged a new way of evaluating projects incorporating both internal and external factors especial uncertainty and risks for a more structured assessment when evaluating performance for a more objective success rate.

Based on data analysis, the following hypotheses were derived for future research:

H0 = Variance in time, quality and cost does not confirm if any value has been derived nor if the project has been successful.

H1 = Positive variance (in time, quality and cost) causes a reduction of uncertainty and risks.

H2 = Negative variance (in time, quality and costs) causes an increase of uncertainty and risks.

H3 = Variance relates directly to the phenomena of uncertainty and risk management.

22.3.2 Research Limitations

- Data associated with project outcomes are rich and qualitative, and conversion into a quantitative form is required analysis or interpretation. This process can be disreputably difficult; hence case studies have been limited to only nine cases for this paper.
- Greater sample size will be required to obtain scientific significance.

22.4 Findings

The outcomes of project assessments were firstly based using the iron triplets and then based on a more flexible model/approach (Fig. 22.2). Refer to the next page for an illustration of how a contemporary review on the same factors linked to uncertainty and influenced the overall success rate by 55% for these cases.

It was clear that benefits and outputs of these projects were mostly ignored during evaluations or simply not assigned significant weight. All these projects derived variances, which resulted in failure based on the iron triplets. However, based on the

Case Study	Project Aim	Project Start Date	Project completion	Cost variance	Project assessment (Risk and Uncertainty)	Assessment, Iron Triplets (Cost, Time, Quality) on completion.	Adopted Review	Variance, Details and Analysis
Edinburgh Trams (1)	To improve accessibility, reduce congestion and promote sustainability (Connect Edinburgh Airport to City centre and development areas).	2007	2014	328million	Cost and Time variance caused a rightwards shift increasing uncertainty and risk levels. Whilst a left shift based on (q)variance reduced uncertainty and risk for successful delivery.	Failed	CV= IUC+Risk QV=DUC+Risk TV=IUC+Risk = Success	Highest customer satisfaction rates in the UK. Overspent, delayed Completed to Specification
Scottish parliament (2)	Building a parliament building for Scotland	1999	2004	414million	An overall increase in uncertainty and risk.	Failed	CV= IUC+Risk QV=IUC+Risk TV=IUC+Risk = Failed	Claimed to be out of context for the land it represents. Overspend, delays, design Not to specification
NHS information IT (3)	Centralised patients e-record system	2003	2013 - ceased existence	3.6billion	An overall increase in uncertainty and risk.	Failed	CV= IUC+Risk QV=IUC+Risk TV=IUC+Risk = Failed	By 2013 only 13 trusts received full patient administration information compared to the ordinal 169. Plague with delays, no inhouse system integration, the unreliability of the data, complaints Not to scope
London Eye (4)	Monument to Mark the start of the century	1998	2000	The cost was declared to be 75million higher than normal construction.	Cost variance caused a rightwards shift increasing uncertainty and risk levels. Whilst time and quality variance led to a leftwards shift based on reduced uncertainty and risk for successful delivery.	Failed	CV= IUC+Risk QV=DUC+Risk TV=DUC+Risk = Success	Highest Tourist Attraction, promising revenue and ROI. Most successful architectural project. to scope specification
London Olympics (5)	Host-Summer Olympic Games	2005	2012	157%	Cost and quality variance caused a rightwards shift increasing uncertainty and risk levels. Whilst a left shift based on (t)variance reduced uncertainty and risk for successful delivery.	Failed	CV= IUC+Risk QV=IUC+Risk TV=DUC+Risk = Success	A potential source of economic income regeneration No to scope
UK Passport Agency (6)	Provision of passport services to British Nationals in the UK, most economically and promptly.	1991		285million	An overall increase in uncertainty and risk.	Failed	CV= IUC+Risk QV=IUC+Risk TV= IUC+Risk = Failed	UK customer satisfaction Index currently at 76.9 or 0.8 points at the lowest level since July 2015. Failed
Portsmouth Spinnaker Tower (7)	Designed as a monument to commensurate millennium celebrations.	2001	2005	24.5million	Overall increase in uncertainty and risk.	Failed	CV= IUC+Risk QV=IUC+Risk TV=IUC+Risk = Failed	Residents had demanded the change of original Arabic design. Over cost, delayed by 5years and not to specifications. No to scope
Channel Tunnel (8)	A railway tunnel connecting England and France.	1988	1994	11.4billion	Cost and Time variance caused a rightwards shift increasing uncertainty and risk levels. Whilst a left shift based on (q)variance reduced uncertainty and risk for successful delivery.	Failed	CV= IUC+Risk QV=DUC+Risk TV= IUC+Risk = Success	Approximately 80million vehicles, 185million Eurostar passengers, goods valued over, £150bn worth of goods, travelled through this tunnel since 1994, including the Olympic Torch and Tour de France. Cost overrun, delayed schedule The variance from the original scope.
Thames Barrier (9)	A Movable barrier system to prevent floodplain.	1974	1984	0.1billion	Cost and Time variance caused a rightwards shift increasing uncertainty and risk levels. Whilst a left shift based on (q)variance reduced uncertainty and risk for successful delivery.	Failed	CV= IUC+Risk QV=DUC+Risk TV=IUC+Risk = Success	Closed 186 times since opening to protect Greater London from floods, next review 2030 subject to climate change, but specialist review reckon can be delayed for 40 Completed yrs. The variance from original scope.

Fig. 22.2 Performance data of case studies based on the iron triangle constraints (Chatelier 2020). V variance, C cost, Q quality, T time, UC uncertainty, R risk, I increase, D decrease

adopted model, variances were considered as interdepend factors for success and not sole determinants. An evaluation was also suggested to be carried out throughout the project life cycle including post project to monitor uncertainty and risk rates for better performance.

22.5 Discussions

22.5.1 The Fallacy of Traditional Project Management Theory

Whilst there are established approaches and guidance made available in the project management domain (BABOK, 2005 and PMI, 2008), they are only acceptable for projects, not megaprojects. Similarly, Mishra et al. (2015) also criticised historic economic techniques used for analysis to construct compatible future cash flows and proposed a framework for addressing uncertainty and risk for TIPs (transport investment projects) for both private and public institutions.

There has been commotion amongst researchers and practitioners that the continuance of project failure problems may be more closely associated with traditional project management (TPM) theory than expected. A well-presented example is critics that TPM centres around the efficiency of outputs based on elements of the iron triangle whilst paying less attention to processes that encourage value or benefit generation (AShurst et al., 2008; Remenyi and Sherwood-Smith, 1998). This paper supports this view; whilst the project environment has evolved over the last 50 years, the principles used to measure performance have stagnated.

Both the analysis of case studies on page 7 and literature demonstrate there is still much controversy regarding the overall evaluation of projects. A clear lack of understanding is evident when practitioners had simply accepted the impossible likelihood of success using the iron triplets whilst disregarding the uncertainty and risks instigated. Authors like Oisen (1971), Barnes (1988) and Weaver (2007) referred to these three factors as 'The iron triangle of project management' because of this strong cohesion in the project management domain.

Several other authors like Baccarini (1999), Cooke-Davies (2002) and Dvir et al. (2003) highlighted the differences between realising success by firstly delivering product specifications (measured against realised benefits of the project) and by secondly successfully managing the project (as per iron triangle constraints). In the context of the latter, a hospital operation can be successful based on the efficiency wrt (with regard to) money spent, but there will be no value derived if the patient does not survive.

The Contemporary Iron Triangle Performance advancing through levels is shown in Fig. 22.3 where the labels "Level 1," "Level 2," etc. are used to characterize uncertainty and risk levels. A project at Level 3 is experiencing high levels of uncertainty and risks (3 or more unknown factors) compared to projects at Level 2 (2xor more unknowns). Each vertex represents the project starting at 0 for the relevant constraint where a variance increase will cause a move to the right side of the triangle and reduction will trigger a move to the left. A variance to the right = increase in uncertainty and risk levels and vice versa.

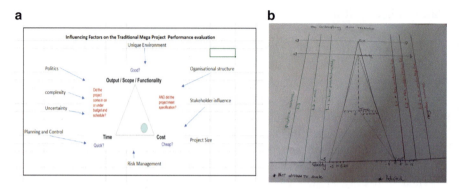

Fig. 22.3 (**a**, **b**) Adopted from iron triangle (Albert, 1969)

The authors of this paper challenge the way megaprojects are being evaluated using the popular 'iron triangle', arguing that the constraints originally shared by Dr. Martin Barnes in 1969 do not reflect the complexity and uncertainty of megaprojects nor the contemporary changing project environments and more specifically cannot be extended for use within the megaproject domain. The use of these constraints is leading misinterpretation of what constitutes success for megaprojects and therefore should be declared obsolete. If not addressed, the issue has the potential to escalate the ambiguity within this dynamic domain where other constraints like stakeholder conflicts, resources levelling and strategic misinterpretation had become major challenges.

22.5.2 Case Study Analysis Based on the Iron Principles (Quality, Time and Costs)

Measuring project performance during its project's life can help deal with scope creep. Dwain Wilcox CIO of Axial Corporations argued 'As we see elements that would contribute to scope creep, we address it in risk management profile because scope creep is affecting the outcome of the project or the budget or resources; it is truly a risk that has to be managed'. He further insisted on classifications to be made by both size and scope for organisations to establish the right tracking and outcomes to pursue. Other researchers also suggested alternatives to quality; most popular interpretation is to replace quality with scope (e.g. Badewi, 2016; van Wyngaard et al., 2012). The paper will, however, retain the original constraints of the iron triangle throughout.

Both the Scottish parliament building and the Portsmouth tower had issues with the original designs which lead to significant time delays. In the case of the Portsmouth's millennium tower meant for residents to celebrate the year 2000, unfortunately, the tower was later renamed as spinnaker due to missing the century milestone. O'Brien (1998) and Ibbs et al. (2001) consider this as common practice for construction projects, causing not only significant variation to the contract duration (Ibbs et al., 1998) but to cost and scope. Regarding the suggested revised model, both uncertainty and risk had increased for all three factors, and therefore the project has been declared a failure.

In the case of the London eye case review, which serves as a good example of the contradictory argument, this paper posits where the immediate outcome of the project was regarded satisfactory due to early completion of 16 months instead of the expected 24 months. Delivered on the specification but because of the cost overrun, the most popular tourist attraction was deemed a failure according to the iron triplets, though revived now as successful based on the model illustration (Fig. 22.2). Both the quality variance and time variance moving inwards towards the centre line of the triangle (reduction of risk and uncertainty) oppose to cost variance moving outwards, leading to an overall positive outcome. The project outcome will therefore be regarded as successful.

Although one could imagine that the Olympics games unlike other major projects share some likenesses and can learn from predecessors, the cost variance would, therefore, be expected to be much lesser compared to other major projects. On the contrary, the London 2012 summer games case revealed the costs at an astonishing 15.0 billion USD and further proved to be the most expensive to date compared to other games like Beijing (2008) who achieved costings of 50% less and an overall variation of 2% compared to other destinations (Leslie-Carter et al., 2009). But Koch-Weser (2013) dims the spotlight by arguing that China has a reputation for lack of reliable economic reporting. Therefore, it is worth observing that clearly with such high variance percentages, there is no obligation from organisers that they have any intention to comply with the principles of the 'iron triangle'. Because of the mitigated risk of non-deliverance at the sacrifice of cost, the project will still be considered successful as per the above-revised model taking into account the benefits to all athletes, hosts and society outweighing that of the costs sacrificed.

Even though it was found that the Olympics are the most costly megaprojects across all industries with an average cost overrun of 156%, followed by IT at 107%, road constructions are the least at 20%. Mishra et al. (2015) highlight that TIPs (transport investment projects) demand long-term maintenance and commitment due to its irrevocable investments but contribute to both direct and indirect costs (Ibbs et al., 1998). Even though the construction industry had made room for variations in contracts like the NEC3/4, performance indicators for the industry seem to lag. The revised model equally presents a reflection on both opportunities and risks—the Olympics could directly impact FDI and boost the economy.

The Channel Tunnel has been dismissed in literature as a permanent burden to the taxpayer and a definite failure. Poor collaboration between the French and English governments meant engineering works and designs caused major delays for this project. On the contrary, this paper argues that this project has successfully exceeded specifications, mitigating safety risks for all using the channel, at the sacrifice of cost and time. It is worth noting that some of the benefits include the £150bn worth of goods including fresh fruits and veggies being transported yearly, consequently playing an integral role in trading with Europe; see case analysis for other benefits.

The Edinburgh trams received many critics about ruining the sight and navigation around the city when constructed, but years later obtaining the highest customer satisfaction rates in the UK, another example that the quality output had outweighed the cost and time variance sacrifice for the projects.

All nine projects discussed had initially been condemned as a failure based on the iron triangle constraints, but it has become apparent that when thoroughly evaluated accounting all outputs, uncertainty and risk mitigation of these projects had already delivered positive returns or have the ability to deliver these in the future. A revised success rate of 55% for these projects most definitely inspires hope for the future of megaprojects.

22.5.3 Risk Management in Complexity and Uncertainty

Lock (2003) highlighted 'The principal identifying characteristic of a project is its novelty. It is a step into the unknown, fraught with risk and uncertainty' (Lock, 2003, p. 4). Project management theory tends to 'solve' this dilemma by proposing a risk management process. Franke (1993), who studied the correlation between risk management and project control, suggested the main task of taking control of the project is by reducing ambiguity whilst simultaneously linking risks associated with project delivery. However, the iron triangle theory contradicts the 'problem-solving method' by constraining the participating performance factors; thus more is unknown causing more uncertainty.

Turner and Zolin (2012), supported by Beier and Heathcote (2010), suggested two ways of coping with risk: (1) avoiding the uncertainties where possible or taking the responsibility to investigate the reason for uncertainties and if this could, in turn, become certain or (2) accept the existence of uncertainties and proactively manage them. Not much is explained by Turner on available methodology to turn the 'uncertain' to 'certain'; the author would, therefore, suggest the third as (3) continuous testing, tracking, reviewing, resolving and accepting of new uncertainties as the project environment changes and fourth as (4) maximising opportunities if viably presented by these uncertainties. Many theoretical authors including Paul Roberts (2013) encourage PMs to report on variance escalated but, however, do not provide any guidance on the root cause analysis for variance and corrective action anticipated. By providing an opportunity for risk and uncertainty to be taken into account when evaluations on project successes are completed. It sets a reflective platform which is most convenient and cost-effective for any root cause analysis of variance to be investigated asap. If managed well, it could reduce both uncertainty and complexity. From this perspective, the complexity appears not only as a problem but simultaneously an opportunity where mutual interdependencies would encourage coordination. It was suggested that the inevitable uncertainties in megaprojects could enable harness benefit of reflexivity, adaptability and exploration of alternative pathways (e.g. Gelatt, 1989).

22.5.4 Misinterpretation Justified: Hirschman's Hiding Hand Theory (HHP)

Following the popularity of the iron triangle, many concluded that it was better to brush off the issues—literature was pointing out about the limitations of the iron law, especially for megaprojects. Alfred Hirschman was, however, keen to address these concerns by introducing the Hiding Hand in 1967. He argued that there is a rough balance in megaprojects: a tendency to underestimate the costs and problems of megaprojects, but similarly the tendency to underestimate the creativity with which people address the costs and problems that arise. Flyvbjerg later condemned this view as 'beneficial ignorance' as simply strategic misinterpretation and lack of accountability (2016).

This paper is not joining the parade on project failure rates but, instead, viewing the iron triangle as a didactic device, intended to communicate the relationship between time, cost and other potential criteria; the authors note that the triplet variations in megaprojects are not only interrelated but correlated with other active variables such as uncertainty risks, complexity, the achievability of requirements and the standard to which deliverables are produced.

However, due to the small sample size and research, a generalisation although apparent is not yet concluded until scientific significance is obtained using a larger sample size.

22.6 Summary and Conclusions

This review of the challenges facing megaprojects when evaluating performance finds still struggling with issues identified and at best only partially addressed for practice. The greater complexity and uncertainty that come with size serve to demonstrate the cumulative impact of megaproject management's contemporary issues.

The findings of this paper conclude that firstly the iron triangle should be declared obsolete or at the least adapted to include more relevant factors as illustrated on page 7. Secondly, there is a clear need for an introduction to a new way of evaluating megaprojects in line with their unique characteristics. Based on both literature and practice, the iron triangle certainly does not promote an accurate holistic view of how projects perform. More guidance is therefore needed for megaprojects to perform. Several hypotheses are proposed based on literature and cases presented for the incorporation of how this research develops.

Once the city is known for leading the world in delivering megaprojects, London's risk management's response with regard to the biggest challenge of its lifetime (COVID-19 pandemic) will not go unnoticed. The general assumption is probably relevant to the public service social act stipulating that projects 'might improve the economic, social and environmental well-being of the relevant area' (Public Services (Social Value) Act, 2012, p. 2).

After the pandemic, the demands for risk management will only escalate with existing projects revaluating the need to save cost and be efficient to create more value, and therefore demanding accuracy when evaluating megaprojects has never been more essential. In this climate, there is an absolute urgent call for megaprojects to perform, but for megaprojects to perform, they need to step away from the iron triangle to reduce uncertainty and risk subsequently. To perform is to take a complex series of actions that integrate skills and knowledge to produce a valuable result. The model presented is therefore a stepping stone towards a holistic framework or strategy needed to address the problems faced by project managers in megaprojects. Developing performance is a journey, and the level of performance describes the location in the journey. Attention is drawn to uncertainty and risk management to help deliver meaningful successful projects for a better society and sustainable economies. This paper is, therefore, now declaring the triangle obsolete and any supporting theories including the HHP.

> How wonderful that we have met with a paradox, now we have hope of making progress (Niels Bohr, 1996)

References

AShurst, C., Doherty, N. F., & Peppard, J. (2008). Improving the impact of IT devel-opment projects: The benefits realization capability model. *European Journal of Information Systems, 17*(4), 352–370.

Atkinson, R. (1999). Project management: cost, time and quality, two best guesses and a phenomenon, its time to accept other success criteria. *International Journal of Project Management, 17*(6), 337–342.

BABOK. (2005). *A guide to business analysis body of knowledge.* International Institute of Business Analysts.

Baccarini, D. (1999). The logical framework method for defining project success. *Project Management Journal, 30*(4), 25–32.

Badewi, A. (2016). The impact of project management (PM) and benefits management (BM) practices on project success: Towards developing a project benefits governance framework. *International Journal of Project Management, 34,* 761–778.

Barnes, M. (1988). Construction project management. *Project Management, 6*(2), 69–79.

Beier, M., & Heathcote, J. (2010). The early stages of project planning. Conference paper Leeds Metropolitan University, International Conference on Strategic Project Management.

Berechman, J., & Chen, L. (2011). Incorporating risk of cost overruns into transportation capital projects decision-making. *Journal of Transport Economics and Policy, 45*(1), 83–104.

Böhle, F. et al. (2016). A new orientation to deal with uncertainty in projects. *International Journal of Project Management 34,*1384–1392.

Boonstra, A., Boddy, D., & Bell, S. (2008). Stakeholder management in IOS projects: Analysis of an attempt to implement an electronic patient file. *European Journal of Information Systems, 17*(2), 100–111.

Bruzelius, N., & Flyvbjerg, B. (2002). W. Rothengatter big decisions, big risks. Improving accountability in megaprojects. *Transport Policy, 9*(2), 143–154.

Bryde, D. J. (2008). Perceptions of the impact of project sponsorship practices on project success. *International Journal of Project Management, 26*(8), 800–809.

Chatelier, et al. (2020). Seeds conference paper, Leeds Sustainability Institute, Leeds Beckett University, School of the Built Environment and Engineering.

Collins, A., & Baccarini, D. (2004). Project success – A survey. *Journal of Construction Research, 5*(2), 211–231.

Cooke-Davies, T. (2002). The real success factors on projects. *International Journal of Project Management, 20*, 185–190.

De Wit, A. (1988). Measurement of project management success. *International Journal of Project Management, 6*(3), 164–170.

Dvir, D., Raz, T., & Shenhar, A. J. (2003). An empirical study of the relationship between project planning and project success. *International Journal of Project Management, 21*(2), 89–95.

Dvir et al. (2006). The future center as an urban innovation engine. *Journal of Knowledge Management, 10*(5), 110–123.

Eisenhardt, K. M. (1989). Building theories from case study research. *Academy of Management Review, 14*, 532–550.

Flyvbjerg, B. (2013). Quality control and due diligence in project management: Getting decisions right by taking the outside view. *International Journal of Project Management, 31*, 760–774.

Flyvbjerg, B. (2014). *Planning and managing megaprojects: Essential readings* (Vols. 1–2). Edward Elgar.

Flyvbjerg, B. (2016). The fallacy of beneficial ignorance: A test of hirschman's hiding hand. *World Development, 84*, 176–189.

Flyvbjerg, B. (2017). *The Oxford handbook of megaproject management* (p. 2). Oxford University Press.

Flyvbjerg, B., Bruzelius, N., & Rothengatter, W. (2003). *Megaprojects and risk: An anatomy of ambition*. Cambridge University Press.

Forbes. (2017). https://www.forbes.com/sites/georgebradt/2017/11/08/why-you-must-take-a-strategic-approach-to-risk-management. (Assessed December 2019).

Frey, T. (2017). *Megaprojects set to explode to 24% of global GDP within a decade.*

Gardiner, P., & Stewart, K. (2000). Revisiting the golden triangle of cost, time and quality: the role of NPV in project control, success and failure. *International Journal of Project Management, 18*, 251–256.

Gelatt, H. B. (1989). Positive uncertainty: A new decision-making framework for counseling. *Journal of Counseling Psychology, 36*(2), 252–256.

Gil, N., & Tether, B. S. (2011). Project risk management and design flexibility: Analysing a case and conditions of complementarity. *Research Policy, 40*(3), 415–428.

He, Q., Luo, L., Hu, Y., & Chan, P. C. (2015). Measuring the complexity of mega construction projects in China-A fuzzy analytic network process analysis, *International Journal of Project Management, 33*(3), 549–563.

Hirschman, A.O. (1969). Development projects observed. *Institute of Development Studies Bulletin, 1*(3), 22–25.

Ibbs, C. W., Lee, S. A., & Li, M. I. (1998). Fast tracking's impact on project change. *Project Management Journal, 29*(4), 35–41.

Ibbs, C. W,. Wong, C. K., & Kwak, Y. H. (2001). Project change management system. *Journal of Management in Engineering, 17*(3),159–65.

Ika, L. A. (2009). Project success as a topic in project management journals. *Project Management Journal, 40*(4), 69.

IPA Annual Report on Major Projects. (2017). https://www.gov.uk/government/publications/national-infrastructure-and-construction-pipeline-2016. (Assessed, April 2020).

Jugdev, K., & Müller, R. (2005). A retrospective look at our evolving understanding of project success. *Project Management Journal, 36*(4), 19–31.

Kaming, P., Olomolaiye, P., Holt, G., & Harris, F. (1997). Factors influencing construction time and cost overruns on high-rise projects in Indonesia. *Construction Management and Economics, 15*, 83–94.

Koch-Weser, I. N. (2013). *The reliability of China's economic data: An analysis of national output*. US-China Economic and Security Review Commission, US Congress.

Leslie-Carter, R., Zou, P. X. W., & Tan, E. X. (2009). Managing international projects: a case study of the "water cube" aquatic centre for Beijing 2008 Olympic games. In: Dainty, A. (Ed) Procs 25th Annual ARCOM Conference, 7–9 September 2009, Nottingham, UK, Association of Researchers in Construction Management, 959–68.

Liu, Z., et al. (2016). Handling social risks in government-driven mega project: An empirical case study from West China. *International Journal of Project Management, 34*(2), 202–218.

Lock, D. (2003). *Project management* (8th ed.)

Merrow, E. (2011). *Industrial megaprojects: Concepts, strategies, and practices for success.* Wiley.

Mishra, S., Khasnabis, S., & Swain, S. (2015). Incorporating uncertainty and risk in transportation investment decision-making. *Transportation Planning and Technology, 38*(7), 738–760.

Mulholland, C., Chan, P. W., & Canning, K. (2019). Deconstructing social value in decommissioning: A case study of industrial heritage at Dounreay. In C. G., & A. K. A. R. Martin Loosemore (Eds.), *Social value in construction.* London: Routledge.

Müller, R., & Turner, R. (2007). The influence of project managers on project success criteria and project success by type of project. *Eur Manag J 25*(4), 298–309.

Müller, R., Geraldi, J., & Turner., J. R. (2012). Relationships between leadership and success in different types of project complexities. *IEEE Transactions on Engineering Management, 59*(1), 77.

Niels Bohr (1996, quote) How wonderful that we have met with a paradox. Now we have some hope of making progress.: Niels Bohr – Place for writing thoughts. published independently by Spotnotebooks (March, 2020).

O'Brien, J. J. (1998). *Construction change orders.* McGraw Hill.

Oisen, R. P. (1971). Can project management be defined? *Project Management Quarterly, 2*(1), 12–14.

Oyegoke, A. S., & AI Kiyumi, N. (2017). The causes, impacts and mitigations of delay in megaprojects in the Sultanate of Oman. *Journal of Financial Management of Property and Construction, 22*(3), 286–302.

Oyegoke, A. S., Awodele, O. A., & Ajayi, S. (2019). Managing construction risks and uncertainties, A management procurement and contracts perspective. eBook Risk Management in Engineering and Construction, Imprint Routledge, 21.

PMI. (2008). *A guide to project management body of knowledge.* Project Management Institute (PMI).

PMI. (2016) . *All rights reserved. "PMI", the PMI logo, "Making project management indispensable for business results." And "Pulse of the Profession" are marks of Project Management Institute, Inc.* Project Management Institute.

Pollack, J. (2007). The changing paradigms of project management. *International Journal of Project Management, 25*(2007), 266–274.

Public Services (Social Value) Act (2012), Public Services (Social Value) Act 2012 (legislation. gov.uk) (Assessed, April 2020).

Remenyi, D., & Sherwood-Smith, M. (1998). Business benefits from information systems through an active benefits realisation programme. *International Journal of Project Management, 16*(2), 81–98.

Savolainen, P., Ahonen, J. J., & Richardson, I. (2012). Software development project success and failure from the supplier's perspective: A systematic literature review. *International Journal of Project Management, 30,* 458–469.

Siggelkow, N., & Rivkin, J. W. (2009). Hiding the evidence of valid theories: How coupled search processes obscure performance differences among organizations. *Administrative Science Quarterly, 54,* 602–634.

Toor, S., & Ogunlana, S. (2010). Beyond the 'iron triangle': Stakeholder perception of key performance indicators (KPIs) for large-scale public sector development projects. *International Journal of Project Management, 28*(3), 228–236.

Turner, J. R., & Zolin, R., (2012). Forecasting success on large projects: developing reliable scales to predict multiple perspectives by multiple stakeholders over multiple time frames. *Project Management Journal, 43*(5), 87–99.

Van der Hoorn, B., & Whitty, S. (2015). Signs to dogma: A Heideggerian view of how artefacts distort the project world. *International Journal of Project Management, 33*(6), 1206–1219.

van Wyngaard, C., Pretorius, J., & Pretorius, L. (2012). Deliberating triple constraint trade-offs as polarities to manage – a refreshed perspective. Proceedings of the 2013 IEEE International Conference on Industrial Engineering and Engineering Management, IEEE Conference Publications: 1265–1272.

Wateridge, J. (1998). How can IS/IT projects be measured for success. *International Journal of Project Management, 16*(1), 59–63.

Weaver, P. (2007). The origins of modern project management. In: *Fourth annual PMI college of scheduling conference, Proceedings mosaic project services*. 15–18 April 2007.

White, D., & Fortune, J. (2002). Current practice in project management – An empirical study. *International Journal of Project Management, 20*(1), 1–11.

Chapter 23
How Calls for New Theory Might Address Contemporary Issues Affecting the Management of Projects

John Heathcote

23.1 Introduction

Several observable issues persist in the contemporary practice in the management of projects. They perhaps serve to demonstrate that there is some unhealthy adherence to a set of practices that can be readily demonstrated to be flawed. These flaws have consequences for the management of projects in praxis, and many of the persistent issues around project management performance can be attached to these issues.

Project management (PM) appears to be a profession that has assimilated a lot of disparate tools, techniques and methods from other fields. It can be seen to contain much from preceding management philosophies in its professional so-called bodies of knowledge. These 'bodies of knowledge' prepared by the UK's Association for Project Management (APM) and the US-based Project Management Institute (PMI) are themselves depositories of a collection of techniques and approaches that have grown almost exponentially over time, for instance, PMI 2000, PMI 2004, PMI 2008, PMI 2013 (PMI, 2013) and the PMI BoK 2017 and the 'bodies of knowledge' from the APM issued in APM 1992, APM 1994, APM 1996, APM 2000, APM 2006, APM 2012 and APM 2019. These professional body approaches have an important impact on the way the profession is conceptualised and place the main emphasis of the management of projects as a matter of following an appropriate toolkit of processes. The notion that project management might be reduced to a series of broad processes appears to have been influenced by quality management tenets but may be attempting to apply a quality management principle to too broad and variable an undertaking as a project. PM as a profession appears to react to

---

J. Heathcote (✉)
School of the Built Environment, Engineering and Computing, Leeds Beckett University, Leeds, UK
e-mail: j.heathcote@leedsbeckett.ac.uk

issues that arise by developing new process solutions, for instance, the growing acknowledgement that 'stakeholders' could dramatically influence projects saw the development of 'stakeholder management', a process which largely mimics a risk management process. This points to PM limiting its reactions to ones that belong to only in the 'process school of thought' as identified by Bredillet (2008). Morris (2009) points out that the positivist approach to project management (PM) is slowly learning it also needs to consider that our PM knowledge is largely socially constructed. Padalkar and Gopinath (2016) conducted a longitudinal systematic review of PM literature and noted that *deterministic* views focusing on the control of planned projects began to include work that *sought explanations* for project phenomena (*explanatory* approaches) which was then joined by *non-deterministic* perspectives on projects developed by Morris and others since 2000. All these perspectives continue and, in parallel, perhaps create a greater confusion.

Writers on PM have noted that the 'lived experience' of those managing projects does not closely match that as presented by texts and professional bodies' publications of process approaches. These approaches can be equated with *deterministic* approaches (Crawford et al., 2006; Morris, 1997, 2010; Svejvig & Anderson, 2015). Crawford et al. (2006) questioned whether project managers could be considered 'reflective professionals' or 'trained technicians' with the tendency towards process-based solutions and the associated training presenting the role of PM as appearing to be about simply running through a set of processes/procedures. Crawford et al.'s (2006) implication is that this would not serve the complexity and challenges of a project in real life.

Acknowledging that there is a gap and an incongruence between practice in the management of projects and the general professional and textbook advice, academia has developed *the rethinking project management* agenda (Svejvig & Anderson, 2015; Padalkar & Gopinath, 2016). This is examined later in the paper, but to set that rethinking in context, a review of persistent issues faced in the management of projects has been considered next, and this precedes a review of suggestions raised by the rethinking PM agenda.

This study utilises a grounded theory approach to bridge the gap between the rethinking agenda and the issues experienced in practice; these differences are referred to in literature, but not explicitly linked to the need for new theory. This study proposes new linkages between contemporary issues and need to evolve *theory for practice*. The paper goes on to induct new theory that offers the potential to address how project and project management need to be reinterpreted to address issues experienced in practice.

23.2 Research Methodology

While *explanatory* research approaches commence with a theoretical position (Yin, 1994), the approach taken for this study contrasts with the *explanatory* approach by taking an *exploratory* approach to arrive at grounded theory. To do this Strauss and

Corbin (1990) and Yin (1994) proposed collecting data and find themes and issues to follow up from that data. This can be translated as theory emerging from the process of data collection and analysis. This study therefore is an inductive approach, arriving at theory. The sampling approach taken with this study includes a selected sample from papers associated with the rethinking project management agenda (Winter et al., 2006), and they are contrasted with a selection of contemporary issues that are repeating issues observable in the management of projects (Morris, 1997, 2009; Flyvbjerg et al., 2009; Gray, 2010; Koskela & Howell, 2002; Morris et al., 2011; Lenfle & Loch, 2010; Morris, 2013; Söderland et al., 2012; Padalkar and Gopinath (2016); Carvalho & Rabechini Junior, 2015; Brady et al., 2012; Crawford et al., 2006) and are selected by the author. The sample is deliberately selected to demonstrate the issues underpinning the difficulties experienced in the management of projects. This sample selection is therefore *purposeful* (Strauss & Corbin, 1990). Any bias introduced by the two samples' purposeful selections is addressed by only arriving at grounded theory, for later hypothesis testing.

The study's research aim then is to identify relationships between the data (see Table 23.1) and goes on to develop questions and hypotheses to test these as advocated by Strauss and Corbin (1990) for these types of studies.

23.3 Literature, Secondary Data

23.3.1 *Contemporary Issues in the Management of Projects*

Defining Success There are several references to failure rates being high in the success of projects, but given that, identifying what success is may be quite variable. Defining a project's success will depend on the perspective you are taking. If the measure of project success is judged by whether it meets the time schedule and cost budget set of the project at its approval to go ahead, few projects manage to achieve this. But where the benchmark of a time and cost plan also incorporates and is revised for any subsequent change requests provides a movable success measure. Whether changes to plan are incorporated might be unclear in any sample. Creating a problem for research where 'project success' is a measure sought as a comparator. Because a better measure of success for the project is to measure whether the intended 'quality' specification has been met, together with the time and cost, these three constraints (time, cost and quality) have formed a lasting, if over simple, idea of project success. Cooke-Davies (2002) points out though that 'time, cost and quality' ought to merely reflect 'project management' success and 'project' success might be something more nuanced. This problem of determining what represents 'project' success arises because the underpinning theory of 'project' remains undefined (Koskela & Howell, 2002). Several candidates for arriving at what represents project success are offered by the APM who suggested project success is represented by meeting 'stakeholder satisfaction'; previously the approved constraints of *time*, *cost* and *quality* specification budgets have been offered, and this remains a very prevalent notion in PM; and more recently the idea of *benefits* has been offered in the development of the *theory of Programme Management*, and this is sometimes used as a substitute for *quality* and otherwise seen as a locally interpreted derivative of the project's host organisation's *strategic intent*.

Aside from the problem of what a *project* is, there lacks a theoretical underpinning for practice, so what a *project* is is uncertain, and consequently determining whether it has been delivered successfully or not becomes a matter of positioning. An additional number of 'issues' persist in the management of projects and form a repeating cadre of causes that the profession struggles to address with any permanency. Attempts are made but those solutions tend to fail in application; this tends to lead to a criticism that the prescribed solution was not properly implemented. For instance, the UK's NAO regularly suggested, following project audits, that more efforts should be producing more accurate planning estimates. This despite an understanding of this difficulty is explained and demonstrated with experimental science in the 1970s. Such a criticism of audited projects assumes that the uncertainty of a project proposal can be accurately estimated and that other factors might serve to create further uncertainty and be either predicted and mitigated or else 'made up for' once slippage has occurred. In fact, the notion that project managers will make up for poor performance in the project and will be addressed by creating better performance in subsequent tasks (the thermostat model) assumes *float* (or *slack*) always exists in budget allocations and that mediocre performance will be the norm. Clearly this is problematic.

The Planning Fallacy Kahneman and Tversky (1977) coined the phrase the *planning fallacy* to describe the effect of a 'hard wired' cognitive error that prevented human beings accurately making planning predictions (which would apply to both cost and schedule). They showed, using experimentally supported evidence, that humans have a natural and largely unavoidable tendency towards 'optimism bias' when planning. Later, Buehler et al. (1994) examined experimental work from the field of human economic behaviour, to show that expertise was no mitigation to the planning fallacy. Because planning and control are at the heart of project management, it is difficult to understand quite why this key flaw does not feature more prominently in PM texts. Dealing with the consequences of the planning fallacy consumes normative PM through the application of more processes to planning activities to achieve the desired control of the project. This 'controlling' interpretation (Koskela & Howell, 2002) of the 'management' function for project management is also referred to as the 'engineering paradigm' (because of the association of planning engineering and then 'doing' it), and yet the planning fallacy predicts that this will not address the issue of plans being optimistic/projects late. The planning fallacy and its attendant uncertainty and optimism biasing effects (for instance, Flyvbjerg et al. (2009) and Gray (2010)) identified optimism in project 'business cases' also creates other issues for the management of projects and although heavily interrelated are not always considered together. Nor is the planning fallacy a commonly understood issue in project management professional training or practice.

Deterministic Planning One important effect of the planning fallacy is that the reliability of the project plan might itself be overstated. Koskela and Howell (2002), Morris et al. (2011), Lenfle and Loch (2010), Morris (2013), Söderland et al. (2012) and Padalkar and Gopinath (2016) all saw the deterministic treatment of the project plan as being important to understanding why the theory of project management is unable to address *theories for practice* in the management of projects. A *deterministic* plan assumption refers to the idea that the plan can be known, and where sufficient effort made in the early stages of planning then the plan will be sufficiently accurate to allow it to be treated as a 'benchmark' (or 'baseline') against which progress and performance rates can be judged. The 'planning fallacy'-related experimentation (Kahneman & Tversky, 1977; Buehler et al., 1994) demonstrates this is unlikely, and any project is likely to encounter a series of risks; some might be anticipated but be of uncertain probability and others unknown and so unidentified. And yet the project management *control paradigm* (Brady et al., 2012; Padalkar & Gopinath, 2016) entirely relies on this idea of a baseline plan, set out in the 'project definition' and to which the 'business case' is usually appended (APM, 2012).

Together with the planning fallacy, and the resultant variability of reality creates uncertainty for projects, the tendency towards deterministic approaches adds a layer of unreality to the management of projects, as project teams, their managers and the contracting suppliers are obliged to create monitoring 'plans' that do not adequately 'model' a more *non-deterministic* (or *'probabilistic'*) reality. It is easy to imagine that in such an environment, fallacious reporting is a necessary situation that project agents are pushed into meeting the expectations of sponsors and client organisation directors who are seeking certainty to reduce the risk, the risk the organisation is exposed to as a result of what might be the 'risky' significant capital investments that many projects represent.

Proprietary project management methodologies and explanatory approaches further embed into normative PM a deterministic assumption. In the UK the APM's encouragement of membership for PMs through a series of proprietary short courses oversimplifies the challenge of PM and presents a hard paradigm emphasis (the 'hard' paradigm refers to PM as a series of processes and data tools and is contrasted with the 'soft' paradigm which emphasises 'social issues, decision making, and values', Pollack, 2007). An analogy for generalised project methodologies might be a recipe that is attempting to be representative of all dishes, from soup to salad! Project methodologies tend to represent single planning strategies (Prince2 offers a 'wave' or 'stage' planning linear approach, with check-listing; DSDM offers a more agile, prototyping approach, with tasks as prioritised to-do lists); some of the methodologies include a consideration for, for example, team dynamics, as with DSDM. The growing popularity of 'agile' approaches, particularly for IT projects, invites flexibility in business cases by allowing for more rapid changes to content; usually this means doing less than was originally planned for. Crawford et al. (2006) saw the adoption of process methods for managing projects as reducing the PM role to one of 'trained technicians', intimating that a more 'reflective practitioner' was what was required in practice.

Managing Risk and Uncertainty Like the planning fallacy, Gigerenzer (2011, 2013) points out the 'risk' is also typically misrepresented, with human beings falling prey to a 'loss aversion' bias they will be unconscious to (Kahneman, 2011; Sutherland, 2007). But risk management (RM) is central to the way that PM presents itself (Brady et al., 2012). Although RM imagines that the uncertainty surrounding project environments can be brought under control through iterative scanning for new risks, seek first to assess them (and score on the basis of probability of occurrence and impact if they do happen), and then mitigate the risk so that they are 'controlled' or lessened. RM also recommends that, while this is happening, project teams prepare 'contingency' plans in case the mitigating efforts do not succeed. In this way the 'risk management process' looks like a 'defence in depth' of known uncertainties and an ongoing attempt to identify new uncertainties. Much work in PM has gone into developing more mathematical based approaches to calculating risk, but these miss the key point that risk is itself a subjective impression and, as Gigerenzer establishes, cannot easily be accessed, and not by 'gut feel and intuition'. Carvalho and Rabechini Junior (2015) showed how PM risk techniques proved ineffective for practice.. Gigerenzer uses large statistical data sets to reveal truths about known risk that are habitually mis-read, but such evidence is not always available to PMs and the evidence suggests that project workers will prioritise their own sense of risk putting this ahead of any data.

While undoubtably still worth doing, the limitations of risk management do not feature in the PM literature (Carvalho & Rabechini Junior, 2015); moreover, the project team could not hope to anticipate more than a few of the key risk items to a project. As it stands it might serve to further maintain a myth that the project can be 'controlled' to a greater degree than is possible in practice.

False Performance Reporting *Adversarial contracted teams, underbidding and business case exaggerations.* Problems with adversarial behaviours leading to contractual conflict between contracted parties on projects are a problem leading to poor teamworking and cooperation and a tendency to lose efficiency in work performance. Long seen as a national industry-wide problem in the UK (Latham, 1994; Egan, 1998), this issue is important because much of project delivery (as a temporary endeavour) relies on the utilisation of contracted services from specialist contracting organisations and their sub-contractors. Heathcote et al. (2019b) provided an explanation for adversarial contracting seeing it resulting from competitive bidding pressures, in which game theory would predict that low bids are made to win tenders and then variation claims are fabricated to recover any underbid, during construction (Heathcote et al., 2019b). Flyvbjerg et al. (2009) and Gray (2010) showed how the client business cases for projects tended to underestimate cost forecasts, to achieve authorisation, a behaviour to which they both attributed the desire for client project managers to achieve authorisation for their projects regardless of the project's actual viability. This has the effect of creating pressures to arrive at a lower contract price. Much effort has gone into attempting to address these pressures; for contract bidding a move towards more 'partnering' approaches between contractors and clients and contractors and their supply chains has been generally accepted and is ostensibly in practice. However, Cheung et al. (2003), Chan et al. (2006), Bresnen (2007), McMeeken (2008), Gadde and Dubois (2010) and Morris et al. (2011) all saw partnering approaches struggling in implementation, with 'partnering' being paid little more than 'lip-service', excluding most sub-contractors from partnering benefits, even [commercial] 'bullying' of supply chain sub-contractors, all of which further reinforces the tenacity of adversarial relationships and an adversarial culture to dominate the behaviour between contracted team members. Heathcote et al. (2019b) suggested that this meant that contactors, who might be significant contributors to a project, would need to be disguising progress from the client. Case study evidence of public sector projects (whose performance is a matter of public scrutiny and so visible), which have gone significantly awry from their original (on their easily measured time and cost budget) estimates, provides a growing body of evidence to support these hypotheses.

This may go some way towards explaining why the project management challenge is frequently referred to as being about the management of uncertainty (Cagliano et al., 2015), and this too alludes to a *deterministic* way of thinking.

Stakeholder Management In general normative PM (APM, 2006) treat 'stakeholders' as risks to be managed, but more recently this management of stakeholders, in acknowledgement of their ability to influence the project, has evolved from 'manage' to 'engage' (Aaltonen & Kujala, 2010) and, consequently, multiple negotiations are necessary throughout the project to avoid stakeholders from using their power to block the process (Van Rijswick & Salet, 2012, cited in Heathcote et al., 2019a). However, without the tools that persist in PM for 'managing', stakeholders reflect a risk management process, with only Heathcote et al. (2019b) suggesting a more nuanced negotiated engagement. This approach from Heathcote et al. (2019b) departs from the preceding 'controlling' approaches to one where their interests must be engaged with and be addressed through negotiation, somewhat moving the paradigm away from *controlling* and towards an approach that seeks opportunities for mutual gain. This change asks for a reinterpretation of how projects should be managed and may not readily fit into the deterministic approach.

The Link to Strategy, Portfolio and Programme Management (and the Attendant Benefits Management) In its infancy in practice, and being highly variable, the theory of programme management alludes to both theories of project and expands it by including benefits (Lycett et al., 2004) (and so *value generation*, Heathcote and Ben Baha, 2019). The theory of programme management, however, makes *benefits management* the responsibility of superordinate roles of programme managers and allocates the delivery of *'product' outputs* to the project teams. This demarcation might be very significant as it firmly retains the role of deterministic planned delivery of output 'products' with the project team and manager and reserves for the role of programme manager and programme team that of managing the *benefits*. Arguably projects are undertaken for the benefits they bring, and so *benefits* are a strategic concern and can be regarded as the intended strategic outcome for the project. Arguably, all projects are strategic in nature then, but the existing theory of programme management separates this issue of benefits from the project team. The application of programme management has been found to be limited (Shehu & Akintoye, 2009) and only available to organisations that can absorb the overhead of a coordinating bureaucracy that is normally associated with it and administered through a PMO (project management office). The emergence of programme management, which, unlike other PM processes, can be solely attributed to the field of project management, seeks to link projects more firmly to the organisational strategy and to provide further control oversight for senior decision-makers in the organisation (programme management is also forecasted as bringing some advantages that might arise from the ability to coordinate between related projects). Heathcote and Ben Baha (2019) found that their respondents were most concerned about the project-strategy link and remained concerned after the installation of programme management approaches, suggesting it is not easily applied in practice. They were able to hypothesise that *H1* 'The systems nature of projects and programmes means that the two should not be treat as separate disciplines, but rather interconnected and overlapping systems'; *H2* 'That without further development, that addresses the control emphasis of some programmes, that programmes may serve to limit the creation of value by projects'.

Summary The control paradigm is based on a reductionist and deterministic approach that reduces project management to an emphasis on related deterministic planning assumptions that cannot be maintained in practice. Project management as a profession, supported by professional bodies, develops greater complexity in methods that seek to address control failures without addressing the underlying conditions that bring these difficulties about. The difficulty in comparing unique project scenarios presents barriers to quantitative testing of the theory that normative PM relies on, further: that underpinning theory is not expressed or identified in PM literature, and academic writers have been obliged to induct theory from observed practice. Padalkar and Gopinath (2016) provide an example of this induction, identifying deterministic, explanatory and *emerging* non-deterministic interpretations in a longitudinal review of PM literature. This study considers next the emergent non-deterministic approaches that seek to create a route to better equipping project management in practice.

23.3.2 The Project Management Rethinking Agenda

Morris (1997, 2009), Crawford et al. (2006), Svejvig and Anderson (2015), Svejvig and Anderson (2015) and Padalkar and Gopinath (2016) acknowledged an incongruence between practice in the management of projects (the 'lived experience') and the general professional and textbook advice, to which academia has developed the project management rethinking agenda. Linked to this is inadequacy of underlying project management theory (Padalkar & Gopinath, 2016) and the need to establish an 'explicit understanding of the theoretical basis of project management'

(Pollack, 2007, p. 266). Padalkar and Gopinath (2016) suggested this confusion comes from 'multiple perspectives, paradigms, methodologies and weak theories' (p. 1316). Pollack (2007), Brady et al. (2012), and Padalkar and Gopinath (2016) proposed research that advocated a need to reconsider project paradigms and explore non-deterministic perspectives and consider the negative impact existing approaches were having on projects. Below four selected contributions to the rethinking agenda are presented:

Koskela and Howell (2002) identified two general theories of project in operation, one being about creation of a product and the *transformational* and *flow* of work to create that product (these theories see projects as the operation by which things are 'built'); they might be an important contribution to the filed by highlighting the theory of project also in action which emphasised *value generation*; this is important because this theory of project sees the project as having a more encompassing (strategic) purpose. The two theories can work together although theories of transformation and flow will always be sub-ordinate to that of the project being about the generation of 'value' (and it should be noted that value is a subjective and perhaps personal interpretation). They also sought to identify the dominant theories for management that PM invoked. Perhaps it is important that Koskela and Howell's theories of project management do not appear to readily account for value generation (Fig. 23.1).

Winter et al. (2006) reporting on 'directions for future research in PM' concluded with Fig. 23.2. It is possible to determine from their conclusions a revising of the theory of project. They note that projects are 'complex' and need to understand that complexity; that projects are 'socially constructed by social processes'; and that this is important because projects are about value creation, concurring with Koskela and Howell's interpretation and having the implication that projects will need to have a broader conceptualisation as a consequence of that 'value' interpretation. Implied in this is that project managers will be involved in other social processes such as negotiation with other interests, and this will impact on how the role is enacted, *away from operating processes* (and *away from collections of processes such as methodologies* perhaps), towards the need for a more *flexible* and *facilitative* role.

Subject of Theory		Relevant theories
Project		Transformational
		Flow
		Value generation
Management	Planning	Management as planning; Management as organising
	Execution	Classical comms theory; Language/Action perspective
	Controlling	Thermostatic model; Scientific experimentation model

Fig. 23.1 Ingredients for a new theoretical foundation of project management (Koskela & Howell, 2002)

Theory ABOUT Practice	
\multicolumn{2}{|c|}{Direction 1}
The **Lifecycle** Model for Projects and PM	Theories of the **complexity** of Projects and PM
FROM: the simple life-cycle based models of projects, as the dominant model of projects and project management. And, FROM: the (often unexamined) assumption that the lifecycle model *is* assumed to be the actual 'terrain' (i.e. that the actual reality is 'out there' in the world).	TOWARDS: the development of new models and theories which recognise and illuminate the complexity of projects and PM at all levels. And TOWARDS: new models and theories which are explicitly presented as only partial theories of the complex 'terrain'.
\multicolumn{2}{|c|}{Implication}
\multicolumn{2}{|c|}{The need for multiple images to inform and guide action at all levels in the management of projects, rather than just the classical lifecycle model of project management, as the main guide to action, (with all its codified knowledge and techniques).   Note: theories ABOUT practice can also be used as theories FOR practice.}
\multicolumn{2}{|l|}{Theory FOR Practice}
\multicolumn{2}{|c|}{Direction 2}
Projects as Instrumental Processes	Projects as Social Processes
FROM: the instrumental lifecycle image of projects as a linear sequence of tasks to be performed on an objective entity 'out there', using codified knowledge, procedures and techniques, and based on an image of projects as temporary apolitical production processes.	TOWARDS: concepts and images which focus on social interaction among people, illuminating the flux of events and human interaction, and the framing of projects (and the profession) within an array of social agenda, practices, stakeholder relations, politics and power.
\multicolumn{2}{|c|}{Direction 3}
Product Creation as the Prime Focus	Value Creation as the Prime Focus
FROM: concepts and methodologies which focus on; *product creation* – the temporary production, development, or improvement of a physical product, system or facility etc. – and monitored and controlled against specification (quality), cost and time.	TOWARDS: concepts and frameworks which focus on *value creation* as the prime focus of projects, programmes and portfolios. Note however: 'value' and 'benefits' as having multiple meanings linked to different purposes: organisational and individual
\multicolumn{2}{|c|}{Direction 4}
Narrow Conceptualisation of Projects	Broader Conceptualisation of Projects
FROM: concepts and methodologies which are based on: the narrow conceptualisation that projects start from a well-defined objective 'given' at the start, and are named and framed around single disciplines, e.g. IT projects, construction projects, HR projects, etc.	TOWARDS: concepts and approaches which facilitate broader and ongoing conceptualisation of projects as being multidisciplinary, having multiple purposes, not always pre-defined, but permeable, contestable and open to negotiation throughout.
\multicolumn{2}{|l|}{Theory IN Practice}
\multicolumn{2}{|c|}{Direction 5}
Practitioners as Trained Technicians	Practitioners as Reflective Practitioners
FROM: training and development which produces practitioners who can follow detailed procedures and techniques, prescribed by project management methods and tools, which embody some or all of the ideas and assumptions of the 'from' parts of 1 to 4.	TOWARDS: learning and development which facilitates: the development of reflective practitioners who can learn, operate and adapt effectively in complex project environments, through experience, intuition and the pragmatic application of theory in practice.

Fig. 23.2 Directions for future research in project management: The main findings of an EPSRC research network (Winter et al., 2006)

Svejvig and Anderson (2015) reviewed selected works that compared classical project management, with rethinking project management, between 1995 and 2010; see Fig. 23.3.

They also perceived a move towards the theory of project being about value creation. They pick out projects as dealing with complexity and 'emerging'; moving away from the control paradigm; and implying a link between value and organisational strategy. They also see the PM role as having to deal with this complexity and make the project about innovation and needing to be adaptive.

In Fig. 23.4 is included Dalcher's (2017) commentary on how he perceives the implications for these emerging changes for projects and project management.

Comparing classical project management with rethinking project management.

Author	Classical Project Management	Rethinking Project Management
Packendorff (1995, p. 328)	Project metaphor: the project as a tool Process: linear, with the phases plan, control and evaluate	Project metaphor: the project as a temporary organization Process: iterative, with the phases expectation setting, actions and learning
Jugdev et al. (2001, p. 36)	Project management: as a set of tools and techniques used to achieve project efficiencies	Project management: as a holistic discipline used to achieve project/program/organizational efficiency, effectiveness and innovation
	Success: measured by efficiency performance metrics	Success: a multidimensional construct measured by efficiency, effectiveness and innovation
	Practice project management: focus on the project details at the operational level and tactically	Sell project management: be an advocate and champion of project management by aligning its value with the firm's strategic business priorities
Winter et al. (2006c, p. 642, original emphasis)	Simple life-cycle-based models of projects, as the dominant model of project and project management with the (often unexamined) assumption that the life-cycle model *is* (assumed to be) the actual terrain	New models and theories that recognize and illuminate the *complexity* of projects and project management, at all levels. The new models and theories are explicitly presented as only *partial* theories of the complex terrain
Shenhar and Dvir (2007, p. 11, original emphasis)	Approach: traditional project management Project goal: completing the job on time, on budget and within the requirements Management style: one size fits all	Approach: adaptive project management Project goal: achieving multiple business results and meeting multiple criteria Management style: adaptive approach, one size does *not* fit all
Andersen (2008, p. 5, 10, 49)	Perspective: task perspective Project definition: a project is a temporary endeavor undertaken to create a unique product, service or result (Project Management Institute, 2004, p. 5) Main focus: execute the defined task	Perspective: organizational perspective Project definition: a project is a temporary organization established by its base organization to carry out an assignment on its behalf Main focus: value creation. Create a desirable development in another organization
Lenfle and Loch (2010, p. 45)	Project type and target: routine execution, target given and defined from above	Project type and target: novel strategic project with a general vision and direction, but detailed goals not known and partially emergent
	Examples of domain of relevance: • Known markets and customer reactions • Known performance drivers of developed systems • Known environmental parameters	Examples of domain of relevance: • New markets and unknown customer reactions • Unknown technology • Complexity with unforeseeable interactions among drivers and variables

Fig. 23.3 Comparing classical project management with rethinking project management (Svejvig and Anderson 2015)

Dalcher identifies people as the subtitle to what can be seem as projects as social systems. His leadership category describes a facilitative approach that concurs with others. His broad contextualisation of the project environment reinterprets the project in the social system, and the strategy of the organisation, and also adds a sustainability perspective by including 'long-term thinking' which also adds to the value imperative. Like some others Dalcher also sees these changes as invoking the need to see projects as innovating.

A new theory of project might be distilled from these works presented in Figs. 23.1, 23.2, 23.3, and 23.4, as one of *creating a valued outcome, as defined through facilitating a negotiation with the (open) social system that comprises internal (strategic) and external interests; acknowledging this will lead to an emergent (multiple) outcome(s) from necessarily multiple (and innovated) possibilities.* Dealing with complexity often means that simplifications are developed that look at component parts of a larger whole. This might be what happened with PM. The rethinking project management agenda reasserts the complexity of the project as an 'open system' (Smyth & Morris, 2007) which leads to multiple interactive variables without any constants. PM has attempted to create temporary constants (of the 'plan', the 'business case' or the 'requirements') but undermined the more wholistic reality of project context as it did so.

Identifying Key Themes [for further advances in project management]
People: Normative PM overlooks considerations for 'people'. Human aspects need to be better addressed stakeholders, motivation, needs discovery, engagement, marketing, influencing, persuading and understanding of users and their role. Examples of such new methods include: *Systemic evaluation of the spheres of influence; Driving stakeholder engagement by role and contribution; Determining the (human) pace of progress ; Repositioning projects as "social endeavours"; Focusing on gatekeepers, customers, client chains, and contractors as stakeholders; Utilizing choice engineering; Fostering project resilience; Considering the methodology of compelling behaviours.*
Leadership: moving from managing to leading. Managing might be about certainty and a control- oriented perspective, while leadership points to a different and more varied skill set. The combination of uncertainty and a greater reliance on a network of participants requires a more organic approach emphasising influence, participation, and collaboration.
Context: Projects rely on situational and contextual factors that managers need to understand. Interacting with projects in complex environments [might] require an awareness of the specific characteristics, including informational, contextual, strategic, geomorphological, geological, environmental, and public perception considerations and a willingness to experiment and adapt.
Strategy: Project management is concerned with the delivery of projects, while projects link strategy and execution. Improving the alignment between strategy and execution requires strategic or organisational or portfolio level engagement from project professionals.
Value: Projects are often created to satisfy strategic needs and objectives and therefore project management is increasingly called upon to deliver benefits and value. However, it is completed projects that satisfy users by subsequently providing benefits, and not project management per se. Benefit realisation and value delivery capability can be linked to projects, but only via a strategic, or organisational, frame of thinking that extends beyond execution.
Long-term thinking: The long-term perspective is often invoked to consider ethics, decision making, return on investment, benefits realisation, value accumulation, decommissioning, extended life cycles and warranty periods. It is here that the distinction between temporal project management (focused on delivery to predefined schedules and budgets), and the sustained outcomes, and even outputs of a project, … Sharing knowledge, resources and talent often requires organisational considerations that extend beyond any single project. Similarly, as project managers are asked to relate to a wider horizon, or adopt an extended life cycle, they enter a different level of conversation about the project and its impacts. They therefore need new ways of reasoning for this type of conversations such as employing new and extended methods of addressing multiple levels of success, timely engagement modes and ultimate project outcomes.
Innovation: […] long term success comes from management innovation. Innovation and experimentation provide essential learning opportunities for validating and improving performance. Moreover, adaptation, trial-and error and resilience enable managers to adjust and respond to the unknown. Innovation can lead to bold new suggestions; in the approach to managing projects, [and in project outcomes].

Fig. 23.4 Adapted from; PM World Journal, The Case for Further Advances in Project Management Vol. VI, Issue VIII—August 2017. Prof Darren Dalcher

23.4 Analysis

Table 23.1 Summary of findings: how might the rethinking agenda's conclusions support issues observable in contemporary management of projects'

Contemporary issues	Rethinking agenda
Planning fallacy	Interpreted as complexity and uncertainty in the rethinking agenda
	Although addressed by avoidance, the agile methodologies seek to accommodate variance and scope and output uncertainty, by simply avoiding commitment to a scope
	A ready 'planning fallacy' solution has been proffered in 'reference class forecasting'
	Plans need to accommodate multiple possibilities and probabilities on outcome Heathcote and Coates (2018) suggest thinking a project plans as 'modelling for analysis'
Proprietary project management methodologies and explanatory approaches	Not suitable for complex environments and open system project challenges. May also make the difficulty of understanding the project challenge worse by causing a comforting distraction

Contemporary issues	Rethinking agenda
False performance reporting, in projects and from contractors	Hard paradigm models need to serve soft system possibility exploration to create innovation possibilities. Hard system artefacts like Gantt charts and cash flow can serve as models to facilitate risk identification and innovation opportunities. Heathcote and Coates (2018) suggest thinking a project plans as 'modelling for analysis'
Hard paradigm emphasis, *in social systems*	
Risk management	
Stakeholder management	Value creation/generation should be the goal of projects, separation of benefits and products between programme/benefit roles, and project roles may not be useful and could be specifically harmful to a value-focussed theory of project, though the value interpretation most definitely brings a strategic focus to the theory of project which is crucial. The role of stakeholders internal (strategy makers/directors) and external (which could be thought of as the 'market') can be acknowledged in interpreting whether value is being added/created
Link to strategy, programme management	
Contradicting theories of project in action. Value generation	

23.5 Discussion and Concluding Hypotheses

The emerging trends identified by authors of the rethinking project management agenda move from deterministic and explanatory to non-deterministic paradigms and might imply that PM is changing, but observable in both practice and education and most notably in a plethora of professional training courses, and is the continued prevalence of the early theories of project that have a continuing pervasive presence. This means that there is a continuing presence of multiple theories which are sometimes incompatible. Underpinning the issue though is that of the 'planning fallacy'. This known fallibility in plan forecasting might need to be acknowledged as an issue connected to the notion that projects are complex. Complexity is not something that recently happened as some PM texts suggest but was always a consequence of uncertainty and probability.

Arising from the findings in Table 23.1, a preliminary new theory of project is proposed: *projects seek to create a valued outcome, as defined through facilitating a negotiation with the (open) social system that comprises internal (strategic) and external interests; acknowledging this will lead to an emergent (multiple) outcome(s) from necessarily multiple (and innovated) possibilities.*

Key to addressing the prevalent contemporary issues in the management might mean that an understanding of the competing theories of project is in operation. It would appear that the open system interpretation of projects as socially constructed value generators better fits and explains the problems experienced by approaches that rely on unexplored, untested and hidden theoretical premises that can be demonstrated to be incomplete. Some contemporary and established project methods, such as the linear development impression created by waterfall proprietary project methodologies, and some programme management systems, such as the separation of project outputs from beneficial strategic outcomes, might be having a detrimental effect on the issues examined here. Approaches to contracting would also need to accommodate solutions to planning fallacies as an antecedent to the better design of contract/procurement strategies.

Some scientific work hints that the approaches identified might lead to more suitable approaches to managing projects: Heathcote and Coates (2018) proposed addressing this dilemma by considering plan development not as deterministic plans of how things are expected to happen, but to see the 'plans' instead as *model[s] for analysis*, methods by which multiple probabilities are anticipated, modelled and variously mitigated to create more valuable outcomes and showed how this approach leads to innovation for better project performance. Coates and Heathcote's (2017) quantitative study provided statistically significant findings to demonstrate that such an approach would lead to time, cost and commercial/market benefits which outperformed the control group. Heathcote, Butlin, and Kazemi (2019) proposed 'interest-based negotiation' as a method for facilitating positive outcomes with stakeholder interests.

References

Aaltonen, K., & Kujala, J. (2010). A project lifecycle perspective on stakeholder influence strategies in global projects. *Scandinavian Journal of Management, 26*, 381–397.
Association for Project Management. (2006). *APM body of knowledge* (5th ed.).
APM. (2012). *APM body of knowledge* (6th ed.). Association for Project Management.
Brady, T., Davies, A., & Nightingale, P. (2012). Classics in project management: Revisiting the past, creating the future. *International Journal of Managing Projects in Business, 5*(4), 718–736.
Bredillet, C. N. (2008). Exploring research in project management: Nine schools of project management research (part 6). *Project Management Journal, 39*(3), 2–5.
Bresnen, M (2007). Deconstructing partnering in project-based organisation: Seven pillars, seven paradoxes and seven deadly sins. *International Journal of Project Management, 25*, 365–374.
Buehler, R., Griffin, D., & Ross, M. (1994). Exploring the 'planning fallacy': Why people underestimate their task completion times. *Journal of Personality and Social Psychology, 67*(3), 366–381.
Cagliano, A., Grimaldi, S., & Rafele, C. (2015). Choosing project risk management techniques. A theoretical framework. *Journal of Risk Research, 18*(2), 232–248. http://ebscohost.com.
Carvalho, M., & Rabechini Junior, R. (2015). Impact of risk management on project performance: The importance of soft skills. *International Journal of Production Research, 53*(2), 321–340. http://ebscohost.com.
Chan, A. P. C., Chan, D. W. M., Fan, L. C. N., Lam, P. T. I., & Yeung, J. F. Y. (2006). Partnering for construction excellence – A reality or myth? *Building and Environment, 41*, 1924–1933.
Cheung, S.-O., Ng, T. S. T., Wong, S.-P., & Suen, H. C. H. (2003). Behavioural aspects in construction partnering. *International Journal of Project Management, 21*, 333–343.
Coates, A., & Heathcote, J. (2017). Measuring the impact of key planning principles on 'Gross Margin'. SEEDS International Conference.
Cooke-Davies, T. (2002). The "real" success factors on projects. *International Journal of Project Management, 20*(3), 185–190.

Crawford, L., Morris, P., Thomas, J., & Winter, M. (2006). Practitioner development: From trained technicians to reflective practitioners. *International Journal of Project Management, 24*, 722–733.

Dalcher, D. (2017). The case for further advances in project management. *PM World Journal, 6*(8).

Egan, J. (1998). *Rethinking construction*. Construction Task Force, London.

Flyvbjerg, B., Garbuio, M., & Lovallo, D. (2009). Delusion and deception in large infrastructure projects. *California Management Review* [Online], *51*(2), 170–193. Available from: [Accessed 27/03/2017].

Gadde, L., & Dubois, A. (2010). Partnering in the construction industry – problems and opportunities. *Journal of Purchasing and Supply Management, 16*, 254–263.

Gigerenzer, G. (2011). *What Scientific concept would improve everyone's cognitive toolkit?* The Edge Question. https://www.edge.org/response-detail/10624

Gigerenzer, G. (2013). Risk Literacy TedZurich availabale online at https://www.youtube.com/watch?v=g4op2WNc1e4.

Gray, B. (2010). The defence strategy for acquisition reform. MOD.

Heathcote, J., & Coates, A. (2018). Illustrating how a systems approach to modelling project plans improved innovation in operations. SEEDS International Conference.

Heathcote, J., & Ben Baha, G. (2019). An Investigation into the gap Between Programme Management theory and Practice. *SEEDS International conference*.

Heathcote, J., Butlin, C., & Kazemi, H. (2019a). Stakeholder management: Proposal for research; Do successful project managers employ 'interest-based negotiation' to create successful project outcomes? In *SEEDS International conference*.

Heathcote, J., Kazemi, H., & Wilson, M. (2019b). Illustrating game playing on construction contracts: The negative impact of procurement strategies. A proposal for research. In *ARCOM international conference Leeds Beckett University*.

Kahneman, D. (2011). *Thinking, Fast and Slow*. UK: Penguin Random House.

Kahneman, D., & Tversky, A. (1977). Intuitive prediction: Biases and corrective procedures. *TIMS Studies in Management Science, 12*, 313–327.

Koskela, L., & Howell, G. (2002). The underlying theory of project management is obsolete. In *Proceedings of the PMI research conference 2002* (pp. 293–302).

Latham, M. (1994). Constructing the team: joint review of procurement and contractual arrangements in the United Kingdom construction industry. London: Crown.

Lenfle, S., & Loch, C. (2010). Lost roots: How project management came to emphasise control over flexibility and novelty. *USA: California Management Review, 53*, 32–55.

Lycett, M., Rassau, A., & Danson, J. (2004). Programme management: A critical review. *International Journal of Project Management, 22*, 289–299.

McMeeken, R. (2008). Egan 10 years on. Available from [Accessed 4/04/2019].

Morris, P., Pito, J., & Sederlund, J. (2011). *The Oxford handbook of project management*. Oxford University Press.

Morris, P. (2013). *Reconstructing project management*. Wiley-Blackwell.

Morris, P.W.G. (1997). *The Management of Projects*. Thomas Telford.

Morris, P.W.G (2010) Research and the future of project management. *International Journal of Managing Projects in Business* [Online] *3*(1), 139—146. Available from [Accessed 13th April 2020]

Padalkar, M., & Gopinath, S. (2016). Six decades of project management research: Thematic trends and future opportunities. *International Journal of Project Management, 34*(7), 1305–1321.

Pollack, J. (2007). The changing paradigms of project management. *International Journal of Project Management, 25*, 266–274.

Project Management Institute. (2013). *A guide to the project management body of knowledge (PMBOK Guide), Fifth edition*. Newtown Square, Pennsylvania.

Shehu, Z., & Akintoye, A. (2009). Construction programme management theory and practice: Contextual and pragmatic approach. *International Journal of Project Management, 27*(7), 703–716. www.sciencedirect.com.

Söderland, J., Geraldi, J., Brady, T., Davies, A., & Nightingale, P. (2012). Dealing with uncertainty in complex projects: Revisiting Klein and Meckling. *International Journal of Managing Projects in Business, 5*(4), 718–736.

Strauss, A., & Corbin, J. (1990). *Basics of qualitative research*. Sage.

Sutherland, S. (2007). *Irrationality*. Cornwall: Pinter and Martin.

Svejvig, P., & Anderson, P. (2015). Rethinking project management: A structure literature review with a critical look at the brave new world. *International Journal of Project Management, 33*, 278–290.

Smyth, H. J., & Morris, P. W. G. (2007). An epistemological evaluation of research into projects and their management: Methodological issues. *The International Journal of Project Management, 25*(4), 423–436.

Van Rijswick, M., & Salet, W. (2012). Enabling the contextualization of legal rules in responsive strategies to climate change. *Ecology and Society, 17*(2), 18.

Winter, M., Smith, C., Morris, P., & Cicmil, S. (2006). Directions for future research in project management: The main findings of a UK government-funded research network. *International Journal of Project Management, 24*, 638–649.

Yin, R. K. (1994). *Case study research: Design and methods*. Sage.

Chapter 24
Can Hard Paradigm Artefacts Support Soft Paradigm Imperatives? An Unpaired Comparative Experiment to Determine whether Visualisation of Data Is an Effective Collaboration and Communication Tool in Project Problem-Solving

Alison Davies and John Heathcote

24.1 Introduction

Project management is now prevalent across all sectors (Maylor, Brady, Cooke-Davies, & Hodgson, 2006; Pellegrinelli, 2011) and has developed beyond product delivery to encompass organisational change, the transformation of businesses and the implementation of strategies (Winter, Smith, Morris, & Cicmil, 2006, p. 638). This "projectification" (Midler, 1995, quoted in Maylor et al., 2006, p. 663) has resulted in organisations adopting numerous PM practices and techniques to deliver change (Maylor, 2010). Nevertheless, despite decades of PM tools, approaches and processes, projects continue to fail (Morris, 2013; Pinto, 2013). Svejvig and Andersen (2015) suggest that classical PM methodologies based upon a hard, deterministic paradigm of certainty and rational decision-making have "remained fairly static in the past" (Koskela & Howell, 2002, cited in Svejvig & Andersen, 2015, p. 278) and have proven inadequate in practice (Cicmil, Williams, Thomas, & Hodgson, 2006). This includes the failure to recognise and focus upon the front-end work to ensure problems are fully identified (Morris, 2013; Pinto & Winch, 2016) and effective decision-making takes place to identify optimum solutions (Shore, 2008; Samset & Volden, 2016), ensuring projects are delivered successfully and benefits realised. Indeed, Samset and Volden (2016, p. 301) add that "agreeing on the most effective solution to a problem and choice of concept need to be dealt with

A. Davies (✉) · J. Heathcote
School of the Built Environment and Engineering, Leeds Beckett University, Leeds, UK
e-mail: j.heathcote@leedsbeckett.ac.uk

as early as possible" as this work at the front-end phase is crucial to project success. Stingl and Geraldi (2017, p. 121) purport that decision-making is "integral to the management of projects", which is supported by Pollack and Adler's (2015, p. 247) research into keyword analysis of 95,000 unique records that revealed, "issues associated with decision making are central to PM research". Research has also consistently identified effective communication as a key factor to project success (Rezvani et al., 2016), and therefore team working and the identification of problems and agreement of solutions are essential; nevertheless effective team working is difficult to achieve (Stingl & Geraldi, 2017). However, behavioural economic theory states that human decision-making is flawed and prone to make the same mistakes repeatedly, such as loss aversion and availability error (Kahneman, 2012; Sutherland, 2013), and "actual decision behaviour deviates strongly from the rational ideal" (Stingl & Geraldi, 2017, p. 121); therefore this poses a key challenge to project success.

Consequently there is a need to consider new ways of thinking about the PM discipline, and Pollack (2007) suggests a soft paradigm is required that focuses upon facilitation, participation and effective communication with stakeholders to address these contemporary issues; it seems the "primary problems of project managers are not technical, but human" (Posner, 1987, quoted in Pollack, 2007, p.270). It appears that project managers are applying "practices that circumvent some of the problems encountered in the classical approach" (Svejvig & Andersen, 2015, p. 286). Hällgren and Södrholm's (2011) research demonstrates that project plans can be used in the hard paradigm model for tracking progress but also can be utilised to enhance understanding and communication within the soft paradigm. Therefore, one approach to tackle the complex nature of decision-making in projects is to summarise the data into infographics. This visual representation of data in a graphic format can enhance understanding and communication as it allows for rapid summarising of data (Bititci, Cocca, & Ates, 2016) to identify themes and issues and can aid effective decision-making through the identification of possible problem and solutions. As a result, this research explores the use of the visualisation of data to address the problem of flawed decision-making, problem-solving and social processes within projects.

24.2 Project Management: Hard and Soft Paradigms

Growing theoretical disquiet (Pollack, 2007) among researchers and criticism of classical PM (Svejvig & Andersen, 2015) by senior practitioners have seen an emergence of a RPM initiative, which identifies five directions for further research to address these concerns and to "connect it more closely to challenges of contemporary project management practice" (Winter et al., 2006, p. 639). The initiative calls for new approaches in "project complexity, social process, value creation, project conceptualisation, and practitioner development" (Winter et al., 2006); this research draws upon the first four directions, in particular. The project complexity direction suggests that a new way of thinking about projects is required, which is currently based upon a dominant literature (Padalkar & Gopinath, 2016) that is predicated upon a deterministic paradigm or "hard systems model" (Winter et al., 2006, p. 640), which centres upon control and certainty. However, RPM does not dismiss the hard paradigm of tools and techniques but suggests a need for considering project conceptualisation in a new light, which encompasses a soft paradigm, focusing upon "people and participation" (Pollack, 2007, p. 270). Indeed, people are fundamental to value creation, and the need to focus upon problem-solving at the front end of projects is now widely recognised (Matinheikki, Artto, Peltokorpi, & Rajala, 2016; Morris, 2013; Samset & Volden, 2016). RPM calls for new approaches and holistic thinking to "assist practitioners at the messy front end" (Winter et al., 2006, p. 645) through problem-solving methods that combine the soft and hard aspects of projects. However, projects are inherently complex, which makes objective decision-making difficult, but one solution could be the use of visualisation of data to facilitate improved group interaction, decision-making and problem-solving. Eppler and Platts (2009) identify three main types of benefits for visual representation: cognitive, social and emotional, which would support the issues identified. Data represented visually would help a project group develop a common view of the information quickly, which would assist them in decision-making in a variety of ways including the identification of constraints, problems and possible solutions. It could be argued that visual representations are already used extensively in projects, such as Gantt charts, but they are used in a standardised format to track project progress and PM literature does not take into account these additional fields of study within practice to aid project success. As a result this research aims to contribute to RPM by exploring the concept that PM should focus on the human decision-making aspect to solve problems and consider whether the use of visual tools helps groups work more effectively together, facilitating better interaction, thus further developing practical application of PM. It may also advance project conceptualisation by examining whether hard paradigm artefacts can support soft paradigm imperatives and establish a conceptual link to replace this paradigmatic separation.

24.3 Visualisation and Cognitive Fit Theory

Visualisation of data is defined as the "collection, transformation and presentation of qualitative and quantitative data" (Al-Kassab, Ouertani, Schiuma, & Neely, 2014, p. 3), in a variety of visual formats including graphs, tables, bar charts, diagrams and infographics. According to Moore (2017), standalone data does not contribute to sense making, but when data is combined in a meaningful way, it generates knowledge that can assist decision-makers in problem-solving. As such visualisation of data in "a methodically developed graphic" (Moore, 2017, p. 130) presents an opportunity for decision-makers to gain "insights, develop understanding, identify patterns, trends or anomalies faster, and promote engaging discussions"

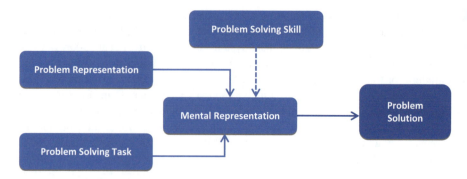

Fig. 24.1 Cognitive fit theory: problem-solving model (Vessey & Galletta, 1991)

(Dasgupta et al., 2015, quoted in Moore, 2017, p. 130). It is a useful tool to present complex information and thus enable more effective group interaction (Moore, 2017).

Vessey and Galletta's (1991) experimental study into the effectiveness of graphs versus tables in problem-solving resulted in development of cognitive fit theory, which draws upon consumer behaviour and behavioural decision theory. Cognitive fit theory (Fig. 24.1), which is the dominant theory within visualisation literature (Engin & Vetschera, 2017; Kopp, Riekert, & Utz, 2018; Teets, Tegarden, & Russell, 2010), suggests that "for the most effective and efficient problem solving to occur, the problem representation and any tools or aids employed should all support the strategies, methods or processes, required to perform the task" (Vessey & Galletta, 1991, p. 64).

According to Vessey and Galletta (1991), if the problem representation (visual data) fits the problem-solving task appropriately, then the problem-solver is able to formulate a mental representation of the problem, which will ultimately generate solutions quicker and more accurately (Teets et al., 2010). If, however, there is a mismatch between the problem representation and task, then cognitive fit will not occur, and the problem-solver will have to transform either the task or representation to allow for problem-solving (Vessey & Galletta, 1991).

Research identified that graphs are "spatial problem representations", which "emphasize information about relationships in the data" (Vessey & Galletta, 1991, p. 67), and tables are "symbolic problem representations, which emphasize information on discrete data values" (Vessey & Galletta, 1991, p. 67). Subsequent research has determined that the choice of visual representation significantly affects decision-making, and selecting the appropriate visual can therefore enhance problem-solving performance, for example, a spatial task would be best supported by graphical information (Moore, 2017; Teets et al., 2010).

In addition, cognitive load also plays an important role within cognitive fit theory and effective decision-making (Phillips, Prybutok, & Peak, 2014). Cognitive load is defined as the "mental effort on the part of the decision-maker" (Phillips et al., 2014, p. 375). If there is a mismatch between task and representation, then problem-solving becomes more difficult as the cognitive load becomes greater. The problem-solver needs to "expend more mental effort" (Phillips et al., 2014, p. 379), which results in the "resource consuming and effortful System 2 being replaced by heuristic methods of System 1" (Engin & Vetschera, 2017, p.96), thus reducing performance. The relevance of the visual representation is therefore important to reduce cognitive load, so that "cognitive resources remain available for deeply processing and understanding the given material" (Kopp et al., 2018, p. 369).

Literature has highlighted a number of benefits in using visualisation of data including "cognitive, social and emotional" (Eppler & Platts, 2009). Visual data improves "synthesis of information" (Bititci et al., 2016, p. 1573) and facilitates pattern recognition and problem-solving. It also allows disparate groups to come together to gain a mutual understanding of issues, and from an emotional perspective, it "creates involvement and engagement, providing inspiration and convincing communication" (Bititci et al., 2016, p. 1573).

The benefits of visualisation as a decision-making tool have already been discussed and are incorporated within Eppler and Platts's (2009) cognitive benefits. Visualisation as a means of knowledge management aids the transfer, sharing and creation of knowledge (Al-Kassab et al., 2014) by "operating as a catalyst for interpretations" (Al-Kassab et al., 2014, p. 6). It could be argued that this function is linked to Eppler and Platt's social benefit as it enhances shared learning. However, care is needed to avoid misrepresentation of data (Bititci et al., 2016) that may focus attention on partial information, thus distorting knowledge (Al-Kassab et al., 2014). Indeed Bresciani and Eppler (2015, p. 1) identified a number of "pitfalls of visual representations" and call for further research. The selection of an appropriate image or visual can ensure complex information is readily comprehensible (Al-Kassab et al., 2014) and is therefore an effective communication medium. Visualisation improves communication and acts as a "collaboration catalyst" (Eppler & Bresciani, 2013, p. 146), and it seems these two elements are intrinsically linked. The suitability and fit of visual data used in this research are discussed in the Methodology section below. It appears that visualisation of data is an effective tool to aid decision-making and problem-solving in projects; however, there is limited literature in this area within the field of PM. Research tends to focus upon 3D visualisation (Jaber, Sharif, & Liu, 2016) often within the construction industry or on strategic decision-making (Killen, 2013; Killen & Kjaer, 2012) in a project portfolio setting. This research therefore contributes to a gap in the PM discipline, particularly relating to the use of visual data to enhance group collaboration and communication to improve problem-solving and decision-making.

24.4 Research Review and Methodology

There is a wealth of literature about the multi-disciplinary nature of PM and analysis of research trends and themes (Kwak & Anbari, 2009; Padalkar & Gopinath, 2016; Pollack & Adler, 2015). Smyth and Morris' (2007, p. 433) epistemological evaluation of project research criticised the "lack of epistemological care taken in the selection and application of research methodologies", particularly relating to a positivist methodology and, for example, the use of contradictory case study approaches. Positivist epistemology explores human and social behaviour (Easterby-Smith, Thorpe, & Jackson, 2015), and this is "typically marked by an experimental design" (Stingl & Geraldi, 2017, p. 124). Stingl and Geraldi (2017) identified three schools of thought regarding behavioural decision-making in projects, and this research draws upon the reductionist school, which assumes decision-makers are cognitively limited (Lovallo & Kahneman, 2003). The methodology builds upon "the experimental approach of psychology and cognitive sciences" (Stingl & Geraldi, 2017, p. 124), and it is rooted in a positivist epistemology. As a result, an unpaired experimental design, with an independent variable (visualisation of data), will be conducted, which is expected to improve the dependent variable (improved collaboration and communication). Analysis of this empirical data will test the hypotheses defined below.

Research question: Does the visualisation of data facilitate improved group interaction?

Research Hypotheses.

H_0 (null hypothesis):	There is no statistically significant relationship between the use of visual data and the facilitation of better interaction in a project group.
H_1 (alternative hypothesis):	There is a statistically significant relationship between the use of visual data and the facilitation of better interaction in a project group.

24.5 Research Method

This study is based upon a postpositivist epistemology that tests whether visualisation of data can improve group communication and collaboration. Similarly, Killen's (2013, p. 804) research explored use of visualisation of data to "support project portfolio decision making" and utilised a controlled experiment with a control group that only had tabular information and two other intervention groups, with differing visual data to determine whether data types impact upon decision-making. Experimental designs are generally conducted within the fields of psychology and economics and, as such, are not a method that is often used in PM (Killen, 2013), providing valuable experience within the discipline (Killen, 2013).

Research was conducted in a Hydra Suite laboratory, which is a laboratory-style facility, designed to create "critical decision immersive simulations" (Leeds Beckett, n.d.). Participants undertook a simulated problem-solving task so that the number of identified problems and solutions was recorded and therefore could be measured. The Hydra Suite laboratory allows for a controlled environment and removes multiple variables so that the focal variable (visual representation) can be tested. This comparative, true experiment (Robson, 2011) will test the difference between treatment conditions, that is control (no treatment) and intervention groups (experiential treatment). The control group was presented with a typical business case, which is in text format, and asked to identify problems and possible improvements to the organisation; the intervention group was provided with the same information, but the independent variable was a visual representation of data. It is predicted that the visual data will enhance collaboration, communication and problem-solving, and the number of outputs will be recorded and measured. In addition, the Hydra Suite laboratory has video recording capabilities so that observation of group interaction also took place.

Research has shown that charts and graphs reduce information overload (Kopp et al., 2018). To allow for participants' cognitive style, which is a preference for how information is processed, a number of visual images were used; this ensures cognitive load is not increased (Engin & Vetschera, 2017). The visual data included a Gantt chart depicting a standard manufacturing process, which also included a bar chart displaying financial information. Leach (2010, cited in Jaber et al., 2016) suggests Gantt charts should be over-layered with budget information to provide a comprehensive overview of business processes. A separate infographic, which represents information from these other two visuals, was also provided. Details of the experiment and outlining how the data was collected are shown below.

Upon arrival at the Hydra Suite laboratory, all participants received a short briefing session about the experiment and randomly assigned to either control or intervention groups; each group was assigned to a separate Hydra laboratory syndicate room to participate in the activity. Each room contained a briefing document, which provided instructions to the participants, for the 20-min group activity: view the documents provided in the room relating to the business case; identify problems with existing processes and possible options/solutions; and assign one person the task of writing the group's ideas on the paper provided. The control group was provided with a text format only, which provided information about a typical manufacturing company. The intervention group was also provided with an independent variable, that is, the visual data. The researcher observed and recorded each group in the Hydra laboratory control room, and an example of an intervention group activity is shown in Figs. 24.2, 24.3, and 24.4.

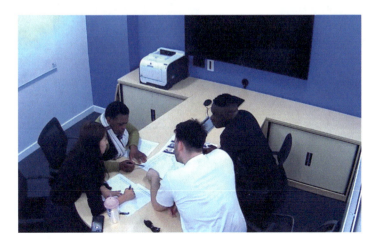

Fig. 24.2 Video snapshot of experiment – intervention group

Fig. 24.3 Intervention group interaction with visual documents

Fig. 24.4 Control group: individual focus on text only documents

24.6 Research Results

Observationally (Fig. 24.2) interaction was evident, facilitated by the pictorial schematics and Gantts. The Hydra Suite laboratory allowed for a recorded observation.

Observationally (Fig. 24.3) interaction was evident, facilitated by the pictorial schematics and Gantts.

Whereas the control groups initially worked through the text only documents individually (Fig. 24.4) and this delayed communication and collaboration to identify problems.

Overall participants in the intervention groups identified the first problem more quickly, which seems to suggest visual representation of data aids decision-making through easier identification of patterns and problems.

Interestingly, one participant in a control group drew a visualisation of the data to help synthesise the information, particularly relating to procurement, which she had identified as an issue. She explained to the group, "It's difficult to think about processes and how they overlap.um. connect. I'm going to jump in and draw a little something for the procurement so I can see…so I can get my head around it". This is important as it suggests the visual aid eased cognitive load and enabled the participants to make sense of the information. As a result, the controlled environment became flawed, as a visual diagram was introduced, and therefore the outputs of this group were reclassified as an intervention group.

Figure 24.5 shows the number of problems and solutions following the redistribution of data, and again the box and whisker plots show a visual difference between the intervention groups (1) and control groups (2).

In this instance an outlier was introduced in the number of problems identified, which skewed the results. Overall, however, findings show that there is a visual difference between the independent variable introduced to the intervention group, within the box and whisker plots. Nevertheless, the probability of results being due

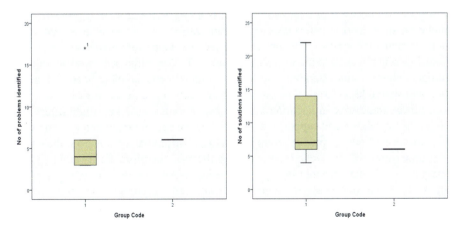

Fig. 24.5 No of problems and solutions identified

to chance remains high. There is concern that behavioural economic experiments "are too often, too easily generalised and interpreted as general human attributes" (Berg, 2014,p. 228), and this is mirrored by Kahneman who admits that it is not "uncommon among psychologists to overestimate the reliability and validity of results based on small samples" (Kahneman, 2011, quoted in Berg, 2014, p. 228). The experiment set out to test whether the visualisation of data improved group interaction, but this research acknowledges on this occasion the number of iterations of the experiment (N- = 7) resulted in insufficient data to arrive at statistical significance and therefore the null hypothesis (H_0) could not be accepted or rejected. Although not conclusive, the additional observational notes do appear to support the alternative hypothesis (H_1). It appears likely that statistical significance could be achieved as the experiment is based on sound methodology if additional iterations were conducted.

24.7 Conclusion

This study tackles a theoretical and contemporary challenge in practice that focuses upon flawed decision-making in projects. Firstly, this study confirmed that cognitive biases and two systems thinking identified and developed within behavioural economic theory affect project success. This is due to a number of biases that hamper effective decision-making which provide an illusion of control and overconfidence in problem-solving, with a predetermined or first solution often being selected for project development. This research has focused upon the reductionist school of behavioural decision-making, which identifies these as predictable and deep-rooted errors. The dominant, hard, deterministic PM paradigm has remained fairly fixed and inadequate in practice, but a change in approach is needed, to address this decision-making issue, as projects become increasingly complex.

Secondly, the study revealed that visualisation of data could offer such a new approach to improved problem-solving, particularly at the front-end problem formulation stage as it transforms complex data into an understandable format. It is dependent upon cognitive fit theory and the need to reduce cognitive load so that the visual data is matched to the problem-solving task. When this occurs, visualisation of data enhances understanding and forges a mutual understanding of the information and shared knowledge, thus improving interaction and decision-making.

Thirdly, there was anticipation that the experimental analysis would reject the (H_0) null hypothesis and support the (H_1) alternative hypothesis to demonstrate that visualisation of data improves group interaction. Unfortunately, however, the study was underpowered and lacked statistical significance; therefore the H_0 (null hypotheses) could not be accepted or rejected. Nevertheless, the box and whisker plots showed a visual difference in favour of the intervention groups, and observations highlighted that there was better collaboration and communication, together with quicker problem identification in the intervention groups. Arguably, further iterations could produce statistical significance, as the research methodology is sound and repeatable.

Finally, this research has determined that projects and project success are about people. People have to manage project complexity and uncertainty and undertake effective problem-solving processes at the neglected front end of projects to ensure they choose the right concept, and this is delivered effectively. Problem-solving and decision-making are intrinsically linked, but fundamentally human behavioural decision-making is flawed, which impacts directly upon project success. Armed with this knowledge, project environments can be changed, and a new holistic approach adopted to minimise these biases. Visualisation of data is used as a project management tool under the traditional dominant hard deterministic paradigm, but there is an opportunity to adopt new perspectives and use visual data to generate knowledge, effective communication and a mutual understanding of problems to aid quality decision-making. Effectively this would bring the two project management paradigms together and in response to Pinto (2013) would stop making our lives more difficult than necessary.

24.8 Limitations

The main weakness and limitation of this study are the number of participants recruited was an insufficient sample size to achieve statistical significance and therefore the (H_0) null hypothesis could not be accepted or rejected. Therefore, results could not be generalised to the population.

References

Al-Kassab, J., Ouertani, Z., Schiuma, G. & Neely, A. (2014) Information visualization to support management decisions. International Journal of Information Technology & Decision-making [Online], 13, (2), pp. 407–428. Retrieved August 4, 2018, from http://www.ebsco.host.com.

Berg, L. (2014) Who benefits from behavioural economics? Economic Analysis and Policy [Online], 44, pp.221–232. Retrieved August 4, 2018, from http://www.ebsco.host.com.

Bititci, U., Cocca, P. & Ates, A. (2016) Impact of visual performance management systems on the performance management practices of organisations. International Journal of Production Research, 54 (6), pp.1571–1593. Retrieved April 4, 2018, from http://www.ebsco.host.com.

Bresciani, S. & Eppler, M. (2015) The pitfalls of visual representations: A review and classification of common errors made while designing and interpreting visualizations. Sage Open [online], 5 (4), pp.1-14. Retrieved July 24, 2018, from http://www.ebsco.host.com.

Cicmil, S., Williams, T., Thomas, J., & Hodgson, D. (2006). Rethinking project management: Researching the actuality of projects. *International Journal of Project Management, 24*, 675–686.

Dasgupta, A., Poco, J., Wei, Y., Cook, R., Bertini, E., & Silva, C. (2015). Bridging theory with practice: An exploratory study of visualization use and design for climate model comparison. *IEEE Transactions on Visualization and Computer Graphics, 21*(9), 996–1014. Quoted in: Moore, J. (2017) Data visualization in support of executive decision-making. Interdisciplinary

Journal of Information, Knowledge & Management [Online], 12, pp.125–138. Retrieved July 4, 2018, from http://www.ebsco.host.com.
Easterby-Smith, M., Thorpe, R., & Jackson, P. (2015). *Management and business research* (5th ed.). London: Sage.
Engin, A. & Vetschera, R. (2017) Information representation in decision-making: The impact of cognitive style and depletion effects. Decision Support Systems, 103, pp.94-103. Retrieved July 29, 2018, from http://ebscohost.com.
Eppler, M. & Bresciani, S. (2013) Visualization in management: From communication to collaboration: A response to Zhang. Journal of Visual Languages and Computing [Online], 24 (2), pp.146–149. Retrieved July 4, 2018, from http://www.ebsco.host.com.
Eppler, M. & Platts, K. (2009) Visual strategizing: The systematic use of visualization in the strategic planning process. Long Range Planning [Online], 42, pp.42–74. Retrieved July 4, 2018, from http://www.ebsco.host.com.
Hällgren, M., & Söderholm, A. (2011). Projects-as-practice—New approach, new insights. In P. Morris, J. Pinto, & J. Söderland (Eds.), *The Oxford handbook of project management*. Oxford: Oxford University Press.
Jaber, K., Sharif, B. & Liu, C. (2016) An empirical study on the effect of 3D visualization for project tasks and resources. Journal of Systems and Software 115, pp.1-17. Retrieved July 29, 2018, from http://ebscohost.com.
Kahneman, D. (2011). *Thinking, fast and slow*. Allan Lane: Penguin Books. Quoted in: Berg, L. (2014) Who benefits from behavioural economics? Economic Analysis and Policy [Online], 44, pp.221–232. Retrieved August 4, 2018, from http://www.ebsco.host.com.
Kahneman, D. (2012). *Thinking, fast and slow*. London: Penguin Books.
Killen, C. (2013). Evaluation of project interdependency visualizations through decision scenario experimentation. *International Journal of Project Management, 31*, 804–816.
Killen, C., & Kjaer, C. (2012). Understanding project interdependencies: The role of visual representation, culture and process. *International Journal of Project Management, 30*, 554–566.
Kopp, T., Riekert, M. & Utz, S. (2018) When cognitive fit outweighs cognitive load: Redundant data labels in charts increase accuracy and speed of information extraction. Computers in Human Behaviour [Online], 86, pp.367–376. Retrieved August 10, 2018, from http://ebscohost.com.
Koskela, L. & Howell, G. (2002) The underlying theory of project management is obsolete. PMI Research Conference 2002, PMI, pp.293-302. Cited in: Svejvig, P. & Andersen, P. (2015) rethinking project management: A structured literature review with a critical look at the brave new world. International Journal of Project Management, 33, pp.278–290.
Kwak, Y., & Anbari, A. (2009). Analyzing project management research: Perspectives from top management journals. *International Journal of Project Management, 27*, 435–446.
Leach, K. (2010) *Project schedule visualization*. Retrieved October 26, 2016, from http://www.steelray.com. Cited in: Jaber, K., Sharif, B. & Liu, C. (2016) an empirical study on the effect of 3D visualization for project tasks and resources. Journal of systems and software, [online], 115, pp.1-17. Retrieved July 29, 2018, from http://ebscohost.com.
Leeds Beckett University (n.d.) Leeds Beckett University facilities [online], Leeds: Leeds Beckett University. Retrieved August 8, 2018, from http://www.leedsbeckett.ac.uk/leeds-business-school/facilities//.
Lovallo, D. & Kahneman, D. (2003) Delusions of success: How optimism undermines executives' decisions. Harvard Business Review [online], 81 (7) July, pp. 56-63. Retrieved November 29, 2018, from http://ebscohost.com.
Matinheikki, J., Artto, K., Peltokorpi, A., & Rajala, R. (2016). Managing inter-organizational networks for value creation in the front-end of projects. *International Journal of Project Management, 34*, 1226–1241.
Maylor, H. (2010). *Project management* (4th ed.). Harlow: Pearson Education Limited.
Maylor, H., Brady, T., Cooke-Davies, T., & Hodgson, D. (2006). From projectification to programmification. *International Journal of Project Management, 24*, 663–674.

Midler, C. (1995). Projectification of the firm: the Renault case. *Scandinavian Journal of Management, 11*(4), 363–375. Quoted in: Maylor, H. et al. (2006) From projectification to programmification. International Journal of Project Management, 24, pp.663–674.

Moore, J. (2017) Data visualization in support of executive decision-making. Interdisciplinary journal of information, Knowledge & Management [Online], 12, pp.125–138. Retrieved July 4, 2018, from http://www.ebsco.host.com.

Morris, P. (2013) Reconstructing project management reprised: A knowledge perspective. Project Management Journal [Online], 44 (5), pp.6–23. Retrieved January 14, 2018, from http://www.ebsco.host.com.

Padalkar, M., & Gopinath, S. (2016). Six decades of project management research: Thematic trends and future opportunities. *International Journal of Project Management, 34*, 1305–1321.

Pellegrinelli, S. (2011). What's in a name: Project or programme? *International Journal of Project Management, 29*, 232–240.

Phillips, B., Prybutok, V. & Peak, D. (2014) Determinants of task performance in a visual decision-making process. Journal of Decision Systems [Online], 23 (4), pp.373–387. Retrieved July 4, 2018, from http://www.ebsco.host.com.

Pinto, J. (2013) Lies, damned lies, and project plans: Recurring human errors that can ruin the project planning process. Business horizons [online], Business Horizons, 56, pp.643–653. Retrieved July 4, 2018, from http://www.ebsco.host.com.

Pinto, J., & Winch, G. (2016). The unsettling of "settled science" – The past and future of the management of projects. *International Journal of Project Management, 34*, 237–245.

Pollack, J. (2007). The changing paradigms of project management. *International Journal of Project Management, 25*, 266–274.

Pollack, J., & Adler, D. (2015). Emergent trends and passing fads in project management research: A scientometric analysis of changes in the field. *International Journal of Project Management, 33*, 236–248.

Posner, B. (1987). What it takes to be a good project manager? *Project Management Journal, 18*(1), 51–54. Quoted in: Pollack, J. (2007) The changing paradigms of project management. International Journal of Project Management, 25, pp.266–274.

Rezvani, A., Chang, A., Wiewiora, A., Ashkanasy, N., Jordan, P., & Zolin, R. (2016). Manager emotional intelligence and project success: The mediating role of job satisfaction and trust. *International Journal of Project Management, 34*, 1112–1122.

Robson, C. (2011). Real world research (3rd ed.). Chichester: Wiley.

Samset, K., & Volden, G. (2016). Front-end definition of projects: Ten paradoxes and some reflections regarding project management and project governance. *International Journal of Project Management, 34*, 297–313.

Shore, M. (2008) Systematic biases and culture in project failures. Project Management Journal [Online], 39 (4), pp.5–16. Retrieved July 20, 2018, from http://www.ebsco.host.com.

Smyth, H., & Morris, P. (2007). An epistemological evaluation of research into projects and their management: Methodological issues. *International Journal of Project Management, 27*, 423–436.

Stingl, V., & Geraldi, J. (2017). Errors, lies and misunderstandings: Systematic review on behavioural decision making in projects. *International Journal of Project Management, 35*, 121–135.

Sutherland, S. (2013). *Irrationality: The enemy within*. London: Pinter & Martin.

Svejvig, P., & Andersen, P. (2015). Rethinking project management: A structured literature review with a critical look at the brave new world. *International Journal of Project Management, 33*, 278–290.

Teets, J., Tegarden, D. & Russell, R. (2010) Using cognitive fit theory to evaluate the effectiveness of information visualizations: An example using quality assurance data. IEEE Transactions on Visualization and Computer Graphics [Online], 16 (5), pp.841–853. Retrieved July 4, 2018, from http://www.ebsco.host.com.

Vessey, I., & Galletta, D. (1991). Cognitive fit: An empirical study of information acquisition. *Information Systems Research [online], 2*(1), 63–84. Retrieved July 20, 2018, from http://www.ebsco.host.com.

Winter, M., Smith, C., Morris, P., & Cicmil, S. (2006). Directions for future research in project management: The main findings of a UK government-funded research network. *International Journal of Project Management, 24*, 638–649.

Chapter 25
Collusion within the UK Construction Industry, An Ethical Dilemma

Joseph Thorp and Hadi Kazemi

25.1 Introduction

Comparatively, the construction industry is one of the most powerful economic drivers in the UK – with the scope to be one of the most prominent and economically available industries worldwide. In 2018, the construction contributed £117 billion to the UK economy, 6% of GDP (Rhodes, 2019). Due to the lucrative nature of the industry, it is increasingly common that businesses and corporations alike do the upmost to generate high revenues, thus maximising profit margins and meeting essential capitalist business objectives. However, due to the competitive nature of markets, individuals or corporations might find themselves conspiring in collusive activities to bypass ethical and legal boundaries and be the beneficiaries of monetary gain. Collusive activities in the construction industry can take various forms. These comprise bidding – via suppression and rotation; market division and sharing; collusion within public and private sectors; as well as collusion deriving from the relationships between clients, main contractors, sub-contractors and suppliers (Shan et al., 2020). The taxpayer within the UK provides substantial funding for public construction projects. But a major talking point is whether the public are aware that they are potentially providing the funds to fuel cartel activities. The Office of Fair Trading (OFT) was responsible for protecting consumer interests throughout the UK. Since its closure in April 2014, however, the question remains over the prevalence of collusion in the UK construction industry and the perception of individuals regarding collusive activities and possible penalties imposed as a result of being involved in such activities. The aim of this study, therefore, is to

J. Thorp · H. Kazemi (✉)
School of Built Environment, Engineering, and Computing, Leeds Beckett University, Leeds, UK
e-mail: h.kazemi@leedsbeckett.ac.uk

develop a comparative study and investigate the potential of collusion in the UK construction industry compared to overseas examples. In addition, this paper will discuss and review the impact of collusion and the risks associated with it and analyse benchmark trends of overseas construction industries and the potential for the issue to occur in the UK.

25.2 Literature Review

Construction industry is a lucrative market, and the fierce competition in the market means organisations do the upmost to generate higher revenues whilst reducing cost deficits, a vital business success strategy. Consequently, there is a potential for unethical and corruptive practice to gain bigger benefits. Chotibhongs and Arditi's (2012) work provides a good example of an opinionated stance on the issue of collusion within the construction industry who refer to collusion as an insidious issue within the construction industry globally. They further suggest that despite a wide acknowledgement about unethical and illegal practice in the industry, there is a lack of detailed research to 'detect collusive bidding'. In their work, Chotibhongs and Arditi (2012) investigate potential methods of detection of collusion – in particular, collusive bidding – in North America and Canada.

Harding and Joshua (2010) share similar stance and come up with the idea of 'business cartels' referring to firms that in the same market control competition between them by fixing prices, sharing markets or rigging tendering procedures. This observation also suggests the concealment of illegal conduct from their most important influencers, the customers. In addition, it is stressed that business cartels must also conceal their illegal identity from non-participants and internal and external watchdogs (Baker & Faulkner, 1993). In another study, Wells (2014) provides an overview of corruption and collusion within the construction industry and refers to construction sector as being 'one of the most corrupt globally'. This view is enhanced by the fact that, in relation to the charts of Transparency International's Bribe Payer's Index, public works and construction often top such charts and are 'perceived as the sector most likely to engage in bribery' (Hardoon & Heinrich, 2011). Furthermore, it is stated that an estimated 20–30% of the value of projects were lost through widespread corruption including collusion. Wells (2014) identifies 13 reasons behind collusion, ranging from a company's size, complexity, uniqueness and the fact that all projects go through many phases which involves various contractual links that, in turn, results in accountability to be dispersed among numerous separate agents. This point sounds extremely valid and provides clarity to the issue as to why it has become so prevalent in contemporary businesses. There is an implied positive correlation here which suggests that as more contractual links and networks of connection are created, accountability of the issue diminishes as it becomes more difficult to find out where such collusion has unfolded. Furthermore, within her view of corruption in construction, Wells (2014) emboldens the collusion to allocate contracts as a major issue.

In comparative terms, there is a level of certainty surrounding the existing evidence to suggest that from previous investigations, cartel and collusive bidding are predominantly originated from developed countries. More specifically, the Integrity Department (INT) of the World Bank reports on the prevalence of collusion in roads sector in a large number of developing countries, including Kenya, Tanzania, Uganda, Cambodia, Philippines, Indonesia, Nepal, Pakistan, parts of India as well as Colombia and Peru (World Bank, 2011). Nevertheless, in comparison to the point that collusion does not always mean higher prices, Dorée (2004) raises the case of contractors involved in collusion in the Dutch construction industry where shielding business against predatory pricing helped the contractors to reduce uncertainty about future workload fluctuations. This is an intuitive argument as it provides a reasonable and thorough evaluation of the notion of collusion with complementarities such as the fact it reduced rivalry and created a more stable and predictable market environment (Wells, 2014). In another example, Jaspers (2018) unfolds bid rigging, corruption, illegal financing of political parties, and the infiltration of organised crime in the local construction industry in the province of Quebec between 1996 and 2011. Moreover, Jones and Kovacic (2019) support the indication of collusive practices in public sector construction projects in the USA. In their view, public procurement systems render companies to distort through both corruption and stable supplier collusion by covering bidding, market and customer allocation or bid rotation. Accordingly, they signify that bid rigging wastes public funds and diminishes public confidence in the competitive process and, within the government, reduces the quantity and quality of vital goods and services.

Within the UK context, and in contrast to the statistical approach adopted by the Chinese and USA government to conclude an obvious prevalence of collusion within construction, Lee and Cullen (2018) provide an insight into professional ethics within quantity surveying sector. According to them, construction industry is believed to be one of the most corrupt industries in the world, and such practices are present in all stages of a project's life cycle by means of tender collusion and other corrupt practices. This view, also acknowledged by Transparency International (2011), follows the general idea set out by Le and Shan (2014) and Jones and Kovacic (2019) that the construction industry is facing problems involving the legality of practices. Furthermore, reviewing the work of Grant Thornton (2013), Lee and Cullen (2018) signify that collusion has been viewed as 'endemic' in the UK construction industry. Where Le and Shan (2014) and Jones and Kovacic (2019) fail to incorporate professional ethics, Lee and Cullen (2018) find this a necessity and develop the argument around the need of codes of conduct, more specifically the *Rules of Conduct*, whereby each member of the RICS (Royal Institute of Chartered Surveyors) is required to abide by these rules. Nonetheless, Bowen et al. (2007) suggest that the use of codes of conduct does not guarantee active adoption by members. Oke et al. (2017) conducted a methodology to examine the causes of collusion in the South African construction industry and revealed that collusive activities are extremely prevalent in road projects within public construction domain. It concluded that the top three causes of collusion within the industry derived from contractor greed, political influence, and poor ethics and corporate governance, the latter supporting the notion of Lee and Cullen (2018) that professional ethics and the ignorance by professionals to Codes of Conduct play a fundamental role in collusion within not just the UK construction industry but the construction industry inter-continentally. Shah and Alotaibi (2018) follows the view that ethical misconduct is evident within the UK construction industry mentioning that collusion and other unethical conducts are prevalent in the industry. However, Shah and Alotaibi (2018) is quick to dismiss the fact that unethical behaviours are not always down to a lack of knowledge on the issue, but rather a piece of the strategy of learning reasonable business or being brought into the practice (Vee & Skitmore, 2003). Accordingly, Shan et al. (2020) classifies unethical practices into four universal actions known as conflicts of interest including unfair conduct, fraud, collusion and bribery. Mueller (2018) agrees with the ethical issues faced within the construction industry and identifies five major steps to improve the ethical stature of construction companies as (1) developing an ethics blueprint, (2) developing a corporate code of ethics in reflection to what is important to the organisation, (3) providing scheduled ethical training to employees on a regular basis, (4) encouraging managers to develop and maintain personal relationships with team members, and (5) engaging with a third-party organisation to enable employees to make anonymous and confidential ethics reports.

In brief, similarities can be drawn up between the UK construction sector and overseas as cover-pricing is a commonality again. Macnamara (2009) states that submitted bids give a misleading impression to clients about the level of competition in the market. This is effective in a way in which it incorporates the element of perception in the identification of the issue within the UK construction industry and pairs legal issues and personal interpretation together. In addition, the role of ethics in construction practice needs to be re-defined as it provides the foundations for decision-making process in construction context.

The following table (Table 25.1) presents a summary of literature review which paves the way to data collection and analysis.

25.3 Research Method

A mixed method combining quantitative and qualitative research allows the incorporation of different elements of each research technique to create an extensive overview and a well-rounded study. As a result, this paper uses a mixed method approach – quantitatively via historical data provided in the literature review and qualitatively using questionnaires and semi-formal interviews. Due to the availability of adequate literature surrounding collusive practices in overseas construction industries and the UK construction industry, there was not a significant emphasis placed on the exploration of the issue. However, there was a focus on the attitudes of professionals within the industry and their respective opinions on the potential

Table 25.1 Summary of literature review

Construction industry	Themes	References
Overseas	The construction industry is the most corrupt industry	Hardoon and Heinrich (2011), cited in Wells (2014) Transparency International (2011), cited in Lee and Cullen (2018)
	Collusion occurs in public sector construction	Jaspers (2018) Le and Shan (2014) Jones and Kovacic (2019)
	The cause of collusion usually involves a level of personal greed	Oke et al. (2017) Ritchey (1990), cited in Vee and Skitmore (2003)
	Despite the many issues it can present, collusion can be beneficial	Wells (2014)
	The effects of collusion can be detrimental to the value and quality of construction projects	Le and Shan (2014) Competition and Markets Authority (2019) Wells (2014) Jones and Kovacic (2019)
UK	Collusion is present within the UK construction industry	Competition and Markets Authority (2019) Macnamara (2009)
	'Cover-bidding/pricing' is the most common form of collusion in the UK construction industry	Competition and Markets Authority (2019)
	Sanctions are usually enforced in the form of financial penalties to construction firms conspiring in collusive activities within the UK	Competition and Markets Authority (2019) Macnamara (2009)

for collusive practices to be prevalent in the UK construction industry at present. To achieve the aim of the study, qualitative data collection was implemented for this research through convenient sampling to gain an insight into the topic. This was followed up by quantitative data analysis to gain an understanding of accumulative figures surrounding trends. In total, three telephone interviews were conducted to retrieve opinionated information, while 22 questionnaires were completed to provide further knowledge of the qualitative and opinion-based information, enabling quantitative statistics to be created from the resulting responses. Further investigation was affected as a result of COVID-19, and research had to be concluded with the available data. Despite the nature of the topic area being somewhat subjective due to its legal boundary, the research plan succeeded in gaining honest answers to help understand the issue in the domestic construction industry.

25.4 Data Analysis

Collected qualitative data from interviews was analysed through different phases of data familiarisation, re-reading the transcripts and inclining similarities in the data set which then enabled the identification of themes as presented for which the data could be represented. Table 25.2 below represents the themes identified in interviews.

The abbreviations shown in the 'activity' column represent the following: ON for 'Opinion', OES for 'Opinion on the Ethical Stance', OPO for 'Opinion on Collusive Practices Overseas and UK comparison' and BS for 'Behaviours and Potential Solutions'.

It would be extremely difficult to ascertain the issue of collusion from the industry in its entirety, but if the correct pool of professional respondents were selected, it would provide an indication of what could be reciprocated at a larger scale. Therefore, following the interviews, a sample of 22 questionnaire responses were collected. Questions were ratified based on the interview results to investigate if they comply with the themes found in literature review. Questions were evolved around the (potential) existence of collusion in construction industry, role of ethics, motivations for collusive practice, and information about conceivable imposed penalties. The responses gained from the questionnaires provided an abundance of data sets and trends. For the purpose of analysis corresponding to the questionnaire, it was most appropriate to select sets of data that showed predominant and most common outcomes. Although majority of the respondents indicated an awareness of the

Table 25.2 Thematic analysis of telephone interviews

Activity	Theme	Reoccurrence of theme	Quote from interview
ON	There is the potential for collusion to occur within the UK construction industry	1,2,3	"If collusion has happened before, then it most likely will happen again" – Interview 2
OES	Collusion is unethical practice and should not be seen within industry	1,2,3	"[collusion] is not ethical because it reduces fair competition in the market. It is bad practice and should not be seen at any level, or in any industry" – Interview 1
OPO	Collusion may be part of a common cultural practice in overseas industries	1,3	"In other construction industries around the world, many will adopt cultural practices to support their methods of construction, so that may influence their approach" – Interview 3
BS	More convincing and complex regulations need to be ascertained as well as market barriers	2,3	"To really take authority and prevent potential collusive practices specific to UK construction, tighter regulations need to be put in place as well as barriers to market entry. It is difficult to suggest the exact barriers that must be implemented, but ones that filter for clean and cooperative future competition" – Interview 2

concept of collusion and unethical behaviour, alarmingly however, nearly a third of respondents were not aware of what collusion is, which could lead to potential ignorance to the issue going forward in industry. This could stem from a lack of education or perhaps the lack of experience in the industry and its potential flaws. However, absolute majority of the respondents believe that collusive practice does not have the potential to be prevalent in today's UK construction industry. Accordingly, it can be assumed that the industry is perceived as becoming more ethical and creates greater provision for fair competition. However, it could be suggested that a weakness with this approach lies by the fact that opinions do not determine the outcome of the industry, with the potential prominence of collusive practice yet to be confirmed. In terms of the regulatory bodies surrounding the governance of fair and competitive practice within the UK construction sector, it could be suggested that companies within the sector are not doing quite enough to educate their employees on such governance. Over half of the respondents were unaware of potential sanctions that can be imposed by the Competition and Markets Authority (CMA) as a result of any collusive practice. Although it can be assumed that many organisations follow the regulations closely, there needs to be a more consistent and transparent knowledge throughout the tiers of organisations to enable them to move forward more efficiently in order to achieve the targets of fair competition identified by the Competition and Markets Authority. Subsequently, the findings from the questionnaires supports the hypothesis that there is potential for collusive practice to occur within the UK construction industry, but its effects may not be as prevalent when compared to previous studies conducted in overseas construction industries. In addition to this, the responses from the interviews suggest a similar trend that collusion has the potential to be prevalent within the UK construction industry in future but prioritises that more sustainable regulations such as tighter market barriers to entry can be put in place to help prevent collusion within the industry. The data analysis techniques used here were useful in portraying and illustrating the views of the respondents in this mixed method of exploratory design. Following (Fig. 25.1) is a snapshot of respondents' responses to some of the questions discussed above.

It was stated within the Introduction of this study that collusion was a significant issue in overseas industries which hampered fair competition across the construction market. It was, therefore, pertinent to gain an insight as to whether the UK construction industry had the potential to reciprocate the wider issues in a growing UK economy.

Collusion in practice may seem somewhat beneficial to the financial operation of organisations, but collusion can have lasting detrimental effects to construction industries that experience its prevalence. Collusion can limit and reduce the value and quality of construction projects (Competition and Markets Authority, 2019; Jones & Kovacic, 2019; Le & Shan, 2014; Wells, 2014). In addition, the risk of conspiring in collusive activities can lead to hefty financial fines, damaging the financial reserves organisations need to be able to operate and compete within markets. In relation to the findings of this study and previously cited literature, it can be suggested here that many people are aware of the risks and impacts that collusion

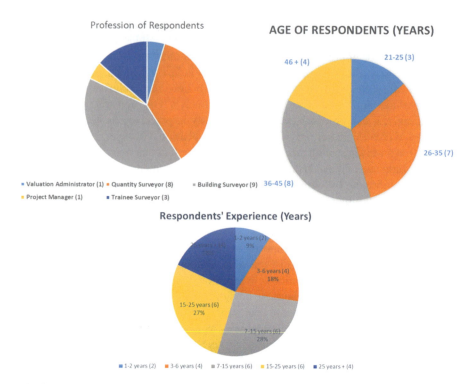

Fig. 25.1 Snapshot of responses to some of the questions

can have, but this may not deter people from conspiring because they can see where profitability can be achieved, even if it is unethical.

Collusion is commonplace within overseas construction industries such as in Canada, China and the Netherlands, with large amount of cases deriving from bid rigging and public procurement of infrastructure. Trends can be derived from the literature to imply that the major cause of collusion within these sectors comes down to greed, more specifically, contractor greed. The significance of this study, though, is the fact that because the Competition and Markets Authority (CMA) is in place to do everything to ensure collusion among other illegal practices does not happen, the potential of collusion being a future issue may not be as large in scale and prominent in comparison to overseas industries. However, the potential is still there. The predominant cause for previous collusion, as seen in literature, tends to involve a level of personal greed. It can be assumed that without the benefits of colluding, construction professionals would not choose to conspire because there is no real incentive. Therefore, a suitable and possible solution going forward could be to educate and promote fair competition and its rewards. Within the exploratory design, it was seen that another potential solution could be to ascertain a robust scheme for preventing collusion that is transparent across all levels of construction. This could entertain the idea of implementing tighter barriers to entry. These findings conclude that the construction industry is not perfect and has areas which need improvement but changing the attitudes of professionals will go a long way in eradicating the potential issue moving forward.

25.5 Conclusion

Providing good and sufficient education to construction professionals will significantly influence how collusion is perceived. The present issue is that there are professionals within the construction industry that are not aware of collusion and how detrimental it can be to fair competition and ethical standards.

A recommendation specific to construction organisations going forward would be to implement a robust education scheme throughout the industry to educate professionals and those wanting to work within the industry about the effects of the potential problem, the regulatory bodies that are in place to provide security and fluidity in industry and the potential prosecutions that can be executed. Contractor greed has been cited as a predominant issue, so recommendations could include foreign bodies safeguarding the contractor side and supply chain to reduce collusion in future. This would have to be done immediately to reduce the effects that collusion can have, and prosecutions need to involve larger financial fines to deter others from following the same unethical and illegal path.

Furthermore, it may be suggested that recommendations can be made to promote fair and ethical competition throughout all global construction industries.

There are gaps within the research that need to be filled in future studies to gain a full understanding of the potential issue. A recommendation for future study would be to select a larger sample size that gives a better and more concise indication of the UK construction industry. Despite the difficultly in execution, selecting a sample of professionals who are at the summit of the hierarchy such as executives may provide a more thorough indication because of their vast awareness and industry expertise but could gain data that is unreliable because of the subjective topic area. One other consideration for future study would be to evaluate regulatory bodies that try to achieve fair competition to recognise their perspectives on the potential issue to allow the study to fully represent its aim and objectives.

References

Baker, W. E. and Faulkner, R. R. (1993). The social organization of conspiracy: Illegal networks in the heavy electrical equipment industry. *American Sociological Review, 58*(6), 837–860.

Bowen, P., Akintoye, A., Pearl. R. and Edwards, P. J. (2007). Ethical behaviour in the South African construction industry. *Construction Management and Economics, 25*(6), 631–648.

Chotibhongs, R. and Arditi, D. (2012). Analysis of collusive bidding behaviour. *Construction Management and Economics, 30*(3), 221–231.

Competition and Markets Authority. (2019). *Construction suppliers accused of colluding to keep prices up*. London: HMSO.

Dorée, A. G. (2004). Collusion in the Dutch construction industry: An industrial organisation perspective. *Building Research and Information, 32*(2), 146–156.

Harding, C. and Joshua, J. (2010). *Regulating cartels in Europe*. New York: Oxford University Press.

Hardoon, D. and Heinrich, F. (2011). Bribe Payers Index 2011. Berlin: Transparency International Secretariat. Retrieved November 27, 2019, from https://issuu.com/transparencyinternational/docs/bribe_payers_index_2011.

Jaspers, J. D. (2018). Business cartels and organised crime: Exclusive and inclusive systems of collusion. *Trends in Organised Crime, 22*, 414–432.

Jones, A. and Kovacic, W. (2019). Fighting supplier collusion in public procurement: Some proposals for strengthening competition law enforcement. [online]. Retrieved October 26, 2019, from https://papers.ssrn.com/sol3/papers.cfm?abstract_id=3375632.

Le, Y. and Shan, M. (2014). Research trend of collusion in top construction journals. In J. Wang, Z. Ding, L. Zou, & J. Zuo (Eds.), *Proceedings of the 17th international symposium on advancement of construction management and real estate* (pp. 1133–1141). Dordrecht: Springer.

Lee, C. and Cullen, D. (2018). An empirical comparison of ethical perceptions among the consultant's quantity surveyor and contractor's quantity surveyor in the UK construction industry. RICS. [Online]. Retrieved October 27, 2019, from https://pdfs.semanticscholar.org/fe7e/a0dca1eed801576c1586d3ac561d6fcfae3f.pdf.

Macnamara, K. (2009). Builders fined and pound; 129m for rigging contract bids. *Independent*, [online]. Retrieved November 10, 2019, from https://www.independent.co.uk/news/business/news/builders-fined-pound129m-for-rigging-contract-bids-1791268.html.

Mueller, S. (2018). Five steps to improve construction ethics. Construction Executive. [Online]. Retrieved March 10, 2020, from https://www.constructionexec.com/article/five-steps-to-improve-construction-ethics.

Oke, A., Aigbavboa, C., Mangena, Z., and Thwala, W. (2017). *Causes of collusion among people in construction*. Presented at the joint CIB W099 and TG59 International safety, health, and people in construction conference, Cape Town, South Africa, 11-13 June 2017. Retrieved November 10, 2019, from https://journals.co.za/docserver/fulltext/jcpmi_v7_n2_a8.pdf?expires=1572870599&id=id&accname=guest&checksum=5BF065B4E9630B4AFA7E653CEAA13B23.

Rhodes, C. (2019). *Construction industry: Statistics and policy*. House of Commons Library Briefing Paper.

Ritchey, D. (1990). Bonded by trust: a new construction ethics committee is formed. *Arkansas Business and Economic Review, 7*(7), 50. Royal Australian Institute of Architects, http://www.raia.com.au/html/coc

Shah, R. K. and Alotaibi, M. (2018). A study of unethical practices in the construction industry and potential preventative measures. *Journal of Advanced Engineering and Management, 3*, 55–77.

Shan, M., Le, Y., Chan, A. and Hu, Y. (2020). *Corruption in the public construction sector: A holistic view*. Singapore: Springer.

Grant Thornton (2013). Time for a new direction, Fighting fraud in Construction. [Online]. Retrieved June 10, 2020, from https://www.grantthornton.co.uk/globalassets/1.-member-firms/united-kingdom/pdf/publication/2013/time-for-a-new-direction.pdf.

Transparency International. (2011). *Bribe Payers Index 2011*. [Online]. Retrieved March 20, 2020, from https://issuu.com/transparencyinternational/docs/bribe_payers_index_2011?mode=window&backgroundColor=%23222222.

Vee, C. and Skitmore, M. (2003). Professional ethics in the construction industry. *Engineering Construction and Architectural Management, 10*(2), 117–127.

Wells, J. (2014). Corruption and collusion in construction: A view from the industry. In T. Soreide & W. Aled (Eds.), *Corruption, grabbing and development: Real world challenges* (pp. 23–34). New York: Edward Elgar Publishing.

World Bank. (2011). *Curbing fraud, corruption and collusion in the roads sector*. Washington, DC: World Bank.

Chapter 26
The Myth of the Post Project Review

Amelia Maya Guinness and John Heathcote

26.1 Introduction

This research considered the perceived paradigm of project learning. It appears that the sphere of Project Management fails to notice the lack of intra-project and inter-project learning and future project success from existing knowledge pools. Observable in practice, people appear to not learn from projects (excepting their own immediate project experiences), (Argyris, 1977; Boud et al., 1985; Dewey, 1938), and yet normative project management sees learning from *post project reviews* and *lessons learned* as a fundamental part of any project lifecycle (Hamilton, 2001; Lock & Scott, 2013; Ronak, 2019; Swanson, 2017). Project Management theory does anticipate learning between projects to be a difficulty and result in a problem for synergy and benefits; this therefore may support a hypothesis that the potential benefits are lost as a result of the bureaucracy in project learning, post project reviews and lessons learned. Kolb's (2015) learning theory, amongst others (Argyris, 1977; Mcleod, 2010; Revans, 2011), suggests some possible barriers to learning between projects, including the absence of *concrete experience*. Furthermore, behavioural economic research also describes the *Pygmalion effect* which would also block learning efforts (Sutherland, 2007). The Pygmalion effect talks about 'self-fulfilling prophecy' (Rosenthal & Jacobson, 2011) and how this could affect learning abilities. Another potentially relevant theory is *communication theory* (Roger & Kincaid, 2019) (Scudder, 2019), which also has input into the idea that people struggle to learn on projects; it suggests that communication is important to learning and is not always performed or utilised in the most effective way to achieve the most beneficial outcomes. Risk is frequently cited as one of the main

A. M. Guinness (✉) · J. Heathcote
School of the Built Environment, Engineering and Computing, Leeds Beckett University, Leeds, UK

contributors to whether a project is successful or unsuccessful; therefore, it may be important to understand how individuals learn from project to project or throughout a project. Project team individuals will have different perceptions towards risks, and this may create animosity and team tension when the project team is trying to establish the potential risks worthy of mitigation. Risks are a subjective and personal perception (Gigerenzer, 2002); consequently, it may be that learning has a correlation with risk perception. It is possible that 'You never really understand a person until you consider things from his point of view… Until you climb inside of his skin and walk around in it' (Lee, 1960). From the conundrum of a deficit of learning arising from a process emphasis on post project learning, this study addresses itself to the research question: 'why and to what extent might intra-project learning be avoided; how does learning transfer fail throughout the course of the project'?

26.2 Research Methodology

This research adopted a mixed method approach of a sequential interview and then closed questionnaire, described by Cresswell (2013) as an *exploratory sequential* design. The researcher collected data from semi-structured open-question interviews and analyse it to induct a pattern and then developed a hypothesis. For the interviews, a sample group of only highly experienced project professionals were chosen; additionally they had experience across a range of projects and so were able to have an expanded notion of what the challenges were, of projects, and saw project learning as being something they personally thought about simply because they were interested in their work and doing it well. Pink theorises that people are motivated generally to develop mastery over their work (Pink, 2010). This seemed important to the research because it allowed for the respondents to call on a longitudinal (Iwh.on.ca., 2015) experience and volume which would probably see each develop their own theories about what was important, and also allow them to have an expansive degree of project experience on which to base this, a potential strength of qualitative interviews (Symon & Cassell, 1998). This inductive research of the open interviews permitted the development of hypothetical possibilities that allowed the building of a more closed deductive-based questionnaire (Cresswell, 2013). Questionnaires are a method that allows the data to be gathered in a structured data format. The follow-on deductive approach allowed for some testing of the hypotheses developed from the interviews and explored the inductive data, as is intended in Cresswell's (Cresswell, 2013) mixed method, exploratory sequential design.

The original question asked in the interviews was intended to be answered by project professionals, so that the information given was applicable to the research question. The interview question was 'what would you say are the three most vital and important pieces of learning your experience has taught you on projects'? This question was designed to be open to interpretation so that the interviewees had the freedom to answer with their own opinions and feelings. This meant that the information that the researcher received gave rich qualitative data, which could be

> *"**Communication**. This is a bit of an old saw I suppose but it does lie at the very heart of what Project Management is about in my opinion and consists on many levels. First and fore-most it is the manager's own ability to completely and fully understand the project's objectives; let's face it if he isn't clear then what hope is there for the project. It is interesting that many managers do not seem to take the time and/or effort to get a full understanding and will often rush into the project start up; "time is money", "deadlines to be met" etc, etc. Instead they should spend time testing their own understanding with the principal stakeholders because that dialogue and relationship which is formed at the outset, will often be the means by which a project can survive and overcome even the most difficult issues as and when they arise. Whilst a project will often be defined and commissioned by a contract the 'real' way in which it will be delivered is through the relationships and the most important one is the one between the manager and the stakeholders. I did have the experience of working on some projects that were relatively leading-edge technology. In one case we built what was, at the time, the largest distributed relational database (very old tech now!) ever implemented in Europe and which was beset by delays as we broached and then solved any number of technical issues. The project survived and was eventually regarded as a major success not just because we got the tech to work but also because of my relationship with the stakeholders. That is not to say that the dialogue from the manager down the pyramid isn't equally important, but the essential difference is that the manager (hopefully) will have the wherewithal to affect some sort of control and resolution in the downward direction. In its simplest context communication is an absolute must. An accurate and frequent reporting up and down the pyramid, which at times can seem to be a chore, even an unnecessary one when measured against the frenetic activity of a project in full flow, is fundamental to a successful project simply because it is the means by which issues are identified and the simple process of having to discuss/think/resolve can reveal issues which may otherwise remain hidden and which could stymie the project's progress."*

Fig. 26.1 Example interviewee response

examined and analysed to find common themes. Below is an example of the response from one of the interviewees (Fig. 26.1). However, an issue with analysing this type of data is that it demonstrates that the researcher must interpret the qualitative information given which may have led to bias data (Symon & Cassell, 1998). Nevertheless, the participants were selected for the interview by the researcher, due to the participants being a targeted sample group who were known to the researcher as being project professionals. This was done purposefully as the researcher required professionals who had significant experience. This inductive interview approach was chosen for the flexibility advantages, making the information given by the participants fluid (Symon & Cassell, 1998).

Communication

This is a bit of an old saw I suppose but it does lie at the very heart of what Project Management is about in my opinion and consists on many levels. First and foremost, it is the manager's own ability to completely and fully understand the project's objectives; let's face it if he isn't clear, then what hope is there for the project. It is interesting that many managers do not seem to take the time and/or effort to get a full understanding and will often rush into the project start up; "time is money", "deadlines are to be met" etc. Instead, they should spend time testing their own understanding with the principal stakeholders because that dialogue and relationship, which is formed at the outset, will often be the means by which a project can survive and overcome even the most difficult issues as and when they arise. Whilst a project will often be defined and commissioned by a contract, the 'real' way in which it will be delivered is through the relationships, and the most important one is the one between the manager and the stakeholders. I did have the experience of working on some projects that were relatively leading-edge technology. In one case, we built what was, at the time, the largest distributed relational database (very old tech now!) ever implemented in Europe and which was beset by delays as we broached and then solved any number of technical issues. The project survived and was eventually regarded as a major success not just because we got the tech to work but also because of my relationship with the stakeholders. That is not to say that the dialogue from the manager down the pyramid isn't equally important, but the essential difference is that the manager (hopefully) will have the wherewithal to affect some sort of control and resolution in the downward direction. In its simplest context, communication is an absolute must. An accurate and frequent reporting up and down the pyramid, which at times can seem to be a chore, even an unnecessary one when measured against the frenetic activity of a project in full flow, is fundamental to a successful project simply because it is the means by which issues are identified and the simple process of having to discuss/think/resolve can reveal issues which may otherwise remain hidden and which could stymie the project's progress'.

26.3 Findings

The Theme Matrix (Table 26.1) is used to organise the research to facilitate the organisation of the information given by the respondents and create a synthesis of data that is more easily analysed (Alhojailan, 2012). The Theme Matrix identifies and divides the information into four umbrella themes which were *communicating objectives and alignment of purpose, failures and bias, not learning as a team and team selection.* The next stage of the research was to develop common themes between the seven participants selected for the interviews, and this was analysed using a Theme Matrix which demonstrated seven prominent similarities to what were interpreted from the interviews to be important aspects of learning on projects. They were:

Clarity and understanding of the purpose of the project.
The benefits of using visual aids and communicating effectively through them.
Building relationships with the stakeholders.
Making sure that people take responsibility for failures on projects.
Not to be too optimistic about potential risks.
Utilising post project review and lessons learned.
Picking the right people for the project – team selection.

The Theme Matrix (Table 26.1) appears to represent a group of similarities that require project managers and project teams to display soft skills as opposed to hard skills (Pollack, 2007) that have been previously seen as the most important aspects for project teams to possess to be successful. The soft skills invoke Gardner's (1983, quoted in Goleman, 1996) theory of 'interpersonal intelligence'; this idea is 'the ability to understand other people: what motivates them, how they work, how to

26 The Myth of the Post Project Review

Table 26.1 Theme Matrix 'showing the themes interpreted from the interview respondents'

	Theme Matrix:						
	Themes: (common themes)						
	Communicating objectives and alignment of purpose			Failures and bias		Not learning as a team	Team selection
Respondents:	Clarity and understanding of the purpose of the project and its objectives	The frequency on projects of a lack of a pre-determined communication plan and not using communication tools such as Gantt charts and other visual aids such as cash flow diagrams and raid logs	Prioritising stakeholder engagement when necessary, farming a relationship with the stakeholders	Not taking responsibility for the failures in a project	Project team members using optimism bias	Not utilising post – Project reviews, lessons learned and not learning due to cutting corners with project closures	Team selection processes, making sure there are the right people for the job, e.g. hiring PMOs
A	Y		Y		Y	Y	Y
B		Y	Y	Y		Y	
C		Y	Y	Y	Y		
D	Y						
E	Y	Y	Y		Y		Y
F						Y	Y
G	Y					Y	Y

work cooperatively with them' (1983, quoted in Goleman, 1996). Therefore, communication appears to be vital to project learning and success. This is supported by an article in the *Sunday Telegraph* (Hepworth, 2017) espousing that 'while we will see many professions replaced by increasingly advanced systems in 2030, the core elements of Project Management will still require human judgement and insight' (Hepworth, 2017).

Explaining the rationale and outputs from each question:

1. *What is your job role?* The first question of the survey is 'what is your job role?'. In this chart, it shows that the majority of the people who responded to the survey questionnaire classified themselves as 'Other' which was 39%. The other choices involved 'Project Management', which was closely behind 'Other' at 34%, 'Programme Management' at 15% and 'Programme Management Office' at 12%. Therefore, 61% are the predefined job roles that the researcher has made individual answers.
2. *How many years' experience do you have?* This second question roughly identifies how old the participants are by how many years of experiences they have; the question is written like this so that the participants don't feel like they must divulge personal information like their exact age. The majority of the participants have 21 years or more experience in addition; 74% of the participants have 12 or more years' experience which implies that the majority of people that responded to the questionnaire were most likely older to be able to have so many years' experience.
3. *What sector do you work in?* The third question tries to identify what background the participants work in; this section has several options. These options include the following: Healthcare, Information Technology, Education, Transport, Local Authority, Engineering, Construction, Retail and Other. The two most common sectors are Information Technology and Other, both at 30% each. Similarly, with the second question, there appears to be many sectors that haven't been included as options; thus, participants have selected Other. Healthcare was selected as the next highest number of participants at 15%; the rest of the options varied between 2 and 8%. This demonstrates that whilst IT, Other and Healthcare are the majority, there is a spread between the alternative options.

Following from the preliminary questions is the statement question from the questionnaire. The question is 'How useful do you believe the following statements to be for project success?'. Subsequent to that are seven statements which the participants had five options to choose from. These options ranged from *not very useful* to *very useful* (using a Likert scale); the participants were asked to choose one of the options for each of the seven statements to the question. The charts start with *not very useful* at the left progressing to *very useful* on the right.

1. *Using Predetermined Communication Plans and Visual Tools such as Gantt Charts, Cash Flow Diagrams and Raid Logs.* This section appears to show the majority of the participants believe that the use of predetermined communication

plans is *useful* for project success. More than half of the respondents regarded predetermined communication plans as either *useful* or *very useful*.
2. *Prioritisation of Stakeholder Engagement and Relationship Building.* There is a clear pattern of respondents who believe that stakeholder engagement and relationship building is *very useful* for projects to be successful, as 85.4% of respondents answered that it was either *useful* or *very useful*.
3. *Taking Responsibility for Failure on Projects.* Like the previous statement, most of the respondents have answered it is *useful* to take responsibility for failures of projects with 81.3% of respondents believing that it is either *useful* or *very useful*.
4. *Clarity and Understanding of the Purpose of the Project and Its Objectives.* The answers to this question were almost unanimous, as almost all of the respondents answered that they believed that it was *very useful* to have clarity and understanding of the purpose of the project so that projects are effective and successful, with 97.9% of respondents answering either *useful* or *very useful*.
5. *Being Optimistic About Potential Risks.* The chart for this question has mixed answers; the majority of respondents (25%) answered in the middle, the neutral answer. The chart appears almost level offering no significant commonalities.
6. *The Utilisation of Post Project Reviews and Lessons Learned.* Most respondents answered that post project reviews and lessons learned were *useful* or *very useful*. This shows a pattern that project professionals believe that post project reviews and lessons learned are beneficial for project success with 60.4% of respondents answering that they are *very useful* and 20.8% that they were *useful*.
7. *Team Selection for a Project.* Team selection also appears to show a pattern that respondents believe that it is useful for project success if project teams are selected appropriately. 87.6% of respondents believed that team selection was either *useful* or *very useful*.

26.4 Discussion

The findings appear to show a pattern that most of the participants responded by saying that the statements to the questions were very valuable. For example, the statement questions in the questionnaire such as 'using predetermined communications plans' and 'the utilisation of post project reviews and lessons learned' had a high percentage of respondents saying that they were very useful for project success. The literature also says that post project reviews and lessons learned are important, for example, Hamilton (2001) who says that lessons learned should be 'used as feedback to the project and used, where applicable, as intelligence in future projects' (Hamilton, 2001), yet they do not say why it is very rare that they are undertaken. In practice, if organisations were using these learning tools to their full advantage, then there might be a significant drop in project failures, risks and issues, as the project team would have documentation to help refer to previous instances of when things either went wrong or right, assuming that this is what they would do. This suggests that there is an issue in that prior learning is not being shared or imparted for the benefit of future projects as people are not inclined to engage in new learning experiences because they believe that they already know what they know (Luft & Ingham, 1955) (Syed, 2015). This is presented in the literature by Gigerenzer (2002) and more philosophically by Taleb (2007) who show that there are 'unknown – unknowns' implying that project professionals do not known that they need to learn because they believe that they already know everything that they need to know on projects for the projects to be successful. Furthermore, the research's intentions were to establish an understanding of project learning, why and to what extent does the learning not appear to function as a productive endeavour.

The primary data sought correlations between the theory and the practise, of project learning, to establish the barriers to learning on projects. The soft skills that appear to be required to learn effectively by project teams' members seem to have similarities to the difference between qualitative data collection and quantitative data collection, and both Dallas (2006) and Buunk and Van Yperen (2019) saw this as a method to improve project decision-making. Additionally, the quantitative data findings appear to have a parallel with Pollack's (2007) *hard skills* (Pollack, 2007), giving the impression that as Project Management progresses, the importance of *soft skills* seems to increase. The hard skills are more tangible to teach and learn from project to project, for example, teaching someone to use Microsoft Project would be a more concrete piece of learning. Due to soft skills being a more abstract idea, it appears that learning these attributes may prove difficult as they are not as concrete as the hard skills, and quantitative data would be, in terms of learning and acquiring knowledge. This is theorised by Kolb's (2015) learning theory; it is easier to learn something if one has a concrete, practical experience of something. One of the research objectives was to find key themes as to why people do not learn throughout projects. The Theme Matrix findings appear to help demonstrate attributions given by the interview respondents as to why people do not learn from preceding project work. The presence or absence of the prescribed hard process methods to which the questionnaire respondents attributed a failure to learn, in themselves, does not appear to be able to influence learning without a more collective and individual desire to re-examine any difficulties experienced on the project. The interview portion of the research demonstrated a significant set of learning from individual (and offline from any project) professionals who were able to reflect on past experience, and both a frustration that this did not happen more collectively at the beginning of subsequent projects, but they also attributed this lack of learning transfer (to others) to the mechanisms in normative project management, the *communication plan* and so, whilst ignoring the fact that these devices, closely associated with project management, had not permitted inter-project learning, in their experience.

However, learning theory can comprehensively explain the barriers to learning from project to project. It attributes learning to the concrete practice of personal experience; learning theory highlights the importance for this type of experience leading individuals to reflectively re-learn. One of the issues with learning theory might be that it does not acknowledge that learning is continuous. If this is the case, then learning theory appears to dismiss the findings for the qualitative data, as the qualitative data. As previously mentioned, the qualitative data from the interviews resulted in a consensus that the most important aspects of learning are not the ones that you can necessarily quantify, for example, 'forming relationships with stakeholders'. This portion of data that was analysed through the Theme Matrix does not align with the theory supposed about learning from Kolb (2015) learning theory, which itself provides a more simplistic perspective, and approaches learning as an individual enterprise.

The Pygmalion effect also disagrees with aspects of the Theme Matrix. One of the umbrella themes analysed was failures and biases in learning. The Pygmalion effect describes the 'self-fulfilling prophecy' (Rosenthal & Jacobson, 2011) which states that if you have positive expectations, then you are more likely to do well (or learn from them). However, the research findings infer that to be able to learn successfully, it is important to take responsibility for failures on projects. Therefore, there is a lack of continuity between what the literature states and what the research findings demonstrate, although it should be noted that the themes arise themselves from the opinions (and biases) of the interviewees, who attribute a self of ownership, and responsibility, as motivators towards the need to learn.

26.5 Conclusion

This study asked: 'why and to what extent might intra-project learning be avoided; how does learning transfer fail throughout the course of the project'? Project management literature discusses how learning is important and that learning tools such as post project reviews and lessons learned are beneficial. However, there are not clear explanations as to why they are useful to a project, nor research that demonstrates its efficacy in practice. There appears to be no direct gratification by using the standardised methods of learning; thus, project professionals dismiss these tools just as an aspect of the bureaucratic system of tick box exercises and form filling. However, the research findings do support the idea that project professionals do believe that learning is important. One of the purposes of this research is to illustrate and understand a gap in the literature that can be understood through the study in order to increase the success of projects, through learning.

This research infers that, in fact, learning is a difficult effort that people avoid as they find it either difficult or they have not learned how to learn. Learning theory from Revans (2011) and Kolb (2015) predicted this, as transferable learning may not be achieved without a 'concrete experience'. However, it appears that one of the major issues of learning transfer is that other factors such as soft skills have not been incorporated into the project learning realm. Several works hint at the importance of soft skills emphasis to raising the barrier to otherwise readily achieved learning insights, including Boud et al. (1985), Dallas (2006), Jenner (2012), Syed (2015) and Buunk and Van Yperen (2019). Understanding the human cognitive challenge at a deeper level (Sutherland, 2007), and the importance of conceptualising projects as social systems (Smyth & Morris, 2007), appears to be what needs to happen before the project teams can be expected to transfer key learning to future projects. In the meantime, seasoned project professions will likely learn individually from their experiences, but struggle to pass this on to new teams they work with.

26.6 Limitations

This paper reports on an original master's degree thesis and consequently time was constrained. With more time, then the sample of participants for the interview, $n = 7$, and for the questionnaires, n = 48, may have been bigger, which would have subsequently enriched the data received from the overall participants. Moreover, the rhetoric questionnaire's questions may have been too leading, together with the sample selection of participants, which were people known to the researcher; this may have made the participants biased towards how they decided to answer on the questionnaire, for instance, through reciprocation (Symon & Cassell, 1998). There may be a need to do more research into different sectors and job roles. This is because there were a high number of participants answering 'Other' to some questions, indicting poor or no understanding of those questions. Thus, there should have been more options to choose from so that the researcher had more of an understanding of the participants' experience, and whether that limited their responses. Knowing more about the participants may help make a better and more informed analysis of the findings. Additionally, one of the questionnaire questions appeared to have a much more level, chart question 5 'Being Optimistic About Potential Risks', depicting that the participants may not have understood the question or the question may be poorly formed. Alternatively, the participants may have answered randomly because they have no definite opinion about the question. Furthermore, the participants for both the interviews and the questionnaires were a sample of participants who were not from a specific location. The participants were project professionals who work in many other countries, for example, France and Switzerland. This may have affected the research findings because it added another variable to the study (Saunders et al., 1997). It may be that in other countries, there are different standard practises in terms of learning, which would mean that the participants had very different experience of a project. The research being a mixed method exploratory approach relied heavily on the interpretations of the interview findings, and these included an interpretation of what the sample of just $n = 7$ had *attributed* lack of inter-project learning to. The themes arrived at included some attributions, which are clearly rooted in normative project management. The potential for further and other interpretation here needs to be examined. In summary, further effort might go on into future studies in this area to eliminate the potential for confounding variables hidden in this sample, such as organisational and other cultural bias, particularly in relation to tools and methods to address the issue of inter-project learning. The authors continue to debate the factors that may contribute to allowing the avoidance of 'known unknowns', that is to say, clearly understood risks to project failure that can be discovered, and understood, prior to committing to a course likely to lead to project failure. This subject offers opportunities for future research.

References

Alhojailan, M. (2012). Thematic analysis: A critical review of its process and evaluation. *West East Journal of Social Sciences, 1*.

Argyris, C. (1977). Double loop learning in organizations. *Harvard Business review*.

Boud, D., Keogh, R., & Walker, D. (1985). *Promoting reflection in learning in reflection turning experience into learning*. London: Kogan page.

Buunk, B. and Van Yperen, N. (2019). *Illusory superiority*. [online] En.wikipedia.org. Retrieved February 13, 2019, from https://en.wikipedia.org/wiki/Illusory_superiority.

Cresswell, J.W. (2013). *Steps in conducting a Scholarly Mixed Methods Study*. DBER Speaker Series. 48. Retrieved from http://digitalcommons.unl.edu/dberspeakers/48

Dallas, M. (2006). *Value and risk management*. Oxford: Blackwell Publishing Limited.

Dewey, J. (1938). *Experience and education*. New York: Simon and Schuster.

Gigerenzer, G. (2002). *Reckoning with risk: Learning to live with uncertainty*. New York: Penguin.

Goleman, D. (1996). *Emotional Intelligence*. London: Bloomsbury Publishing Plc.

Hamilton, A. (2001). *Managing projects for success*. London: Thomas Telford Ltd..

Hepworth, P. (2017). Project Management will be at the Centre of how we work in 2030. *Sunday Telegraph*.

Iwh.on.ca. (2015). *Cross-sectional vs. longitudinal studies | Institute for Work & Health*. [online]. Retrieved February 4, 2019, from https://www.iwh.on.ca/what-researchers-mean-by/cross-sectional-vs-longitudinal-studies

Jenner, S. (2012). *Managing benefits—Optimizing the return from investments*. TSO.

Kolb, D. (2015). *Experiential learning* (2nd ed.). Upper Saddle River, NJ: Pearson Education.

Lee, H. (1960). To Kill a Mockingbird.

Lock, D., & Scott, L. (2013). *Gower handbook of people in project management*. London: Gower Publishing Limited.

Luft, J. and Ingham, H. (1955). *Johari window*. [online] En.m.wikipedia.org. Retrieved January 11, 2019, from https://en.m.wikipedia.org/wiki/Johari_window.

Mcleod, S. (2010). *Kolb learning styles*. [online]. Retrieved January 6, 2019, from http://www.simplypschology.org/learning-Kolb.html.

Pink. D. H. (2010). *Drive: The surprising truth about what motivates us*. Canongate.

Pollack, J. (2007). The changing paradigms of project management. *International Journal of Project Management, 25*, 266–274.

Revans, R. (2011). *The ABC of action learning*. Farnham, Surrey: Gower Publishing Limited.

Roger, E. and Kincaid, D. (2019). *BizComm: The convergence model of communication*. [online] prezi.com. Retrieved February 14, 2019, from https://prezi.com/4fbibqjku0bi/bizcomm-the-convergence-model-of-communication/

Ronak, C. (2019). *Got Projects Going Over Budget?* [online] Project Times. Retrieved January 9, 2019, from https://www.projecttimes.com/articles/got-projects-going-over-budget.html.

Rosenthal, R. and Jacobson, L. (2011). *The Pygmalion effect and the power of positive expectations*. [video] Retrieved January 25, 2019, from https://www.youtube.com/watch?v=hTghEXKNj7g.

Saunders, M. Lewis, P. Thornhill, A. (1997). *Research methods for business students*. Financial Times. Pitman Publishing.

Scudder, S. (2019). *Communication theory - meaning and examples*. [online] Managementstudyguide.com. Retrieved January 25, 2019, from https://www.managementstudyguide.com/communication-theory.htm.

Smyth, H. J., & Morris, P. W. G. (2007). An epistemological evaluation of research into projects and their management: Methodological issues. *International Journal of Project Management, 25*(4), 423–436.

Sutherland, S. (2007). *Irrationality*. London: Pinter Martin Ltd..

Swanson, I. (2017). *Edinburgh Tram Inquiry: How did Edinburgh's trams become a fiasco?* [online] Edinburghnews.scotsman.com. Retrieved January 26, 2019, from https://www.edinburghnews.scotsman.com/news/transport/edinburgh-tram-inquiry-how-did-edinburgh-s-trams-become-a-fiasco-1-4550298.

Syed, M. (2015). *Black box thinking*. London: John Murray.

Symon, G., & Cassell, C. (1998). *Qualitative methods and analysis in Organisational research: A practical guide*. New York: Sage.

Taleb, N. (2007). *The black swan*. London: Penguin Group.

Chapter 27
The Tendency Towards Suboptimal Operational Planning

Thomas Price and John Heathcote

27.1 Introduction

Both projects and project management are seen as key elements within organisations across all sectors (Maylor et al., 2006; Pellegrinelli, 2011). Despite this, a study conducted by the National Audit Office (2016) found that 34% of projects were unsuccessful in delivering their planned outcomes and failed to deliver any value. This viewpoint may be the reason that, in literature, there is considerable focus on value as a basis of project success (Association for Project Management, 2018; Laursen & Svejvig, 2016). However, throughout literature, there is much confusion as to what is seen as project value. In the British Standards BS EN 12973:2000, value is seen as satisfactory needs over the use of resources, which may be heavily linked to the 'iron triangle' (the idea that projects are constrained by three predicable constraints: time, cost and quality) cited measures of project success (Bryde, 2008). Similarly, the APM have described value as 'value = stakeholder satisfaction/ resources required' (APM, 2005) before changing their view to 'value = benefits and key stakeholder requirements/resources used' (Association of Project Management, 2012). This seems to follow the 'control and monitoring' paradigm of the project management theory, basing project success around the scope, budget and time parameters, with little regard to benefits realisation and value creation (Midler, 1995). The PMI (2016) suggest that project value is the benefits created for the project stakeholder or beneficiary as a result of successful project completion. However, Charter (2006), in a paper for the PMI, suggested that project organisations will only see a big difference in how much value is received when the company steps back and thinks through its business processes and empowers their workers with a certain set of tools. This may highlight the complexity in

T. Price (✉) · J. Heathcote
School of the Built Environment and Engineering, Leeds Beckett University, Leeds, UK

determining a project's value, as value could be determined in many ways (Lechler, 2010). This suggests that value is a concept that is not absolute and can be influenced by stakeholders' perceptions and interest (Heathcote, 2017). These perceptions may in turn be linked to the wide view of the definition of projects and project management found in literature (Project Management Institute, 2004; Young, 2006; APM, 1993; Marray, 2006; Kerzner, 2006), which changes in each sector and thus so does the approach to and vision of value. This multidimensional view of value as a concept may suggest that value can be considered as a balance of quality, resource and risk (Goodpasture, 2001). This misinterpretation of value may be linked to the project managers' dominant ways of thinking (Winter et al., 2006).

Commonly, in both project management literature and practices, determining what represents value is an unresolved issue. Where the value debate might lead for an individual project, improving the … 'V = benefits/cost (IVM, 2019)' equation… this can be achieved by either improving 'benefit' (or 'functionality') or alternatively by reducing the cost of delivering the same or more benefit. Therefore, making a process more efficient may deliver the same value at a reduced cost or adding cost and may result in further benefits and more value than planned, though the latter is lacking evidence in literature, which led to several assumptions which this study set out to explore: is value not being fully exploited in projects; do project run with inefficiencies; can planning and visualisation tools benefit projects; and what are the factors that potential block value from being realised?

27.2 Research Method

Research Question: Is there potential latent value and inefficiency in projects?

Research Aim: The overall aim of this research is to investigate the potential for latent value lost to inefficiency of planned delivery in a series of project case studies. The study tested whether projects in practice appear to under-exploit the use of project management visualisation tools to help unlock potential value and reduce inefficiencies to the delivery sequence.

The limitation of this study may be that the research is not being collected in a quantitative/deductive approach, which statistically test hypothesis drawing conclusions from data analysis (Rovai et al., 2014). However, the method of this study is qualitative; thus, it allows for 'constructing theory' based on opinion and perception, resulting from first-hand practice and open to interpretation (Ghauri & Gronhaug, 2005). However, to construct any hypotheses, a thematic coding method commonly used in qualitative study must be undertaken to help identifying and analysing report themes and patterns within data (Braun & Clarke, 2006).

27.3 Literature Review

The debate around what represents value, and the process of implementing it, may highlight that the issue within projects may be the reason projects in practice appear to under-exploit the use of project management visualisation tools to help identify inefficiencies and unlock potential value within the planned implementation. Therefore, Green and Sergeeva (2018) suggest a shift from project success being evaluated based on the 'iron triangle' to the origin of through-life value creation does not comprise a 'paradigm shift', although it does highlight the importance of better decision-making and the vast tools a project manager may need to call upon to influence problem-solving. This need for focus on problem-solving and value creation is now widely recognised at the front end of project management (Matinheikki et al., 2016; Morris, 2013; Samset & Volden, 2016).

The Project Management Institute believed that rational decisions lead to better project outcomes (Heffernan, 2015), whereas problem-solving identifies the problem and selects the most appropriate decisions to achieve the optimal outcome (Hicks, 1997). In literature, it is agreed that both are coupled and influence each other heavily (Koppenjan & Klijn, 2004; Lee, 1997). Thus, it is commonly agreed that that it is worthwhile spending time on defining the problem as it can help recognise a clear solution (Maylor, 2010; Kerzner, 2011). Although it is seen that making the correct decision is key to project management, many project managers are unwilling to try to improve the quality of their decisions (Goodwin & Wright, 2004), which may lead to starting projects that should not have been started (Flyvbjerg, 2016). This may be linked to the lack of research about the problem formulation stage and its link to later problem-solving (Choo, 2014). As a consequence, several issues occur which may be linked to irrational decision-making, such as the availability error, optimism bias, sunk cost error, planning fallacy, tactical exaggeration and the first idea error (Kahneman & Tversky, 2000; Sutherland, 2007). However, Hanselman (2013) suggests project managers may not make the most of the problem-solving process, which may highlight a 'Lack of Process' in practice. This may be the reason why project management literature provide several improvement methodologies to give structure, such as 'Value Management', 'Lean', 'Six Sigma' and Goldratt's (1990) 'Theory of Constraints'. Drawing on these tools to help problem-solving may help rationality, creativity, risk identification, problem-solving (Crawford, 2006), value creation (Ingason & Shepherd, 2014), project leadership, strategy and innovation (Dalcher, 2017).

Tidd and Bessant (2009) recognise this need for applying value through there '4Ps' innovation model as a process of turning ideas into reality and capturing value from problem-solving. Similarly, literature often sees 'process innovation' as the first step in creating value (Drucker, 1995; Tidd & Bessant, 2009), as it is frequently the easiest to implement and may also result in 'product innovation', 'position innovation' and 'paradigm innovation'. This may be the reason that in literature and practise process improvement is highly discussed and implemented. Commonly used methodologies are Value Management (VM), Lean, Six Sigma (SS), and

Theory of Constraints (TOC). However, each varies in their method and areas where improvement may be realised. It is often common practise in a Value Management framework to provide a structured decision-making process, by facilitating a stepped approach and how the project team will achieve them (VM SiG, 2017). The implementation of the VM structure is commonly seen as a method of collecting information, based on the problem and value; prioritising information, to form the project objectives; creative thinking, to help stimulate idea; and prioritising options, balancing objectives with available resources (Kelly et al., 2004). The APM Value Management Special Interest Group (VM SiG, 2017) summarises this by suggesting a facilitated approach that helps suspend judgement, and boosts non-typical lateral thinking, to avoid typical decision-making errors. This may benefit the idea of the 'availability error' (Sutherland, 2007) in avoiding the 'first good idea error' by allowing for the generation of several possible solutions. This in turn may suggest that visualising the direction of a project at the development stage opens latent value previously unseen. However, there is little in literature to suggest these broad decision-making processes are considered further down the project cycle, when it comes to planning the correct project to undertake. Text around 'value planning' and 'value analysis' are primarily concerned with project cost rather than adding value (Hayles et al., 2010) or whether further creativity is needed. In summary, to implement a VM approach, an opportunity or problem must be identified, and this then supports effective project management by developing understanding, emphasising project purpose, developing scope, reducing cost and improving value. While *Lean* has often been promoted as the new change and improvement method, particularly as a cost reduction process (Achanga et al., 2006; Bicheno, 2004), many researchers (de Treville & Antonakis, 2006; Hines et al., 2004; Hopp & Spearman, 2004; Paez et al., 2004; Shah & Ward, 2003) suggest that the method should be understood on two levels: the strategic level of understanding value and the operational tools of eliminating waste (Hines et al., 2004). Womack and Jones (1994) suggest that this first principle is about understanding that value is arguably more important than reducing waste. Thus, finding the 'value stream' is a sequential process all the way from raw material to final customer (Bicheno, 2004). Similarly, Six Sigma approach is also heavily based on statistics (Pepper & Spedding, 2010), with a systematic approach to develop process improvement (Ahmed et al., 2013; Madhani, 2016; Pyzdek, 1999). This technique focuses heavily on problem definition by examining the organisation's planning and controlling of current projects and providing practical knowledge on managing the value, schedule and resources (Kumar, Nowicki, Ramirez-Marquez & Verma, 2008). Thus, this analysis involves project life cycle, work breakdown structure (WBS), network diagrams, scheduling techniques and resource allocation decisions. To identify process improvements by analysing this data to determine the key process inputs and in turn optimise the outputs, and also controls the improvement process to ensure it is sustained (Foster, 2010; Pulakanam & Voges, 2010; Starbird & Cavanagh, 2011). To help improve the current baseline, many academics have suggested the implementation of project management tools at each stage of the DMAIC process. Anthony suggests the section of the DMAIC at which the implementation of project management tools may be beneficial, which may suggest that problem-solving tools help visualise the latent value in an improvement-based methodology. Goldratt's (1990) idea, the Theory of Constraints (TOC), suggests that every process has at least one bottleneck that impedes the system from reaching better performance in terms of purpose. This is a contrasting view to other improvement process like Lean and Six Sigma, which are improvement approaches that were initially developed to save resources within the process (Naor et al., 2013). Within literature, TOC is generally seen as a combination of philosophy, concepts, principles and tools, used to maximise the performance of any system by identifying, managing and controlling restrictive factors that limit project implementation and performance (Dettmer, 1995; Lee, 1997; Ricketts, 2008). Others have emphasised that, although every system may have at least one constraint, the system does not become infinitely stronger as the constraint simply migrates to a different area (Goldratt, 2003; Rahman, 1998). This may be because constraints that effect the process can be both external (not physical) and internal (physical) (Goldratt, 2003; Goldratt & Cox, 2002). He suggests that to successfully do this, project managers must make three generics decisions while dealing with constraints: what to change, what to change to and how to cause the change.

Literature highlights that the visualisation of data is a useful tool to collect, transform and present both qualitative and quantitative data (Al-Kassab et al., 2014; Moore, 2017). These tools typically include images, videos, sketches, Gantt charts, network diagrams, regular tables, prototyping and other interactive features, which can help enhance the process by making information more transparent and understandable (Sindiy et al., 2013). Studies carried out by both Bauer and Johnson-Laird (1993) and Larkin and Simon (1987) explain that visual representation is superior to verbal representations and can further benefit the analysis of information. Patterson et al. further suggest that, without this process to assist understanding, a greater effort level is required to obtain information. Heathcote and Coates' (2018) research showed that greater performance might be attributed partly to the ability to visualise the projects' complexity and thereby allowing the project manager to coordinate the project better. This means that all stakeholders within a project could benefit from focusing their attention to analytical or creative thinking via information visualisation (Reddivari et al., 2013; Sibbet, 2013). Heathcote and Coates (2018) added that visualisation should be treated as a process, stimulating the imagination and creative thinking, leading to finding the best solutions for existing problems. Further research has suggested that visualisation through Gantt chart-type modelling of the planned operation by individuals or groups can benefit decision-making, efficiency and value within a project (Davies, 2018; Heathcote, 2017). Similar to aspects of the other improvement process considered in this literature review, the Gantt enables users to produce a 'mental map' of the project to assist in stopping errors, although it may be argued that it can also unlock latent value. The Gantt is seen as a key planning tool and those in literature suggest project management is closely associated with planning (Morris, 1997; Dvir et al., 2003; Gardiner, 2005; Maylor, 2010). Therefore, a way of visualising the planned process of a project may be important in providing the project manager with the skills to conceptualise the complexity of the project (Gupta et al., 2016) and develop rational decisions to progress the work (Wood & Heathcote, 2009). This may in turn help improve from the current baseline. The improvement methods previously discussed in this thesis show the importance of mapping a process to understand the variations, waste and bottlenecks. Therefore, visualisation of the plan through the use of a Gantt may help the identification of the 'critical path' (tasks that do not have float time), by developing a more accurate way of reviewing logic linked tasks (Dvir et al., 2003). Heathcote and Coates (2018) add to this, suggesting that the feasibility of visualising Gantt charts could be reduced to two key aspects: the work breakdown structure (WBS) and the treatment of the logic links between tasks. This method may further reduce the complexity of information and prevent overloading people and compromising decision-making (Levitin, 2015). It can be drawn from the above that project visualisation is in solving the decision-making problems during the whole project cycle. This visualising of the plan allows for modelling and analysis of the problem definition, to help identify a solution similar to other improvement processes. The following literature can be summarised by suggesting problem-solving and decision-making are key factors in the success and realisation of value in a project, and hence many organisations are taking on board several of the improvement processes discussed to try and improve the current percentage of failed projects (National Audit office, 2016). That there are several constraints identified which the improvement processes seen in literature look to deal with in order to improve the process and add value, these can be summarised as follows: remove risk, reduce waste, reduce variation, reduce cost/resource, remove bottlenecks, parallel working, add benefits and add value. However, it has been identified that these methodologies have several limitations when applied individually, which may be the reason it has become common in project management to combine these methodologies and techniques to gain the desired outcome.

27.4 Findings

The following set of illustrations include the detail of the problems identified and the subsequent changes proposed for each of the case study projects. Because they are filled with detail, this makes them difficult to read in detail. However, this should not be particularly important to the paper; here we seek to simply present an illustration of the general amount of problems that could be identified and then the changes that are possible to the original plans. Of note is that the project cases have been able to be improved on in terms of shortening the schedule required. Further descriptions in the findings show how additional benefits could be realised in terms of commercial and market/benefit advantages arising from the challenges of the initial plan of work.

27.4.1 Project Case Study 1: Water Pump Manufacturing Project

This case study follows the typical process of manufacturing water control pumps for a regional energy supplier (Fig. 27.1).

Visualising the typical current case has shown several constraints within the process. These can be summarised as follows: waste (customer response period 5 days; tender evaluation period 5 days; design consideration 5 days), bottlenecks (design 12 days; procurement 9 weeks; manufacturing 34 days), variation (client 2 weeks) and risk (business strategy 10 days or project lost) (Fig. 27.2).

Through the visual analysis process, it has been highlighted that there were common constraints, from which it is possible to suggest the feasible solution modelled in above. This modelled improvement resulted in removal of tender process, reduced time for procurement and manufacturing, more predictable completion times, product innovation and reduced costs. Alongside these results shown by the interventions, the process now only takes 97 days to complete the project, meaning that the process is faster and cheaper and provides a more client focus approach.

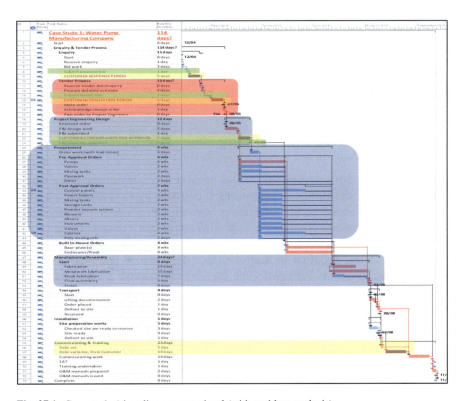

Fig. 27.1 Case study 1 baseline process visual (with problem analysis)

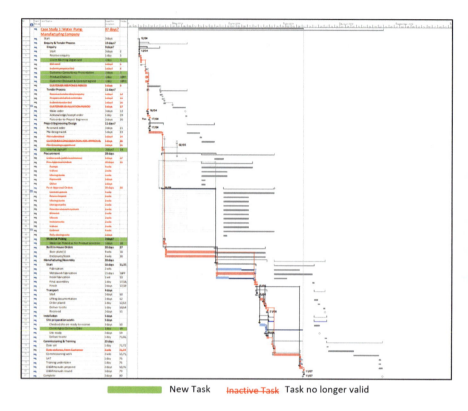

Fig. 27.2 Case study 1 solution analysis visual Gantt

27.4.2 Project Case Study 2: Care Home Construction Project

This case study follows a typical client-side quantity surveying organisation's project process for the demolition and new construction of a larger residential care facility, as well as the relocation of residents until the build is complete. This meant that delays in the programme cost the client more in compensation to those being moved and loss of revenue from new residents (Fig. 27.3).

It may be summarised from the analysis of the current process that there are several constraints which are causing latent value from being produced, summarised as follows: waste (appointing team 11 weeks), bottlenecks (planning process 25 weeks; client approval and operational mobilisation 14 weeks), variation (design VM review 4 weeks; tender VM review 2 weeks) and Risk (implementation of build 99.6 weeks) (Fig. 27.4).

Using the Gantt to visualise the baseline of this project has helped identify ways to reduce time for design and construction, increase quality of specialist tasks, reduce client compensation to residents, realise revenue benefit sooner, increase cost and add further value to the client. The solution suggests that the process becomes more focused on allocating more resources to specialist and lengthy tasks

27 The Tendency Towards Suboptimal Operational Planning

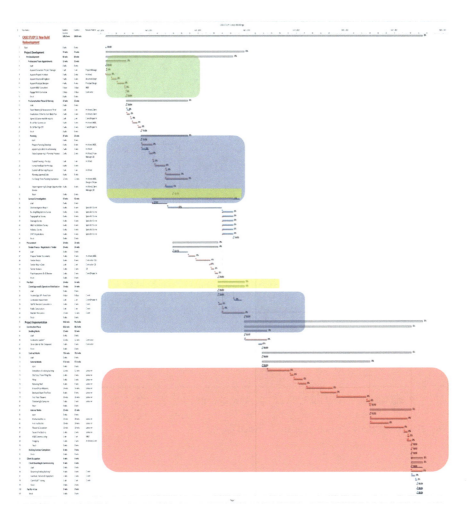

Fig. 27.3 Case study 2 baseline process visual (with problem analysis)

to ensure they are completed to the highest quality sooner, reducing bottlenecks and risks to client and other stakeholders. It can also be seen that if the organisation can reduce the amount spent on relocation and generate new earnings faster than planned, further value will be realised.

27.4.3 Project Case Study 3: FF&E Furniture Project

Furniture consultation, design, procurement and project management service to a wide range of organisation, purchasing FF&E, from initial design concept to installation on site. However, it can be seen from the way in which projects are developed

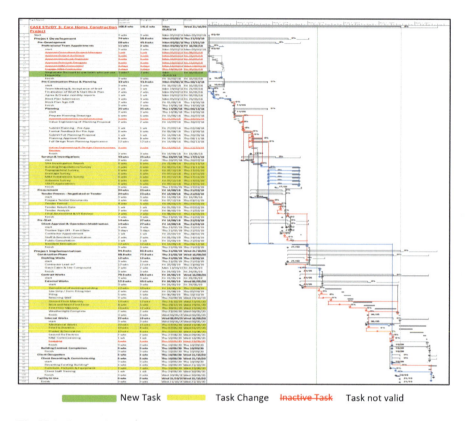

Fig. 27.4 Case study 2 solution analysis visual Gantt

and implemented that the approach has become complacent and known issues often affect project time and time again (Fig. 27.5).

The current process Gantt of a typical project carried out by the furniture organisation shows several problems which require improvement, summarised as follows: waste (client tender review 21 days; client process of order 15 days; awaiting meetings 32 days), bottlenecks (design 29 days; manufacturing 52 days), variation (client revisions 22 days) and risk (bidding process 96 days or lost; storage 31 days; third-party conflict) (Fig. 27.6).

Using the Gantt as a method of analysis and a prototyping solution helps identify ways to reduce time for design, procurement and installation, bring forward project completion, reduce costs, reduce lost revenue, communicate more effectively and reduce known risks. As well as the results shown by the interventions, the process now only takes 140 days to complete the project, saving 103 days. This means that 1.7 projects can now be completed in the same time as the current process, which may lead to an increase in revenue and market share. The customer also receives the additional value of discount, clear objectives from the start and earlier completion.

27 The Tendency Towards Suboptimal Operational Planning

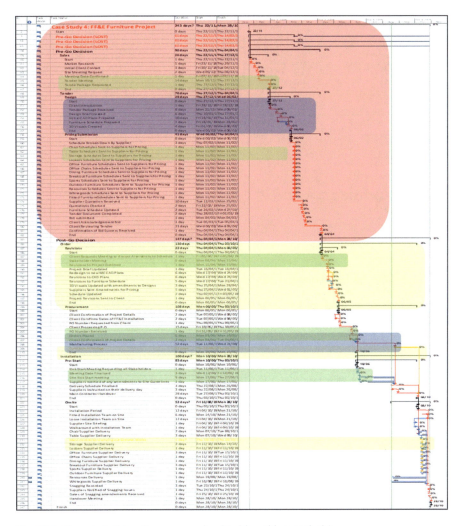

Fig. 27.5 Case study 3 baseline process visual (with problem analysis)

27.4.4 Project Case Study 4: IT Software Procurement Project

This case project relates to a public sector procurement project for the tender of a new CRM system. The current system and support are at the end of their contractual period and must be either extend or re-tendered. In this case, the organisation has chosen to tender under the deadline that the implementation process must be almost complete before the end date of the current system (Fig. 27.7).

From the visual findings, it can be summarised that there are several constraints within the current process, which prevent further benefits and value from being realised. These are as follows: waste (awaiting tender bids 30 days; internal due

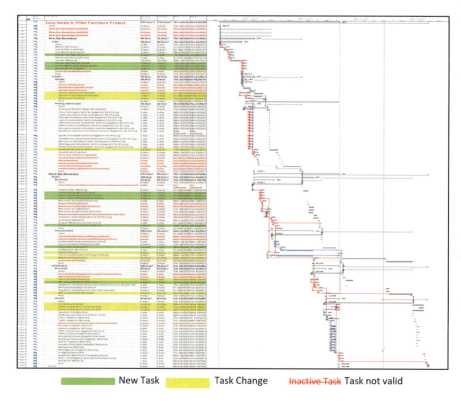

Fig. 27.6 Case study 3 solution analysis visual Gantt

diligence 30 days; tender evaluation 5 days), bottlenecks (tender preparation 29 days), variation (tender variation unknown) and risk (missing milestone; tender strategy; system implementation unknown).

27.4.5 Possible Solution

Using the Gantt to visualise the baseline of this project has helped identify ways to reduce time for procurement and implementation and increase quality of specialist solution. This model does this by changing the buying strategy of the host organisation to one that removes the tender process risk by applying a R&D testing approach to find the best end user value solution via negotiation. In addition to the above, the solution is implemented faster than previously planned, with less constraint influencing the process. This results in a method that is focused on the finding of value through analysis of the marketplace, delivering the best system for the client and thereby resulting in product and paradigm innovation (Fig. 27.8).

27 The Tendency Towards Suboptimal Operational Planning 361

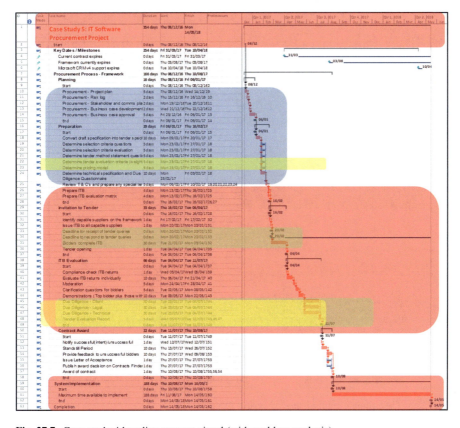

Fig. 27.7 Case study 4 baseline process visual (with problem analysis)

27.5 Summary

From the case studies analysed, the following findings can be summarised from the visualisation of mapping the current process using the Gantt chart, and several limitations have been highlighted, similar to those found in research objective 2. These are summarised in the Table 27.1 below:

From the modelling and identification of constraints, it was possible to use the visual model as a creative method of problem-solving to test the feasibility of possible solutions. Latent value found in the cases was achievable through the action summarised in the Table 27.2 below:

In the analysis and testing of the findings seen in the tables above, it is possible to suggest the potential achievable gains related to the time seen in Table 27.3. It is also possible to see the significant latent value for both the project organisations and their clients.

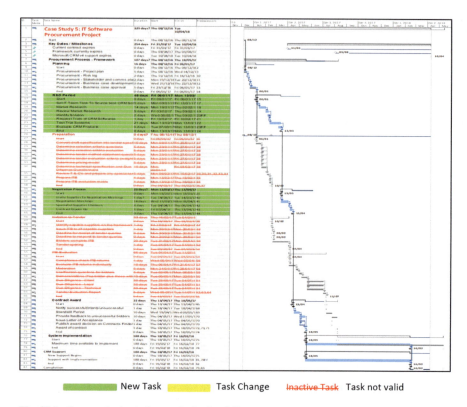

Fig. 27.8 Case study 4 solution analysis visual Gantt

Table 27.1 Summary of constraint findings

Constraints	Case study 1	Case study 2	Case study 3	Case study 4
Waste	Customer response period Tender evaluation Design evaluation	Appointing team	Tender review Process of order Awaiting meetings	Awaiting tender bids Due diligence Tender evaluation
Bottlenecks	Design bottleneck Procurement bottleneck Manufacturing bottleneck	Planning process Client approval and operational mobilisation	Design bottleneck Manufacturing bottleneck	Tender preparation
Variation	Variation cause waste and risk Client variation	Design VM review Tender VM review	Client revision process	Tender variation Due diligence
Risk	Tender/business strategy	Implementation of build	Tender strategy Storage of goods Third-party conflict	Missing milestone Tender strategy Implementation
Current duration	114 days	168.8 weeks	243 days	354 days

27 The Tendency Towards Suboptimal Operational Planning 363

Table 27.2 Summary of feasible solutions

	Case study 1	Case study 2	Case study 3	Case study 4
Possible solution	Removal of tender process Free consultancy service and discount for direct business Standardised designs Product innovation Better client communication	Team appointed using relationship and negotiation Resources add to help planning bottleneck Residents relocated sooner Evaluating given back to experts Resources add to help implementation	Removal of tender process Discount applied No design bottlenecks Client variations less likely Adding in communication Contract and sign-off process	Buying strategy of the organisation to one that removes the tender process risk by applying a R&D testing approach to find the best end user value solution via negotiation.
New duration	97 days	141.6 weeks	140 days	329 days

Table 27.3 Potential gains in process implementation

	Case study 1	Case study 2	Case study 3	Case study 4
Current duration	114 days	168.8 weeks	243 days	354 days
New duration	97 days	141.6 weeks	140 days	329 days
Percentage gain	**15%**	**17%**	**43%**	**8%**

These potential gains suggest that the process of visual modelling for analysis helped identify the root cause of the problem while also highlighting feasible solutions in the projects examined.

27.6 Conclusion

This conclusion provides an explanation on how this research determined the influence of the use of visualisation tools, in this case the Gantt chart, to support the identification of potential latent value and inefficiencies in a project, including recommendations for project practitioners. This study challenges a contemporary issue, both within project literature and practise, regarding the reasons as to why projects do not deliver their planned outcomes (National Audit office, 2016) and the latent value (Goodpasture, 2001) that remains unrealised. There are indications that projects are open systems (Smyth & Morris, 2007), open to the marketplace and open to the organisations that host them. This study confirms that, to motivate innovation in a project plan process, there is a need for an external and compelling motivation (such as a competitive challenge or company loss) to spark any internal willingness to change. If the general unwillingness to innovate/improve is not fully addressed, it will act as a barrier to innovation (Baird, 1997; Lorenz et al., 2012; Hanselman, 2013). Hence, the concept can be drawn that innovation barriers are as prevalent in projects as they are in organisations.

This study revealed that client-led design avoids the advantages of more collaboration with supplier expertise, which may be a consequence of ill-considered procurement strategies. Thus, the application of standardised and cost reduction-focused procurement strategies pursued 'out of procedural habit', or procurement rule interpretation, is an important barrier to value-adding opportunities within projects. Consequently, it can also be hypothesised that client value and value interpretations are not typically analysed in most project supply arrangements. This failure to explore value interpretations leads to a loss of potential value in favour of simple interpretations of what the project is required to achieve. From this thesis, it can also be seen that some project management-related approaches, such as Lean and Six Sigma, with their focus on correcting inefficiency, similarly overlook value-adding opportunities for projects. This study has also indicated that, in practise, the unavailability of modelled baseline plans results in lost value-adding opportunities, simply because those opportunities cannot be readily visualised. This study has highlighted that project managers tend to attribute more advantages to not modelling the plan with any rigour, over the advantages of modelling the plan with rigour, which may be linked to cognitive bias or barriers to learning and innovation within the host organisation. This study established that the absence of even basic network planning in projects robs the project manager and team of a critical tool in dealing with complexity (even the simplest of plans can rapidly become complex and require a Gantt-type model to allow for effective and efficient coordination of interlinked tasks).

Therefore, it could be agreed that modelling and visualisation allow for more complex examination of problems and opportunities that arise from solving these problems. An inability to see what is happening in the complexity of the process means that some of the issues common to the case studies and identified in the literature are readily identified in the modelling process used in the study. Consequently, it can also be theorised that Gantt chart modelling allows for the root cause analysis of problems in project processes (this is further supported by the findings of Davies (2018)). As a result of this and following the process undertaken in this study, it can also be seen that greater plan-task-coordination results in faster projects at a reduced cost. From this conclusion, several hypotheses can be drawn:

Hypothesis 1: *That modelling for analysis of the Gantt chart in projects allows for the identification of the root cause of the problem and helps build a clear problem definition.*

Hypothesis 2: *Visualisation can aid rational decision-making as it allows complex investigation of the problem and provides data that would have been overlooked.*

Hypothesis 3: *That project managers are currently underestimating the results of the facilitation of visual analysis of their current baseline, resulting in lost value-adding opportunities, which may be the cause of the high rates of failure in project deliverables.*

Hypothesis 4: *Projects are vulnerable to innovation barriers, and this may be the reason that organisations are not exploiting the use modelling as an analysis tool to innovate in the project plan process.*

A number of recommendations can be made for improving the current approach to projects. Firstly, the visualisation of data could benefit practitioners by unlocking their creativity, allowing for better problem-solving and decision-making, which is an integral part of project management. Secondly, the visualisation of data could benefit practitioners by helping find the root cause of constraints within their operation. Thirdly, project managers should encourage the project team and organisation to exploit the use of visual modelling for analysis in order to innovate and add further value. Finally, the evidence in the thesis suggests that practitioners should not view the Gantt chart solely as a planning and control tool but utilise it as a method of testing feasible solutions.

References

Achanga, P., Shehab, E., Roy, R., & Nelder, G. (2006). Critical success factors for lean implementation within SMEs. *Journal of Manufacturing Technology Management, 12*, 460–471.

Ahmed, S., Manaf, N. H. A., & Islam, R. (2013). Effects of lean six sigma application in healthcare services: A literature review. *Reviews on Environmental Health, 28*(4), 189–194.

Al-Kassab, J., Ouertani, Z., Schiuma, G. & Neely, A. (2014) Information visualization to support management decisions. International Journal of Information Technology & Decision-Making [online], 13, (2), pp. 407-428. Retrieved August 5, 2019, from http://www.ebsco.host.com.

APM. (2005). *Body of knowledge* (5th ed.). Totton: Hobbs the Printers Ltd.

APM. (1993). Directing change: A guide to governance of project management. High Wycombe, UK: Association for Project Management.

Association for Project Management (2018), Value Management Special Interest Group (SIG). Retrieved July 5, 2019, from https://www.apm.org.uk/group/apm-value-management-specific-interest-group.

Association of Project Management. (2012). *The APM body of knowledge* (6th ed.). APM: Buckinghamshire.

Bauer, M. I., & Johnson-Laird, P. N. (1993). How diagrams can improve reasoning. *Psychological Science, 4*(6), 372–378.

Baird, I. C. (1997). Imprisoned bodies—free minds: Incarcerated women and libertatory learning. In The 38th Annual Proceedings of the Adult Education Research Conference (pp. 13–18). Stillwater: Oklahoma State University.

Bicheno, J. (2004). *The new lean toolbox: Towards fast, flexible flow*. England: Picsie Books.

Bryde, D. (2008). Perceptions of the impact of project sponsorship practices on project success. International Journal of Project Management, [online] 26(8), pp.800–809. Retrieved July 8, 2019, from https://www.sciencedirect.com/science/article/abs/pii/S0263786307001810.

Braun, V., & Clarke, V. (2006). "Using thematic analysis in psychology". Qualitative Research in Psychology. *3*(2), 77–101.

Charter, M. (2006). *The project management value question: In search of value at the expense of delivering it.* Paper presented at PMI® global congress 2006—North America, Seattle, WA. Newtown Square, PA: Project Management Institute.

Choo, A. (2014). Defining problems fast and slow: The U-shaped effect of problem definition time on project duration. Production and Operations Management 23(8), pp.1462–1479. Retrieved July 7, 2019, from https://onlinelibrary.wiley.com/doi/abs/10.1111/poms.12219

Crawford, J. K., (2006). "The project management maturity model". Information Systems Management, *23*(4), 50–58.

de Treville, S., & Antonakis, J. (2006). Could lean production job design be intrinsically motivating? Contextual, configurational and levels-of-analysis issues. *Journal of Operations Management, 24*(2), 99–123.

Dalcher, D., (2017). The case for further advances in Project Management. PM World Journal. Vol (VI) issue (VIII), august 2017.

Davies, A. K. (2018). *Can hard paradigm artefacts support soft paradigm imperatives? An unpaired comparative experiment to determine whether visualisation of data is an effective collaboration and communication tool in project problem solving.* Msc Strategic Project Management Dissertation: Leeds Beckett University, UK.

Dettmer, H. W. (1995). Quality and the theory of constraints. *Quality Progress, 28*(4), 77–81.

Dvir, D., Raz, T., & Shenhar, A. (2003). An empirical analysis of the relationship between project planning and project success. *International Journal of Project Management, 21*, 89–95. https://doi.org/10.1016/S0263-7863(02)00012-1.

Drucker, P.E. (1995), Managing for the future. Oxford: Butterworth-Heinemann.

Flyvbjerg, B. (2016). The fallacy of beneficial ignorance: A test of Hirschman's hiding hand. *World Development, 84*, 176–189.

Foster, T. (2010). *Managing quality – Integrating the supply chain* (4th ed.). Hoboken, NJ: Pearson Prentice Hall.

Goldratt, E. (1990). *The theory of constraints*. New York: North River Press.

Goldratt, E. M. (2003). *Production the TOC way with simulator*. New York: North River.

Goldratt, E. M.; Cox, J. (2002). Manual da Teoria das Restrições. Porto Alegre: Bookman. (2002) The goal, New York: North River Press.

Goodpasture, J. (2001). *Managing projects for value*. Vienna, VA: Management Concepts.

Goodwin, P., & Wright, G. (2004). *Decision and control*. Hoboken, NJ: Wiley.

Green, S. and Sergeeva, N. (2018). Value creation in projects: Towards a narrative perspective. International Journal of Project Management 37(5), pp.636–651. Retrieved July 19, 2019, from https://e-tarjome.com/storage/panel/fileuploads/2019-07-15/1563181562_E12206-e-tarjome.pdf.

Gupta, S., Dumas, M., McGuffin, M. J., and Kapler, T. (2016). Movement slicer: Better Gantt charts for visualizing behaviours and meetings in movement data. In *Pacific visualization symposium (PacificVis), 2016 IEEE* (pp. 168-175) IEEE.

Ghauri, P., Gronhaug, K. (2005). Research Methods in Business Studies: A Practical Guide 3rd Ed. Harlow, Pearson Education Ltd.

Gardiner, P. (2005) Project management: a strategic planning approach. Basingstoke: Palgrave Macmillan.

Hanselman, A. (2013). *Overcoming the barriers to innovation—How do you measure up?* [online] Retrieved July 9, 2019, from http://www.andyhanselman.com.

Hayles, C. S., Graham, M., & Fong, P. W. S. (2010). Value management as a framework for embedding sustainable decision making. *Proceedings of ICE – Municipal Engineer, 163*(1), 43–50.

Heathcote, J. (2017). *Projects need to add value: Not just spend money.* (SEEDS) 2016 Conference proceedings at the third international conference. LSI Publishing.

Heathcote, J., & Coates, A. (2018). *Illustrating how a systems approach to modelling project plans improved innovation in operations* (SEEDS international conference 2018). Leeds Sustainability Institute: Leeds Beckett University.

Heffernan, M. (2015). Willful blindness—How projects go wrong (but could get a lot better!). In *Keynote session presented at PMI® global congress 2015—EMEA, 11–13 may.* London: UK.

Hines, P., Holweg, M., & Rich, N. (2004). Learning to evolve - a review of contemporary lean thinking. *International Journal of Operations & Production Management, 24*(10), 994–1011.

Hicks, M. (1997) Problem solving in business and management: hard, soft and creative approaches. London: International Thomson Business Press.

Hopp, W., & Spearman, M (2004). To Pull or Not to Pull: What Is the Question? Manufacturing & Service Operations Management. *6.* 133–148. 10.1287/msom.1030.0028.

IVM. (2019). *What is Value Management—Value for Money* | The Institute of Value Management. [online] Retrieved July 31, 2019, from https://ivm.org.uk/what-is-value-management.

Ingason, H. T., and Shepherd, M. M. (2014). "Mapping the Future for Project Management as a Discipline–For more Focused Research Efforts". Procedia-Social and Behavioral Sciences, vol. 119, pp. 288–294.

Kahneman, D., & Tversky, A. (Eds.). (2000). *Choices, values, and frames.* New York: Cambridge University Press.

Kelly, J., Male, S., & Graham, D. (2004). *Value Management of Construction Projects.* Oxford: Blackwell.

Kerzner, H. (2011). *Project Management metrics, KPIs, and dashboards: A guide to measuring and monitoring project performance.* New York: International Institute for Learning.

Kerzner H (2006) Essentials of strategic PM. 9th Ed.

Koppenjan, J. F. M., & Klijn, E. H. (2004). *Managing uncertainties in networks: A network approach to problem solving and decision making.* London: Routledge.

Kumar, U. D., Nowicki, D., Ramírez-Márquez, J. E., & Verma, D. (2008). On the optimal selection of process alternatives in a six sigma implementation. International Journal of Production Economics, 111(2), 456–467.

Larkin, J., & Simon, H. (1987). Why a diagram is (sometimes) worth ten thousand words. *Cognitive Science, 11*(1), 65–100.

Laursen, M., & Svejvig, P. (2016). Taking stock of project value creation: A structured literature review with future directions for research and practice. *International Journal of Project Management, 34*(4), 736–747.

Lechler, T. (2010). The project value mindset of project managers. In *Paper presented at PMI® research conference: Defining the future of Project Management, Washington, DC.* Newtown Square, PA: Project Management Institute.

Lee, J. (1997). Design rationale systems: Understanding the issues. IEEE Expert, [online] 12(3), pp.78–85. Retrieved July 5, 2019, from https://ieeexplore.ieee.org/abstract/document/592267.

Levitin, D. (2015). The organized mind: Thinking straight in an age of information overload_ penguin.

Lorenz, R. et al. (2012). Barriers to service innovation—Perspectives from research and practice.

Madhani, P. M. (2016). Six sigma deployment in supply chain management: Enhancing competitiveness. *Materials Management Review, 12*(6), 31–34.

Marray. R (2006). Starting out in project management. 2nd ed.

Matinheikki, J., Artto, K., Peltokorpi, A., & Rajala, R. (2016). Managing inter-organizational networks for value creation in the front-end of projects. *International Journal of Project Management, 34*(7), 1226–1241.

Maylor, H., Brady, T., Cooke-Davies, T., Hodgson, D. (2006). "From projectification to programmification". *International Project Management Journal, 24*(8), pp. 663–674.

Maylor, H (2010) Project management, 4th ed., Essex: Pearson Education Limited.

Midler, C. (1995). "Projectication of the Firm: The Renault Case". *Journal of Management, 11*(4), pp.363–375.

Moore, J. (2017) Data visualization in support of executive decision-making. Interdisciplinary Journal of Information, Knowledge & Management. 12: 125-138. Retrieved July 19, 2019, from http://www.ebsco.host.com.

Morris, P. (2013) Reconstructing project management reprised: A knowledge perspective. Project Management Journal, 44 (5): 6-23. Retrieved January 14, 2018, from http://www.ebsco.host.com

Morris, P.W.G. (1997) "The Management of Projects" Thomas Telford. London.

Naor, M., Bernardes, E. S., & Coman, A. (2013). Theory of constraints: Is it a theory and a good one? *International Journal of Production Research, 51*(2), 542–554.

National Audit Office. (2016). "Delivering major projects in government: a briefing for the Committee of Public Accounts".

Paez, O., Dewees, J., Genaidy, A., Tuncel, S., Karwowski, W., & Zurada, J. (2004). The lean manufacturing enterprise: An emerging sociotechnological system integration. *Human Factors and Ergonomics in Manufacturing & Service Industries, 14*, 285–306. https://doi.org/10.1002/hfm.10067.

Pepper, M. P. J., & Spedding, T. A. (2010). The evolution on lean six sigma. *International Journal of Quality & Reliability Management, 27*(2), 138–155.

PMI. (2016). *The high cost of low performance how will you improve business results*. Retrieved from Newton Square, PA.

Pulakanam, V., & Voges, K. E. (2010). Adoption of six sigma: Review of empirical research. *International Review of Business Research Papers, 6*(5), 149–163.

Pyzdek, T. (1999). *The complete guide to six sigma*. Tucson, AZ: Quality Publishing.

Pellegrinelli, S., (2011). "What's in a name: project or programme?" . *International Journal of Project Management, 29*, pp.232–240.

Project Management Institute. (2004). A guide to the project management body of knowledge (PMBOK® guide) (3rd ed.). Newtown Square, PA: Project Management Institute, Inc.

Rahman, S. U. (1998). Theory of constraints: A review of the philosophy and its applications. *International Journal of Operations and Production Management, 18*(4), 336–355.

Reddivari, S., Rad, S., Bhowmik, T., Cain, N., & Niu, N. (2013). Visual requirements analytics: A framework and case study. *Requirements Engineering, 19*, 257–279. https://doi.org/10.1007/s00766-013-0194-3.

Ricketts, J. A. (2008). *Reaching the goal: How managers improve a services business using Goldratt's theory of constraints*. Boston, MA: IBM Press, Prentice Hall-Pearson.

Rovai, A. P., Baker, J. D., & Ponton, M. K. (2014). Social Science Research Design and Statistics. Chesapeake, VA. Watertree Press LLC.

Samset, K., & Volden, G. (2016). Front-end definition of projects: Ten paradoxes and some reflections regarding project management and project governance. *International Journal of Project Management, 34*, 297–313.

Shah, R., & Ward, P. T. (2003). Lean manufacturing: Context, practice bundles and performance. *Journal of Operations Management, 21*(2), 129–149.

Sibbet, D. (2013). *Visual leaders: New tools for visioning, management, and organizational change*. New York: Wiley.

Sindiy, O., Litomisky, K., Davidoff, S., & Dekens, F. (2013). Introduction to information visualization (InfoVis) techniques for model-based systems engineering. *Procedia Computer Science, 16*, 49–58. https://doi.org/10.1016/j.procs.2013.01.006.

Smyth, H. J., & Morris, P. W. G. (2007). An epistemological evaluation of research into projects and their management: Methodological issues. *International Journal of Project Management, 25*(4), 423–436.

Starbird, D., & Cavanagh, R. (2011). *Building engaged team performance: Align your processes and people to achieve game-changing business results*. New York: McGraw-Hill.

Sutherland, N. (2007). *Irrationality*. 2nd ed. Pinter & Martin.

Tidd, J., & Bessant, J. (2009). *Managing innovation*. West Sussex: John Wiley & Sons.

VM SiG. (2017). *APM VM SiG AGM: An overview VM*. [Unpublished]. Leeds Beckett University.

Winter, M., Smith, C., Morris, P., & Cicmil, S. (2006). Directions for future research in project management: The main findings of a UK government-funded research network. *International Journal of Project Management, 24*, 638–649.

Womack, J., & Jones, D. T. (1994). From lean production to the lean Enterprise. *Harvard Business Review, 72*(2), 93–104.

Wood, S, Heathcote, J (2009), Supply chain management: Providing a systems perspective for project managers, Unpublished.

Young, T. L (2006). Successful Project Management. 2nd ed. London UK: Trevor L Young.

Part VII
Smart Digital Innovation

Chapter 28
Challenges of Projects Supporting Smart Cities' Development

Zahran Al-Hinai, John Heathcote, and Hadi Kazemi

28.1 Introduction

Cities and towns are inhabited by more than 50% of the world's population, which means more than 3.5 billion people live in cities and towns. By 2030, this number is expected to reach about 5 billion (United Nations Population Fund, 2016). People move to urban places seeking better life these places provide to their inhabitants (Monzon, 2015), and some of which are acting as centres of innovation and creativity (Kitchin, 2013).

The rapid increase of cities' population, and their concentration in cities, can bring benefits if managed well, since cities are dynamic places, and they rely on the flow of people, resources, ideas and global connections. However, cities also generate a wide range of problems which might be difficult to tackle as they grow in size and complexity (Monzon, 2015). These problems and challenges might affect different sectors such as economic, social and environmental sectors. For instance, increasing pressure on city services like transport and healthcare could result in further effects on other sectors of smart cities. Nevertheless, cities should respond to people's needs through sustainable solutions for social and economic and environmental aspects (Albino et al., 2015). Accordingly, the concept of smart cities has emerged to tackle these challenges and provide better living style. The term 'smart city' is rather broad and ambiguous, with no agreed definition or consensus on how cities should approach the agenda. However, it is apparent that all definitions agree on the basic functionality of smart cities in terms of providing a creative atmosphere

---

Z. Al-Hinai · J. Heathcote · H. Kazemi (✉)
School of Built Environment, Engineering, and Computing, Leeds Beckett University, Leeds, UK
e-mail: h.kazemi@leedsbeckett.ac.uk

for better living and establish a more responsive and resilient environment to address any issues/problems that smart cities face.

Despite the fact that smart cities strive to deal with challenges that face cities in general, the development of smart cities itself encounters challenges and obstacles. This raised deliberation on how project management could help addressing these challenges in such kind of environment. Although project management faces contemporary issues in general, and the development of such cities might add new issues to them, these contemporary issues could be seen as opportunities which might lead to more effectiveness in managing projects. Moreover, investigating how projects should be approached in this type of environment might give a chance to enhance and improve the way that projects are being perceived in such cities. This arguably might improve project management performance in general, since the statistics on project performance shows that there is a significant percentage of projects which are underperforming. For instance, the National Audit Office (2016) states that 34% of public sector projects in 2015 were not delivered successfully, although the 'project success criteria' need to be clearly described. Nonetheless, this could be considered as an indicator that projects encounter challenges, complexities and issues that restrain them from achieving the intended value.

On the other hand, despite the divergences of perspectives on whether projects deliver product or value – although both concepts exist together, a clearer differentiation might be required in specific projects to enhance the governance of projects – there has been an emerging trend for the value perspective (outcomes) instead of product perspective (outputs) to be placed at the centre of a project (APM, 2012; Heathcote, 2016). The outcomes of this paper lean towards the value perspective in this equation. Accordingly, the aim of this paper is to investigate the ways in which the current project management approaches could address the challenges of smart city projects. The paper examines the suitability of existing project management methods and how the current practice of developing projects could address the challenges of smart cities' projects. Eventually, this paper aims to propose a model for smart cities' project development within the capabilities of existing project management techniques.

28.2 Literature Review

The term 'smart city' is relatively recent, although the concept has been discussed since the 1990s when 'it was used in order to signify how urban development was turning towards technology, innovation and globalization' (Gibson cited in Siuryte & Davidaviciene, 2016, p. 255). Since then, numerous definitions have been given due to the 'fuzzy concept' of the term, the use of which is not always consistent (Albino et al., 2015) and the result is confusion and uncertainty in defining the boundaries and scope of the concept (Centre for Cities, 2014). It has been argued that the definitions of smart cities have different dimensions and concentrations, where some concentrate on broader aspects, while others focus on specific subjects centric on citizens, technology and data. Accordingly, Nam and Pardo (2011) categorise smart cities into three different dimensions of technology, citizens and community. Comparatively, Siuryte and Davidaviciene (2016) also believe that smart cities should combine all the mentioned aspects together: technology, particularly 'Information and Communication Technologies (ICTs)', 'smart citizens' – citizens who are able to utilise technologies and participate in public life – and interconnection and integration of different elements of such cities, for instance, transportation, healthcare, administration, etc. which help getting a dynamic and active smart city. Nonetheless, despite the ideal perspectives that the concept of smart city can bring as elaborated previously, such cities face complex challenges in different aspects. Within this, there exists a lack of clarity in terms of content, targeted objectives and policy tools driving the economic success (Campbell, 2014) due to the 'fuzzy concept' of the term and inconsistencies in using the concept (Albino et al., 2015). According to Monzon (2015), the main objectives of smart city projects should be on solving their problems efficiently and improve the sustainability and quality of life for residents. However, solving city challenges is extremely difficult as the system of a city is inherently interrelated as 'system of systems', where a change in one city system might consequently affect another system. This type of chain effects on the system would require more rigorous and systematic approaches for problem-solving.

28.2.1 Technology/Infrastructure

There is a wide recognition amongst different authors regarding the importance of technologies as a key component of the smart cities. Albino et al. (2015) suggest that 'smart' should essentially contain 'intelligence', as smartness is realised only when an intelligent system adapts itself to the users' needs. Similarly, Nam and Pardo (2011) indicate that technologies should be utilised to mainly meet the needs of users, and also be user-friendly through affordable and available devices that could be used anytime and anywhere (Albino et al., 2015; Nam & Pardo, 2011). Availability of sufficient funds is always an issue for such initiatives, but the long-term monetary advantage has created enough justification for these investments for investors. Nevertheless, business models for rolling out smart technologies are still underdeveloped (Centre for Cities, 2014) which in turn causes uncertain financial return and restricts the availability of finance from private sector (Hirst et al., 2012). In addition, despite the potential impact that technology can provide in resolving issues facing smart cities, simply embedding expensive technologies is clearly a misunderstanding of the concept (Monzon, 2015) which may lead to consequent side effects. This presents itself in the form of marketisation by giant tech companies, which in turn raises privacy and protection of sensitive data concerns from the end user's perspective. It is often recognised that the degree of accessing and sharing inhabitants' personal information in smart cities is a rather important issue (Mason, 2015).

28.2.2 Human Dimension

Khatoun and Zeadally (2016) state that smart cities should provide encouraging environment that inspires and motivates their inhabitants to share knowledge amongst themselves through empowered technologies. Consequently, an emerging creative class shall be encouraged through developing an appropriate environment where social and intellectual capital empowers establishing relationships and connecting people (Albino et al., 2015). This, as Haque (2012) points out, then puts the 'smartness of citizens' as the focal point of any model for smart cities' development which aims to enhance the human capital welfare. Notwithstanding, technologies have no core value without people, even if that technologies provide high level of quality and services (Siuryte & Davidaviciene, 2016). In addition, increasing citizen participation is also difficult, where cities and private sector face difficulties to increase citizens' participation in the smart agenda beyond the committed few. As a result, providing platforms to engage citizens is not enough, and these should be capable to effectively participate to raise issues of corruption besides participating in solving such issues (Mason, 2015).

28.2.3 Institutional Dimension

Governance, regulations and policy are the institutional factors vital for effective connections between people and ICTs in smart cities (Siuryte & Davidaviciene, 2016). The active cooperation in a harmonious way amongst people, technologies and government and policy are essential for designing and running a smart city (Nam & Pardo, 2011).

Smart cities are state run and therefore effective connections between people and ICTs in smart cities are based on institutional factors that depend on governance, regulations and policy (Siuryte & Davidaviciene, 2016), and any paucity in these factors might cause issues. Introducing new policies has always been a challenge for governments. Therefore, in the case of smart cities, there is a risk of wrong policies being introduced as disruptive and technologies are normally embedded which are by nature unpredictable (Sissons & Thompson, 2012). Furthermore, creating systems that pursue more effective modes of governance in such cities causes tension and threats amongst the citizens in terms of stifling of rights to privacy, freedom of expression and confidentiality. Therefore, breaking down silos and joining different sectors by better coordination integrating different interests and stakeholders for better collaboration are required (Centre for Cities, 2014) in addition to framing projects in a multi-stakeholder and municipally based partnerships (Monzon, 2015).

28.3 Research Review

Smart city initiatives appear to raise some challenges for the management of projects that will bring about the 'smart city', and these can be added to the contemporary issues facing project management in general. As a response to these issues and complexity in current methodologies, there is a suggestion to move forwards towards different methodologies that give more concentration to value, such as 'project strategy' and 'agile' methodology. In essence, concepts such as 'programme management' have the potential for tackling some of the challenges as seeking better coordination amongst projects across a programme might lead to targeting common strategic goals and sharing resources amongst them, and therefore, gaining greater benefits (De Hertogh et al., 2006). In addition, integrating projects and programmes would lead to gain benefits as deliverables of projects empower the deliverables of programme and create change capability that allows realising the long-term benefits. This would mainly satisfy the needs of end users' by generating tangible benefits. Lycett et al. (2004, p. 290) illustrate the fundamental goals of programme management by categorising them twofold: 'efficiency and effectiveness goals' and 'business focus goals'. Principally, such management aspects, if taken integrally, would lead to improved efficiency and effectiveness of management and overall governance structure of programmes and inherent projects.

28.4 Research Method

This paper is aimed to test the applicability and validity of project management approaches to address new kind of modern projects apart from the typical ones that have traditionally been raised in projects' environment. The main research question, therefore, is to identify the main challenges that smart cities, as a new kind of developments, would impose on their deriving projects. Hence, research question and research design, while compatible against each other, need to address this issue parallelly. However, research design might differ based on the nature of research question and how it might reach to answer the question (Saunders et al., 2015). As a result, this research adopted a qualitative/inductive method by utilising a 'scoping review' research design, aiming at generating new hypotheses. Scoping review, according to Dijkers (2015), aims to map the key concepts underpinning a research area and the main sources and types of evidence available. This process is normally useful when a particular field/topic is complex or has not been reviewed comprehensively before. Accordingly, this research has considered utilising a scoping review, due to the lack of scientifically based evidence in the fields of project management and smart cities, because one area is rather new and the other area does not have a history of looking at qualitative approaches. In this sense, this research has conducted 'Evidence Mapping' and extended to doing a 'Scoping Review' as illustrated previously, in combination with a partial 'Systematic Review'. But these cannot be considered 'high-quality evidence' since it does not exist for project management or for smart cities. Furthermore, as a way of summarising and structuring this research methods as per the research 'onion' process presented by Saunders et al. (2015), this research follows an interpretivism philosophy using inductive approach and in particular grounded theory strategy via secondary data sampling. The idea is to map the existing project management methodologies against challenges of innovative and modern initiatives of smart cities and test their capabilities in tackling those challenges.

28.5 Discussion

Examining how smart cities' challenges can be assisted by project management might need to consider the modular level inside project management concept which includes tools – soft skills paradigm, value management, negotiation, benefit realisation and stakeholder engagement – which could be implemented to address these challenges (Fig. 28.1). Table 28.1 also illustrates more details on gathering the smart city challenges, proposed solutions from literature, project management methods and possible modular project solutions.

Despite the abundance and divergences in defining 'smart cities', most definitions have common meeting points supported by what have been explored throughout this research. To this end, it is argued that solving problems and challenges faced by cities – sustainability (social, economic and environmental) which aims to provide better living – would be achieved by implementing smart technologies, cooperation and engagement of people and integration with policies and regulations (government). As such, in the context of smart cities, these three dimensions of technology, citizens and government seem to be essential. Technologies might need to be adapted in a way that provides services which meet the needs of users, are designed to be user-friendly through affordable and available devices and could be used regardless of time and location (Albino et al., 2015; Nam & Pardo, 2011). This would not be achieved without collaboration and engagement of inhabitants. Thus, their creativity, education and diversity seem to be required (Nam & Pardo, 2011).

In addition, the role of the government through governance, regulations and policies would enhance getting effective connections between people and ICTs in smart cities. Providing platforms that engage people to interact with public life in the city and participate in decision-making which would have direct impact on their lives would result in prosperity, transparency and democracy. This will, in the context of city and wider society, reduce the role of the government because of decisions being made by people.

Consequently, it is evident that chosen project methodologies can address a number of challenges faced by smart cities, if not all, especially within the three broad dimensions of technology, citizens and government. In fact, the concepts of programme management and project strategy seem to be the most effective and suitable approach that might be used in this kind of environment.

28 Challenges of Projects Supporting Smart Cities' Development

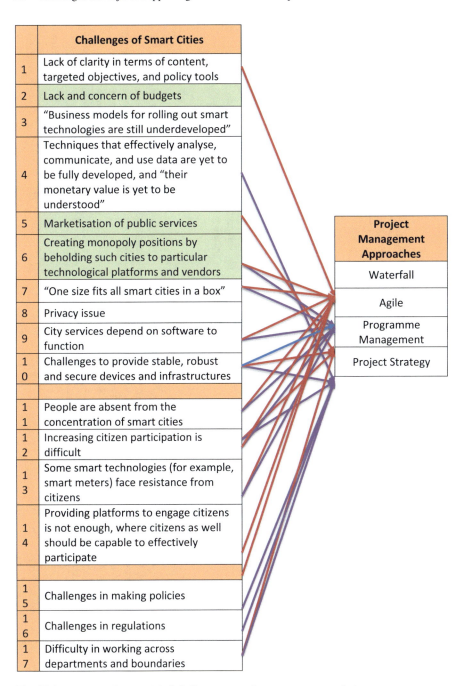

Fig. 28.1 *Mapping of smart cities' challenges to project management techniques*

Table 28.1 How smart cities' challenges can be assisted by project management

	Smart city challenges	Description	Literature proposed solutions	Project management methods	Possible modular project solutions
1	Lack of clarity in terms of content, targeted objectives; and policy tools	Lack of clarity in terms of content, targeted objectives; and policy tools driving the economic success	The scope, focus, objectives and expectations should be clear since the beginning and shared by all partners involved. It should be continuously re-evaluated and modified if necessary (van Winden et al., 2016)	Programme management	– Stakeholder engagement – Value management – Interest-based negotiation – Innovation – Creativity and lateral thinking
				Identifying a set of benefits which then can be adopted to policies:	
2	"Business models for rolling out smart technologies are still underdeveloped"	Lack of business models cause uncertain financial return. Besides long payback periods restrict the availability of financing from private sector	"Documents a clear and measurable strategic purpose for the project, linked to overall strategic and programme-level objectives for the city"	Project strategy	– Value management – Stakeholder engagement – Innovation – Creativity and lateral thinking
				Through value case could demonstrate the business model	
4	Privacy issue	Too much access to inhabitants' personal information and too much control to their digital activities, where "digital traces" are left behind them while using the Internet even in social media, which allows the "tech giants" gain the benefits	"Plan for privacy, security and resilience from the start: it is vital to address these issues from the initial discovery phases of the project" (British Standards Institution, 2014, p. 22)	Programme management and project strategy	– Value management – Benefit realisation – Stakeholder engagement – User involvement – Innovation – Creativity and lateral thinking

(continued)

	Smart city challenges	Description	Literature proposed solutions	Project management methods	Possible modular project solutions
5	Challenges to provide stable, robust and secure devices and infrastructure	Such cities are exposed to viruses, crashes, glitches and security hacks that can cause havoc	"Establishment of National Cyber Security and Smart City defence operation centre as a single point of defence to monitor and attend any incidents attacking city services", besides providing advises for public and private sector (Al-Shidhani, 2016)	Programme management and project strategy	– Value management – Benefit realisation – Stakeholder engagement – Innovation – Creativity and lateral thinking
6	Increasing citizens' participation is difficult, where cities and private sector face difficulties to increase citizens' participation in the smart agenda beyond the committed few	Perhaps this might be due to many reasons, for example, people not having the skills to use the Internet or not having access to it, especially amongst old people and low-income communities	Required user involvement in a multi-layered and ongoing process: • "The degree to which user involvement is necessary depends on the type and goal of the project… Different user types have to be approached in different ways" (van Winden et al., 2016, p. 109) • "Creating awareness requires a different approach from changing behavior or testing user acceptance for a new solution prior to upscaling" (van Winden et al., 2016, p. 109)	Programme management and project strategy	– Value management – Benefit realisation – Interest-based negotiation – Stakeholder engagement – User involvement – Innovation – Creativity and lateral thinking
7	Providing platforms to engage citizens is not enough; the citizens as well should be capable of to effectively participating	For instance, raising problems of corruption besides participating in treating it. Therefore the 'smartness' of the citizens' is essential to get an effective smart city	• Education, awareness and training (Al-Shidhani, 2016)		

(continued)

Table 28.1 (continued)

	Smart city challenges	Description	Literature proposed solutions	Project management methods	Possible modular project solutions
8	Challenges in making policies	Since such cities embed technologies, some of which are disruptive and unpredictable; the governments might introduce the wrong policies (Sissons and Thompson, 2012). These policies along with a range of practices and instruments need to be complemented with data and technologies:	'Proactive policies and regulations' before the problems take place might be one of the solutions (Al-Shidhani, 2016):	Programme management and project strategy	– Value management – Benefit realisation – Interest-based negotiation – Stakeholder engagement – User involvement – Innovation – Creativity and lateral thinking
9	Challenges in regulations	Creating systems that pursue more effective modes of governance in such cities causes tension and threats amongst the citizens in terms of stifling of rights to privacy, freedom of expression; and confidentiality:	The balance of benefits of data analytics with individual and societal rights is required to maintain trust in government and democracy, particularly when so much of data are processed by corporate systems (Kitchin, 2013). Therefore, making sure that common standards and regulations are in place is essential (Sissons and Thompson, 2012):	Programme management and project strategy Regulation should follow the market Reducing the role of government	– Value management – Benefit realisation – Interest-based negotiation – Stakeholder engagement – User involvement – Innovation – Creativity and lateral thinking
10	Difficulty in working across departments and boundaries	Data is rarely shared amongst them and coordination is seldom	"Smart cities need to develop new operating models that drive innovation and collaboration across these vertical silos'" (British Standards Institution, 2014, p. 3)	Programme management and project strategy	– Value management – Benefit realisation – Stakeholder engagement – User involvement – Innovation – Lateral thinking

In addition, projects should concern themselves about outcomes, benefits and value, where it should be value-centric rather than product-centric (Winter & Szczepanek, 2008). Considering that value is subjective, and has different perspectives and interpretations, much attention should be given to how the value is identified, who identifies it and how to evaluate it. Therefore, the project strategy approach might have the capacity to overcome this potential disconnect between programme and projects, since it has 'strategic' rather than 'methodological' approach to managing projects. Correspondingly, project is value-centric which aims to deliver valued outcomes. Furthermore, this approach provides a particular way of looking at the project as a strategic problem that needs to be solved creatively to bring positive change.

However, these methods/approaches on their own might not be enough and constitute accurate work and are not holistic. The environment of smart cities seems to be stimulating and attractive for this type of tools, although it might not be empowered in the same extent in all cities. However, the two approaches to smart city initiatives (top-down and bottom-up) might demonstrate the capacity of some of these tools. As a result, the smart city challenges might lead to a change in the way of developing and perceiving projects. Apparently, in the challenging environment in smart cities, the drivers are going to be more bottom-up than top-down, so top-down polices might try influencing what people do, but it is more likely to be a very innovative and entrepreneurial environment. Therefore, projects need to focus on these primary concerns.

28.6 Conclusion

As discussed, capabilities of existing project management approaches would address some of the major challenges imposed by smart cities. However, as these approaches cannot be holistic and all-embracing, a simple and flexible 'model' for smart cities' project development can be proposed. The model is concerned with having a fundamental approach that involves 'benefit realisation', engagement (stakeholder and people), innovation (through innovative tools, for instance, hackathon) and soft skills in addition to value/benefit equation that drive the development approach through 'methodologies' and strategy (Fig. 28.2).

This model could employ the two approaches ('top-down' and 'bottom-up'). The 'top-down' is embodied when the city has a strategy, and this strategy will be implemented through programme and project strategy, involving the modular level and skills in all stages. On the other hand, 'bottom-up' approach starts with people at the modular level and raises the creativeness and involvement of citizens to participate in solving issues and generate creative ideas, allowing projects to be merged in a programme, which then shapes a strategy. All projects might be considered to be strategic in nature, and each single project should contribute in delivering a strategy. Nonetheless, both approaches could be used harmoniously and could lead to better results.

Fig. 28.2 Smart cities' project development model

Therefore, in this kind of environment, it might be realised that the way projects are being approached might need to move forwards to more strategic thinking, and more value-centric through less processes, focusing on social interaction amongst people while enhancing soft skills and deeper entrepreneurial perceptions and adapting new technologies.

Arguably, this is what project management in general might need to concern itself about in any case.

References

Albino, V., Berardi, U. and Dangelico, R. (2015). Smart cities: Definitions, dimensions, performance, and initiatives. *Journal of Urban Technology, 22*(1), 3–21.

Al-Shidhani, A. (2016). *Leading cyber-threats for smarter urban communities. [online]*. YouTube. Retrieved April 18, 2018, from https://www.youtube.com/watch?v=Xkp9Bv7OiAc

APM. (2012). *APM body of knowledge* (6th ed.). APM: Buckinghamshire.

British Standards Institution (2014). *PAS 181 – Smart cities framework*. BSI Group. [Online]. Retrieved from www.bsigroup.com

Campbell, M. (2014). *Smart cities? Not for me! [online]*. PRAXIS. Retrieved January 6, 2018, from http://professormikecampbell.wordpress.com

Centre for Cities. (2014). *Smart Cities - Centre for Cities. [Online]*. Retrieved January 6, 2018, from http://www.centreforcities.org/publication/smart-cities/

De Hertogh, S., Vandenbroecke, E., Vereecke, A. and Viaene, S. (2006). A multi-level approach to Programme objectives: Definition and managerial implication. In *Vlerick Leuven Gent Management School Working Paper Series 2006–11*. Vlerick Leuven Gent Management: School.

Dijkers, M. (2015). What is a scoping review? *Centre on Knowledge Translation for Disability and Rehabilitation Research, 4*(1), 1–5.

Haque, U. (2012). *Surely there's a smarter approach to Smart Cities? [Online]*. Retrieved from http://www.wired.co.uk/article/potential-of-smarter-cities-beyond-ibm-and-cisco

Heathcote, J. (2016). *Projects should add value: Not just spend money! [online]*. Retrieved December 15, 2017, from https://www.apm.org.uk/blog/projects-should-add-value-not-just-cost-money/

Hirst, P., Hummerstone E., Webb, S., Karlsson, A., Blin, A., Duff, M., Jordanou, M. and Deakin, M. (2012). JESSICA for smart and sustainable cities, Horizontal study. Report published by the European Investment Bank.

Khatoun, R. and Zeadally, S. (2016). Smart cities: Concepts, architectures, research opportunities. *Communications of the ACM, 59*(8), 46–57.

Kitchin, R. (2013). The real-time city? Big data & smart urbanism. *GeoJournal, 79*(1), 1–14.

Lycett, M., Rassau, A. and Danson, J. (2004). Programme management: A critical review. *International Journal of Project Management, 22*(4), 289–299.

Mason, P. (2015). *We can't allow the tech giants to rule smart cities [online]*. UK edition. Retrieved January 6, 2018, from https://www.theguardian.com/uk/commentisfree

Monzon, A. (2015). *Smart Cities Concept and Challenges: Bases for the Assessment of Smart City Projects*. Published in the 2015 International conference on smart cities and green ICT systems (SMARTGREENS), 20–22 May, Lisbon, Portugal.

Nam, T. and Pardo, T. (2011). Proceedings of the 12th annual international digital government research conference: *Digital government innovation in challenging times*. June 12–15, Maryland. New York, Association for Computing Machinery.

National Audit Office (2016). Delivering major projects in government: A briefing for the Committee of Public Accounts. National Audit Office.

Saunders, M., Lewis, P. and Thornhill, A. (2015). *Research methods for business students* (7th ed.). England: Pearson Education.

Sissons, A. and Thompson, S. (2012). *Three dimensional policy. Why Britain needs a policy framework for 3D printing*.The Work Foundation, Big Innovation Centre, Lancaster University.

Siuryte, A. and Davidaviciene, V. (2016). An analysis of key factors in developing a smart city. *Mokslas: Lietuvos Ateitis, 8*(2), 254–262.

United Nations Population Fund. (2016). *Urbanization* [Online]. Retrieved April 18, 2018, from https://www.unfpa.org/urbanization

van Winden, W., Oskam, I., van den Buuse, D., Schrama, W. and van Dijck, E. (2016). *Organising Smart City projects: Lessons from Amsterdam*. Amsterdam: Hogeschool van Amsterdam.

Winter, M. and Szczepanek, T. (2008). Projects and programmes as value creation processes. *International Journal of Project Management, 26*, 95–103.

Chapter 29
I Spy with My Little Eye: Improving User Involvement in Elderly Care Facility Design through Virtual Reality

Abdul-Majeed Mahamadu, Udonna Okeke, Abhinesh Prabhakaran, Colin A. Booth, and Paul Olomolaiye

29.1 Introduction

The healthcare industry according to Lim and Tang (2000) has been ascribed as one of the fastest-growing industries in the service sector. Deloitte (2019) report on global healthcare outlook ties the healthcare industry's teeming growth to the world's ageing and growing population, greater prevalence of chronic diseases and exponential advancement in innovative digital technologies.

The provision of healthcare facilities to meet global standards and evolving demands has been met with increased complexities and rigorous challenges at every stage of procuring the design, construction, operation and maintenance of these facilities. Contributing factors to the multifaceted concern range from stakeholder (physician, nurse practitioners, physicians' assistants, specialists, etc.) participation and satisfaction, energy consumption, fit for purpose versus ideal for use arguments and the effect of facility design on patient's status of physical and psychological recovery.

Ulrich (2000, 2001) emphasises the impact of environmental design in improving medical outcome, propagating the theory of supportive design, with growing research supporting this concept (Dalke et al., 2006; Laursen et al., 2014). According to Leung et al. (2019), interior building design features and layout have a significant impact on elderly patients' physical health, psychological wellbeing, social relationships and cognitive functioning. As a result of this, it is recommended that there is more user involvement in the design process to ascertain user's preferences as well as acceptance of design at an early stage.

A.-M. Mahamadu (✉) · U. Okeke · A. Prabhakaran · C. A. Booth · P. Olomolaiye
Faculty of Environment and Technology, University of the West of England (UWE), Bristol, UK
e-mail: Abdul.Mahamadu@uwe.ac.uk

29.2 User Engagement in Healthcare Facility Design

Sharma et al. (2017) define patient or user engagement in healthcare as the active partnership between patients, families, caregivers and other stakeholders, working together to inform quality improvement initiatives in healthcare delivery (Khodyakov et al., 2017). Stakeholder or user engagement is considered a critical criterion in the design of any facility, more so a healthcare facility, and while researchers deliver compelling ethical rationale that supports patient engagement, there also seems to be a common consensus on the limitations of how best to go about user engagement. Khodyakov et al. (2017) further note that a gap often exists between the intentions to involve stakeholders and the actual engagement of stakeholders.

Approaches such as community-based participatory research (CBPR) and evidence-based research (EBR) pull at improving patient outcomes by the active involvement of community members, organisational representatives and researchers in all aspects of the research process (Israel et al., 1998). Kim et al. (2018) cite initiatives in the United States, The Precision Medicine Initiatives, which seek comprehensive user engagement through the life cycle of health research.

Traditional methods of engagement involve focus groups, in-depth interviews, surveys, email communication, conference calls, patient home visits, patient advisory boards, deliberative sessions and consensus-building techniques, relying primarily on in-person or small-group interaction (Deverka et al., 2012; Domecq et al., 2014; Kim et al., 2018). For Carman et al. (2016), this mode of engagement is plagued by logistical barriers, huge expenses, potential selection biases, ethical concerns and legal impediments. Lavallee et al. (2014) in proffering a solution to above challenges highlights the impact of communication technologies (online surveys, email communication, conference calls or webinars) in facilitating more heterogeneous user engagement and participation, though this may be perceived as impersonal and shallow in the depth of engagement (Kim et al., 2018).

Thus, in order to facilitate more meaningful user engagement, there is the need to engage technologies and strategies that are far-reaching and still able to attain meaningful user participation. This is even more important in the building design scenario given the need for communicating technical details in the most user-friendly manner.

29.3 Virtual Reality and User Involvement in Healthcare Facility Design

Although there is user involvement in some cases, traditional approaches to user engagement often rely on 2D drawings and documentation including architectural renderings of viewpoints and paper models (Lin et al., 2018). However, this process is characterised by misunderstandings and information gaps and has often proven futile as end-users fail to fully understand the concept and content of these drawings. Primarily, this is because they lack engineering knowledge and experience to interpret these highly technical pieces of 2D data (Lin et al., 2018). Recently, the concept of Building Information Modelling (BIM) is viewed as a key enabler of collaborative communication as well as design visualisation. In healthcare facility design, BIM-enabled technologies rely on object-oriented modelling, thus can allow end-users to visualise spaces and provide effective feedback that can help designers better understand and interpret user requirements (Okada et al., 2017). This process involves 3D visualisation, but on 2D interfaces, hence the need for technologies that provide near-real-life or immersive capabilities (Kang & Kuncham, 2014).

Evidence from many researchers echo two common deficiencies in practices associated with healthcare facility design, including the difficulty experienced by medical practitioners and stakeholders in understanding and interpreting design concepts (communication) and the discrepancy between the rendered design images and the finished construction product (simulation) (Huang et al., 2017; Lin et al., 2018). Researchers have therefore developed other innovative processes to combat some of these deficiencies.

For example, Huang et al. (2017) presented BIM Visualisation and Interactive System (BIM-VIS), a real-time rendering system that integrates BIM with a game engine and VR technologies to achieve more realistic scenes in the design of healthcare facilities, with the option of a wireless gamepad for navigation. This tool is however semi-immersive VR and relies on several projection screens to display BIM model views. This system thus relies on seral pieces of hardware and significantly difficult to set up.

Lin et al. (2018) presented a Database-supported VR/BIM-based Communication and Simulation (DVBCS) similar to the BIM-VIS concept. This system integrates BIM with a game engine and VR technologies in a semi-immersed VR environment, with the option of a gamepad for navigation. The core distinction with this approach is the dismantling of the file-based BIM model into BIM elements, which are saved as database BIM elements of the BIM model, usable for advanced simulations and communication.

Kang and Kuncham (2014) presented a BIM Computer Aided Virtual Environment (BIM CAVE) system based on Autodesk Navisworks to display the BIM 4D construction sequence. The system initially included three screen walls and three computers to project the rendered images in real time. This was later modified to include nine screen walls to enhance virtual reality and user experience. Hilfert and Koning (2016) have suggested the need for more cost-effective approaches that require less hardware.

The above developments have significantly improved concerns of communication and simulation plaguing healthcare facility design (Hilfert & Koning, 2016; Lin et al., 2018), and while the issue around hardware cost may cause hesitation among stakeholders, the new generation of VR tools offer a more cost-effective solution. There is a need for tailored VR solutions. Furthermore, the above developments have focussed on hospital design rather than care facilities. It is worth noting that care facilities are used mostly by elderly users, and thus there is a need for exploring the most suitable VR systems in this scenario as well as identifying design requirements. In order to address this, the current study explores safe and fully immersive virtual reality (VR) communication systems for capturing user requirements and preferences during healthcare facility design. The feasibility of using this to engage elderly care facility users is explored. Furthermore, the functionality that can allow for more effective use of VR tools on the care facility user scenario is explored.

29.4 Proposed Approach

A two-stage approach was followed. The first was the development of the prototype VR system using a direct prototyping methodology. This is presented in Fig. 29.1. This was subsequently tested through demonstration and interviews with healthcare and design practitioners.

In the first phase, Building Information Modelling (BIM) software (Autodesk, 2019) was utilised as the primary modelling tool to develop libraries of elderly care design objects. A fully functional parametric model was developed using Autodesk Revit with a focus on interior design elements as well as furniture, fixtures and finishing. Filmbox (.fbx) file formatting was used to improve model fidelity and visuals (Simmons, 2014). The interactivity was achieved through a game engine,

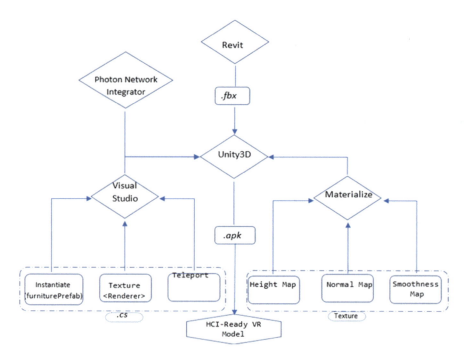

Fig. 29.1 Details of the interactive VR development workflow

Fig. 29.2 Screenshots of scenes in the interactive care facility design user engagement VR system

Unity3D (Unity3D, 2019), in conjunction with FBX library purposely used to ensure faster file exchange in the workflow. FBX format allowed lightweight parametric model where only relevant metadata for the proposed tool was extracted. Unity3D further supported virtual visualisation and interactivity through its intuitive tools and object-oriented programming (OOP). The interactivity built into the model related mainly to interior design manipulation for visualising different options of interior decor including furniture, building fixtures, finishes and positioning of objects. C-Sharp (C#), an object-oriented programming (OOP) language was used for the development of the custom tools and user interface for enabling human-computer interaction (HCI). The VR system developed also allowed users to outline preferences through an interactive form. The system was developed as a fully immersive VR simulation and was trialled on two headsets, Oculus Go and HTC Vive. Furthermore, third-party texture enhancement applications were used in conjunction with the gaming engine (i.e. Unity3D). The outputs from the visualisation are presented in Fig. 29.2. The user interfaces included in virtual tablet (Fig. 29.3) are used for object manipulation through hand gestures which are read by motion sensors connected to the VR headset.

29.5 Testing of Proposed System

The study relied on a quasi-experiment and qualitative interviews approach to test the application's usefulness, practicability and potential risks (Prabhakaran et al., 2018). The participants were engaged in a direct-prototyping format where their feedback was used in improving the system as the research went along (Eid, 2015). Each participant had the opportunity to test applications in a fully immersive and interactive VR environment using two different headsets. This was followed by interviews and transcription of interview responses. A sample of seven ($n = 7$) experts with experience in healthcare, healthcare building design and VR development were thus interviewed. These professionals were chosen because of their understanding of user requirements in the research domain. This was to allow for their input for improving the system before a more tailored and safe approach can

Fig. 29.3 Screenshots of virtual tablet user interface for object manipulation and interaction in the VR system

Table 29.1 Profile of participants

Interviewee	Experience in healthcare	Participant reference
Research fellow – Elderly care	4	A
Healthcare practitioner – Nurse	3	B
Healthcare practitioner – Nurse	3	C
Healthcare practitioner – Carer	10	D
Architect	10	E
Healthcare building design expert	20	F
Healthcare project manager	6	G

be designed for elderly users themselves. The profile of the interviewees is presented in Table 29.1. A thematic analysis of the interview responses was performed in order to systematically document participant's views in relation to the potential usefulness, safety considerations and elderly-specific user requirements. This aided the proposition of future considerations for research in this area.

29.6 Findings and Discussion

Based on the thematic analysis, it was concluded that the proposed system was very effective in communicating design intent as well as the elicitation of user requirements. The key findings are discussed in the themes below.

29.6.1 Elderly Friendly Functionality

The key themes identified were aesthetics, isolation and multiplayer functionality, navigation ease and multisensory experience. The following were identified as the most critical user engagement functions which can enhance the experience of elderly users: variations in a colour contrasting, wall surface and finishing fidelity, lighting, furniture selection and layout, scenery and location of windows and ability to test the nostalgic value of design features [Interviewee A, B, C, D, E, F, G]. This concurs with previous studies that have highlighted the importance of fidelity and visuals in VR immersion (Hilfert & Koning, 2016; Johnny et al., 2014; Leung et al., 2019; Lin et al., 2018). In addition to aesthetics, the VR experience was identified to be isolating for elderly [Interviewee A] (Hodge et al., 2018). This could be improved by multiplayer VR-gaming functionality that allows several people to experience the same scene at the same time. The multiplayer approach to exploring designs is less isolating especially if other participants can be viewed as an avatar. Furthermore, this multiplayer functionality could allow participation of more able-bodied persons or carers who can take control of navigation rather than elderly users who may have disabilities or dexterity issues. The significance of multiplayer functionality cannot be overemphasised at a time where social distancing is critical due to the impact caused by the COVID-19 pandemic. Teleportation functions were viewed as more appropriate than walking navigation. Teleportation was also viewed as a safer option for the elderly who could sit on a chair and be stationary while navigating to different scenes in the virtual environment [Interviewee A, B, C, D, E, F, G]. Furthermore, this approach is regarded as safer with walking navigation viewed to have more propensity of causing dizziness [Interviewee A, B, C, D, E, F, G]. In the opinion of participants, incorporation of multisensory experiences would improve the usefulness of VR in the elderly user scenario [Interviewee A, D, E, F, G]. The use of sound instructions was highlighted as advantageous. Intuitive functions and voice command were identified more suitable for elderly users, thus reducing the need for user orientation. Gesture recognition for interactivity was viewed as a limitation in the elderly care scenario. The outlined safety concerns are not unusual given the delicate nature of elderly care and requirements for the design of their living spaces (Hodge et al., 2018; Nichols & Patel, 2002).

Non-immersive technologies also appear more suitable in elderly care scenarios from the experiments in this study [Interviewee A, F]. Some of the main concerns for elderly users were mobility, tripping, likelihood of falls and potential dizziness when fully immersive VR is used. These have also been reported as VR risks in other application scenarios (Nichols & Patel, 2002). However, it is also established that in order to provide the right level of rendering, the system will require higher specification hardware to set up. High-fidelity mixed-reality (MR) system thus seems to be the best alternative although related technologies are not as advanced yet (Prabhakaran et al., 2018). Interviewees also wanted an approach that allows saving of user preferences such that different choices and configurations of design can be compared while the tool is being used [Interviewee A, B, C, D, E, F, G].

29.6.2 Desirable Features of the Virtual Building Design

The key themes identified were improving design options and incorporating standards. Interviewees thought that design rules regarding safety and other health-building standards should be incorporated in the system as additional information to guide choices [Interviewee A, F, G]. Such a system should be built into the interactivity to help users to dynamically view the impact of their choices and implications on meeting building, and design regulations as well as standards (Lin et al., 2018). This includes incorporating functionality for assessing navigation, social distancing features, infection and cross-contamination risks associated with alternative designs and fixtures. This cannot be overemphasised in view of the need to follow new standards and approaches in healthcare facility design in response to COVID-19 pandemic (Waite & Pitcher, 2020). Participants also suggested the incorporation of a larger object library consisting of home adaptation equipment with 'non-clinical looking' design options. The digital objects' libraries that relied on the VR simulation should be connected to a live database of available design components to improve a variety of choices in real time. Similar limitations have been highlighted in previous experiments and development of VR for participatory design (Hilfert & Koning, 2016; Lin et al., 2018).

29.7 Conclusion

This research explored how to enhance healthcare stakeholder's involvement in the design of healthcare facilities. A novel interactive virtual reality (VR) communication system for capturing user requirements and preferences during healthcare facility design was developed and tested. Based on pilot tests, it has been established that interactive immersive technology (VR) can be used to engage users in care facility design. The current generation of fully immersive VR tools, however, support care professionals more than the elderly end-users or patients themselves. The research found that safety issues hamper the deployment of fully immersive VR for the elderly users who may have navigation challenges as well as hindered spatial awareness which could lead to trips and falls. Teleportation functions built into the application were found to be very useful. From the findings of this study, it is therefore suggested that fully immersive experiences are supported by multiplayer (multi-user) functionality as well as teleportation functions for user navigation. Multiplayer functionality could address potential issues of loneliness in the virtual world as well as allow more physically abled participants to coordinate controls while an elderly person is immersed in the same experience. Furthermore, use of such functionality will also help in the maintenance of social distancing while enabling elderly user participation in healthcare facility design without exposing them to the risk of infection given their vulnerability. It was also suggested that semi-immersive methods are tested to ascertain their relative advantage over fully immersive solutions in the context of safety for elderly users. Although semi-immersive solutions like augmented and mixed reality could potentially be safer in terms of visibility of surroundings, they often provide less visual quality when compared to fully immersive VR. Future work will aim at developing a high-fidelity mixed-reality version of the tool and also incorporate conversational interaction including voice command systems to achieve more intuitive interactivity.

References

Autodesk. (2019). *What You can do with Revit*. Retrieved February 15, 2019, from https://www.autodesk.co.uk/products/revit/architecture.

Carman, K. L., Maurer, M., Mangrum, R., Yang, M., Ginsburg, M., Sofaer, S., Gold, M. R., Pathak-Sen, E., Gilmore, D., Richmond, J., & Siegel, J. (2016). Understanding an informed Public's view on the role of evidence in making health care decisions. *Health Affairs (Project Hope) [online], 35*(4), 566–574.

Dalke, H., Little, J., Niemann, E., Camgoz, N., Steadman, G., Hill, S., & Stott, L. (2006). Colour and lighting in hospital design. *Optics and Laser Technology [online]., 38*, 343–365.

Deloitte. (2019). *Global health care outlook: Shaping the future [online]*. London: Deloitte LLP. Retrieved from file://nstu-nas01.uwe.ac.uk/users3$/mt2-adebayo/Windows/Downloads/gx-lshc-hc-outlook-2019.pdf.

Deverka, P. A., Lavallee, D. C., Desai, P. J., Esmail, L. C., Ramsey, S. D., Veenstra, D. L., & Tunis, S. R. (2012). Stakeholder participation in comparative effectiveness research: Defining a framework for effective engagement. *Journal of Comparative Effectiveness Research [online]., 1*(2), 181–194.

Domecq, J. P., Prutsky, G., Elraiyah, T., Wang, Z., Nabhan, M., Shippee, N., Brito, J. P., Boehmer, K., Hasan, R., Firwana, B., Erwin, P., Eton, D., Sloan, J., Montori, V., Asi, N., Dabrh, A. M., & Murad, M. H. (2014). Patient engagement in research: A systematic review. *BMC Health Services Research [online]., 14*(89), 1–9. Accessed 20 June 2019.

Eid, M. (2015). *Requirement gathering methods*. Retrieved July 20, 2018, from https://www.umsl.edu/~sauterv/analysis/F2015/Requirement%20Gathering%20Methods.html.htm.

Hilfert, T., & Koning, M. (2016). Low-cost virtual reality environment for engineering and construction. *Visualisation in Engineering [online]., 4*(2), 1–18.

Hodge, J., Balaam, M., Hastings, S. and Morrissey K. (2018). *Exploring the design of tailored virtual reality experiences for people with Dementia*. Conference on Human Factors in Computing Systems (CHI) 2018, April 21–26, 2018, Montréal, QC, Canada.

Huang, C.Y., Yien, H.W., Chen, Y.P., Su, Y.C. and Lin, Y.C. (2017). Developing a BIM-based visualisation and interactive system for healthcare design. In: *Proceedings of the 34th international symposium on automation and robotics in construction* [online]. Taipei, Taiwan. International Symposium on Automation and Robotics in Construction (ISARC). doi:10.22260/ISARC2017/0051.

Israel, B. A., Schulz, A. J., Parker, E. A., & Becker, A. B. (1998). Review of community-based research: Assessing partnership approaches to improve public health. *Annual Reviews Public Health [online]., 19*, 173–202. Accessed 20 June 2019.

Johnny, W. K. W., Skitmore, M., Buys, L., & Wang, K. (2014). The effects of the indoor environment of residential care homes on dementia suffers in Hong Kong: A critical incident technique approach. *Building and Environment, 73*, 32–39.

Kang, J, Kuncham, K. (2014) BIM CAVE for 4D Immersive Virtual Reality. *Proceedings of the Creative Construction Conference*, Prague, Czech Republic, 2014. Creative Construction Conference 2014 [online]. Retrieved June 18, 2018, from http://2015.creative-construction-conference.com/wp-content/uploads/2015/01/CCC2014_J_Kang.pdf.

Khodyakov, D., Stockdale, S. E., Smith, N., Booth, M., Altman, L., & Rubenstein, L. V. (2017). Patient engagement in the process of planning and designing outpatient care improvements at the veteran administration health-care system: Findings from an online expert panel. *Health Expect [online]., 20*(1), 130–145. Accessed 20 June 2019.

Kim, K.K., Khodyakov, D., Marie, K., Taras, H., Meeker, D., Campos, H.O. and Ohno-Machado, L. (2018). A novel stakeholder engagement approach for patient-centered outcomes research. *Medical Care* [online]. 56(10 Supplementary 1), pp. 41–47. [Accessed 20 June 2019].

Laursen, J., Danielsen, A., & Rosenberg, J. (2014). Effects of environmental design on patient outcome: A systematic review. *Health Environments Research and Design Journal [online]., 7*(4), 108–119. Accessed 17 June 2019.

Lavallee, D. C., Wicks, P., Alfonso-Cristancho, R., & Mullins, C. D. (2014). Stakeholder engagement in patient-centered outcomes research: High-touch or high-tech? *Expert Review of Pharmacoeconomics & Outcomes Research [online].*, *14*(3), 335–344. Accessed 20 June 2019.

Leung, M., Wang, C., & Chan, I. Y. S. (2019). A qualitative and quantitative investigation of effects of indoor built environment for people with dementia in care and attention homes. *Building and Environment, 157, 15*(2019), 89–100.

Lim, P. C., & Tang, N. K. H. (2000). The development of a model for total quality healthcare. *Managing Service Quality: An International Journal [online].*, *10*(2), 103–111. Accessed 17 June 2019.

Lin, Y. C., Chen, Y. P., Yien, H. W., Huang, C. Y., & Su, Y. C. (2018). Integrated BIM, game engine and VR technologies for healthcare design: A case study in cancer hospital. *Advanced Engineering Informatics [online].*, *36*, 130–145. Accessed 17 June 2019.

Nichols, S., & Patel, H. (2002). Health and safety implications of virtual reality: A review of empirical evidence. *Applied Ergonomics, 33*(3), 251–271.

Okada, R. C., Simons, A. E., & Sattineni, A. (2017). Owner-requested changes in the design and construction of government healthcare facilities. *Procedia Engineering [online].*, *196*, 592–606. Accessed 17 June 2019.

Prabhakaran, A., Mahamadu, A.M., Mahdjoubi, L. and Manu, P. (2018). An approach for integrating mixed reality into BIM for early stage design coordination. In: *The 9th International Conference on Engineering Project and Production Management (EPPM)*, Cape Town, South Africa, 2018. UWE Research Repository [online]. Retrieved June 17, 2019, from http://eprints.uwe.ac.uk/37844.

Sharma, A. E., Knox, M., Mleczko, V. L., & Olayiwola, N. (2017). The impact of patient advisors on healthcare outcomes: A systematic review. *BMC Health Services Research [online].*, *17*(693), 1–14. Accessed 18 June 2019.

Simmons, T. (2014). *Which format is better-FBX Or OBJ?* Retrieved February 15, 2019, from http://aecobjects.com/2014/10/which_format_is_better/.

Ulrich, S. R. (2000). Effects of healthcare environmental design on medical outcomes. In: *Design & Health – The Therapeutic Benefits of Design*. Proceedings of 2nd International Congress on Design & Health, Karolinska Institute Stockholm, Sweden, June 2000. Academia [online]. Retrieved June 17, 2019, from https://www.academia.edu/696899/Effects_of_Healthcare_Environmental_Design_on_Medical_Outcomes.

Ulrich, S. R. (2001). Effects of interior design on wellness: Theory and recent scientific research. *Journal of Healthcare Interior Design [online].*, *3*, 97–109. Accessed 17 June 2019.

Unity3D, (2019). *Unity3d.* Retrieved February 16, 2019, from https://unity3d.com/.

Waite, R. and Pitcher, G. (2020) How will Covid-19 change the design of health facilities? *Architects Journal.* Retrieved June 15, 2019, from https://www.architectsjournal.co.uk/news/how-will-covid-19-change-the-design-of-health-facilities/10047080.article.

Chapter 30
Smart Enterprise Asset Management

Mark Lenton, Dave Lister, Jim Garside, Richard Pleace, Gary Shuckford, Simon Roberts, Tim Platts, Paul Redmond, Christopher Gorse, Bashar Alhnaity, and Ah-Lian Kor

30.1 Introduction

According to ISO (2014), the ISO 55000-2014 standard describes asset management as 'a coordinated activity of an organisation to realise value from assets'. Typically, assets are managed with information in silos due to the lack of system interoperability. This could result in user frustration, suboptimal decisions and even chaos without a holistic perspective of the entire asset and its complete lifecycle. In this paper, we adopt an ecosystem approach to address the inadequacies of the current siloed approach for asset management. We present our proposed smart

---

M. Lenton · P. Redmond (✉)
SRO Solutions Ltd, Eccles, UK
e-mail: Paul.Redmond@srosolutions.net

D. Lister
IAconnects Technology Ltd, Beckhoff Automation, Henley-on-Thames, UK

J. Garside
1st Horizon Surveying & Engineering Ltd, Barnsley, UK

R. Pleace
SiteDesk Ltd, Cheshire, UK

G. Shuckford
Edenbridge Healthcare Ltd, Wakefield, UK

S. Roberts
Mercateo UK Ltd, London, UK

T. Platts
TP Professional Services, Sherburn in Elmet, UK

C. Gorse · B. Alhnaity · A.-L. Kor
Leeds Beckett University, Leeds, UK

enterprise asset management (SEAM) approach, and its ecosystem of solutions as well as partner organisations. This paper describes the ecosystem in the context of a smart government building (within the built environment – note due to NDA, no specific reference can be made) to demonstrate the interoperability across a diverse range of innovative technologies and work in an asset management that encompasses the entire lifecycle. A smart government building has been chosen as a use case for this paper because of the UK Government's commitment to build new buildings and improve maintenance for older buildings. This provides timely opportunities to operationalise plans for digital buildings and estates that will improve lifecycle performance, environmental sustainability and digital transformation and traceability and reduce backlog maintenance. Building and estate digitisation, sustainability and efficiency will be central to the approved plans of all government building projects, though unfortunately many stakeholders have not fully considered/understood the benefits nor addressed how they will meet the required Treasury Green Book (e.g. evidenced-based 'fit for purpose') (HM Treasury, 2018), transparent and measurable outputs needed to identify what success looks like and therefore how initiatives can and should be continuously improved. However, our ecosystem of partners are working together on a more sophisticated digitisation, environmental sustainability, traceability and accountability across various programmes to improve inefficient building, estates and operational performance. In so doing, we need to address several challenges listed below:

- Challenge 1: Technical Challenge
- For the smart government building, the ecosystem of partners will utilise some of the latest interoperable and innovative technologies and seek to introduce improvements in day to day, real-time operational capability as well as providing key re-usable data for analysis to support longer term strategic 'Smart' Building and Operations objectives. As part of a planned 'think big – start small – scale fast' approach, this process will provide the platform which can be extended to support the proposed new builds and the existing estate.
- Challenge 2: Cultural Challenge
- Current approaches to estates, digital systems and data are diverse, siloed and poorly coordinated and lead to massive inconsistency, duplication and wasted effort and do not longitudinally maximise the value of information towards ensuring sustainable, efficient and resilient performance. Communication barriers (lack of openness and transparency) within and across internal teams, stakeholders, data types and technologies increase errors and costs and cause delays in data transmission and rework – this can occur during build phases and more importantly in operation. Processes are challenged with lack of openness to participation and inclusiveness to all stakeholders (individual groups such as procurement, finance, soft and hard FM teams). Systems are constantly challenged when attempting to maintain consistency, reliable evidenced-based information, integrated and holistic outcomes for 'clients' and continuous improvement across asset lifecycles. There is lack of transparency, often resulting from social and sometimes localised political interference and a lack of digital automation,

which leaves asset and operational management vulnerable to inertia and prevents stakeholders from being open to communicate and share information and best practice freely. Improvements through new ways of working based on digitised, integrated, interoperable and holistic thinking offered by the ecosystem of partners will allow all stakeholders the opportunity to participate and have transparency through a single source of the truth over the connected, sustainability, digital and efficiency programmes proposed.

- Challenge 3: Sustainability Challenge and Data Re-usability
- Underpinned by the 'smart' and interoperable approach the ecosystem partners offer, the aim is to develop a data collection and management structure that supports the delivery and operation of resilient, sustainable, new and existing built environment facilities whilst ensuring the principles of the Green Building Council publication (Net Zero Carbon Buildings (UKGBC, 2019)) and the suggestions set out by the digital advisors and supported by academic rigour. A sustainability strategy will be developed in line with a digital/data strategy that will provide operational information and foster knowledge for effective and efficient delivery and operation.

This paper is organised as follows: Section 30.1, Introduction; Section 30.2, Literature Review; Section 30.3, Methodology; Section 30.4, Smart Enterprise Asset Management Ecosystem; and Section 30.5, Conclusion.

30.2 Literature Review

The notion of smart asset management, which is rooted in asset management (AM), focuses on the technology as a driving process to enhance the captured information to assist in making strategic decisions (Urmetzer et al., 2015). Thus, asset-owning organisations rely on AM performance strategies to excel in the competitive business environment. The overarching aim of our proposed SEAM is to improve AM strategies in an organisation, using cutting-edge technologies. Moreover, SEAM has more to offer, such as a robust structure to validate and improve asset performance by collecting and incorporating reliable asset information in strategic business decision-making (Shah, et al., 2009). The needs of a clear definition of SEAM and a delineation of its implementations have dramatically increased as a result of the potential for enhanced asset analysis and performance possibilities in any business structure. Recent studies have shown the current use of SEAM in the industry is not documented, and with only few research and white papers have referred to some form of development and classification of smart assets. Several other studies have shown variations in implementing SEAM extensively across different organisations. Thus, in these interpretations of SEAM, a common ground needs to be identified. This calls for the need of a concise definition of SEAM to avoid conflicting terminologies, and establish a basis for future research. It is in these conflicting terminologies that various concepts have already been established in existing literature (Bughin et al., 2010; Lampe & Strassner, 2003; Liyanage & Langeland, 2009; Nel & Jooste, 2018). Although these concepts are acknowledged separately, they relate to the central idea of technological entities management. These separate ideas also need to be addressed to establish a foundation for further development of SEAM.

In recent decades, global businesses operate in an era of technological performance which sees overwhelmingly diverse variations in electronic goods and technology affordances. Market competition is a driving factor for organisations wanting to acquire and at the same time offer cutting-edge innovative expertise (Wang et al., 2014). According to literature, there is a business need for implementing SEAM (see business outcomes in Fig. 30.2). Therefore, it is imperative for academics and industrial researchers investigating SEAM and its implementation to first and foremost establish the implementation prerequisites. As a result of information availability through interconnected communication structures, researchers have established a clear link between technological/business performance management and electronic convergence (Nel & Jooste, 2018). Thus, such a link unravels the potential for businesses to offer comprehensive and meaningful information across multiple platforms, rendering knowledge (and business intelligence) more accessible to various stakeholders in the organisation. With such advancement of technical capabilities, organisations could fully exploit the creative use of a portfolio of technologies to create added value and competitive advantage. The general benefits of implementing asset management strategies are to reduce operational cost, improve profitability and improve competitive edge in the marketplace (Aberdeen Group, 2009; Aberdeen Standard Investments, 2018). However, it is imperative for the portfolio of technologies to be interoperable so that resources could be well coordinated and managed to reduce costs (e.g. manufacturing, etc.) (Nel & Jooste, 2018). The emphasis here will be the use of technologies to monitor and manage physical assets (through shared essential information) in optimal as well as optimised conditions, thereby delivering predetermined targets in Quality of Service (QoS – high performance), Quality of Effectiveness (QoE – excellent user satisfaction) and Quality in Sustainability (QiS – environmental friendliness).

To reiterate, it is evident that asset managers are at the helm of innovative technologies where strategic, tactical as well as operational decisions about assets need to be appropriately addressed to exploit their technological potential. The subsequent sections will discuss how these diverse technologies could be seamlessly integrated and the benefits of our proposed ecosystemic approach to SEAM.

30.3 Methodology

In this section, we shall discuss smart enterprise asset management and digital twin methodologies. The smart management macro-level methodology is depicted in Fig. 30.1a, b and discussed in Subsection 30.3.1, whilst the micro-level discussion of each component in smart enterprise asset management is in Subsection 30.3.2

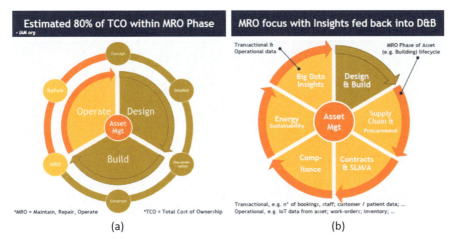

Fig. 30.1 Smart enterprise asset management lifecycle

Fig. 30.2 Digital twin and smart enterprise asset management

and Table 30.1. Detailed discussion of the digital twin methodology is also in Subsection 30.3.2.

30.3.1 Smart Enterprise Asset Management Lifecycle

ISO 55001:2014 (BSi, n.d.) is a framework for asset lifecycle management which aims to achieve the best return from assets and reduce the total cost of ownership (TCO). It encompasses financial, organisational and physical assets. Such a standard helps reduce asset ownership-associated risks, improve customer satisfaction through provision of quality-assured products and services, invoke new business

Table 30.1 Smart enterprise asset management ecosystem innovative modules

Ecosystem modules	Description	Data sources	Benefits	Beneficiaries
3D (and/or 2D) scan and BIM digital twins	Dimensionally accurate 3D replication of the built environment and M&E assets; sufficient to: navigate as if walking the building, architect changes, provide training, locate critical fixed / mobile assets and visualise other critical materials/ assets within, including near real-time traffic light status for monitored critical assets	2D drawings 3D scans 3D models and data schemers, e.g. Revit, IFC, etc. Standard or manufacturers' product 3D images and data sheets	(i). Improve collaboration and handover from design and build to operate lifecycle phase (ii). Data from existing BIM models can be pulled into asset management solution (iii). Using 3D models (digital twins) to interact with assets and locations in an intuitive manner (iv). Quick and easy retrieval of asset or location information by navigating 3D model and selecting item of interest (iv). Easily access functionality of asset management, CAFM/CMMS solution	Designers Architects Builders Contractors Developers

Ecosystem modules	Description	Data sources	Benefits	Beneficiaries
IoT connected assets	Sensing capability either 'as built' or enhanced 'retrofit' for key assets/spaces, to support data creation for measurement and reporting of desired outcomes: Staff experience, availability, performance, sustainability, compliance, utilisation, safety and efficiency	Sensor data gathered to meet client requirement BMS files type of sensor feeds data to information repository, e.g. footfall sensor, occupancy sensor, air quality, power usage, etc.	Benefits of an integrated IoT strategy are: (i). Easy 'plug and play' style deployment and commissioning of most IOT sensors (ii) ability to join up data sources from different assets, e.g. BMS HVAC + room occupancy and temperature (iii). Operations – Reactive, predictive and proactive maintenance for improving asset availability and uptime, remote monitoring/metering and trigger event on threshold breach (iv). Provide insights - optimise/maximise asset performance via trends measurement, SLAs improvement and TCO reduction, and data provision to ensure fit for purpose and evidence-based insights (v). Health and safety/compliance – Provide insights into workers' Well-being, working environment condition (noise, CO_2, air quality, temperature), heart rate and fatigue detection, proximity/danger area and PPE detection	Property consultants, administration, designers and architects looking, maximising and optimising location (space) or assets Facilities management for evidence-based insights for third-party SLA measure and control Engineers and operations ensuring easy deployment and visibility of asset performance, availability and ability to continuously improve

(continued)

Table 30.1 (continued)

Ecosystem modules	Description	Data sources	Benefits	Beneficiaries
Asset management and database management	Support key asset specification and performance information; sufficient to locate, understand condition or origin, monitor status, measure performance, report for service, service, assess spares requirement, record repair activity, replace and retire	COBie file Existing BMS, CAFM or EAM system Manufacturer data sheets Commissioning records Operating and maintenance procedures Service providers Service history record Product/asset data templates	(i). A scalable world-class CMMS/CAFM solution with unique integrated BIM, IoT, e-procurement and potential blockchain interfaces (ii). Central and easy access to information related to assets and locations: Name, reference, criticality, associated contracts and SLAs, spares, documentation, plans, etc. (iii). Schedule work orders based on asset and process criticality and appropriate and available resource (iv). Manage, track, measure and continuously improve operations based on work order data (v). Auditable adherence for H&S and compliance regulations and guidelines. Record for audit purposes. Avoid corporate exposure and liability. Measure and continuously improve supplier contract (vi). Service levels. Align service levels with asset and underpinning process criticality, costs to deliver and negotiations (vii). Create digital twin. Improve collaboration throughout whole asset lifecycle. Intuitive visualisation and navigation of asset and location information for faster decision and accuracy (viii). Ensure right product, right price and at right lead-time based on right criteria (criticality, compliance, sustainability, etc.) (ix). Simplify procurement process and reduce cost of procurement. Reduce number of suppliers. Reduce and ensure just-in-time inventory. Improve uptime (c.50% downtime due to parts not available) (x). Ability to integrate with 'shared ledger' capabilities across end-to-end process enabling a real-time auditable trail that can be shared between customer and third parties to ensure SLAs, compliance, pricing, contractual agreements, traceability, etc., are controlled, respected and improved	Site operations and engineering Finance and purchasing departments Facilities management Property consultants, administration, designers and architects
	Storing critical asset performance data, merging and analysing with other associated or external data to form evidence and insights to support complex decision-making, e.g. sustainability, space repurposing, capital projects, supplier negotiation, etc.	Revit data/scan data Construction data Commissioning data Maintenance data Sensor data Performance data 'Client' data Climate data Other external information		

Ecosystem modules	Description	Data sources	Benefits	Beneficiaries
Supply chain and e-procurement management	Capability to store and order product and service supplier 'best value' contract information, stores information and inventory monitor and gather performance against SLA to support QoS, renegotiation and sustainability goals	COBie file Finance/procurement ERP system Procurement contract/supplier database Asset performance data Asset service history Stores inventory data	(i). Ensure auditability of right product, right price and right lead-time based on right criteria to ensure uptime, compliance, health and safety, SLAs and target KPIs (ii). Just-in-time inventory – Planned and IoT predictive maintenance to ensure parts delivered on time, in stores and reserved for jobs (iii). Improve uptime: c.50% downtime due to parts not available (iv). Optimise procure-to-pay for materials, spares and consumables. Single order for multiple suppliers and single monthly invoice (v). Improve supplier performance, management and consolidation, including performance reporting (vi). Traceability of parts and materials ordered, why they were ordered and for which asset and job	Finance: Procurement, payment teams – Consolidated orders, consolidated suppliers, monthly invoices, automated/semi-automated ordering process (remove non-value add tasks) Operations teams – Ensuring right product, right time and right price

(continued)

Table 30.1 (continued)

Ecosystem modules	Description	Data sources	Benefits	Beneficiaries
Blockchain and auditability/ compliance	Provides audit trail and ascertain compliance. Capability to provide immutable evidence of supplier product traceability, service performance or contract compliance by automatically registering (writing) transaction steps into a shared ledger (blockchain) which cannot be edited or changed	Any system that records relevant transactional activity	(i). Ensure auditability and traceability (ii). Ensure visibility and compliances (iii). Enhance supply chain management through monitoring of supply chain activities (iv). Facilitate integrated and distributed ledger for production management (v). Shared information system (vi). Fosters distributed and certified systems (vi). Timeliness of delivery (vii). Prove authenticity and provenance of goods	Producers Forwarder Broker Consumer
	Provides distributed shared ledger that is consensually shared and synchronised across multiple sites, geolocations, accessible by multiple users. Provides secured fund transfer	Blockchain and distributed ledger technology Smart contracts for digital certification and distributed SLA management	(i). Confidence and improvements in services: Measure actual performance of third parties and suppliers SLAs vs expected (ii). Invoking potential penalties on immutable evidence if SLAs not respected (iii). Reduction in time and effort in any 'validation' within this process, e.g. PO – Invoice payment record reconciliation (everything on immutable ledger, so need only sample checking) (iv). Potential reduction in insurance premiums, e.g. auditable evidence of compliance, health and safety; ensure genuine products and certified operators used to complete the work order (v). Reduce procurement overheads and improve buyer and maintainer productivity	Operations Finance Buyers Third parties

Ecosystem modules	Description	Data sources	Benefits	Beneficiaries
Big data analytics and insights	Collect and analyse big data from e-procurement and IoT sensors to provide evidence-based insights for end customers, suppliers and partners. This will provide support for decision-making and informed actions to improve QOS, productivity, cost-effectiveness, sustainability, health and safety, compliance, asset performance, etc.	Sensor data Relevant system data	(i). The ability to provide data analysis solutions covering a wide range of requirement, skill sets and cost and provide data sources and platform to trigger events and visualise data for front-line teams (ii). Merging transactional (client, procurement) and operational (IoT, asset performance) data to produce evidence-based insights and triggering events or visualising data aimed to improve building or service performance, worker care and experience (iii). Detect and predict anomalies and improve building, asset or workforce efficiency (iv). Facilitate complex decision-making by applying machine learning and artificial intelligence (v). Applying business rules quickly, easily and effectively to ensure events are triggered at right time for right team (vi). Reduce unplanned downtime and boost asset and equipment longevity, yield, quality and effectiveness	All teams operations, finance, procurement, architects and designers, etc. – As data sources and insights could potentially cover performance and optimisation improvements for assets, locations, buildings and services – All those listed above

acquisition and support international business growth. The asset management lifecycle is iterative. According to PECB (2016), the four stages in asset management lifecycle are planning, acquisition, operation cum maintenance and, finally, disposal. However, in this paper, we propose a three-phase asset management (see Fig. 30.1a). The phases and their sub-phases are design concept initiation followed by design specification and documentation that will feed into the subsequent phase, build (involves construction based on specification) and operate (MRO – maintain, repair, operate and asset refurbishment). In this lifecycle, it is estimated that 80% of TCO is within the MRO sub-phase.

Figure 30.1b is an extension of Fig. 30.1a where MRO is a major part of the entire lifecycle and its focus is to feed back insights into the design and build phase. The insights are afforded through big data analytics of relevant data relating to transactions (e.g. smart building context – room booking, usage, relevant business/health transactions, etc.) and operations (for the same smart building context – IoT data, environmental monitoring, building performance, etc.). The sub-phases that subsume under MRO phase are supply chain and e-procurement; contracts and service level management/agreement, SLM/A; compliances; energy and sustainability.

30.3.2 Digital Twin

A digital twin is a digital representation or virtual replicas of a real-world entity or system (e.g. physical objects, process, organisation, person or abstraction) (Gartner, n.d.; Networkworld, 2019). Data from multiple digital twins can be aggregated to model real-world systems comprising composite entities, operations or processes (e.g. city, factory, etc.) (Gartner, n.d.). Additionally, they could be exploited to build simulations or IoT platforms (e.g. Azure Digital Twins) to explore optimisation as well as what-if scenarios (Networkworld, 2019) for prediction and gain deeper insights into drivers of better products/services to enhance customer experience (Microsoft, 2020). Gartner (2019) conducted a recent IoT implementation survey and found that 75% of organisations implementing IoT already use digital twins or plan to within a year. Digital twins have been deployed for asset management (Arc Advisory Group, 2020) and integrated into asset management lifecycle to create value and support decision-making (Macchi et al., 2018). It is viewed that assets could be rendered 'smarter' by means of digital twinning (DNV.GL, n.d.). In this paper, we discuss how multiple digital twins could be integrated into smart enterprise asset management lifecycle to create business values (see Fig. 30.2). Gartner (2019) maintains that it is necessary for multiple digital twins to be integrated, and here, we shall discuss multiple digital twins (for a diverse range of areas of interest) that are relevant to smart enterprise asset management:

1. **BIM Integration:** BIM focuses on an entity's design and construction (e.g. a building). On the other hand, a digital twin models the entity's operations and

interactions with other entities within a system (adapted from IoTforall, 2019). This is aligned to Gartner's notion of the next-generation digital twins which include users, processes and behaviours. Such related data is collated through the deployment of sensors (discussed in point iii).
2. **3D Model Scanning:** Point clouds could be created via photogrammetry or LiDAR (light detection and ranging) technologies (Vercator, n.d.). The captured point cloud data are used to make 3D scans of relevant assets (e.g. 3D laser scan) that are subsequently fed into the development of a virtual model, easily accessible as a digital twin (Hannovermesse, 2018).
3. **Connecting Assets:** Connected assets via sensors provide useful information (through data analysis discussed in point iv) on how they perform in real time, and to prevent serious accidents through potential failure prediction (CIOB, 2019).
4. **Data Analysis:** In order to realise the full promise of digital twins, it is essential to integrate systems and data across the entire organisational ecosystems (Deloitte, 2020). Real-time data will be analysed and fed into simulations to clearly understand what-if scenarios, predict outcomes accurately and trigger appropriate actions or events to manipulate the physical world (ibid).
5. **Asset Performance Management:** Asset performance management (APM) involves the deployment of sensors to collect real-time performance and condition data for assets and analysed (using AI and machine learning algorithms). Such collected and systematically analysed data are inputs into digital twins that assume the form of realistic and interactive 3D models (Walters, 2016) that provide evidence-based decisions about the assets (Negi, 2019). According to Walters (2016), APM provides the 'power of combining all systems into one that can deliver actionable intelligence', and this facilitated through the convergence of engineering (electronics, and instrumentation), operational (sensor and controls) and information (software, hardware and systems) technologies. A framework for setting targets (with metrics) in APM measurement have been developed (Green et al., 2016).

As listed in Fig. 30.2, the positive business outcomes of integrating multiple digital twins into the entire lifecycle management of smart assets are improved critical asset uptime, energy efficiency, improved productivity, reduced procurement overheads, drive predictive maintenance, improve health and safety compliance and optimise asset performance. The integration of digital twins for end-to-end whole lifecycle asset management ought to be highly prioritised to afford migration and integration of solutions as customer requirements grow. The key benefits of integrating multiple digital twins (3D scan, BIM integration and connected assets via IoT technologies) with asset management for a smart building use case are (1) precise (accurate to 5 mm) and up-to-date 3D model of building to support and expedite architectural and redesign activity; (2) highlight key assets and locations, colour-coded asset status, asset specification data and their real-time performance; (3) quick and easy retrieval of asset or location information in an intuitive 3D model; (4) improve collaboration between different internal and external teams covering asset whole lifecycle; and (5) spatial and contextual information readily available for assets and locations to assist with health and safety compliance, highlight access restrictions and plan work ahead of site visits. Their beneficiaries are the following:

(1) Administrators, designers and architects who need to plan site alterations, to repurpose areas within the facility and/or to introduce temporary works.

(2) Engineering and maintenance teams by providing ready contextual access to asset and location data for faster corrective and more timely preventive actions as well as assisting with identification and scheduling of predictive tasks. Visual information also helps optimise visits by remotely assessing on-site requirements to avoid wasted visits and by assisting with the identification of nearby assets or locations that can be attended to in one visit.

(3) Operation team to manage staffing based on demand (footfall) and visually control, manage and report energy consumption and CO_2 emissions.

(4) Purchasing team to allow visual reference to assets and locations when identifying what is needed and by intuitive access to specifications to enable accurate ordering of spares and/or replacements.

30.4 Smart Enterprise Asset Management Ecosystem

According to NCA (n.d.), in the natural world, biodiversity and ecosystems afford numerous benefits (also known as 'ecosystem services') to mankind, whilst UNEP (n.d.) view healthy ecosystems and a rich biodiversity vital for proper functioning of cities. Gartner (2020) has developed a digital ecosystem framework to analyse digital ecosystems and understand how the various elements work together within the ecosystem. Inspired by these, we draw a parallel for a smart enterprise asset management ecosystem in the built environment context (depicted in Fig. 30.4 and discussed in Subsection 30.4.3). Such an ecosystem requires diverse yet complementary technologies, and it aims to break down technological silos to create a collaborating and interconnected multidisciplinary system for the ultimate goal of interoperable as well as seamless ecosystem service provision. Our proposed smart enterprise asset management ecosystem demonstrates the essential characteristics of jointness (yet modular), coordinated, scalability and interoperability.

30.4.1 Smart Enterprise Asset Management System Ecosystem: Joined-Up Digital Strategy

UK Department for Transport (n.d.) views the importance a joined-up digital strategy for consistent system operations through the adoption of innovative 'agile' approach for the design of services and platforms. The benefits would be user experience enhancement, efficiency maximisation and cost reduction. Central customer-related information coordination would provide a joined-up cross-ecosystem view that would be more reliable and comprehensive. Figure 30.3a depicts the six different technologies that are essential for a built environment-related smart enterprise

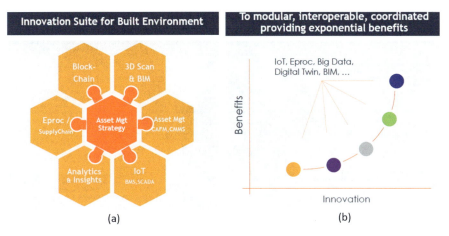

Fig. 30.3 (**a**) Innovation suite for smart enterprise asset management; (**b**) Benefits of the innovation suite

asset management: 3D scan and BIM, asset management, IoT, big data analytics and insights, supply chain and e-procurement and blockchain. Figure 30.3b depicts the exponential benefits afforded when these six modular technologies are well coordinated, and interoperable to function as an ecosystem. To reiterate, all the depicted components of the smart enterprise asset management ecosystem will have their individual digital strategy. However, the overarching need is to have a joined-up digital strategy for the ecosystem to provide seamless quality services that will ensure an optimised trade-off amongst Quality of Service (QoS – high system performance), Quality of Effectiveness (QoE – excellent user experience) and Quality in Sustainability (QiS – reduced environmental impact).

30.4.2 Smart Enterprise Asset Management: Ecosystem Solutions for a Smart Building Use Case

Our proposed smart asset management ecosystem (known as smart enterprise asset management system, SEAM, as depicted in Fig. 30.4) comprises six integrated modules (note: 3D model and BIM are considered as separate).

To exploit an end to end, interoperable 'smart' enterprise asset management [SEAM] system would typically require the five modules, their associated data and ecosystem partners (tabulated in Table 30.1).

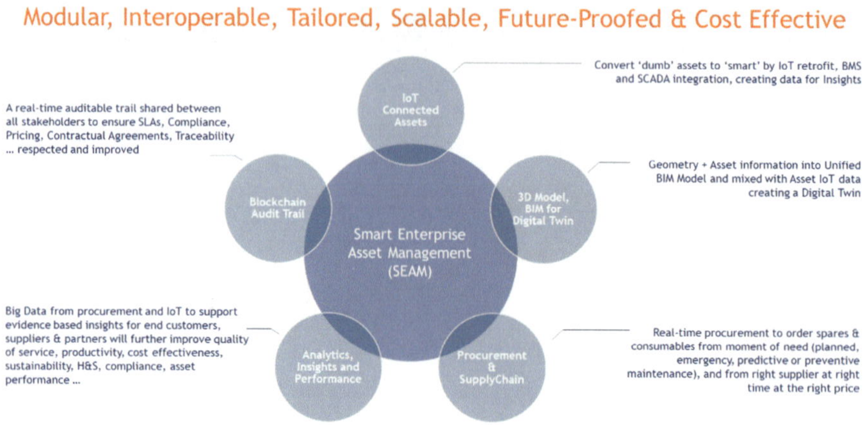

Modular, Interoperable, Tailored, Scalable, Future-Proofed & Cost Effective

Fig. 30.4 Interoperable, scalable and future-proofed smart enterprise asset management solution

Table 30.2 Smart enterprise asset management ecosystem of partners

Module	Partners
Digital twin from 3D scan/BIM model	1st horizon, SiteDesk
Project delivery managed under a BIM level II framework	TP professional services
Core asset/building management control platform	SRO solutions
Asset and location monitoring and IoT sensing	IAC, Beckhoff
Big data analytics and insight	EdenBridge, LBU
Integrated and streamlined e-procurement capability	Mercateo
Tracking and tracing suppliers and supply using shared ledger	SRO solutions

30.4.3 Smart Enterprise Asset Management and Ecosystem of Partners

The end-to-end digital asset management capability is unique as it brings together open and non-proprietary usage of some of the latest digital technologies in a way that will ensure a building and its assets can operate effectively and efficiently to support the workforce delivering the services to the end clients in an optimal way. These outputs are all immediate- to medium-term benefits and do not include the academic outputs. The six modular and interoperable components that underpin the ecosystem value proposition and associated proposed partners (for a smart building use) case are tabulated in Table 30.2.

Through the introduction of an ecosystem of subject matter experts from industry and academia with the ability to track record of delivering digital innovation, sustainability outcomes, a smart building (e.g. government building) will be able to create the foundations of a unique 'end-to-end' approach to full asset, lifecycle and sustainability management. This will be delivered by integrating open and interoperable services such as 3D scanning, digital twins, BIM/visualisation, Internet of

Things (IoT) sensors, predictive maintenance and automated maintenance repair and operation procurement. The ecosystem, working with a government building, has the capability to improve overall productivity in line with the government's digital and sustainability objectives, by using interoperable modules that can be combined to deliver end-to-end, efficient and operational support. These modules can integrate with existing systems (e.g. building management systems), creating visually live IoT-driven data monitoring, i.e. 'digital twins' of a building/asset, allowing failures to be predicted and actioned via automated service request management, including contract/spares management, stores control and e-procurement services *before services* are impacted, e.g. critical plant assets such as HVAC, lighting, power and waste management, fridge temperature, etc. The wider benefits include the ability to track building/department footfall/usage statistics which can then be married with people flow data to identify if the department/building can be optimised. Smart asset manangement system (Redmond, 2017) and mobile tools that digitise the assets (i.e. digital twin) provide evidential data that will simplify and increase the ability to reliably control the management of buildings and assets, to enable robust decisions to be made on how and when they can be used, with the assets – i.e. engineering, spaces, desks, lifts, fridges, rooms, theatres, etc., the entire asset if required – being a complete digital twin.

30.4.4   Summary of Smart Enterprise Asset Management

Our proposed smart enterprise asset management ecosystem provides re-usable and interoperable data that will be made available for different profiles and business outcomes (see Fig. 30.5). All pertinent data is aimed to be re-usable and stored in 'big data' environment to be viewed through lens of different profiles seeking

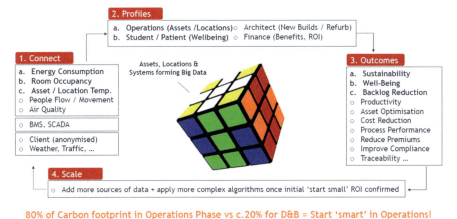

Fig. 30.5 Re-usable and interpoerable data for different profiles and outcomes

different business outcomes (e.g. operator for productivity gains; finance for energy reduction; etc.). To ensure cost-effectiveness and ROI, it is planned to include additional data sources in different phases. Once smart data management is implemented, it is anticipated that predictive maintenance and autonomous decision-making can be developed for key sections of the asset (such architecture being developed with BIM, FM and digital twins in mind). Initially, data collection architecture design for the critical energy and performance systems will be prioritised to reduce energy and improve operation of the asset. The aim is to capitalise on both historic and real-time performance data so that evidence-based decisions can be made on future projects. Central to effective operation is avoiding duplication of practice and unnecessary resource use. The effective utilisation of performance data will radically change understanding of asset performance, enhancing whole lifecycle management.

30.5 Conclusion

The USP and innovation of our proposed smart management ecosystem stems from eight organisations sharing the same vision of an end-to-end, interoperable, smart building solution. Our 'think big – start small – scale fast' approach provides a cost-effective solution that mitigates risk for customer. This offering will help democratise adoption of innovative technology such as IoT, data analytics, digital twin, etc., delivering exponential benefits as modular solutions are joined up. Our initial work with a government building has seen the ecosystem offering for digital sustainability/smart building being integrated into a repeatable template for future capital projects.

References

Aberdeen Group. (2009). *Asset performance management: Aligning goals of CGOs and maintenance managers.* November 2009. Retrieved June 26, 2020, from http://www.forpoint.co.nz/wp-content/uploads/2013/04/Infor-EAM-WP_Asset-Performance-Management-Aberdeen-Group.pdf.

Aberdeen Standard Investments. (2018). *Strategy Guide: Global Absolute Return Strategies Portfolio.* Retrieved June 26, 2020, from https://funds.standardlifeinvestments.com/uk/ifa/GARS_Strategy_Guide.pdf.

Arc Advisory Group. (2020). *Digital twins for strategy guide: Asset management.* Retrieved June 26, 2020, from https://www.arcweb.com/blog/digital-twins-asset-management.

BSi. (n.d.). *ISO 55001:2014, Optimize your assets and improve performance.* Retrieved June 27, 2020, from https://www.bsigroup.com/en-GB/Asset-Management/.

Bughin, J., Chui, M., and Manyika, J. (2010). Clouds, Big Data, and Smart Assets: Ten tech-enabled business trends to watch. McKinsey Quarterly. 56(1): 75–86. Retrieved June 26, 2020, from https://www.mckinsey.com/industries/technology-media-and-telecommunications/our-insights/clouds-big-data-and-smart-assets-ten-tech-enabled-business-trends-to-watch

CIOB. (2019). *Potential to use BIM data in digital twins is being overlooked*. Retrieved June 26, 2020, from https://www.bimplus.co.uk/analysis/potential-use-bim-data-digital-twins-being-overloo/.

Deloitte. (2020). *Digital Twins: Bridging the Physical and Digital*. Retrieved June 26, 2020, from https://www2.deloitte.com/uk/en/insights/focus/tech-trends/2020/digital-twin-applications-bridging-the-physical-and-digital.html.

DNV.GL. (n.d.). *Making your Asset Smarter with the Digital Twin*. Retrieved June 26, 2020, from https://www.dnvgl.com/article/making-your-asset-smarter-with-the-digital-twin-63328.

Gartner. (2019). *Gartner Survey Reveals Digital Twins are Entering Mainstream Use.*. Retrieved June 26, 2020, from https://www.gartner.com/en/newsroom/press-releases/2019-02-20-gartner-survey-reveals-digital-twins-are-entering-mai#:~:text=Gartner%20defines%20a%20digital%20twin,business%20operations%20and%20adding%20value.&text=%E2%80%9CWe%20see%20digital%20twin%20adoption%20in%20all%20kinds%20of%20organizations.

Gartner. (2020). *The Gartner digital ecosystem framework: How to describe ecosystems in the digital age?* Retrieved June 26, 2020, from https://www.gartner.com/en/documents/3979306/the-gartner-digital-ecosystem-framework-how-to-describe-.

Gartner. (n.d.). *Gartner Glossary - Digital Twin*. Retrieved June 26, 2020, from https://www.gartner.com/en/information-technology/glossary/digital-twin.

Green, D., Masschelein, S., Hodkiewicz, M., Schoenmaker, R., & Muruvan, S. (2016). Setting targets in an asset management performance measurement framework. In Proceedings of 2016 International Conference on Quality, Reliability, Risk, Maintenance, and Safety Engineering (QR2MSE 2016) 2016 World Congress on Engineering Asset Management (WCEAM2016): July 25–28, 2016, Jiuzhaigou, Sichuan, China. Retrieved June 26, 2020, from https://repository.tudelft.nl/islandora/object/uuid:f113746e-cc22-4a6d-b452-cf8ac3fddf68?collection=research.

Hannovermesse. (2018). *Point clouds make digital twin accessible*. Retrieved June 26, 2020, from https://www.hannovermesse.de/en/news/news-articles/point-clouds-make-digital-twins-accessible.

IoTforall. (2019). *Digital Twins vs. Building Information Modelling (BIM)*. Retrieved June 26, 2020, from https://www.iotforall.com/digital-twin-vs-bim/.

ISO. (2014). *ISO 55000:2014: Asset management—Overview, principles and terminology*.

Lampe, M., Strassner, M. (2003). The potential of RFID for moveable asset management, proceedings of the 5th international conference on ubiquitous computing (UbiComp), 12-15th Oct, 2003, Seattle, USA. Retrieved June 26, 2020, from https://www.alexandria.unisg.ch/21557/.

Liyanage, J.P., and Langeland, T. (2009). Smart assets through digital capabilities. *Encyclopedia of Information Science and Technology*. second edition IGI global, pp. 3480-3485, doi: https://doi.org/10.4018/978-1-60566-026-4.ch553.

Macchi, M., et al. (2018). Exploring the role of digital twin for asset lifecycle management. *IFAC-PapersOnLine, 51*(11), 790–795. https://doi.org/10.1016/j.ifacol.2018.08.415.

Microsoft. (2020). *What is Azure Digital Twins?* Retrieved June 26, 2020, from https://docs.microsoft.com/en-us/azure/digital-twins/overview.

NCA. (n.d.). *US Global Change Research Program: Ecosystems and Biodiversity*. Retrieved June 27, 2020, from https://nca2014.globalchange.gov/highlights/report-findings/ecosystems-and-biodiversity.

Negi, R. (2019). Experience in asset performance management analytics for decision support on Transmission & Distribution Assets. *Proceedings of IEEE PES Asia-Pacific power and energy engineering conference (APPEEC)*, 1-6 Dec, 2019, Macao, doi: https://doi.org/10.1109/APPEEC45492.2019.8994622.

Nel, C., and Jooste, J. (2018). A policy framework for integrating smart asset management within operating theatres in a private healthcare group to mitigate critical system failure. International Journal of Condition Monitoring and Diagnostic Engineering Management, *155*.

Networkworld. (2019). Retrieved June 30, 2020, from https://www.networkworld.com/article/3280225/what-is-digital-twin-technology-and-why-it-matters.html.

PECB. (2016). *4 key stages of asset management lifecycle*. Retrieved June 27, 2020, from https://pecb.com/article/4-key-stages-of-asset-management-life-cycle.

Redmond, K. (2017). *Smart asset management, BIM, the internet of things and energy*. Retrieved June 26, 2020, from https://www.pbctoday.co.uk/news/bim-news/smart-asset-management-bim/30275/.

Treasury, H. M. (2018). *The green book*. Retrieved June 26, 2020, from https://assets.publishing.service.gov.uk/government/uploads/system/uploads/attachment_data/file/685903/The_Green_Book.pdf.

UK Department for Transport. (n.d.). *Digital Strategy*. Retrieved June 27, 2020, from https://assets.publishing.service.gov.uk/government/uploads/system/uploads/attachment_data/file/49475/dft-digital-strategy.pdf.

UKGBC. (2019). *Net zero carbon buildings: A framework definition*. Retrieved June 26, 2020, from https://www.ukgbc.org/wp-content/uploads/2019/04/Net-Zero-Carbon-Buildings-A-framework-definition-print-version.pdf.

UNEP. (n.d.). *Biodiversity and ecosystems*. Retrieved June 27, 2020, from https://www.unenvironment.org/explore-topics/resource-efficiency/what-we-do/cities/biodiversity-and-ecosystems.

Urmetzer, F., Parlikad, A. K., Pearson, C., & Neely, A. (2015). Design considerations for engineering asset management systems. In J. Amadi-Echendu, C. Hoohlo, & J. Mathew (Eds.), *9th WCEAM research papers* (Lecture notes in mechanical engineering). Cham: Springer. https://doi.org/10.1007/978-3-319-15536-4_22.

Vercator. (n.d.). *LiDAR vs Point Clouds: Learn the Basics of Laser Scanning, 3D Surveys and Reality Capture*. Retrieved June 26, 2020, from https://info.vercator.com/blog/lidar-vs-point-clouds.

Walters, A. (2016). Connecting information, engineering and operational technologies: Asset performance management provides the power of combining all systems into one. *Plant Engineering, 70*(9), 83–85.

Wang, C., Bi, Z., & Xu, L. D. (2014). IoT and cloud computing in automation of assembly modeling systems. *IEEE Transactions on Industrial Informatics, 10*(2), 1426–1434. https://doi.org/10.1109/TII.2014.2300346.

Chapter 31
Strategic Management of Assets and Compliance through the Application of BIM and Digital Twins: A Platform for Innovation in Building Management

Gav Roberts, Lee Reevell, Richard Pleace, Ah-Lian Kor, and Christopher Gorse

31.1 Introduction

BIM was introduced in the early 2000s to support building design for architects and engineers (Sacks et al., 2018). They focused on the improvement of preplanning and design, clash detection, visualisation, quantification, costing and data management. In addition, recently, there has been the appendage of basic functionalities, such as energy analysis, structural analysis, scheduling, progress tracking and jobsite safety (Volk et al., 2014).

The current use of BIM focuses on life cycle stages to maintenance, refurbishment, deconstruction and end-of-life considerations. BIM is now an integral part of a business's building asset management strategy, and maintains that an integration of BIM into asset management will ascertain accurate as well as up-to-date asset information. A benchmark in best practice asset management using BIM processes (i.e. BIM asset management Kitemark) has been developed by BSI (Sacks et al., 2018).

When BIM is used for facilities management in new buildings, clear benefits are reported, e.g. regarding improved information flows and project management, risk mitigation and positive return on investments. In many existing buildings, incomplete, obsolete or fragmented building information is predominating. Missing or obsolete building information might result in ineffective project management,

---

G. Roberts (✉) · L. Reevell (✉)
Halton Housing, Widnes, UK
e-mail: Gav.Roberts@haltonhousing.org; Lee.Reevell@haltonhousing.org

R. Pleace
SiteDesk Ltd, Cheshire, UK

A.-L. Kor · C. Gorse
Leeds Beckett University, Leeds, UK

© The Author(s), under exclusive license to Springer Nature Switzerland AG 2022
C. Gorse et al. (eds.), *Climate Emergency – Managing, Building, and Delivering the Sustainable Development Goals*, https://doi.org/10.1007/978-3-030-79450-7_31

uncertain process results and time loss or cost increases in maintenance, retrofit and remediation processes (Ashworth et al., 2019), as existing buildings often lack as-built documentation due to omitted updating. BIM implementation in existing buildings has benefits of improved documentation management, clearer information on maintenance of warranty and service information, assessment and monitoring, energy and space management, emergency management and retrofit planning.

Various digital tools for building capture and auditing are available, such as 2D/3D geometrical drawings, tachometry, laser scanning or automatic locating of images. If building documentation is inadequate for maintenance or deconstruction processes, capturing and surveying techniques with different qualities are applied to audit and gather the existing buildings' characteristics. The functionality-related level of detail and the corresponding data capturing technique influence all following steps of BIM creation and its associated effort.

Although on the one hand, implementation of BIM both in new and existing buildings induces profound changes of processes and information flows, on the other hand, it accrues considerable advantages. It can enhance project management and risk mitigation or limit costs and duration of facilities management. It might also affect sustainability ratings and certifications. It could be possible to monitor energy consumption, wastewater and maintenance costs. It would be achievable to illustrate environmental effects of the building and to verify and monitor consumption and emission values. It would also validate data management, maintenance schedules and equipment warranties with respect to deterioration and cause-effect relationships. When a building reaches its end of life, there is the ability to consider its recyclability on a component level. With the addition of the latest technologies including IOT, sensors, automated compliance and future trends of automated capture, onsite progress tracking, measurements and monitoring can be shown through cloud computing solutions depicting building information and live transformations.

This can make serious improvements to the way buildings are managed. We aim to show the business advantages and prove cost, time and efficiency savings whilst depicting the environmental factors and how we can improve them.

Notwithstanding those benefits listed, it is essential that the business need and the stages for implementation are addressed and communicated to all parties, if BIM processes are to be effectively implemented (Wang et al., 2018). Whilst BIM Level 2 is claimed to be widespread, it is often poorly delivered (Attrill, 2020). There is a need to clearly define the process and expectations, and also to identify the barriers to the execution of BIM.

31.2 Drivers for an Integrated BIM Asset Management Model

Wang et al. (2015) have discussed how BIM could be exploited to support fire safety management. A key driver for such a change is the introduction of new fire safety legislation for properties with more than one dwelling which has resulted from the

Hackitt Report (Hackitt, 2018; Ministry of Housing, Communities and Local Government, 2020) into the Grenfell Tower fire. The proposed Bill (UK Government, 2020) is currently at the stage 3 Committee stage in the House of Commons and will likely result in the introduction of new regulations which will extend the Fire Safety (Regulatory Reform) Order 2005; this will be supported by new and comprehensive Buildings Safety legislation. A draft version of the Bill will likely appear this summer, and the contents are expected to further clarify the responsibility of building owners, i.e. extending fire risk assessments to include the whole of the building and additional responsibilities for the design and construction of new build properties. Other elements of fire safety will likely include compartmentation and fire mitigation measures. The solution would demonstrate a fire safety model, including records of who fitted what where and when, the materials used, time/date for replacement and other auditing checks. The advantages would include insight into when maintenance, service and checks need performing. We would have easily accessible records of building fabric and internal wall structures, etc. Integration of BIM within asset management would help appropriately address regulatory compliance (Sacks et al., 2018).

Studies are embracing BIM at the knowledge and decision-making level, yet few capture practical and meaningful application. The potential application of a BIM knowledge management system during operation and maintenance has been recognised; however, with the exception of Wang et al. (2015), few considered the importance of fire compliance. The following case study describes the method adopted to provide an integrated BIM asset management model, with a clear need to ensure effective fire compliance.

31.3 Case Study: Process-Interaction Flow – An Integrated BIM Asset Management Model

The process-interaction flow for an integrated BIM asset management model is proposed and the schematic depicted in Fig. 31.1. The process makes it easy to find the building and identify the part of the building or the asset to which the data is linked. The chronology of data linked to the asset can then be found.

The list of interactions is as follows:

- Centred around a web-based portal for the buildings, a 3D model is developed with the ability to be on focus in areas of interest.
- Live information about each individual component is displayed on demand.
- Pop up of alerts for upcoming scheduling information.
- Click on relevant location/asset for compliance documentation and certification.
- Data is sourced from external datasets, and it feeds back to the dashboard creating a full live building model.

Fig. 31.1 Schematic and flow process for asset data management

31.4 Digital Twin and BIM-Enabled Asset Management

Innovation is the successful exploitation of new ideas, and by partnering with Sitedesk and ecosystem partners to introduce a small-scale model, we demonstrated the value of BIM to the business. Lu et al. (2020) proposed integrating the digital twin concept into smart asset management. Thus, future aspirations include an integrated digital twin, for life cycle BIM- enabled asset management for all multiple occupancy and single occupancy buildings (note: the number of buildings currently within the group holdings exceeds 7000).

Illustrated in Fig. 31.2 are the key drivers following industry best practice and positive business outcomes when creating an integrated digital twin based on four sources of data within an asset management ecosystem. They are asset (or survey) data, BIM-related data, IoT sensor data and performance-related data.

The integrated digital twin will undergo an iterative growth as proof of concepts are verified and validated. To reiterate, the aim of this integrated digital twin is to improve workflows and demonstrate an improved fire compliance and fire safety system. It evidences compliance requirements set out in current legislation with quick access to certification and direct links to our management information system. It will reduce management overhead costs and most importantly to make customers safer in their homes. The prototype demonstrates the foundations of a BIM system that facilitates intelligent data lead decisions and focus resource into areas that need them the most.

31 Strategic Management of Assets and Compliance through the Application of BIM... 419

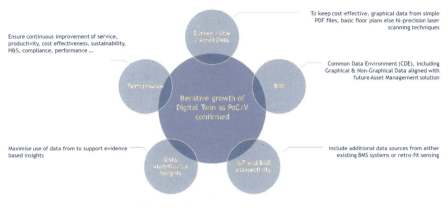

Fig. 31.2 Digital twin for BIM-enabled asset management

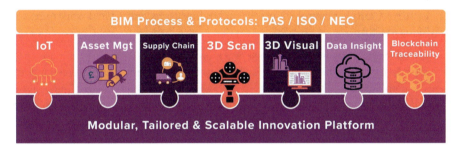

Fig. 31.3 BIM-enabled asset management ecosystem (components required)

31.5 Smart BIM-Enabled Asset Management Ecosystem

Figure 31.3 depicts the components of a smart BIM-enabled asset management ecosystem (within BIM process and protocols framework and standards – PAS 1192–3: 2014; ISO 55000: 2013; NEC contracts). They are IoT sensing, asset management, e-procurement and supply chain, 3D BIM and visualisation, data insights, blockchain and traceability system.

Figure 31.4 presents an integrated schematic of an end-to-end, joined-up, modular and scalable digital framework to provide room for future organic growth. Relevant components are data GPS, data analysis (e.g. using big data analysis to provide insights), data for management and control (e.g. building management system for automation; SCADA; asset management) and data visualisation (through dashboards) and acquisition (through IoT sensors and SCADA).

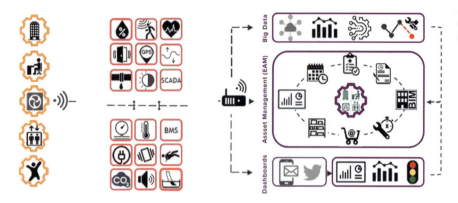

Fig. 31.4 Smart asset management solution

A list of functionalities for our proposed smart asset management systems is as follows:

1. **User Interface:** A web-based solution that everyone in the business can have access to and provide them with the information from any platform.
2. **Control Centre:** Control for offline and online assets and related work orders.
3. Management:
    (a) Manage, measure and optimise work orders and staff.
    (b) Manage compliance, health and safety ensuring right qualified resource to complete the job.
    (c) Manage, measure and optimise third-party contracts and pull data from existing models and visualisations and interact from within an asset management window.
    (d) Service Level Management:

    - Details of KPIs and service level management for ops teams, suppliers and critical assets; the system will directly link to e-procurement for spares and consumables.
    - Service level management for operations, suppliers and assets: SLAs vs costs of service.

    (e) **Inventory Management:** To ensure inventory and reduce procurement overheads.
4. **Procurement:** Procurement model would ensure right product, right price and right lead- time based on right criteria – to ensure uptime, compliance, H&S, SLAs and target KPIs.
5. **Data Analytics:** Directly linked into data analytics displaying energy consumption and carbon footprint with evidencing positive improvements.
6. **Predictive Maintenance:** Planned and IoT-enabled predictive maintenance to ensure parts delivered on time and in stores and reserved for jobs.

7. Performance Improvement:

   (f) Improve uptime and optimise procure-to-pay for materials, spares and consumables.
   (g) Improve supplier performance, management and consolidation.

8. Auditability, Traceability and Transparency:

   (h) Real-time auditable trail shared between customer and third parties to ensure SLAs, compliance, pricing and contractual agreements respected and improved.
   (i) Smart contracts – reduce procurement overheads and optimise P2P process via a traceable end-to-end process.
   (j) Compliance – auditable trace of genuine products and certified operators (insurance and auditors).
   (k) Transparency between customer and suppliers to increase loyalty – one version of truth.

31.6 Methodology

The pieces of the puzzle illustrated (in Fig. 31.4) will create a smart asset management framework, by bringing each piece of the puzzle systematically together. At each stage, the process can be assessed, determining if the benefits have been realised and the business advantage remains. The road is mapped out creating a digital vision that is assured through performance. Our roadmap will have three-phased approach namely: *Start Small*, with the ability to *Scale Fast* whilst *Thinking Big* (see Fig. 31.5).

To achieve the target of an integrated digital twin for end-to-end whole life cycle asset management, an agile methodology (Rigby et al., 2016) will be adopted. It will be underpinned by three principles: *Start Small*, *Scale Fast*, and *Think Big*. Risk aversion will be demonstrated to evidence its benefits. This is supported by the 'Try then Buy' ethos (Coghlan, 2017) and Minimum viable product (MVP) (see Technopedia, 2020) concept. A key premise behind the idea of MVP is that you produce something of value of which to demonstrate as a solution to the business problem. This may be no more than a landing page, or a service with an appearance of automation, but which is fully manual behind the scenes. Seeing what people do with respect to a product is much more reliable than asking people what they would do. The primary benefit of an MVP is so we can gain an understanding of the business' requirements without fully developing the product. Proving the model beforehand will help staff understand the advantages of BIM which can be expanded upon.

Following the MVP approach we have chosen a vehicle as an analogy (see Fig. 31.6). We first begin to build a test scooter to prove we need wheels and demonstrate its benefits. However, our ultimate aim is to make a self-driving car.

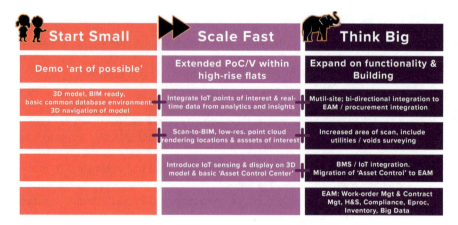

Fig. 31.5 Three-phased approach for smart asset management

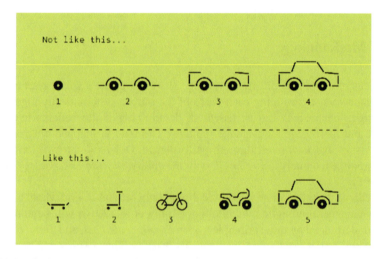

Fig. 31.6 A vehicle analogy: from building a test scooter to a self-driving car

31.7 Use Case: Churchill Mansions

The process of moving to a digital twin solution started by incorporating available 2D and 3D data for Churchill Mansions (see Fig. 31.7) into BIM models which Sitedesk made available as a digital twin – Sitedesk is the BIM visualisation component of the Ecosystem offering, and it provides a simple and easy-to-use environment through which users can interact with 2D and 3D digital twins and also access the underlying power within the various elements of the Ecosystem. Initially Sitedesk's own common data environment (CDE) was used to underpin the collaborative process.

Fig. 31.7 Churchill Mansions, Runcorn, Cheshire, WA7 1DH

The building has 11 floors with 44 apartments and the documentation for this building is limited. Using the plans shown below (Fig. 31.8 (a)–(c); Fig. 31.9), Sitedesk has created the building as a 3D model to use for the visualisation and contains the addition of the extra asset details not in the original drawings.

31.7.1 Phase 1: Start Small

By interacting with the digital twin of Churchill Mansions, Halton Housing were able to identify key components and areas which came under the umbrella of the revised regulatory requirements. Once identified Sitedesk enabled Halton Housing to monitor the workflow and store all associated documentation to provide a complete record of what had been undertaken to comply with the legislative requirements.

The benefits Halton Housing experienced as a result of using Sitedesk are:

- Higher quality data controlled within a central system rather than being kept in an unstructured way within multiple systems.
- Improved productivity through simple, intuitive interaction with the digital twin to locate and report on assets/locations of interest.
- Improved monitoring and adherence to health and safety and compliance obligations by having centralised data with auditable operations and intervention tracing.
- Ability to access the digital twin and all its associated documentation using mobiles, tablets and laptops.
- Being able to make the most of the variable quality of data available for sites being controlled – some sites have 2D documentation only for example.

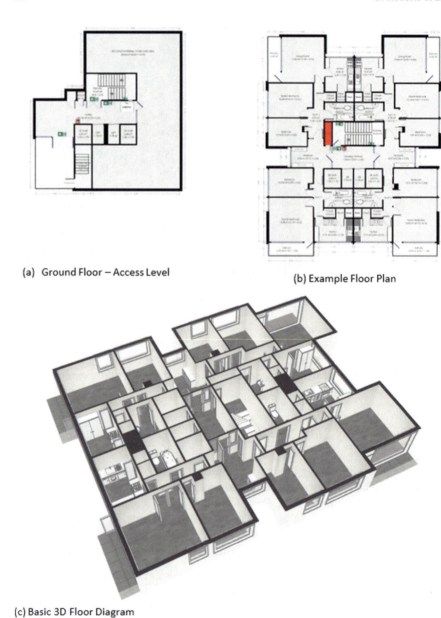

(a) Ground Floor – Access Level

(b) Example Floor Plan

(c) Basic 3D Floor Diagram

Fig. 31.8 Churchill Mansions plans. (a) Ground floor – access level. (b) Example floor plan. (c) Basic 3D floor diagram

Fig. 31.9 Example of a single apartment

In the future, as Halton Housing's requirements grow to necessitate deploying other sophisticated components within the Ecosystem, the data within Sitedesk can be used to seed these other components to provide a straightforward migration path to the further advanced functionality that is available throughout.

Working with Sitedesk on phase 1 demonstrated the art of what is possible by creating the first piece of the puzzle (highlighted in Fig. 31.10); now the prototype has been developed, and it is in its testing phase, working on functionality with the business partners and stakeholders.

The goals (achieved) for phase 1 are to:
- Demonstrate the art of possible.
- Build the 3D model.
- Create a basic database environment.
- Construct a navigation system.
- Record business critical assets.
- Create framework for collaboration and management of risk.
- Record assurance and compliance.
- Support cultural and change through training and mentoring.

Advantages found are:
- Ease of information access.
- Auditing strategies.
- Business acceptance.
- Traceability.
- Compliance.

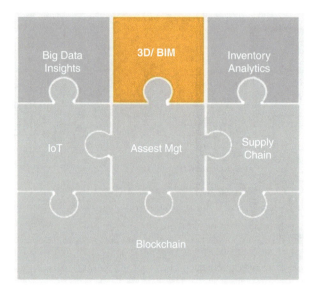

Fig. 31.10 3D/BIM

31.7.2 Phase 2: Scale Fast

Phase 2 is the next step and the foundations have been put in place to start this expansion with the goals of improving collaboration and integration of known assets during the life cycle phase of the building. Using the 3D models previously created allows interactions with assets and locations in an intuitive manner to be able to have quick and easy retrieval of asset condition, service records and location information automatically by connecting the model to data sources. As depicted in Fig. 31.11, it adds more pieces to the puzzle expanding functionality, performance and automation.

- Extended PoC within the high-rise flats.
- Integrate IoT points of interest.
- Generate and demonstrate real-time data.
- Scan to create point cloud renderings of locations and assets of interest.
- Introduce IoT sensing and display on 3D model.
- Add a basic asset control centre.
- Record supply of recorder building materials and associated traceability.

Advantages
- Ease of information access.
- Auditing strategies.
- Business acceptance.
- Traceability.
- Automated compliance.

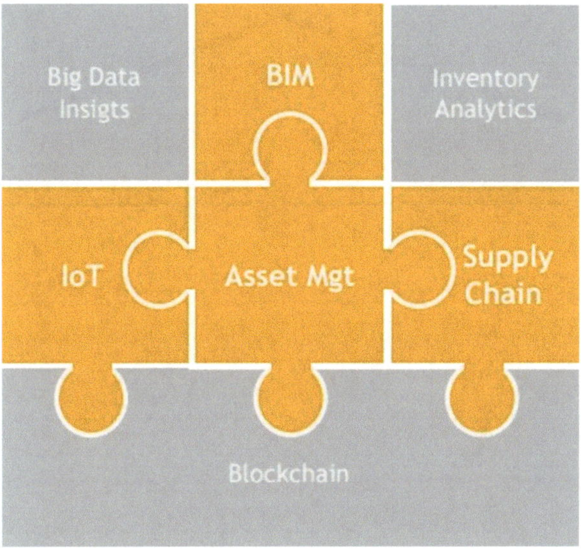

Fig. 31.11 BIM, IoT, supply chain and asset management goals

- Real-time data from overlaying IoT devices for increased productivity.
- Automatic ordering of faulty parts.
- Boiler house monitoring.

31.7.3 Phase 3: Think Big

With a constant stream of live data from people, sensors and devices, the model can encompass all effecting factors. Looking at the final pieces of the puzzle and bringing them online will change the way the buildings are not only managed but lived in as well. Asset management can be challenging. With phase 3, it would be possible to offer comprehensive solutions for managing physical assets on a common enterprise platform for many built forms. Now we are talking about the building as single living asset (see Fig. 31.12).

Advantages
- Operations productivity and KPIs.
- Improve asset uptime.
- Optimise asset usage and performance.
- Improve energy sustainability.
- Gain insights from connected assets.
- Reduce P2P overheads and cost of spares.
- Reduce inventory and consolidate suppliers.
- Improve service level management and introduce smart contracts.

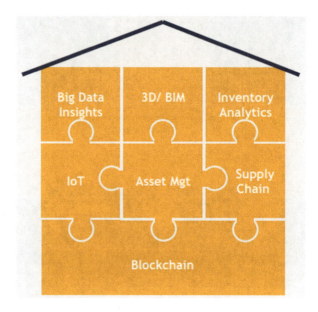

Fig. 31.12 Building as a single living asset

- Compliance and health and safety: apply, trace and report.
- Improve employee working environment.
- Procurement integration.

31.8 Conclusion: Future Development

With phase 1 completed and the business desire to continue to the next phase, the changes to building life cycle management are profound. By expanding to a constant stream of live data from people, sensors and devices, the model can offer comprehensive solutions for managing assets using Building Information Modelling to develop the digital twin of Churchill Mansions and improve strategic management of assets and compliance. This is the start of a platform for future innovations in building management in the social housing sector.

It is clear that so many organisations face challenges when transforming to digitised practice. Prior research evidences the confusion that some organisations face, as they engage with the BIM framework (Winfield & Rock, 2018). The initial steps taken in this case study arose out of a desire to bring about improved performance and capability to the business and also as a result of needing to adequately address the changes to building regulations and fire safety compliance. The phase adoption will allow the business and board to witness the changes, financial implications and benefits.

Whilst the industry is aware of BIM and the potential benefits of digitisation, the transformation within the sector remains slow (Sacks et al., 2018; Volk et al., 2014). The initial steps taken demonstrate that it is essential that the desire to disrupt and digitise the sector should be met with a clear business need. The regulatory changes and recent events such as the COVID-19 epidemic are bringing about a need to rationalise and improve data collection, communication and rapid response processes. Whilst previously acknowledged (Ashworth et al., 2019), the need for collaborative platforms, advancing the way information is delivered and managed, is clearly evident. This project has documented the step change and phases adopted through a case study. The work and research is ongoing, with further reflections to be shared.

References

Ashworth, S., Tucker, M. P., & Druhmann, C. K. (2019). Critical success factors for facility management employer's information requirements (EIR) for BIM. *Facilities, 37*(1/2), 103–118.

Attrill, R. (2020). Issues to be addressed with current BIM adoption prior to the implementation of BIM Level 3. Proceedings In: Scott, L and Neilson, C J (Eds.), *35th Annual ARCOM Conference, 2–3 September 2020*, On-line, UK. Association of Researchers in Construction Management.

Coghlan, C. (2017). Try, then buy? Amazon's move is part of a shopping trend. *The New York Times*. Retrieved June 26, 2020, from https://www.nytimes.com/2017/07/05/fashion/try-before-you-buy-amazon.html.

Hackitt, J. (May 2018). Building a safer future: Independent review of building regulations and fire safety. *Final report*. Retrieved June 26, 2020, from https://www.gov.uk/government/publications/independent-review-of-building-regulations-and-fire-safety-final-report.

Lu Q., Xie X., Heaton J., Parlikad A.K., Schooling J. (2020). From BIM Towards Digital Twin: Strategy and Future Development for Smart Asset Management. In: Borangiu T., Trentesaux D., Leitão P., Giret Boggino A., Botti V. (Eds.), *Service oriented, holonic and multi-agent manufacturing systems for industry of the future*. SOHOMA 2019. Studies in computational intelligence, vol 853. Springer, Cham, doi:https://doi.org/10.1007/978-3-030-27477-1_30.

Ministry of Housing, Communities and Local Government. (2020). Final Report of the Expert Group on Structure of Guidance to the Building Regulations. Retrieved June 26, 2020, from https://www.gov.uk/government/publications/final-report-of-the-expert-group-on-structure-of-guidance-to-the-building-regulations.

Rigby, D. K., et al. (2016). *Embracing agile*. Harvard Business Review, May 2016, Retrieved June 26, 2020, from https://hbr.org/2016/05/embracing-agile.

Sacks, R., et al. (2018). *BIM handbook: A guide to building information modeling for owners, designers, engineers, contractors, and facility managers* (3rd ed.). New Jersey: Wiley.

Technopedia. (2020). *Minimal Viable Product (MVP)*. Retrieved June 26, 2020, from https://www.techopedia.com/definition/27809/minimum-viable-product-mvp.

UK Government. (2020). *Fire safety bill: Explanatory notes*. Retrieved June 26, 2020, from https://publications.parliament.uk/pa/bills/cbill/58-01/0121/en/20121en.pdf.

Volk, R. Stengel, J. and Schultmann, F. (2014) Corrigendum to "building information modelling (BIM) for existing buildings—Literature review and future needs". *Automation in Construction, 43*, 106–127.

Wang, H., Meng, X., McGetrick, P. J. (2018). Incorporating knowledge of construction and facility management into the design in the BIM environment. In: C. Gorse, & C.J. Neilson, (Eds.), *34th*

Annual ARCOM Conference. 3–5 September 2018, Belfast, UK. Association of Researchers in Construction Management, 806–815.

Wang, S., Wang, W., Wang K. and Shih Y. (2015). *Applying building information modelling to support fire safety management, automation in construction*. 59, February 2015, doi: https://doi.org/10.1016/j.autcon.2015.02.001.

Winfield, M and Rock, S (2018). *The Winfield Rock report, overcoming the legal and contractual barriers of BIM*. UK: UK BIM Alliance.

Chapter 32
BIM Education Through Problem-Based Learning Exercise: Challenges and Opportunities in an Inter-Professional Module

Shariful Shikder

32.1 Introduction

The incorporation of BIM is changing the traditional method of working within the architecture, engineering and construction (AEC) industry. BIM is a broad term that describes the process of creating and managing information produced during the lifecycle of a project using 3D information modelling and other digital tools for collaboration and information management. The tools allow professionals from multiple disciplines to share and view design information in a structured way using the common data environment (CDE) (NBS, 2016). The key principle of BIM is to promote a collaborative working culture with a seamless flow of information between stakeholders. BIM has become an integral part of the UK construction industry since the enforcement of the government's BIM mandate in 2016. However, one of the biggest challenges of industry-wide BIM adoption is the lack of BIM knowledge and expertise (Georgiadou, 2019; Kassem et al., 2015). Hence, with the changing scenario of the construction industry, higher education institutions also face the pressure to update their curriculum and education methods.

However, it may be challenging to include BIM education within the compact construction management, surveying and engineering curriculum (Becerik-Gerber et al., 2011). Since the concept of BIM emerged during the 2000s, the method and content of BIM education have been subject to research by many authors in higher education institutions (Alwan et al., 2015; Boeykens et al., 2013; Bozoglu, 2016; Gledson & Greenwood, 2016; Mathews, 2013; Zhao et al., 2015). The project-based education is widely used in construction and engineering education, and recently

S. Shikder (✉)
School of Built Environment, Engineering and Computing, Leeds Beckett University, Leeds, UK
e-mail: s.h.shikder@leedsbeckett.ac.uk

the method is also adopted in BIM education (Bozoglu, 2016; Leite, 2016). However, the rapid development of software and hardware technology makes the transition from the traditional approach to adopt the BIM process more challenging, and few early authors indicated there is a lack of knowledge about BIM pedagogy (Wang & Leite, 2014). Hence, the need to evaluate existing teaching methods and exploring opportunities for the future is necessary.

Existing literature suggests that problem-based learning (PBL) and/or cooperative learning is applied as a teaching method for BIM education. In PBL, students are supposed to achieve a common goal while working in small groups. The common goal for each group may be to deliver a design project, feasibility report or BIM execution plan, etc. The assessment methods include primarily observing and analysing collaborative decision-making process using digital tools.

Based on the constructivist learning principles (Jonassen, 1994), a problem-based learning (PBL) technique is applied in an inter-professional studies (IPS) module to explore opportunities for BIM education. Unlike few other case studies from the literature review (Boeykens et al., 2013; Leite, 2016; Zhao et al., 2015), the IPS exercise is not dedicated specifically for BIM education; however, the aim was to embed BIM education in a construction-related group project (e.g. design proposal and a feasibility report).

32.2 Aim and Objectives

This study aims to identify the prospect of inter-professional group projects (problem-based learning exercise) as a method of educating Building Information Modelling (BIM) among AEC students.

The following objectives are set to achieve the above aim:

- Identify the students' existing knowledge of BIM (method: PollEV survey).
- Identify collaboration methods between group members and use of digital tools for the design/proposal development (method: observation, group-based discussion, PollEV survey).
- Identify limitations and challenges of the inter-professional group projects in delivering BIM education (method: literature review and data analysis).
- Explore further opportunities for improving BIM education through inter-professional studies module (method: literature review and results analysis).

32.3 Research Methods

Problem-based learning exercise in the inter-professional studies (IPS) module:
Students from four disciplines were involved in this module. They include final year undergraduate students (Level 6) of Architectural Technology (AT), Building

Surveying, Quantity Surveying and Project Management courses. They worked in small groups (five to six members) and aimed to develop a schematic design proposal and a feasibility report for a given site.

Data collection and analysis methods: This study employed a mixture of quantitative and qualitative methods, which includes two questionnaire surveys along with qualitative discussion and observation of group activities. Data were collected in various formats for 8 weeks as shown in Fig. 32.1.

Questionnaire surveys were conducted in week 2 to develop a general understanding of the students' BIM knowledge at the beginning of the term (see Fig. 32.1). Another survey was conducted again in week 9 to gather their knowledge of BIM and experience of collaborative working. *Group-based discussion* and *observation* were conducted to identify their understanding of the BIM process, group collaboration and the use of BIM-specific tools. Throughout the period, the author also observed group members' involvement in the collaborative decision-making process. And finally, *assessment* and *observation* took place during the groups' final presentations, which provided an opportunity to assess their understanding of relevant standards and compliance criteria. A schematic representation of the adopted research method is shown in Fig. 32.1.

32.4 Results

Findings from the week 2 survey show that most of the students had a basic understanding of BIM; however, their knowledge was insufficient to understand the BIM process. Figures 32.2, 32.3, and 32.4 present the results of three questions asked to the students. The first question was "What is BIM?", where most of the students were able to identify that it was a process. However, when advanced questions were

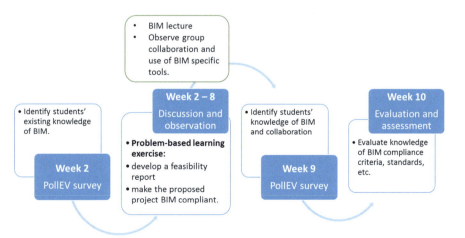

Fig. 32.1 Research flow diagram

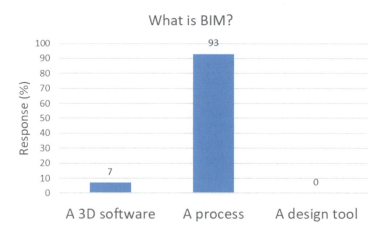

Fig. 32.2 Result of the question "What is BIM?"

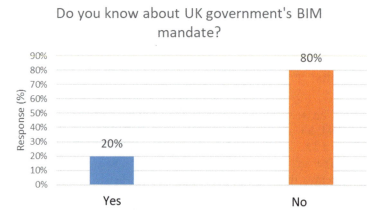

Fig. 32.3 Result of the question related to the UK government's BIM mandate

asked related to governments' BIM mandate and information management, most of the students were not aware of them. The results show that most of the students did not know about common data environment (CDE) and were not aware of UK governments' BIM mandate (IPA, 2016). These findings indicate that they were not aware of the key process and principles of BIM.

Another survey was conducted in week 9, and between these periods, there was a seminar related to BIM, which covered relevant BIM principles, process and standards. The students were also encouraged to conduct research and make their project proposal BIM compliant. As expected, it was found in week 9 that most of the students are well aware of the BIM process and governments' BIM mandate, including some understanding of the common data environment (CDE). However, questions related to group collaboration provided mixed answers.

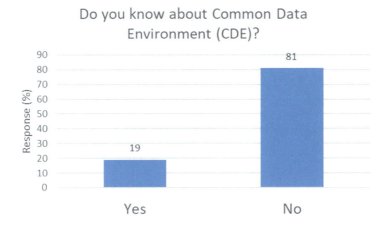

Fig. 32.4 Results of the question related to common data environment (CDE)

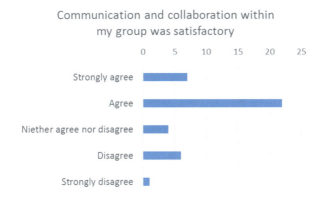

Fig. 32.5 Students' response to communication and collaboration between group members

Students agreed that communication and collaboration within the group members were satisfactory (Fig. 32.5), and it was found that most of the students used real-time text messaging apps (e.g. WhatsApp, Facebook messenger, etc.) to communicate with group members (Fig. 32.7). However, a considerable portion (40%) did not agree that the group exercise improved their learning of a collaborative working environment (Fig. 32.6). An explanation of the above result may be a portion of students were too much focussed on individual performances than group collaboration. A question related to document sharing methods also revealed that the most of them used online messaging services (e.g. WhatsApp) to share files; however, they also used cloud-based file sharing platforms like Dropbox or OneDrive (Fig. 32.8).

The observation of group activities identified that not all group members were equally competent in using BIM-specific tools. In most of the groups, only

Fig. 32.6 Students' response to the learning experience of the collaborative working environment

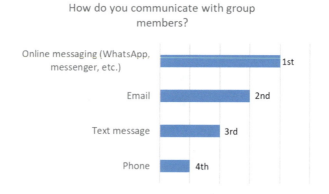

Fig. 32.7 Communication tools between group members

Architectural Technology students used specific BIM authoring tools (Autodesk Revit) for their design development. The work started by printing out the site plan; most of the groups started with sketches and discussion. It was found site plan was converted to CAD drawing, and AT students took responsibility to start drawing up their proposal, where QS/BS students later participated in the costing of the project. Ideally, the CAD drawing could have been taken to a BIM-specific software and further developed. Although some AT students utilised Revit to produce their drawings and models, none of the project managers, BS or QS used any BIM-specific software.

Although the final assessment shows the groups were aware of relevant standards to make the project BIM compliant, collaborative group activities were not intuitive to follow the principles of BIM. Among the 11 groups, most of the groups did not discuss their collaborative working practice as a part of the BIM adoption.

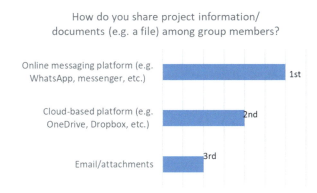

Fig. 32.8 Document sharing between group members

32.5 Discussion

It was evident that the PBL method was successful in developing a feasibility report. Such an approach is expected to work as a mean of transferring theoretical knowledge to workplace practice (Mathews, 2013).

Although the students developed a better awareness of the BIM process, the collaborative use of BIM application lacked within the group exercise. One of the reasons may be their group collaboration using BIM principles and the use of BIM-specific software/tools was not mandatory; hence, they were more concentrated in elements which involved formal assessment. Collaboration is a key principle of the BIM process (Kumar, 2015), and students should be encouraged to utilise digital tools adequately to maximise collaboration. The students must be equipped with adequate tools to support this idea, for example, the CDE, specific software, lab facilities, etc.

Students may not want to learn new software in a short period, and previous experience of using BIM software was found highly beneficial in this context; this finding is also evident from existing literature (Zhao et al., 2015). Separate direct supervision based on disciplines may be necessary to improve the use of BIM-specific software/tools in such a PBL setting (Perrenet et al., 2000).

The understanding of BIM principles and process was also not equal among all group members. Each group member was assigned with a specific task/client concerns, and the cumulative group presentation showed they were able to deal with the specific problems. One of the group members was responsible for developing the BIM compliance strategy for each group, and that particular group member was more involved in BIM-related studies. It should be noted that the adoption of BIM principles within the group exercise was not assessed; hence, the students did not feel any necessity to incorporate such principles/practice.

Teaching BIM may be challenging for various reasons. Leite (2016) highlighted a few challenges of teaching BIM. Firstly, the students need to understand the concept of BIM correctly. It's not a simple software to support any individual discipline. Hence, understanding the collaboration and information management process is as important as learning a software. Because of the rapid change of software and hardware technology, it may be possible that their knowledge of specific software can be obsolete in the near future; hence, self-directed learning should be emphasised. Furthermore, BIM is still emerging and the students should be able to imply critical thinking about the BIM process.

One of the key elements of project-based learning is all tutors need to be clear about the objectives or goals of the module. Hence, incorporating BIM requires all tutors are thinking in the same level and expect similar output (Perrenet et al., 2000).

32.6 Conclusion and Recommendations

The study adopted a mixed approach to identify the BIM education prospect through an inter-professional module. In summary, it is found that the inter-professional module with a PBL setting has prospects of educating BIM principles and process among architecture and construction students. The findings of this study are also supported by the existing literature (Boeykens et al., 2013; Zhao et al., 2015). However, unlike other examples from the literature review, this inter-professional module was not specifically dedicated to educating BIM. The study also identified and discussed several challenges of BIM education in a PBL setting. Apart from technical challenges (infrastructure/CDE/ software, etc.), the discussion highlighted the importance of specifying learning outcomes of the module and each tutors' commitment and engagement to incorporate BIM education.

The findings of this study highlighted some of the key issues that should be considered in the future. Based on the empirical research findings, the following recommendations are proposed to improvise BIM education through the inter-professional module:

- Introduction to BIM-specific software in earlier semesters needs to be focussed so students can apply their expertise in this module and improve their BIM competency. This would provide them with a hands-on experience of applying theoretical knowledge to a practice like environment.
- Students should be encouraged to use BIM-specific software packages in the IPS module.
- Simulating a collaborative BIM environment requires infrastructure (e.g. common data environment and relevant software, etc.), which should be addressed.
- To maximise the benefit of the collaborative working process using BIM principles, all relevant disciplines should establish desired learning outcomes; school-wise awareness is necessary to achieve this aim.
- Because of the rapid changes of software and hardware technology, it may be possible that the knowledge on specific software can be obsolete in the near future; therefore, self-directed learning should be emphasised (Leite, 2016).

- Future studies should include in-depth research including industry professionals and academics to determine specific industry requirement regarding BIM skills and specific pedagogic approach for higher education institutions.

References

Alwan, Z., Greenwood, D., & Gledson, B. (2015). Rapid LEED evaluation performed with BIM based sustainability analysis on a virtual construction project. *Construction Innovation, 15*(2).

Becerik-Gerber, B., Gerber, D., & Ku, K. (2011). The pace of technological innovation in architecture, engineering, and construction education: Integrating recent trends into the curricula. *Journal of Information Technology in Construction, 16*, 411–432.

Boeykens, S., De Somer, P., Klein, R., & Saey, R. (2013). Experiencing BIM collaboration in education. *Computation and performance – Proceedings of the 31st eCAADe conference, 2*(2), 505–513.

Bozoglu, J. (2016). Collaboration and coordination learning modules for BIM education. *Journal of Information Technology in Construction, 21*(June), 152–163.

Georgiadou, M. C. (2019) An overview of benefits and challenges of building information modelling (BIM) adoption in UK residential projects. Construction Innovation, 19(3), pp298–320.

Gledson, B. J., & Greenwood, D. J. (2016). Surveying the extent and use of 4D BIM in the UK. *Journal of Information Technology in Construction, 21*(April), 57–71.

IPA. (2016). *Government construction strategy 2016-2020*. The Infrastructure and Projects Authority, UK.

Jonassen, D. H. (1994). Thinking technology: Toward a constructivist design model. *Educational Technology, 34*(4), 34–37.

Kassem, M., Kelly, G., Dawood, N., Serginson, M., & Lockley, S. (2015). BIM in facilities management applications: A case study of a large university complex. *Built Environment Project and Asset Management, 5*(3), 261–277.

Kumar, B. (2015). *A practical guide to adopting BIM in construction projects*. Scotland: Whittles Publishing.

Leite, F. (2016). Project-based learning in a building information modeling for construction management course. *Journal of Information Technology in Construction, 21*(April), 164–176.

Mathews, M. (2013). BIM collaboration in student architectural technologist learning. In *Architectural technology: Research & Practice* (pp. 213–230). Oxford: Wiley.

NBS. (2016). *What is BIM?* Retrieved April 4, 2019, from https://www.thenbs.com/knowledge/what-is-building-information-modelling-bim.

Perrenet, C., Bouhuijs, P. A. J., & Smits, J. G. M. M. (2000). The suitability of problem-based learning for engineering education: Theory and practice. *Teaching in Higher Education, 5*(3), 345–358.

Wang, L., & Leite, F. (2014). Process-oriented approach of teaching building information modeling in construction management. *Journal of Professional Issues in Engineering Education and Practice, 140*(4), 04014004.

Zhao, D., McCoy, A. P., Bulbul, T., Fiori, C., & Nikkhoo, P. (2015). Building collaborative construction skills through BIM-integrated learning environment. *International Journal of Construction Education and Research, 11*(2), 12.

Chapter 33
Investigating the Trinity Between Sustainability and BIM-Lean Synergy: A Systematic Review of Existing Studies

Hafize Büşra Bostancı, Onur Behzat Tokdemir, and Ali Murat Tanyer

33.1 Introduction

Environmental problems cause the abuse of natural resources and the release of GHG (greenhouse gases) into the atmosphere; therefore, these problems have become the reality that the world has to face today at many fields. Within this context, creating a more sustainable environment has become a phenomenon among the researches and practitioners. Particularly, when the impacts of the construction industry for the environment are considered, the reflection of sustainability on AEC industry has reached the peak to gain more sustainable built environment by minimizing waste and increasing the rapid and precise production. Within this perspective, the cutting-edge technologies have paved the way for constructing more sustainable buildings by controlling and monitoring the whole process from design to construction. The development of CAD technologies has led the development of Building Information Modeling (BIM) technologies that provide any kind of information about buildings. The main benefit of BIM is to make the whole process from conceptual design to physical production of buildings easier for all stakeholders of AEC industry. Garber (2014) states that "BIM enables the whole design and construction groups to digitally arrange the complex building process before the construction" (Garber, 2014, p. 14). Thus, BIM technologies have become a significant matter for sustainable construction by controlling every process to eliminate wastes and to reduce rework, even for more complex structures. However, as Alwan et al. state (2017, "the strategic vision for sustainability needs to be strengthened for inefficient logistics and communication" (Alwan et al., 2017). Sacks et al. (2010) have

---

H. B. Bostancı (✉) · A. M. Tanyer
Department of Architecture, Middle East Technical University, Ankara, Turkey

O. B. Tokdemir
Department of Civil Engineering, Middle East Technical University, Ankara, Turkey

indicated this approach by expressing the relation between BIM and lean construction. It is clarified that both lean construction, which is a cognitive approach for project and construction management, and BIM, a transformative information technology for the construction industry, are two significant innovations leading to profound shifts in AEC industry (Sacks et al., 2010). The synergies between these two concepts could help to reform productivity, efficiency, and quality for sustainable development (Arayici et al., 2011). Hence, the attention on these three concepts has started to increase among researchers and practitioners who want to implement these synergies to design the design process. Within this point of view, this paper focuses on a bibliometric review for the relationship between sustainability and BIM-Lean synergy by analyzing the research fields to identify a knowledge-based conceptual structure and to enable a social network for future studies.

33.2 Conceptual Review

33.2.1 Building Sustainability

The modern concept of sustainable development is introduced in the report "Our Common Future," commonly recognized as Brundtland Report (1987), released by *United Nations World Commission on Environment and Development* in 1987 (Brundtland, 1987; Tolba & Biswas, 1991, p. 29). In this report, sustainable development is described as "paths of human progress that meet the needs and aspirations of the present generation without compromising the ability of future generations to meet their needs" (Brundtland, 1987; Tolba & Biswas, 1991, p. 29). Thus, the interest for sustainable development has started to increase to reduce the detrimental effect of any kind of products from a plastic to a huge-complex building by minimizing potential negative environmental impacts since the Brundtland Report has been published. In this context, the concept of sustainability has also gained importance in the construction industry. The foundations of the term "sustainable construction" have been laid to define the responsibility of the construction industry to achieve sustainability at the International Conference on Sustainable Construction, whose main goal is to evaluate "sustainable construction" or "green construction." At this conference, Kilbert (1994) claimed that sustainable structure means "creating a healthy environment by using resource-efficient, ecologically-based principles" (Hill & Bowen, 1997; Kilbert, 1994). Therefore, minimizing and eliminating the effects of the construction activities by controlling, monitoring, and managing each process has become the main goal of construction industry because of being the top industry among others, such as transportation and infrastructure, which consumes energy mostly, decreasing natural resources and raw materials. According to the data of *Organisation for Economic Co-operation and Development* (OECD), the construction industry consumes nearly 40% of primary energy (OECD, 2003); furthermore, *World Business Council for Sustainable Development* (WBCSD) has reported that 75–85% of the energy consumed in buildings is consumed during use and maintenance (WBCSD, 2008, p. 48). In this sense, creating a more sustainable built environment has become a priority among the researchers and practitioners to save the natural sources and to reduce the impact of climate change as well as to prevent it.

33.2.2 Building Information Modeling (BIM)

Building Information Modeling (BIM) is a system developed by innovative software companies and universities within the scope of programs that mature from object-based parametric modeling. It is widely used in AEC professions as it is an application that transforms building design and process standards. BIM allows the whole design and production group to digitally arrange the complex construction process before the structure is physically produced by leading the designer to find "many more aspects of the project in the initial dimensioning phase using complex computer graphics tools" (Garber, 2014, p. 14). BIM continues to maintain its importance as a technology that reflects and emphasizes building processes and influences not only how buildings are built (efficiency and operations) but also how buildings are designed to generate and analyze building models. Moreover, it enables the production of a model that contains all the geometric and nongeometric data of the building and its surroundings by processing and calculating data in a coordinated and interactive manner. Besides, BIM ensures that the whole process, taking place in the ongoing cycle from design to demolition, is carried out interactively by stakeholders, and promises to integrate all types of data needed into a single file or model; therefore, it causes a paradigmatic change in the AEC sector through these features (Kumar, 2015, p. 23). In consequence, use of BIM, leading a more sustainable globe, changes the way of constructing for a better built environment by (i) reducing waste occurring during design and construction process due to creation of simultaneous studies among stakeholders, (ii) ensuring all kind of information from the quantity of material to the energy consumption of a building during operation and maintenance, and (iii) providing simple solutions for manufacturing of components of complex structures in order to prevent the production errors.

33.2.3 Lean Construction

The background of the lean concept, developed by Toyota Engineer Taiichi Ohno, traces back to the 1960s. The term refers to the main concepts of *Toyota Production System* (TPS) that aims to improve efficiency by eliminating waste occurring during production, with two main pillars: (i) *just-in-time flow* (JIT) and (ii) autonomation. In this point, the lean concept concentrates on minimizing waste to develop value in order to meet customer demand and maintain profitability (Carvalho et al., 2010;

Womack et al., 1991). Huge problems, such as uncertainty of safety conditions, waste of time between workflows, and large amount of construction waste that the construction industry has to deal with, enable the industry to find ways to increase efficiency. As Ballard and Howell (2003) states, Koskela (1992) has introduced the new concepts and techniques in order to explore and apply them to the construction industry (Ballard & Howell, 2003; Koskela, 1992). Furthermore, Koskela (1999) and Howell (1999) have indicated to consider the construction industry as production one by expressing the idea of "to compress the cycle time by eliminating non-value-adding time" and have introduced the main theoretical principles to apply into construction industry (Howell, 1999; Koskela, 1999). Thus, lean thinking has started to be implemented in the construction industry, and it can actually be considered for the construction industry as a new approach in all processes from design and production to workforce and supply chain. Sacks et al. (2010) state that lean construction refers to the adaptation and application of lean principles for construction industry to take attention for reducing waste and enhancing efficiency by maintaining continuity in improvement (Sacks et al., 2010). This could be achieved by reducing variability, increasing reliability and productivity, and developing both workflow and overall efficiency (Francis & Thomas, 2019). As all employees dominate the process during the workflow, the whole process can be enhanced. Consequently, as Pinch (2005) states that the lean thinking is "the slimmer, the more streamlined, the better" (Pinch, 2005), it could be the best approach in order to transform the construction industry by improving the activities hidden within the industry.

33.3 Research Methodology

The rise in the number of academic publications has led the researchers to understand and organize their studies in a different manner because of the challenge in order to accumulate knowledge. Within this context, fulfilling a successful literature review is the main step for sustaining a study. Rousseau (2012) has defined the literature review containing contextual information for a study (Rousseau, 2012, p. 347) as synthesizing alternative solutions by understanding the problem and judging how the results of these solutions are going to benefit in a special context (Rousseau, 2012, p. 52, Aria & Cuccurullo, 2017). Thus, different ways for quantitative and qualitative researchers have been developed. One of these approaches is "bibliometric analysis" that was firstly used in 1969. Between others, bibliometric analysis has the ability to implement a comprehensive, straightforward, and reproducible analysis method focused on the statistical measurement of science, thanks to providing a more objective and reliable analysis (Aria & Cuccurullo, 2017). Therefore, the research methodology of this study is a systematic review based on a bibliometric analysis in order to take attention to the studies focusing on the synergy among BIM-Lean-Sustainability. Within the perspective of this study, data for analysis is extracted from the Scopus database because it contained most of the

Table 33.1 Selected keywords for research

Topics	Keywords search options
BIM	"Building information modelling" OR "building information modeling" OR BIM
Lean	Lean OR "lean construction" OR "lean thinking"
Sustainability	Sustainability OR sustainable OR green OR environmental

higher-ranked journals. First of all, keywords are defined in order to do useful research (Table 33.1). Then, publications containing these three concepts are searched in Scopus (Fig. 33.1).

The research containing the publications between 2011 and 2020 shows that 53% of the results are conference papers, 31% of the results are articles, and 16% of the results are reviews and books (Fig. 33.1a). Besides, it is clarified that there are ten articles within these years, as seen in Fig. 33.1b. The numbers of articles have a peak in 2018; on the other hand, it can be said that the interest for evaluating the relation between BIM, Lean, and Sustainability has decreased by 2019. For this reason, in order to find the trilogy between these three concepts, new research is provided in Scopus by matching the concepts in pairs. In this context, research continues as BIM-Lean, BIM-Sustainability, and Lean-Sustainability. Since the studies carried out at the intersection of three concepts are encountered between the years of 2011 and 2020, the year range in the rest of the study is also determined as 2011–2020. In addition, the research is limited by the studies in the fields of Engineering and Environmental Science and by the studies in English. When the articles studied between these years are examined, values appearing in Fig. 33.2 have been obtained.

The information obtained in this sortation is exported in BibTex format and analyzed separately with "*bibliometrix* R-package and Biblioshiny software package." The focus of the bibliometric analysis is on co-word analysis, which is the only method that uses the most important words or keywords to study the conceptual structure of a research area. Then, co-author analysis, identifying the authors to study the social structure and collaboration networks, and citation analysis, proving the counts to measure the similarity between authors and publications (Aria & Cuccurullo, 2017), are focused.

33.4 Outcomes

The results exported from Scopus are processed in order to address a scientific mapping tool (data update time May 22, 2020). The analysis is limited to the top 20 keywords and the top 10 authors (because there are 10 publications for trinity). As mentioned, because of the limited results for BIM-Lean-Sustainability research topics, the study is carried out in four stages in order to address the synergy between these as in pairs.

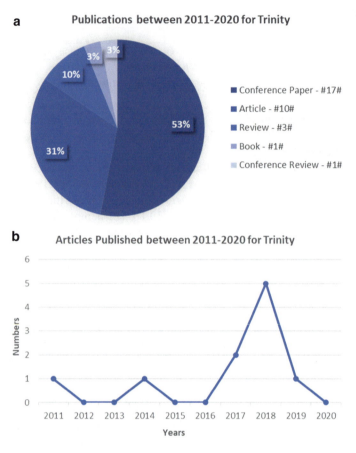

Fig. 33.1 (**a**) The result among publications in Scopus database. (**b**) The numbers of articles published between 2011 and 2020 for trinity

33.4.1 Stage 1: "BIM-Lean" Research

The number of the results according to the research parameters is 71 articles within the perspective of this stage. This stage is contacted with "("building information modelling" OR "building information modeling" OR BIM) AND (lean OR "lean construction" OR "lean thinking")" keywords (Table 33.1). Figure 33.3 shows the top authors' production over the years and the most cited authors.

As it is seen in Fig. 33.3, although the most productive author is seen as Koskela, the most cited one becomes Arayıcı. When the publications are analyzed, it is figured out that there are publications provided by Arayıcı and Koskela collaboration. Besides, the publications, that Koskela isn't the first author for, can affect this duality.

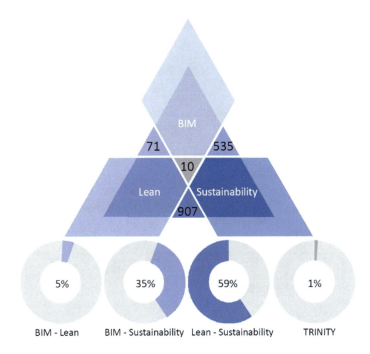

Fig. 33.2 The diagram of the number of results for both pairs and trinity

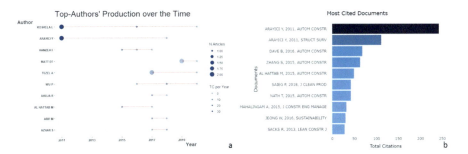

Fig. 33.3 (**a**) The authors' production over the years. (**b**) The most cited authors (Stage 1)

Figure 33.4 presents the most co-occurred keywords. These analyses are conducted according to both keywords plus and authors' keywords. The top 20 keywords are evaluated within the scope of the analysis.

The biggest keywords show the biggest impact. "Architectural design" keyword has the highest impact among the publications according to the topics (Fig. 33.4a). On the other hands, if the most co-occurrence authors' keywords are evaluated, it is seen that "lean construction" keyword has the highest impact (Fig. 33.4b). Besides, it can be stated that keywords for two separated analyses vary from each other.

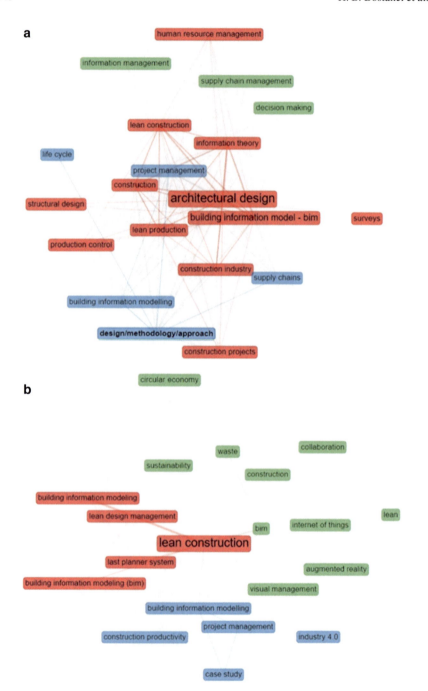

Fig. 33.4 (**a**) Most co-occurred keywords among the topics of publications. (**b**) Most co-occurred authors' keywords (Stage 1)

33.4.2 Stage 2: "BIM-Sustainability" Research

The number of the results according to the research parameters is 535 articles within the perspective of this stage. The interest for these topics is more than the first stage. This stage is contacted with "("building information modelling" OR "building information modeling" OR BIM) AND (sustainability OR sustainable OR green OR environmental)" keywords (Table 33.1). The top authors' production over the years and the most cited authors are shown in Fig. 33.5.

As it is seen in Fig. 33.5, the authors change among being the most productive and the most cited. However, it can be easily understood, particularly from Fig. 33.5a, that the interest for the "BIM-Sustainability" integration is larger than the interest for the "BIM-Lean" when these stages are compared. Besides, the most co-occurred 20 keywords according to both keywords plus and authors' keywords are provided in Fig. 33.6.

As in the first stage, "architectural design" keyword has the highest impact among the publications according to the topics. Furthermore, it can be said that the attention for terms related to energy and efficiency has increased when the keywords are evaluated (Fig. 33.6a). On the other, if the most co-occurrence authors' keywords are evaluated, it is seen that "BIM" and "green buildings" keywords have the highest impact (Fig. 33.6b), and it can be stated that keywords for two separated analyses are related with each other due to concern about energy efficiency in terms of sustainability.

33.4.3 Stage 3: "Lean-Sustainability" Research

The number of the results according to the research parameters is 907 articles within the perspective of this stage. The research interest related with these topics is the highest one among others because lean is generally integrated with sustainability. The third stage is contacted with "(lean OR "lean construction" OR "lean thinking") AND (sustainability OR sustainable OR green OR environmental)" keywords

Fig. 33.5 (**a**) The authors' production over the years. (**b**) The most cited authors (Stage 2)

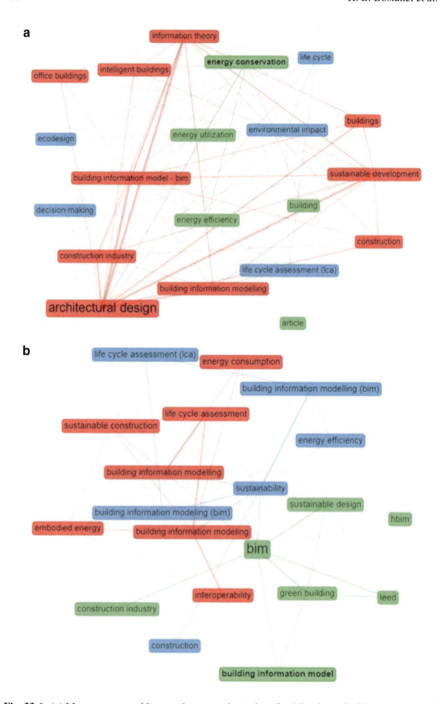

Fig. 33.6 (**a**) Most co-occurred keywords among the topics of publications. (**b**) Most co-occurred authors' keywords (Stage 2)

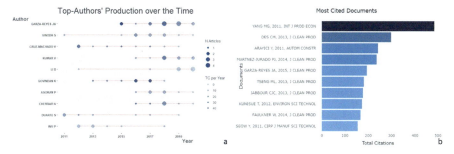

Fig. 33.7 (**a**) The authors' production over the years. (**b**) The most cited authors (Stage 3)

(Table 33.1). The findings about the most productive and the most cited authors are given in Fig. 33.7.

Figure 33.7 shows the change among the authors for being the most productive and the most cited; besides, when the change is taken into consideration, it can be said that the productivity of the top authors has started to increase after 2015 (Fig. 33.7a). Figure 33.8 provides the information about the most co-occurred 20 keywords according to both keywords plus and authors' keywords.

In this stage, the keyword having the highest impact among the publications according to the topics is interestingly "article," but it is followed by the relation between "lean production, manufacture, environmental impact, and sustainable development" (Fig. 33.8a). Furthermore, when the keywords plus among publications are evaluated, the keywords have a big variety from chemistry to gender that can cause a confusion. On the other hand, analysis for the most co-occurrence authors' keywords shows that "lean, sustainability, and lean manufacturing" keywords are the most preferred ones by authors to address the study (Fig. 33.8b).

33.4.4 Stage 4: "BIM-Lean-Sustainability" Research

The number of the results according to the research parameters is ten articles within the perspective of this stage which address the direction of the study. This last stage is contacted with "("building information modelling" OR "building information modeling" OR BIM) AND (lean OR "lean construction" OR "lean thinking") AND (sustainability OR sustainable OR green OR environmental)" keywords (Table 33.1). Since it is mentioned before, the analysis of trinity "BIM-Lean-Sustainability" frames the study because of limited publications within this field. The findings about the most productive and the most cited authors are presented in Fig. 33.9.

The productivity within these topics is not as much as others as it is seen in Fig. 33.9a; it almost has started after 2017 as understood by the number of publications. On the other side, the value of the most cited authors is not too efficient because of ongoing interest for these topics. Lastly, Fig. 33.10 provides the information about the most co-occurred keywords.

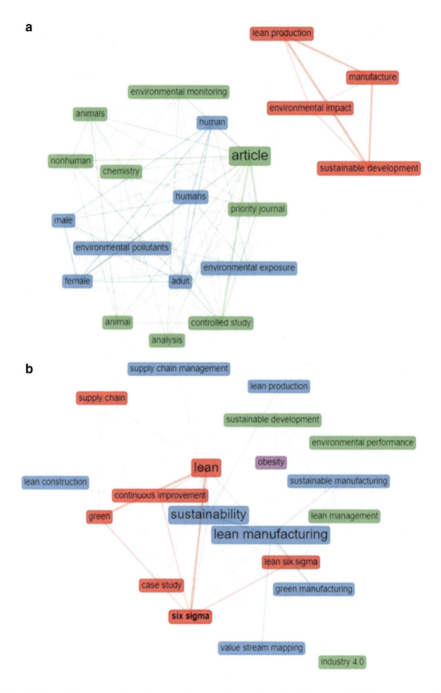

Fig. 33.8 (**a**) Most co-occurred keywords among the topics of publications. (**b**) Most co-occurred authors' keywords (Stage 3)

Fig. 33.9 (**a**) The authors' production over the years. (**b**) The most cited authors (Stage 4)

In the last stage, the keyword having the highest impact among the publications according to the topics is "architectural design," and it is followed by "construction industry" (Fig. 33.10a). The keywords within this stage have a common point among other stages. Moreover, the most co-occurrence authors' keywords analysis has specific keywords such as "Roanoke, Virginia, North America, etc." even though the keywords "BIM, sustainability, and lean construction" involve bigger part of the analysis. It can be interpreted that these keywords form the contents of the relevant studies because of the disconnectedness with the blue side of the figure (Fig. 33.10b).

33.5 Gap in the Field: Emergence of Synergy

Sustainability terms are a driving force for construction industry, so it is important that all actions in the construction industry comply with this terminology. The paradigmatic transformations, seen in the building industry from the use of BIM technologies to lean principles adopted to the construction industry, have led to consider the future of the globe within the context of sustainability. Besides, BIM can facilitate lean construction practices, and most of the lean principles implemented into the construction industry have common functionalities in BIM. Thus, BIM-Lean synergy encourages many aspects such as providing better visualization and information flow among stakeholders and minimizing error by directing the decision-making process for maintenance (Saieg et al., 2018). Therefore, both lean construction and BIM are closely related to sustainability. This is why the number of publications is more as it is proved in the previous section. On the other side, no matter how many studies between binary groups are, the number of studies containing these three terms together is very few. As it is stated by Lekan et al. (2018), galvanizing the idea of BIM and Lean is going to produce a powerful framework that could be helpful in developing qualities leading to sustainable development (Lekan et al., 2018).

The last step for analysis of the fourth stage is evaluated for the consideration for BIM-Lean-Sustainability integration. Three-fields plot – Sankey diagram, which is developed to visualize the energy distribution in a city by exemplifying quantitative

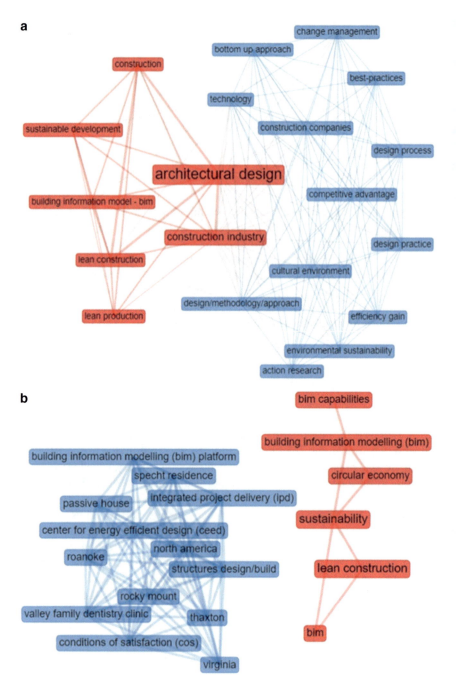

Fig. 33.10 (**a**) Most co-occurred keywords among the topics of publications. (**b**) Most co-occurred authors' keywords (Stage 4)

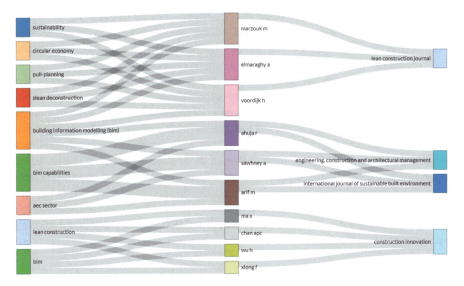

Fig. 33.11 Three-fields plot – Sankey diagram

information about flows (Riehmann et al., 2005), is presented in Fig. 33.11 in order to take attention to the relation between the most co-occurred keywords, the most cited authors, and the sources.

In Fig. 33.11, the left-hand side column shows the keywords, the middle column shows the authors and the right-hand side column shows the sources. Since it is seen in Fig. 33.11, the most used keywords are "building information modeling, BIM capabilities, lean construction, and BIM" followed by "sustainability" used less. The number of authors using these three keywords together is not very high. However, it seems that the authors mostly focus on the relationship between BIM-Sustainability and BIM-Lean. It can be concluded that Marzouk M., Elmaraghy A., and Voordijk H. have articles focusing on sustainability, BIM, and pull-planning. In other respect, other authors are seen as focusing on either BIM capabilities or integration of BIM-Lean for their articles. Thence, it can be extrapolated that even though the research interest is high in binary groups, the interest in associating these three words is not so much. It is thought that the interest in the trinity should increase in order to provide a sustainable future; and it is desired to draw attention to the gap in this field for future studies.

33.6 Conclusion

In this study carried out with a systematic review, the three concepts that have undergone a paradigmatic change in the construction industry are focused on. In the bibliometric analysis made to examine the relationship between BIM, Lean, and

Sustainability terms, it is seen that the authors give priority to Lean-Sustainability terms in their publications due to the adaptation of the lean principles to the construction industry. In addition, the use of BIM, which offers advantages in various fields from material selection and energy-efficient building construction to energy management and optimization of construction processes, is considered to be the leading factor in ensuring the sustainability of the construction industry. Furthermore, the fact that the use of BIM offers technological opportunities in the implementation of lean principles has also increased the interest in this bilateral relationship since 2010. However, although bilateral relations have been established very strongly, few studies focusing on the relationship between these three concepts in the literature draw attention. The main purpose of this study is to visualize the gap in this field with the concept of "trinity," which presents the relation between BIM, Lean, and Sustainability. It is aimed to provide a dynamic platform for future studies with the analyses which are conducted to prove that the direction of a study is the selected keywords.

References

Alwan, Z., Jones, P., & Holgate, P. (2017). Strategic sustainable development in the UK construction industry, through the framework for strategic sustainable development, using building information modelling. *Journal of Cleaner Production, 140*, 349–358.

Arayici, Y., Coates, P., Koskela, L., Kagioglou, M., Usher, C., & O'Reilly, K. (2011). Technology adoption in the BIM implementation for lean architectural practice. *Automation in Construction, 20*(2), 189–195.

Aria, M., & Cuccurullo, C. (2017). Bibliometrix: An R-tool for comprehensive science mapping analysis. *Journal of Informetrics, 11*(4), 959–975.

Ballard, G., & Howell, G. (2003). Lean project management. *Building Research & Information, 31*(2), 119–133.

Brundtland, G. (1987). *Our common future: Report of the 1987 world commission on environment and development* (pp. 1–59). Oslo: United Nations.

Carvalho, H., Azevedo, S. G., & Cruz-Machado, V. (2010). Supply chain performance management: Lean and green paradigms. International Journal of Business Performance and Supply Chain Modelling, 2(3–4), 304–333.

Francis, A., & Thomas, A. (2019). Exploring the relationship between lean construction and environmental sustainability:-a review of existing literature to decipher broader dimensions. *Journal of Cleaner Production, 119913*.

Garber, R. (2014). *BIM design: Realising the creative potential of building information modelling.* New York: Wiley.

Hill, R. C., & Bowen, P. A. (1997). Sustainable construction: Principles and a framework for attainment. *Construction Management & Economics, 15*(3), 223–239.

Howell, G. A. (1999). What is lean construction-1999. *Proceedings IGLC, 7*, 1.

Kilbert, C. J. (1994) Final Session of *First International Conference of CIB TG 16 on Sustainable Construction*, Tampa, Florida, 6–9 November.

Koskela, L. J. (1992). *Application of the new production philosophy to construction* (72nd ed.). Stanford: Stanford University.

Koskela, L. J. (1999). Management of production in construction: A theoretical view. *Proceedings of the 7th Annual Conference of the International Group for Lean Construction.*

Kumar, B. (2015). *A practical guide to adopting BIM in construction projects*. London: Whittles Publishing.

Lekan, A., Oluchi, E., Faith, O., Opeyemi, J., Adedeji, A., & Rapahel, O. (2018). Creating sustainable construction: Building informatics modelling and lean construction approach. *Journal of Theoretical and Applied Information Technology, 96*(10), 3025–3035.

Organization for Economic Cooperation and Development [OECD]. (2003). *Environmentally sustainable buildings: Challenges and policies*. Paris, France: OECD Publications Service.

Pinch, L. (2005). Lean construction. *Construction Executive, 15*(11), 8–11.

Riehmann, P., Hanfler, M., & Froehlich, B. (2005). Interactive sankey diagrams. *IEEE symposium on information visualization, 2005*. INFOVIS 2005., 233–240.

Rousseau, D. M. (Ed.). (2012). *The Oxford handbook of evidence-based management*. Oxford: Oxford University Press.

Sacks, R., Koskela, L., Dave, B. A., & Owen, R. (2010). Interaction of lean and building information modeling in construction. *Journal of Construction Engineering and Management, 136*(9), 968–980.

Saieg, P., Sotelino, E. D., Nascimento, D., & Caiado, R. G. G. (2018). Interactions of building information modeling, lean and sustainability on the architectural, engineering and construction industry: A systematic review. *Journal of Cleaner Production, 174*, 788–806.

Tolba, M. K., & Biswas, A. K. (1991). *Earth and us: Population, resources, environment, development*. Oxford: Butterworth-Heinemann.

Womack, J. P., Jones, D. T., Roos, D., & Technology, M. I. of. (1991). *The machine that changed the world: The story of lean production*. Harper Collins.

World Business Council for Sustainable Development [WBCSD]. (2008). *Business realities and opportunities*. Geneva: WBCSD.

Part VIII
Sustainable Practice

Chapter 34
Management Policy for Laboratory Electronic Waste from Grave to Cradle

Mercy Chinenye Iroegbu, Samantha Chinyoka, Tariro Ncube, and Salma M. S. Alarefi

34.1 Introduction

Electronic waste (e-waste) is growing at an alarming rate around the globe specifically in the developed countries. It encompasses electronic and electrical devices such as washing machines, personal computers, consumer electronics, cell phones, printed circuit boards, power supplies, etc. that have been discarded by users (Sastry & Ramachandra Murthy, 2012). The reasons for discarding such items include, but not limited to, obsolescence of the technology used, end of life and a faulty component within the device that cannot be replaced (Osibanjo & Nnorom, 2008a). In some cases, especially in academic and research institutions, the electronic equipment used in experimental activities may not be possible for reuse leading to disposal. Generally speaking, the main sources of e-waste are government offices, public and private sectors, households, manufacturing industries and academic and research institutions (Chatterjee & Kumar, 2009).

Until recently, the amount of e-waste generated was insignificant such that it was possible to use landfills for disposal of such equipment without causing harm to the environment (Kang & Schoenung, 2004). According to the Global E-waste Monitor (Shevchenko et al., 2019), 44.7 million metric tonnes of e-waste were generated in 2016, and this is expected to rise to 55.2 million metric tonnes by 2021, which is equivalent to 6.8 kg of e-waste for each inhabitant per annum (Jian & Shanshan, 2010). If electronic waste is not effectively managed and recycled, future generations will be deprived of electronic devices.

M. C. Iroegbu · S. Chinyoka · T. Ncube · S. M. S. Alarefi (✉)
School of Electronics and Electrical Engineering, University of Leeds, Leeds, UK
e-mail: S.Alarefi@leeds.ac.uk

Environmental policies and legislation on waste management in most countries are designed based on the principle of extended producer responsibility (EPR) (Osibanjo & Nnorom, 2008a). Promising initiatives by European and some Asian countries have been implemented to entail manufacturing companies to reclaim the products at end of their useful life for recycling. In that regard, the EU has introduced a policy known as the Waste Electrical and Electronic Equipment Directive 2002/96/EC that is designed specifically to improve the environmental performance of electrical and electronic waste (Osibanjo & Nnorom, 2008a). Although the internal generation of e-waste in some developing countries particularly in Africa is not much of a concern, importation of used electronic products from the developed countries contributes significantly to e-waste in these countries (Park et al., 2017). That is because most of the developing countries have limited safeguard, legislation and policies for proper disposal of the e-waste; hence, they end up in informal dumpsites which are not monitored by the governments (Kiddee et al., 2013). In such places, informal primitive extraction methods are employed exposing both the environment and the workers to harmful substances.

Existing e-waste management policies are mainly aimed at reducing the product's impact on the environment thought analysis carried at designing, production processes and disposal of out-of-use products (Baldé et al., 2017). There is however a lack of legislation in the optimisation and management of e-waste from educational institutions. This can pose great challenges to the development of sustainable education particularly in developing countries. In fact, academic institutions in developing countries are characterised by inadequate laboratory equipment due to limited funds to purchase and maintain lab equipment. This paper, therefore, presents findings of qualitative study that aim to assist policy makers with the redeployment of electronic and electric equipment. The findings and recommendations are aimed to stimulate interest in the development of e-waste management policy to benefit institutions in developing countries, whereby discarded electronic equipment from academic institutions in developed countries can be deployed to developing countries. Not only would such policy help reduce e-waste but would also contribute to sustainable education development in developing countries. In setting the scene, a critical review of adverse effects of e-waste and existing management policies is presented. The methodology adopted to conduct the online study is also outlined. As well as discussion of the results, recommendations and direction for further research are presented.

34.2 Literature Review

34.2.1 Problem Statement

E-waste contains valuable metals such as gold and silver, which when recovered can be reused as raw materials to manufacture new devices (Hameed, 2012; Ongondo et al., 2011). Despite existing technology and infrastructure for recycling e-waste, the recycling rates in developing countries remain modest. That is mainly due to the high costs associated with the process. Only about 10% of the e-waste generated is recycled, not to mention that which is commonly disposed of in landfills (Chatterjee & Kumar, 2009). In fact, most e-waste is exported to developing countries, where cheap labour and weak environmental laws offer great potential for profit realisation (Oteng-Ababio, 2012; Puckett et al., 2018). Although the intrinsic value of e-waste makes its trade a lucrative business, the harmful effects of the toxic elements it contains are an ecological and health concern. Developing countries lack adequate infrastructure to effectively manage e-waste such that informal recyclers use unhealthy, primitive methods to process the e-waste.

The open-air incineration of device encasings to recover metals releases toxic fumes which exacerbate global warming and affect human health. Also, the incorrect disposal of batteries leaches lead (Pb) into the soil which eventually contaminates underground water. In Guiyu, China's largest e-waste hub, approximately 80% of the children suffer from lead poisoning (Hameed, 2012; Kiddee et al., 2013). Table 34.1 presented a summary of some of the health hazards of primitive recycling of e-waste.

In Agbogbloshie, Ghana, the world's largest e-waste dumpsite, informal e-waste recyclers, which include scrap dealers, employ children from impoverished families to work as e-waste pickers and dismantlers (Daum et al., 2017; Oteng-Ababio, 2012). These e-waste child workers are not aware of their rights such that they are exploited and paid unfair wages. The direct exposure of these children to toxic environments without personal protective equipment (PPE) or medical compensation shortens their life span. Although the threat to human health is more apparent in developing countries, the mismanagement of e-waste is a tragedy of the commons which has global ramifications.

Table 34.1 Health effects of toxic elements in e-waste

Source of e-waste	Toxic element	Health effect
Cables and plastic encasing of electronic equipment	Dioxin – Released from the burning of polyvinyl chloride (PVC) plastic	• Acute respiratory infection • Immune system impairment
Chip resistors and semiconductors	Cadmium (cd)	• Neural damage • Teratogenic • Accumulates in the kidneys and liver
Soldering in printed circuit boards (PCBs)	Lead (Pb)	• Kidney failure • Impairs child-brain development • Damages the central nervous system
Relays and switches	Mercury (hg)	• Skin lesions • Chronic brain damage

34.2.2 E-Waste Management

A regulatory framework that includes e-waste management policies has been put in place to protect human health and the environment against the harmful effects of e-waste. The policy framework is based on the extended producer responsibility (EPR) approach which allocates significant responsibility for recycling to producers (manufacturers) (Magalini et al., 2019).

The producers are required to take back electronic devices that have reached the end of their useful life from the consumer, for purposes of recycling or reuse, that is, either to produce new products or repair for extended useful life. The main aim of this approach is to reduce e-waste at the source by incentivising producers to produce eco-friendly products that have an increased useful life (Magalini et al., 2019). However, a lack of collection points and a lack of consumer awareness of recycling are limitations to this "take-back" system (Srivastava & Sharma, 2015).

34.2.3 Existing Policies and Legislations

Waste Electrical Electronic Equipment (WEEE) Directive 2002/96/EC

This is a directive by the European Union (EU) which requires manufacturers to recycle and dispose of their products in an eco-friendly manner using the take-back system (Ongondo et al., 2011). The directive is based on the EPR policy approach, with the primary aim of preventing the generation of e-waste. Additionally, it promotes reuse, recycling and recovery of e-waste to reduce its disposal, thereby improving environmental performance (Magalini et al., 2019).

Restriction of Hazardous Substances (RoHS) Directive 2002/95/EC

The RoHS directive restricts the use of hazardous substances such as lead, cadmium, mercury, etc. in the manufacturing of electrical electronic equipment (EEE) (Ongondo et al., 2011). Any EEE that contains hazardous substances higher than the accepted levels will not be allowed on the market. The objective of this directive is to reduce the impact of e-waste on the environment by encouraging the use of biodegradable and eco-friendly materials to manufacture EEE.

Energy-Using Products (EUP) Directive 2005/32/EC

The EUP directive sets eco-design requirements for energy-using products within the internal EU market. The objective of the directive is to ensure sustainable development by increasing energy efficiency whilst protecting the environment (Ongondo et al., 2011).

Basel Convention

The Basel Convention restricts transboundary movements of hazardous wastes except where they are perceived to be following the principles of environmentally sound management (Magalini et al., 2019). Thus, the convention requires that the government of an importing country must be given prior notice of any proposed import of certain hazardous EEE.

Literature reveals that there exists a legal loophole in this convention such that there have been reports of increased e-waste dumping where functional equipment is mixed with obsolete and non-reparable equipment which ends up in landfills (Shagun & Arora, 2013). This has led to the proposal of the Basel Total Ban to ensure that hazardous e-waste is not exported to developing countries for any purpose, even for recycling (Oteng-Ababio, 2012). Additionally, treaties such as the Bamako Convention have been enacted to prohibit imports and control the movement of hazardous wastes within Africa. A comprehensive discussion of the treaty is given in (Magalini et al., 2019).

34.2.4 Limitations of Existing Policies

Although many countries have adopted the reviewed policies, the problems of e-waste continue to escalate. There is still an increase in the dumping of e-waste in developing countries. In a 2018 report released by the Basel Action Network (BAN), the UK was found to be the worst offender in Europe for illegal e-waste exports (). These illegal transboundary movements are achieved through the exploitation of the gaps in the existing policies (Ongondo et al., 2011; Singh et al., 2013).

The policies do not consider the whole life cycle of a product such that there is no effective enforcement of supervision up to the end of the product's useful life. Furthermore, there are no specific regulations on e-waste in most African countries to complement the policies (). Other sustainable e-waste management policies must be enacted to work in concert with the existing policies.

34.2.5 Policy Proposal: Management Policy for Laboratory E-Waste

Literature analysis shows that e-waste management is mainly affiliated with health, resource depletion and environmental concerns. Although the concern for resource depletion addresses sustainable development, there is a research gap on the contribution of e-waste management to sustainable education.

In the developed world, universities can easily replace faulty laboratory equipment due to the ease in the procurement of equipment and the motivation to keep up with the fast-changing technological advancements. This results in the stockpiling of out-of-use equipment, some of which are repairable. Although several electronic devices are covered under the existing WEEE directive, there is no existing policy that specifically focuses on the recycling of e-waste of engineering laboratories in universities.

In developing countries, particularly in Africa, the procurement of laboratory equipment in universities is difficult due to economic constraints. Research in (Osibanjo & Nnorom, 2008b) shows that information and technology development in such countries depend on second-hand devices. Most of these devices are imported without prior functionality testing such that they cannot be purposefully reused, which results in a lack of academic equipment. This paper therefore proposes a management policy for e-waste from engineering laboratories of universities in developed countries, whereby out-of-use electronic equipment such as computers, oscilloscopes, signal generators, integrated circuits, etc. are channelled to developing countries for learning purposes.

Recovery and reuse of laboratory equipment would improve the practical learning experience of students in developing countries, as they learn the repair and maintenance of faulty equipment through laboratory activities. Improving maintenance through repair will extend product life and reduce e-waste. Thus, the policy aims to ensure sustainable education and a global transition towards achieving a circular educational economy through sustainable resource and e-waste management practices.

34.2.6 Barriers to Proposed Policy

The growing scepticism about exportation of e-waste to developing countries impedes the social acceptance of the proposed policy. Many conservationists have been advocating for a ban on total imports of e-waste, arguing that e-waste dumping by developed countries has been ongoing under the disguise of donations to "bridge the digital gap" (Osibanjo & Nnorom, 2008a, 2008b; Shagun & Arora, 2013). In 2010, the Government Ministry of Environment and Forests in India imposed a ban on all imports of second-hand electronic devices, to address illegal transboundary movements (Kiddee et al., 2013).

The lack of government support in developing countries can stifle the transformational potential of the proposed policy. That in addition to the hindrance to sustainability policies and incentives, anticipates an awareness deficiency on the contribution of e-waste management towards the achievement of sustainable education. This alongside the bureaucracy and absence of local legislation can lead to ambiguous responsibilities in the distribution of the deployed laboratory equipment. This risks corruption where the deployed equipment can be misdirected for individual gain.

In addition to that, there is clear information deficit on the status of academic engineering laboratories in developing countries. The latter makes it difficult to assess the reality of the technical needs of the end users which is a key consideration in the formulation of the proposed e-waste management policy.

34.3 Research Method

Research presented in this paper is both explanatory and exploratory in nature. It is explanatory as it attempts to identify the gaps in existing e-waste management policies to suggest a better and complementary policy. It is also exploratory as it aims to assess the social acceptance of the suggested policy.

Indeed, for a policy to be effective, the needs and concerns of the end users must be assessed and addressed. To ensure this, a quantitative research method was adopted as appropriate for assessing the social acceptance of the proposed laboratory e-waste management policy by developing countries. A survey was designed in the form of multiple-choice questions with a mixture of close- and open-ended questions. The close-ended questions were formulated using rating scaling to ascertain the awareness level of individuals on e-waste management in academic institutions. Open-ended questions were used to elicit opinions and insights on the reuse of laboratory e-waste from developed countries.

Although the survey was open to the general public, the main target was students and academic staff from educational institutions. Responses from students were crucial in assessing their practical learning experiences, whilst responses from academic stuff were crucial in identifying the challenges in the procurement of laboratory equipment and their willingness to adopt out-of-use laboratory equipment from developed countries. The survey was conducted using an anonymous online form created using the Google Forms platform. The survey was conducted across certain parts of Africa, namely, Nigeria, Zimbabwe, Kenya, South Africa, Malawi, Ghana, Libya, Tunis and Algeria.

34.4 Research Results and Analysis

From the conducted survey, 152 responses were received. However, 7 responses were rendered invalid as the feedback forms had empty fields. Therefore, analysis of the data was done based on 145 responses.

The questionnaire started by examining the age and education level of participants to ascertain whether the respondents are major stakeholders in the education sector. The results collected indicate that approximately 95.1% of the respondents are major stakeholders in the education sector, thus giving credence to the information received from this survey. Approximately 83.4% of the respondents are within the age limit of 25–44 years, which ascertains that the survey feedback is from mature, yet young, persons who are competent enough to give valid opinions.

In the assessment of the respondents' awareness of e-waste and the impact of effective e-waste management, approximately 99.3% seem to acknowledge the role of e-waste management for the development of sustainable engineering learning environment. Despite the reported lack of e-waste management policies in most educational institutions, 93.7% of the respondents are aware of the concept of e-waste management.

Whilst nearly two-quarters of respondents (i.e. 42.8%) faulted the policy management of e-waste disposal in their institutions, a third (i.e. 32.9%) claimed the e-waste management was adequate. In that regard, around 21% of the participants were undecided, meaning that adequate sensitisation should be made at such institutions on how e-waste is managed. As expected, the high majority of the participants (i.e. 64.8%) claimed that the deployment of relevant out-of-use laboratory equipment would benefit their institutions. Although about a tenth of the remaining (9.6%) are indifferent, a quarter of the participants are against the idea.

From the survey, 42.7% of the respondents thought that their institutions are adequately equipped. Approximately 37.2% ascertain that their laboratories are under-equipped, whilst 20% are not certain whether there is adequate equipment or not. The results also reveal that most of the laboratories are not adequately equipped due to financial constraints as a greater percentage of the respondents expressed difficulties in acquiring the equipment. Only 27.6% of participants have institutions that can easily acquire laboratory equipment (Fig. 34.1).

From Fig. 34.2, more than three quarters of the respondents (~89.7%) agree that increasing the duration of the laboratory session whilst optimizing the size of laboratory groups is important in enhancing the learning experience of engineering students. To further investigate the factors that impact practical experience of learners, the participants were invited to rate the significance of the following: availability of lab equipment, effective space allocation and technical support. Whereas more than two-quarters acknowledged and prioritised the provision of adequate lab equipment (42%) and experienced laboratory technicians (i.e. third of the proportion), less than a quarter have rated the space allocation to have an impact on the practical learning experience of engineering students (Fig. 34.3).

34 Management Policy for Laboratory Electronic Waste from Grave to Cradle   469

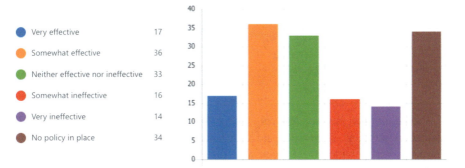

Fig. 34.1 Rating of existing laboratory e-waste (i.e. resistors, inductors, PCBs, PCs, power supplies, oscilloscopes, etc.) management policy

Fig. 34.2 Responses to the significance of laboratory session duration and in enhancing the learning experience of engineering students

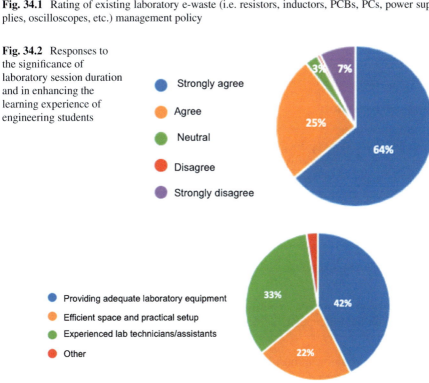

Fig. 34.3 Responses to the key aspects to be considered in improving the practical learning experience of engineering students

From the feedback received through the open-ended questions, individuals expressed concerns about the proposed laboratory e-waste management policy as they argue that it is a ploy to justify a deliberate attempt by the developed countries to dump their e-waste in African institutions. In order to gain a better insight into the key factors to be considered for an effective laboratory e-waste management policy, the participants were asked to rate from the options tabulated in Table 34.2.

Table 34.2 Respondents' opinion on conditions to be fulfilled before the deployment of the equipment

Electronics must be in repairable conditions	Electronics must be safety checked pre and post deployment	Electronics must have at least 5 years of useful life	Learning activities must be included	Staff training must be included
19%	28%	17%	14%	22%

The results reveal that some respondents are willing to accept the policy provided the equipment are reusable and informative learning activities can be included. From the data obtained from the survey, the deployment of out-of-use laboratory equipment would be beneficial to academic institutions in developing countries in Africa and if done in good faith would lead to achieving the grave-to-cradle concept for the circular educational economy. However, for the policy to be effective, the concerns raised by the survey participants must be considered in the formulation of the policy framework.

34.5 Discussion

To enhance social acceptance of the laboratory e-waste management policy, the following discussion addresses the concerns of the stakeholders from the developing countries which were highlighted in the survey. Furthermore, the discussion highlights the opportunities for this policy that address the previously discussed policy barriers.

34.5.1 Policy Opportunities

To address the concerns of e-waste dumping, this paper proposes that the policy must require the following conditions to be fulfilled before the deployment of the equipment:

1. Electronic equipment must be in repairable condition – confirmatory testing of equipment must be done to check its functionality and safety.
2. Electronic equipment should have at least 5 years of useful life – the equipment must be evaluated using a set of global standards.
3. The certified equipment must be branded to enable tracking to its intended destination.
4. Learning activities must be included – complementary training of staff in developing countries must be facilitated for them to acquire technical prowess.

Thus, the policy must clearly define the stakeholders with the financial responsibility for the collection, testing and deployment of the equipment. The policy must also include a deposit refund system for the purchase of laboratory equipment by universities, to promote recycling. Advanced recovery fees (ARF) paid upon purchase will be refunded when collecting the out-of-use equipment. The ARF must be affordable to establish a competitive recycling market.

The previously discussed extended producer responsibility (EPR) policy approach places sole responsibility for recycling on the producers, such that there exists a gap in knowledge and awareness of e-waste handling and management among communities. This paper therefore suggests an integrated system in which the consumers, producers, legislators and policy makers are included in the e-waste management process. This will make all stakeholders aware of the activities and services of e-waste, thereby instilling a culture of unified responsibility for global sustainable development.

Educational institutions in developed countries must be encouraged to adopt an e-waste management policy for the handling of outdated laboratory equipment. Curricula on sustainable education through e-waste management can be introduced in universities to actively engage students. Collaborative learning can be facilitated to enable the transfer of skills and knowledge. For example, students from the developed countries who have had some learning experience of the outdated equipment can create videos to explain the cause behind faulty equipment, which will aid the students in developing countries when repairing the equipment.

Producers of the laboratory equipment must be responsible for providing manuals with instructions for a safe repair. The policy can adopt the EPR approach to mandate manufacturing companies to collaboratively set up repair and maintenance hubs in institutions where their equipment is deployed. These hubs can serve as research and development centres where students inform producers on improved product design through the assessment of learning outcomes. This creates a potential for designing safer and sustainable products with multiuse life, which are easy to disassemble for reuse.

Exchange programmes that facilitate the sharing of theoretical knowledge and practical skills between educational institutions and e-waste workers must be considered. E-waste workers who engage in primitive dismantling and recovery of metals possess useful skills for the disassembly of devices, mastered through heuristic learning. Together with university students, they can devise safer methods for disassembly and recreation of new devices. This form of sustainable inclusive education will mitigate child labour exploitation at e-waste dumpsites. If financial support is provided, the elderly e-waste workers can be formally employed in pilot projects for repairing electronic devices. This will decrease the proliferation of the informal recycling sector and in turn extenuate the environmental footprint of e-waste mismanagement.

34.5.2 *Recommendations*

The following are recommendations to the stakeholders in e-waste management policy making, for the success of the proposed grave-to-cradle e-waste management:

1. Enact an e-waste management policy framework that focuses on the deployment of outdated laboratory equipment to developing countries for learning purposes.
2. Implement a global standards system that certifies the equipment for export.
3. Improve public perspective through technology awareness campaigns on the contribution of e-waste management towards a circular educational economy.
4. Develop and introduce sustainable e-waste management curricula in educational institutions to promote sustainable education.
5. Aid the active participation of universities through public acknowledgement of donations and awarding accolades that contribute to the university's accreditation.
6. Finance the collection, certification, branding, deployment and tracking of laboratory equipment through a revolving fund from ARF (advanced recovery fees).
7. Engage manufacturers to develop technology training and capacity building programmes in developing countries to avoid the abandonment of the deployed equipment.
8. Legislators must put in place more stringent policies and laws to encroach corruption.

34.5.3 Further Research

For the policy to be holistic, further research is to be carried out to assess the willingness of universities in the developed countries to participate in the proposed laboratory e-waste management system. Also, there is a need to conduct empirical inquiries on the amount of out-of-use equipment piled in the universities. Further study and research on the required law and policy reformations to support the proposed laboratory e-waste management policy must be conducted.

34.6 Conclusion

As well as studying the adverse effects of e-waste, the study has critically reviewed existing e-waste management policies. It was highlighted that existing policies have not been entirely effective due to existing loopholes. Also, the study identified a research gap on the contribution of e-waste management to sustainable education, whereby a lack of academic equipment in educational institutions of developing countries can be improved by reusing out-of-use equipment from universities in the developed countries.

To address this, a management policy for laboratory e-waste is suggested based on the outcomes of a qualitative survey that examined the social acceptance of this policy by stakeholders in developing countries. Findings suggested that most educational institutions in the surveyed countries have inadequate academic equipment and are in favour of the deployed equipment given that it fulfils a set of safety and quality check requirements. The certification and branding of the laboratory equipment prior to deployment highlighted in the proposed laboratory e-waste management policy addresses the growing concerns about illegal transboundary movements and e-waste dumping in developing countries. The opportunities for the policy that are presented in this paper show that if adopted, the policy will not only be effective in managing e-waste in educational institutions but will also reduce the proliferation of the informal recycling sector. Although this suggested policy might not be the panacea to solving the problems of e-waste, it could complement the existing policies to mitigate the environmental footprint of e-waste whilst also contributing towards the achievement of a sustainable, circular educational economy.

References

Baldé, C. P., Forti, V., Gray, V., Kuehr, R., Stegmann, P. (2017). *The global E-waste monitor-2017*. United Nations University (UNU), International Telecommunication Union (ITU) & International Solid Waste Association (ISWA), Bonn/Geneva/Vienna. [Online]. Retrieved from https://www.itu.int/en/ITU-D/Climate-Change/Documents/GEM%202017/GlobalE-waste%20Monitor%202017%20.pdf

S. Chatterjee and K. Kumar Effective electronic waste management and recycling process involving formal and non-formal sectors International Journal of Physical Sciences, vol. 4, no. 13, pp. 893–905, 2009. [Online]. Retrieved from http://www.academicjournals.org/ijps.

Daum, K., Stoler, J., & Grant, R. J. (2017). Toward a more sustainable trajectory for E-Waste Policy: A review of a decade of E-Waste research in Accra, Ghana. *International Journal of Environment Research and Public Health, 14*, 2. https://doi.org/10.3390/ijerph14020135.

Hameed, S. A. (2012). Controlling computers and electronics waste: Toward solving environmental problems. *2012 International Conference on Computer and Communication Engineering (ICCCE)*, Kuala Lumpur, pp. 972–977, doi: https://doi.org/10.1109/ICCCE.2012.6271361.

Jian, L., & Shanshan, Z. (2010). *Study e-waste management based on EPR system*. Presented at the 2010 International Conference on E-business and E-government.

Kang, H.-Y., & Schoenung, J. M. (2004). Used consumer electronics: a comparative analysis of materials recycling technologies. *IEEE International Symposium on Electronics and the Environment, 2004*. Conference Record. 2004, Scottsdale, AZ, USA, 2004, 226–230. https://doi.org/10.1109/ISEE.2004.1299720.

Kiddee, P., Naidu, R., & Wong, M. H. (2013). Electronic waste management approaches: An overview. *Waste Management, 33*(5), 1237–1250. https://doi.org/10.1016/j.wasman.2013.01.006.

Magalini, F., Khetriwal, D. S., & Kyriakopoulou, A. (2019). *E-waste policy handbook*. Africa Clean Energy Technical Assistance: Nairobi. Retrieved from https://www.ace-taf.org/wp-content/uploads/2019/11/ACE-E-Waste-Quick-Win-Report20191029-SCREEN.pdf.

Ongondo, F. O., Williams, I. D., & Cherrett, T. (2011). How are WEEE doing? A global review of the Management of Electrical and Electronic Wastes. *Waste Management, 31*(4), 714–730. https://doi.org/10.1016/j.wasman.2010.10.023.

Osibanjo, O., & Nnorom, I. C. (2008a). Overview of electronic waste (E-waste) management practices and legislations, and their poor applications in the developing countries. *Resources, Conservation and Recycling, 52*(6), 843–858. https://doi.org/10.1016/j.resconrec.2008.01.004.

Osibanjo, O., & Nnorom, I. C. (2008b). The challenge of electronic waste (E-waste) Management in Developing Countries. *Waste Management & Research, 25*(6), 489–501. https://doi.org/10.1177/0734242X07082028.

Oteng-Ababio, M. (2012). Electronic waste Management in Ghana- Issues and Practices. *Sustainable Development*, 149–166. https://doi.org/10.5772/45884.

Park, J. K., Hoerning, L., Watry, S., Burgett, T., & Matthias, S. (2017). Effects of electronic waste on developing countries. *Advances in Recycling & Waste Management, 02*, 02. https://doi.org/10.4172/2475-7675.1000128.

Puckett, J., Brandt, C., & Palmer, H. (2018). *Holes in the circular economy: WEEE leakage from Europe*. Seattle: Basel Action Network (BAN). Retrieved from http://wiki.ban.org/images/f/f4/Holes_in_the_Circular_Economy-_WEEE_Leakage_from_Europe.pdf.

Sastry, S. V. A. R., & Ramachandra Murthy, C. V. (2012). Management of E-waste in the present scenario. *International Journal of Engineering and Technology, 4*(5), 543–547. https://doi.org/10.7763/ijet.2012.V4.428.

Shagun, A. K., & Arora, A. (2013). Proposed solution of e-waste management. *International Journal of Future Computer and Communication, 2*(5), 490–493. https://doi.org/10.7763/IJFCC.2013.V2.212.

Shevchenko, T., Laitala, K., & Danko, Y. (2019). Understanding consumer E-waste recycling behavior: Introducing a new economic incentive to increase the collection rates. *Sustainability, 11*, 9. https://doi.org/10.3390/su11092656.

Singh, M., Kaur, M., & John, S. (2013). E-waste: Challenges and opportunities in India. *2013 IEEE Global Humanitarian Technology Conference: South Asia Satellite (GHTC-SAS), Trivandrum, 2013*, 128–133. https://doi.org/10.1109/GHTC-SAS.2013.6629902.

Srivastava, R., & Sharma, D. (2015). Factors affecting e-waste management: An interpretive structural modeling approach. *2015 Fifth international conference on communication systems and network technologies*, Gwalior, 2015, pp. 1307–1312, doi: https://doi.org/10.1109/CSNT.2015.158.

Chapter 35
The Place of Urban Forestry in our Viable Urban Futures: A Cosmetic or a Metaphysic?

Alan Simson

35.1 Introduction

The twenty-first century is the urban century. It has been forecast that urban areas across the world will have expanded by more than 2.5 billion people by 2050, and the UK will not be exempt from this. In the 35 years between 1970 and 2005, the UK population grew by just over five million people. In the 11 years from 2005 until 2016, there was also an increase of another five million people, and it has been estimated that the UK population will increase by a further eight million by 2035 (ONS - Office for National Statistics, UK Government, 2019). Although the large cities will continue to play an important role in this population increase, new technologies are diminishing the advantages of scale, and strengthening the trends of decentralisation. The majority of this growth therefore is likely to be in existing, medium-sized towns – or Second Cities as Foroohar (2006) has called them – rather than the big cities, which will expand the concept of the polycentric city region significantly.

As time moves on, the qualities of cosmopolitan life, formally to be found mainly in the big cities, are likely to become democratised and decentralised, particularly in our post-Covid-19 world. Thus, suburbanism will again become very much alive, as the real growth in jobs, working from home and increasing population are likely to take place there. This has been influenced by the fact that the 'urban utopias' that have been promoted for so many years by urban designers from the times of Ebenezer Howard, Frank Lloyd Wright and Le Corbusier have not really turned out to be as 'utopian' as was promised. So perhaps it is time to make suburbanism work better, rather than just fight it. As Kotkin (2006) has argued, suburbs should be made

---

A. Simson (✉)
School of Electronics and Electrical Engineering, University of Leeds, Leeds, UK
e-mail: alan.simson@btopenworld.com

into small, cultural, social and economic centres of their own, and, without a doubt, urban forestry has a central role to play in this.

35.2 Discussion

Human beings are very sociable animals, and are drawn towards towns and cities to live. Our well-being and often our livelihoods depend upon the many services provided by healthy, natural ecosystems in and around our towns and cities, but, as urban areas rapidly expand, land use planning all too often proved to be inadequate in dealing with this. It fails to appreciate and take into account the multitude of benefits that nature, specifically trees and urban forestry, can provide, a failure that can result in serious environmental, economic and social and cultural problems.

Although there are some very attractive and successful cities and areas of cities in the world, the quality of urban life in general, particularly in our post-industrial urban areas, has not reached the utopian qualities promised, and is in fact already deteriorating, especially in the less affluent areas. If the current approach to urban design is maintained, the speed and scale of future urbanisation will increase both the environmental and health problems for urban dwellers, problems that we know are often made worse by a lack of contact between people and the natural world. Research has shown however that the presence of trees in urban areas – the urban forest – can significantly improve the quality of life there (Simson, 2017). In a hundred years' time, we may well look back on our present towns and cities that are bereft of greenery as our equivalent of the Victorian slums.

The term 'urban forest' is sometimes deemed to be an oxymoron – how can you have a forest in an urban area? Such an argument presupposes that the word 'forest' relates to trees, but it could be argued however that the word 'forest' stems from the Latin word 'foris', meaning 'out of doors', 'the outside'. Thus, the urban forest is really the 'urban out of doors', and includes all aspects of urban and peri-urban green space, both public and private.

Edwards Deming once said that 'Without data, you're just another person with an opinion' (Edwards Deming, 1982). The research into urban forestry has both increased and got more sophisticated over recent years, so now we have that data, and can prove the quantifiable benefits that the urban forest can bring to our urban areas (Dwyer et al., 1992; Turner-Skoff & Cavender, 2019; Tyrväinen et al., 2005). We can, for example, prove that the urban forest:

- Improves our health and well-being.
- Improves children's learning abilities.
- Provides focal points to improve social cohesion.
- Promotes and retains inward investment and job creation.
- Increases property values.
- Improves air quality.
- Promotes biodiversity.

- Limits the risk of flooding.
- Cools our towns and cities.
- Offsets carbon emissions.
- Makes us drive more safely.

The United Nation's New Urban Agenda, launched and adopted by the UN in 2016, emphasised that urban forests and trees can significantly help to address the challenges of greatly increasing urbanisation (United Nations, 2016). In particular, urban forestry has been promoted as being particularly effective in helping to deliver many of their Sustainable Development Goals (SDG), particularly SDG 11, which focused upon making cities inclusive, safe, resilient and sustainable. This was taken further in a Call for Action, which was launched at the First World Forum on Urban Forests, held in Mantova, Italy, in 2018 (United Nations Food and Agriculture Organisation (FAO), 2018).

The UN's vision suggested that towns and cities were a remarkable creation. Nowadays, most of us live in towns and cities, and live our lives through networks of relationships with each other and with our physical environment, something very much demonstrated recently by the reaction of so many communities in many countries when dealing with the issues associated with the Covid-19 virus. These relationships help to create the character and identity of the urban and suburban landscape, and the urban forest is a critically important aspect of that identity. The urban forest helps to mark the passing of time and opens a window for us to observe the cycle of nature, which is also the cycle of our daily life.

Thus, we should deploy urban forestry to make our expanding towns and cities:
- Greener – towns and cities need their urban forest, and people need green spaces. The colour green is often associated with well-being and positive feelings. It's the colour of balance, harmony and growth, which is why it's the interior colour of Leeds Beckett University's Broadcasting Place building. Green cities are strategic for urban communities and will play a key role in our futures (Konijnendijk, 2018). Greening our towns and cities involves all aspects of urban life, not only ecology but the economy, culture, architecture, psychology, education, health and society. Urban forestry is the backbone of green infrastructure and is the key actor in the theatre of our urban landscapes. It is no longer a 'cosmetic', but a 'metaphysic' – a first principle of viable green towns and cities.
- Healthier – we usually feel good when we are in the urban forest. Evidence shows us that the presence of trees has a positive influence upon our health and well-being (Donovan, 2017). Kondo et al. (2020) have proved that by increasing the tree canopy in Philadelphia, obesity, morbidity and mortality have decreased, especially in the lower socioeconomic areas where canopy cover tends to be lowest. The presence of trees makes the urban places seem safer, which results in an increase in physical activities. Thus, we should think of the urban forest as a green physician, improving our health both directly and indirectly and at little expense. We know, for example, that in 2015, urban forestry in the UK saved the

NHS over £1 billion by helping to reduce the impact of air pollutants on urban people (ONS - Office for National Statistics, UK Government, 2015).
- Happier – most people appreciate trees and are happier when they are in green spaces. Urban dwellers use the urban forest for relaxation, either alone or in groups, and for social and cultural events. Many communities support the planting of trees, as well as the conservation of existing trees, in both the well-off and the less well-off parts of the city. The urban forest can also provide natural classrooms for the education of children, and there has been a significant increase in the UK in the setting up of forest schools.
- Cooler – on hot summer days, the urban forest is a breath of fresh air. The right species of trees can also mitigate the thermal extremes of the built-up urban environment much more effectively than air conditioning, and this role will become increasingly to the fore as the issues of climate change escalate. Trees can assist retail businesses too. We know that when the temperature of our streets rises to around 30 °C, which it does more often than people think, shopping becomes less pleasant, and people tend to do their shopping on line. Tress can help to reduce street temperature by between 4 and 8 °C, thus making it more comfortable for pedestrians, and a more comfortable environment in which to spend their money.
- Wilder – biodiversity has an intrinsic value and represents a key element of any landscape, including cities. Indeed, cities have got to step up to the plate, and engage more constructively with reconciliation ecology. Habitat fragmentation is the biggest threat to the conservation of wildlife and natural ecosystems in urban areas. Climate change will stimulate the migration of species, and the urban forest can provide the 'stepping stones' that allow species to migrate to more diverse habitats. Diversity also concerns human communities, and the urban forest can be fundamental for maintaining local identity and maintaining cultural traditions. It can help to create significant landscapes, with particular symbolism that preserves a cultural diversity that characterises most fast-changing cities. In Vancouver, for example, female Ginkgo biloba trees (maidenhair trees) have been planted in the streets of the Chinese Quarter. Female Ginkgo trees are not usually planted in urban areas, as the ripe fruits give off an offensive odour in the autumn when crushed underfoot. However, in Vancouver, the Chinese residents collect the fruits and make useful potions out of them, which is a traditional pastime back in China. The author has been involved in a similar project in the Chinese part of the City of Prato in Italy. Designers can however sometimes overdesign urban green spaces, and it is now known that people sometimes relate better to more 'natural' design than to a formal one.
- Cleaner – a well-managed and healthy urban forest can help to maintain and improve air quality and water quality in and around our towns and cities. The right species of trees in cities can improve the quality of the air through a variety of chemical reactions, and by capturing and holding air pollutants (Conway & Vander Vecht, 2015). By decreasing air temperature, trees also help to reduce the use of air conditioners, which in turn saves energy use and polluting emissions.

The urban forest can also contribute greatly to the sustainable management of water and water resources.
- Wealthier – investing in urban forestry can be a viable strategy for the sustainable creation of jobs, increasing income and boosting local green economies, and it can also attract – and retain – investment, businesses and tourism. The positive effects of the urban forest on our mental and physical health can also generate significant public savings, reducing stays in hospital and speeding recovery times. Urban forestry is often a more affordable approach to urban development than the more traditional, predictable, grey approaches, thus representing a more cost-effective option in addressing the challenges of urban expansion.
- Safer – climate change, rapid urbanisation and high-density and growing urban populations are increasing the vulnerability of our cities. Many urban communities face potential risks to health, well-being and livelihoods. Urban forestry can offer opportunities to restore degraded, neglected and abandoned land, and remediate degraded soils. It can also help to minimise damaging run-off during periods of intense rainfall, and thus reduce the severity of flooding. Urban forestry has positive social effects and increases community safety by attracting people to meet and socialise, play sports and relax.

Thus, for our urban futures to be viable, it is increasingly important that proactive policies, strategic planning and legislation are harmonised so that we can make the most of deploying urban forestry into urban planning internationally, nationally, regionally and locally (Moffat, 2016). However, promoting the concept of urban forestry often faces the challenge of balancing the requirements of green spaces with the compactness of urban development, and this in turn can cause disharmony between the competing approaches to urban design, as promoted by New Urbanism and Landscape Urbanism. New Urbanism defines compactness as 'the agglomeration of urban activities, functions and residents with the physical proximity and continuity of increasing density' (Burton, 2000). Some go further (e.g. Duany & Talen, 2013), arguing that incorporating landscapes and urban forestry into cities threatens the efficiency of urban transportation systems and may lead to more dispersed patterns of suburban and urban sprawl.

Landscape Urbanism on the other hand, as promoted by Waldheim (2016) and others, promotes a much more ecologically based urban design as essential for our towns and cities to become more resilient – essentially towns and cities that are 'ready for anything' – politically, socially, economically and environmentally. The potential exists however for a fruitful collaboration between these two approaches to viable urban design futures, as both concede that a robust, resilient urban forestry strategy, as depicted by the Call for Action, is vital for sustaining the liveability of our towns and cities. Many cities, particularly in Europe, such as Amsterdam and Berlin, have been attempting for some time to incorporate 'nature' into the the design of their built environments, developing a range and a variety of ideas and strategies for expanding their urban forests (Beatley, 2012).

35.3 Conclusion

Human beings have had a long, deep, cultural relationship with their trees, their woodlands and their landscapes – a relationship that transcends national cultures and religions, and which sits as an equal alongside our scientific, economic, ecological and spiritual relationships. Some people take the view that now most of us are urban based, these relationships have much less prominence, and we're more interested in matters such as the price of our food or the fuel for our vehicles. Nothing could be further from the truth however, and you only have to engage with communities on these issues to discover that these ancient links are still there, although they do need to be nurtured. The overwhelming evidence from research and scientific literature suggests that investing in trees is not only an investment in meeting the UN's Sustainable Development Goals, but is ultimately an investment in a better world. This suggests that we should no longer refer to 'green infrastructure' as a specific infrastructure – it should be seen as an equal to grey and blue infrastructure, as no infrastructure can any longer be delivered successfully as a separate entity – we are now dealing with 'critical infrastructure'. Robust and resilient urban forestry will sustain the liveability of our towns and cities, and thus urban forestry is no longer a 'cosmetic', a bolt-on to urban design to soften the hard, inhuman urban developments that are all too often part of urban expansion, but it is a 'metaphysic' – a first principle of the successful planning, design and delivery of our viable urban futures.

References

Beatley, T. (2012). *Green urbanism: Learning from European cities*. Washington, DC: Island Press.

Burton, E. (2000). The Compact City: Just or just compact? A preliminary analysis. *Urban Studies, 37*(11), 1969–2006.

Conway, T., & Vander Vecht, J. (2015). Growing a diverse urban Forest: Species selection decisions by practitioners planting and supplying trees. *Landscape and Urban Planning, 138*, 115.

Donovan, G. (2017). Including public health benefits of trees in urban forestry decision making. *Urban Forestry and Urban Greening, 11*, 20.

Duany, A., & Talen, E. (2013). *Landscape urbanism and its discontents: Dissimulating the sustainable city*. Gabriola Island: New Society Publishers.

Dwyer, J., McPherson, E., & Schroeder, H. (1992). Assessing the benefits and the costs of the urban Forest. *Journal of Arboriculture, 18*(5).

Edwards Deming, W. (1982). *Out of a crisis*. London: MIT Press.

Foroohar, R, (2006). *Unlikely Boomtowns*. Newsweek July 3–10 2006. pp. 50–62.

Kondo, M, Mueller, N, Locke, D, Roman, L, Rojas-Ruela, D, Schinasi, L, Gascon, M & Nieuwenhuijsen, M. (2020). *Health impact assessment of Philadelphia's 2025 tree canopy cover goals*. The Lancet Planetary Health.

Konijnendijk, C. (2018). *The Forest and the City – The cultural landscape of urban woodland* (Future City 9) (2nd ed.). Cham: Springer.

Kotkin, J. (2006). *Building up the burbs*. Newsweek 03-10 July 2006, pp. 80-81.

Moffat, A. (2016). *Communicating the benefits of urban trees*. New York: Taylor & Francis.

ONS - Office for National Statistics, UK Government. (2015). Retrieved from https://www.ons.gov.uk

ONS - Office for National Statistics, UK Government. 2019). Retrieved from https://www.ons.gov.uk

Simson, A J, (2017) A landscape and urbanist perspective on urban forestry. Chapter 15 in The Routledge Handbook on Urban Forestry. Eds. Ferrini, F, Konijnendijk, C & Fini, A. Routledge, London.

Turner-Skoff, J, & Cavender, N. (2019). *The benefits of trees for livable and sustainable communities, in plants, people and planet.* Wiley Online Library.

Tyrväinen, L, Pauleit, S, Seeland, K, & de Vries, S. (2005). Benefits and uses of urban forests and trees. In *Urban forestry and urban greening* (pp. 81-144). Springer.

United Nations. (2016). *New Urban Agenda.* Retrieved from www.habitat3.org/the-new-urban-agenda

United Nations Food and Agriculture Organisation (FAO). (2018). *Call for action.* Retrieved from www.fao.org/forestry/urbanforestry/en

Waldheim, C. (2016). *Landscape as urbanism: A general theory.* Princeton: Princeton University Press.

Chapter 36
Illicit Crops, Planning of Substitution with Sustainable Crops Based on Remote Sensing: Application in the Sierra Nevada of Santa Marta, Colombia

Hector Leonel Afanador Suárez, Gina Paola González Angarita, Leila Nayibe Ramírez Castañeda, and Pedro Pablo Cardoso Castro

36.1 Introduction

Remote sensing methods, such as satellite imagery, are commonly used in environmental research due to their low current cost and even their free availability at satisfactory time and space resolutions, in addition to their easy processing with the map algebra tools of the GIS software (do Valle Júnior et al., 2019). Crop mapping and updating spatial information allow spatiotemporal monitoring of crops, contributing to the planning and managing of water resources, predicting yields, and adapting agriculture to climate change (Food and Agriculture Organization [FAO], 2003). The identification of crops, their location, and their extension are key properties for agricultural planning and best-use proposals.

These remote sensing techniques – and other similar techniques – have been used for the detection and monitoring of illegal crops worldwide, with emphasis on the analysis of time series, GIS and GPS software, and the use of SPOT and Landsat images as documented by Lisita et al. (2013), Armenteras et al. (2013), and Davalos et al. (2011).

In the case of Colombia, illicit crops represent a big problem in different areas of the country because they affect the environment where they are found, generating conflicts in the social, economic, and environmental fields. Being the coca cultivation the one generating the most dramatic impact, and which cultivation in Colombia

---

H. L. A. Suárez · G. P. G. Angarita
School of Environmental Engineering, Universidad Libre, Bogotá, DC, Colombia

L. N. R. Castañeda
School of Industrial Engineering, Universidad Libre, Bogotá, DC, Colombia

P. P. C. Castro (✉)
Business School, Leeds Beckett University, Leeds, UK
e-mail: P.P.Cardoso-Castro@leedsbeckett.ac.uk

© The Author(s), under exclusive license to Springer Nature Switzerland AG 2022
C. Gorse et al. (eds.), *Climate Emergency – Managing, Building, and Delivering the Sustainable Development Goals*, https://doi.org/10.1007/978-3-030-79450-7_36

since 2013, has increased at an average rate of 45% per year, from 48,000 ha in 2013 to 146,000 ha in 2016 (UNODOC, 2017).

In a historical context, Colombia became a producer and exporter of marijuana in the 1960s, cultivated in the Sierra Nevada de Santa Marta and in the Serranía del Perijá (Díaz & Sánchez, 2004). Since then, in the northern area of the Sierra Nevada de Santa Marta, there is still a presence of illicit crops according to the United Nations Office on Drugs and Crime (UNODOC, 2017). The environmental and social problems in this area of the Sierra Nevada de Santa Marta have been generated by decades of deforestation of a large strip of forests for planting illicit crops, in a practice in which these crops are camouflaged in the middle of the wild/native vegetation, but which can be located through satellite photos (Pérez Gutiérrez, 2001).

In relation to the management of illicit crops in Colombia, different approaches have been explored, moving from manual eradication and aerial fumigation with glyphosate to more comprehensive and integral approaches involving policies of social and economic inclusion, development of infrastructure, and crop substitution by crops with alternative/substitute local economic value. The manual eradication proved to be effective – but costly and slow – in mountain areas if combined with economic inclusion and effective law enforcement. However, more recent policies are promoting the use of glyphosate despite the controversy around it in terms of efficacy in the long run and health and ecosystem cost (Solomon et al., 2007). Regarding the call for a more comprehensive approach, the USAID report 2009 (Felbab-Brown et al., 2009) advocated for the implementation of comprehensive interventions in the affected territories, giving particular importance to the crop substitution – by crops with similar local economic value – followed by the delivery of infrastructure to facilitate the socioeconomic inclusion of the communities involved.

To support the design of policies and field activities for the management of illicit crops in Colombia, the monitoring of coca crops has been supported with the interpretation of medium resolution satellite images and the validation of data obtained by aerial reconnaissance (UNODOC, 2017). Similarly, a mapping of the types of crops and identification of the area of greatest cultivation was carried out – as previously reported in a study in Morocco – using data provided by the NDVI and Landsat 8 index for a highly fragmented and intensive agricultural system. The results of the study demonstrated that Landsat 8 satellite images can allow to achieve annual inventories of crops in irrigated, highly fragmented, heterogeneous, and intensive agricultural systems (Ouzemou et al., 2018).

Also, from a spectroradiometer field, spectral signatures of coca crops have been captured according to their phenological status: juvenile, vegetative, and mature (Ángel, 2012). Furthermore, the use of statistical models has allowed estimating the yield of the crops under study from agroclimatic variables of the areas where they are found (Ramírez & Potes, 2019). This study has some limitations related to time lapses (years) of some satellite images collected from a free server as these were obtained from a free server where time lapses (years) for which the analysis could not be performed due to the presence of clouds in this area. In addition, the armed conflict and active drug trafficking in the area imposed another limitation to carry out fieldwork to contrast the analysis carried out with images from remote sensors.

Within this context, this work was carried out with the purpose of searching different sustainable crop planning alternatives in the Sierra Nevada de Santa Marta, where there are crops for illicit use through the analysis of environmental indices (vegetation, humidity, temperature, and precipitation, among others) which will allow knowing the quantity and quality of vegetation, bodies of water, and other properties necessary for decision-making.

36.2 Research Review and Methodology

The project is developed in the Sierra Nevada de Santa Marta, specifically in the department of Magdalena and La Guajira, in the jurisdiction of the municipalities of Santa Marta Dist., esp. Riohacha and Dibulla, and three Regional Autonomous Corporations CORPAMAG, CORPOCESAR, and CORPOGUAJIRA, as can be seen in Fig. 36.1, the northern areas that have been historically affected by the current expansion of illicit crops.

The Sierra Nevada de Santa Marta is a triangular territory of the Andes mountain range, located in the northeast of Colombia, around the geographical coordinates 10° 49'N and 73 ° 39'W. The north flank borders the Caribbean Sea from the flat and arid lands of the south of the La Guajira peninsula to the surroundings of the city of Santa Marta, at the mouth of the Manzanares River (Organización Colparques, n.d.). The study area is of interest not just for the historical cultivation of illicit crops but also for its biodiversity, including some protected areas, challenging topographic conditions, and socioeconomic importance that involves native aborigine populations, settlers, and a diversity of agro-economic activities ranging from small farms to agro-industrial developments.

In this study, we used the vegetation index as an analysis tool to identify the vegetation cover (Ouzemou et al., 2018). The NDVI processing was developed from geoprocessing satellite images with ArcGIS Pro software ver 2.4.1, which allows conducting different processes with Landsat satellite images obtained from the USGS (United States Geological Service) Earth Explorer platform (https://earthexplorer.usgs.gov/). Table 36.1 shows the periods chosen to carry out the investigation and identification of each of the Landsat images used in this study.

The NDVI was performed through the processing of the bands of the satellite images, for Landsat 5 band 3 (red) and 4 (near-infrared) and Landsat 8 band 4 (red) and 5 (near-infrared) by the following equation (Zaitunah et al., 2018):

$$NDVI = \frac{NIR - RED}{NIR + RED} \qquad (36.1)$$

where:
NIR: near-Infrared.

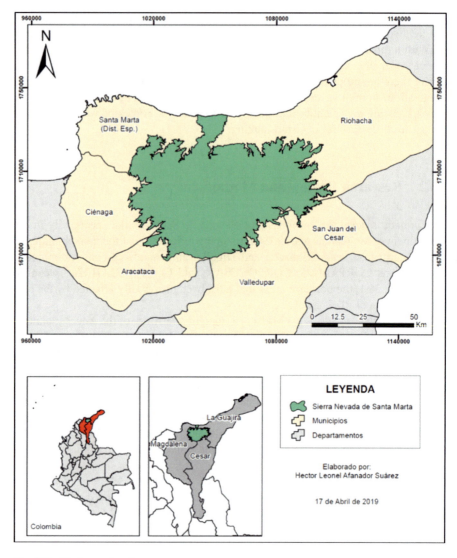

Fig. 36.1 Study area location

RED: reflectivity of the red band.

The formulation of the simulation model of this research considers information related to the crops available to plant in the municipality of Santa Marta and Dibulla. The definition of each of the parameters used is described below.

Set:

$$i \in I : Crops\ ready\ for\ seeding$$

Decision variable:

36 Illicit Crops, Planning of Substitution with Sustainable Crops Based on Remote…

Table 36.1 Landsat images used in the project

Date DD/MM/AAAA	Landsat image ID
20/12/1997	LT05_L1TP_008052_19971220_20161228_01_T1_B4
20/12/1997	LT05_L1TP_008052_19971220_20161228_01_T1_B3
20/07/1997	LT05_L1TP_009052_19970720_20161230_01_T1_B4
20/07/1997	LT05_L1TP_009052_19970720_20161230_01_T1_B3
30/06/1998	LT05_L1TP_008052_19980630_20161223_01_T1_B3
30/06/1998	LT05_L1TP_008052_19980630_20161223_01_T1_B4
17/03/1998	LT05_L1TP_009052_19980317_20161226_01_T1_B3
17/03/1998	LT05_L1TP_009052_19980317_20161226_01_T1_B4
26/12/1999	LT05_L1TP_008052_19991226_20161217_01_T1_B3
26/12/1999	LT05_L1TP_008052_19991226_20161217_01_T1_B4
15/01/1999	LT05_L1TP_009052_19990115_20161220_01_T2_B3
15/01/1999	LT05_L1TP_009052_19990115_20161220_01_T2_B4
29/01/2001	LT05_L1TP_008052_20010129_20161212_01_T1_B3
29/01/2001	LT05_L1TP_008052_20010129_20161212_01_T1_B4
05/02/2001	LT05_L1TP_009052_20010205_20161212_01_T1_B3
05/02/2001	LT05_L1TP_009052_20010205_20161212_01_T1_B4
11/09/2007	LT05_L1TP_008052_20070911_20161111_01_T1_B3
11/09/2007	LT05_L1TP_008052_20070911_20161111_01_T1_B4
18/09/2007	LT05_L1TP_009052_20070918_20161112_01_T1_B3
18/09/2007	LT05_L1TP_009052_20070918_20161112_01_T1_B4
04/01/2015	LC08_L1TP_008052_20150104_20170415_01_T1_B4
04/01/2015	LC08_L1TP_008052_20150104_20170415_01_T1_B5

(continued)

Table 36.1 (continued)

Date DD/MM/AAAA	Landsat image ID
27/11/2015	LC08_L1TP_009052_20151127_20170401_01_T1_B4
27/11/2015	LC08_L1TP_009052_20151127_20170401_01_T1_B5
30/12/2018	LC08_L1TP_008052_20181230_20190130_01_T1_B4
30/12/2018	LC08_L1TP_008052_20181230_20190130_01_T1_B5
05/12/2018	LC08_L1TP_009052_20181205_20181211_01_T1_B4
05/12/2018	LC08_L1TP_009052_20181205_20181211_01_T1_B5

Source: Author, 2019

Parameters:
A_i : Amount of area to sow for the crop i [ha]

$Rend_i$: Crop performance i [ton / ha]

$Precio_i$: Crop value i [\$ / ton]

$Costos_i$: Crop cost i [\$ / ha]

A_{disp} : Area available for the crop [ha]

Inv_{disp} : Available investment [\$]

E_i : Estimated proportion for cultivation i.

The mathematical programming model has the following structure:
Objective function:

$$\max Z_{utilidad} : \left[\sum_{i:1}^{n} (Rend_i \times Precio_i \times A_i) - (Costos_i \times A_i) \right], \quad (36.2)$$

Restrictions:

$$\sum_{i:1}^{n} (Costos_i \times A_i) \leq Inv_{disp}, \quad (36.3)$$

$$\frac{A_i}{A_{disp}} \geq E_i \, \forall i \in I, \quad (36.4)$$

$$\sum_{i:1}^{n} \frac{A_i}{A_{disp}} = 1, \qquad (36.5)$$

$$\sum_{i:1}^{n} A_i \leq A_{disp}, \qquad (36.6)$$

$$A_i \geq 0, \qquad (36.7)$$

Equation (36.2) refers to the calculation related to the maximum utility that can be achieved by each of the selected crops, which meet the agroclimatic characteristics of the area, Eq. (36.3) limits the investment capacity of the takers of decision, Eq. (36.4) allows to establish proportionality conditions for different crops in the same available area, and restrictions (36.5) and (36.6) allow the full use of the available area and limit its use according to its availability.

36.3 Research Results

The NDVI has been implemented over different periods in order to identify changes in the vegetation cover of the study area. The result of this indicator varies from −1 to 1, where the minimum value reflects bare soil, arid terrain, urban areas, and/or rock on maps up to values close to 0 and values greater than 0 show soils with healthy and vigorous vegetation (Islam et al., 2016).

From the index range of 0.36 to 0.51, a selection was made by attributes to know the places in the different years of study where these values were present, as can be seen in Fig. 36.2.

In 1998, it can be seen (Fig. 36.2) a marked change in color of the NDVI (red to green) and coca crops (purple, area of coca crops in 2017 according to the Ministry of Justice; white, coca crops of the study year) between the municipalities of Santa Marta and Dibulla. This is because of the selection of a satellite image from another month concerning the municipality of Dibulla.

Coca cultivation from 2001 to 2007 decreased significantly in the Santa Marta area and a little in the municipality of Dibulla. However, coca crops increased for 2015 and 2018 compared to 2007. For 2018, it is evident that the area of coca crops is not completely white, which represents a decrease in these crops.

In addition to the colors, it is also possible to analyze the histograms found on the maps. From 1997 to 2007, an uneven amount of the NDVI values is observed, compared to the years 2018 and 2015, where the values of the graphs tend to be constant.

Since there is still a presence of coca crops in the study area, the viability of alternative crops to replace them is studied. According to Agronet (2019), ten sustainable crop alternatives are selected for the area. Table 36.2 shows that only four of the ten crops chosen meet the conditions of relative humidity, precipitation, and temperature that occur in the area.

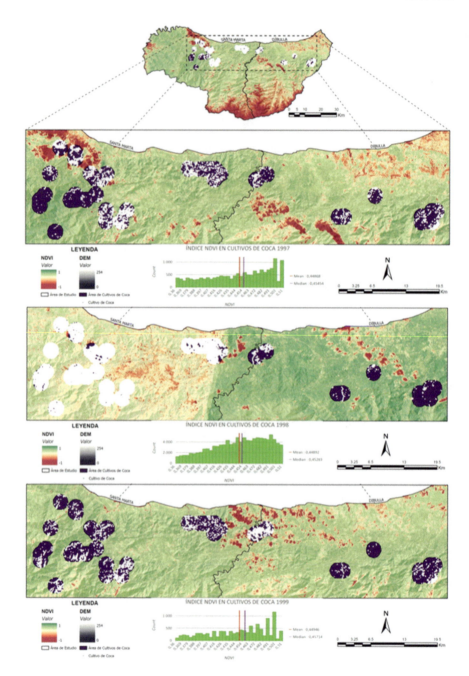

Fig. 36.2 NDVI in coca crops in the years 1997, 1998, 1999, 2001, 2007, 2015, and 2018. Note in purple the coca crops as reported by the Ministry of Justice, and the more recent coca crops from images of this study in white

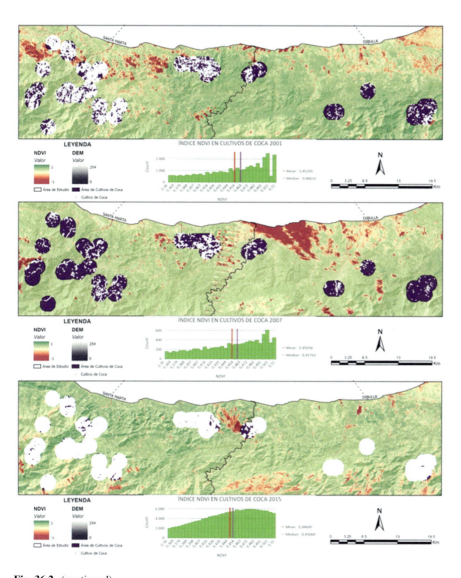

Fig. 36.2 (continued)

The crops that meet the above conditions are mechanized rice (1), beans (5), plantain (9), and cassava (10). These crops are entered into the crop selection model as shown in the following Table 36.3.

The most useful crop is cassava, with an area of 300 ha, followed by mechanized rice and plantain (150 ha and 50 ha, respectively). This significant difference in utility occurs due to the difference in the yields of each crop. The yield of cassava cultivation is 12 ton/ha, compared to 4 ton/ha of mechanized rice and 5.39 ton/ha of plantain (Agronet, 2019). In this analysis, results for the cultivation of beans are not

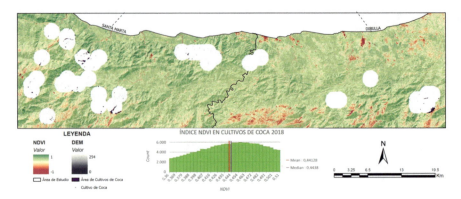

Fig. 36.2 (continued)

Table 36.2 Ideal crops for the study area

No.	Crops	Variable of decision HR	Variable of decision PP	Variable of decision T	Satisfy
1	Mechanized rice	1	1	1	Yes
2	Banana	1	1	0	No
3	Coffee	0	1	0	No
4	Citrus fruit	1	0	1	No
5	Beans	1	1	1	Yes
6	Traditional corn	1	0	0	No
7	Mango	1	0	0	No
8	Palm oil	0	0	1	No
9	Plantain	1	1	1	Yes
10	Cassava	1	1	1	Yes

Source: Author, 2019

Table 36.3 Model crop selection result

No	Crops	Income (USD)	Total cost (USD)	Planting area (ha)	Utility (USD)
1	Mechanized rice	385,845.9	255,671.8	150	130,174.1
2	Beans	–	–	0	–
3	Plantain	183,370.0	111,810.9	50	71,558.8
4	Cassava	1,798,800.0	724,680.2	300	604,645.3

Source: Author, 2019

provided because its utility is negative and does not comply with the restrictions of the crop selection model. The utility of the bean crop is very low in the study area compared to the other crops, being 0.87 ton/ha (Agronet, 2019).

Furthermore, a determining variable in the model is the amount of investment available. According to the data in the table above, an investment of USD 1,092,160.6

(sum of total costs) must be made in order to plant these crops in the area proposed by the simulation model. Not having the total investment amount for planting the crops, the data in the model variable can be updated; however, the available area could not be planted entirety.

36.4 Conclusion

Satellite images – from the US Geospatial Geological Survey database – are a good tool to visualize the multi-temporal change in coca cultivation and to extract information about the vegetation cover of the area analyzed. ArcGIS Pro software is able to process images from that database to carry out the required analysis to provide an assertive interpretation of the issues studied in the area analyzed.

The NDVI was useful to know not only the state of the vegetation cover but also the location of the coca crops, knowing their rank in this index. According to the study carried out, the range is between 0.36 and 0.51; and the most affected areas are north of the Sierra Nevada de Santa Marta between the municipalities of Santa Marta and Dibulla. Therefore, this document provides a critical cartographic base of plant cover, location, and variation of coca crops in the study area during different years.

For instance, from data analyzed in this work, according to the parameters of relative humidity, precipitation, temperature, and utility, rice, plantain, and cassava should be sown, cassava being the crop with the best match with the geophysical attributes of the area studied, and the highest potential to generate higher economic performance to the population in the affected areas.

The results of this research, adding to the generation of new knowledge in the management of the vegetation cover, can provide information to support the creation of guidelines for the formulation of land use policies at a regional, national, and international scale, related – but not limited – to the identification of illicit crops and its substitution. In this context, through the implementation of the methods tested in this work, the case for sustainable crops can be more accurately justified and presented as a solution to the management of substitution of illicit crops.

References

Agronet. (2019). *Principales cultivos por área sembrada.* [Online]. Retrieved January 15, 2020, from https://www.agronet.gov.co/Documents/MAGDALENA_2017.pdf.

Ángel, Y. B. (2012). *Metodología para identificar cultivos de coca mediante análisis de parámetros red edge y espectroscopia de imágenes.* [Online]. Retrieved February 17, 2020, from http://www.bdigital.unal.edu.co/7566/1/7795080.2012.pdf.

Armenteras, D., Rodríguez, N., & Retana, J. (2013). Landscape dynamics in northwestern Amazonia: An assessment of pastures, fire and illicit crops as drivers of tropical deforestation. *PLoS One, 8*(1), 115. https://doi.org/10.1371/journal.pone.0054310.

Davalos, L., Bejarano, A., Hall, M., Correa, H., Corthals, A., & Espejo, J. (2011). Forest and drugs: Coca-driven deforestation in tropical biodiversity hotspots. *Environmental Science and Technology, 45*(4). https://doi.org/10.1021/es102373d.

Díaz, A. M. and Sánchez, F. (2004). *Geografía de los cultivos ilícitos y conflicto armado en Colombia.* [Online]. Retrieved January 15, 2020, from https://economia.uniandes.edu.co/component/booklibrary/478/view/46/Documentos%20CEDE/470/geografia-de-los-cultivos-ilicitos-y-conflicto-armado-en-colombia.

do Valle Júnior, R. F., Siqueira, H. E., Valera, C. A., Oliveira, C. F., Sanches Fernandes, L. F., Moura, J. P., & Leal, F. A. (2019). Diagnosis of degraded pastures using an improved NDVI-based remote sensing approach: An application to the Environmental Protection Area of Uberaba River Basin (Minas Gerais, Brazil). *Remote Sensing Applications: Society and Environment, 14*, 20–33. https://doi.org/10.1016/j.rsase.2019.02.001.

Felbab-Brown, V., Jutkowitz, J. M., Rivas, S., Rocha, R., Smith, J. T., Supervielle, M., & Watson, C. (2009). *Assessment for the implementation of the United States Government's support for plan Colombia's illicit crop reduction components.* Washington, DC: USAID.

Food and Agriculture Organization [FAO]. (2003). *World agriculture: Towards 2015/2030.* London, UK: Earthscan Publications Ltd..

Islam, K., Jasimuddin, M., Nath, B. and Nath, T. K. (2016). Quantitative assessment of land cover change using Landsat Time Series Data: Case of Chunati Wildlife Sanctuary (CWS), Bangladesh. *International Journal of Environment and Geoinformatics, 11*, 112. doi:https://doi.org/10.30897/ijegeo.306471

Lisita, A., Sano, E., & Durieux, L. (2013). Identifying potential areas of *Cannabis sativa* plantations using object-based image analysis of SPOT-5 satellite data. *International Journal of Remote Sensing, 34*(15), 5409–5428. https://doi.org/10.1080/01431161.2013.790574.

Organización Colparques. (n.d.) *Sierra Nevada de Santa Marta.* [online]. Retrieved February 17, 2020, from http://www.colparques.net/SIERRA.

Ouzemou, J.-E., El Harti, A., Lhissou, R., El Moujahid, A., Bouch, N., El Ouazzani, R., & El Ghmari, A. (2018). Crop type mapping from pansharpened Landsat 8 NDVI data: A case of a highly fragmented and intensive agricultural system. *Remote Sensing Applications: Society and Environment, 11*, 94–103. https://doi.org/10.1016/j.rsase.2018.05.002.

Pérez Gutiérrez, C. (2001). Cultivos ilicitos acaban con la Sierra Nevada. *El Tiempo*, 1st august, 2001. [online]. Retrieved February 20, 2020, from https://www.eltiempo.com/archivo/documento/MAM-452849.

Ramírez, L., & Potes, S. (2019). Estimación del rendimiento del cultivo de Passiflora Edulis (Maracuyá) a partir de modelos estadísticos. *Inventum, 14*(26), 33–42. https://doi.org/10.26620/uniminuto.inventum.14.26.2019.

Solomon, K. R., Anadón, A., Carrasquilla, G., Cerdeira, A. L., Marshall, E. J., & Sanin, L.-H. (2007). Coca and poppy eradication in Colombia: Environmental and human health assessment of aerially applied glyphosate. *Reviews of Environmental Contamination and Toxicology, 190*, 43–125. https://doi.org/10.1007/978-0-387-36903-7_2.

UNODOC. (2017). *Monitoreo de territorios afectados por cultivos ilícitos 2017.* doi:ISSN: 2011-0596.

Zaitunah, A., Samsuri, A., Ahmad, A. G., & Safitri, R. A. (2018). Normalized difference vegetation index (ndvi) analysis for land cover types using Landsat 8 oli in besitang watershed, Indonesia. *IOP Conference Series: Earth and Environmental Science, 126*, 10. https://doi.org/10.1088/1755-1315/126/1/012112.

Chapter 37
Bibliometric Study on Particle Emissions of Natural and Alternative Building Materials

Nana Benyi Ansah, Emmanuel Adinyira, Kofi Agyekum, and Isaac Aidoo

37.1 Introduction

Pollution characterised by building construction material emissions has become a hot spot of study due to its accompanying environmental and health complexities. This research aims to conduct a structured literature review on the subject of particle emissions of natural and alternative building materials. The high demand for construction materials is dependent on consumption of raw materials (Lassio and Naked, 2016). The difficulty of determining the pattern of toxins emitted from the use of building construction materials by the built environment professional has led to the use of several toxic materials which are worthy of attention. Some of these materials are legally accepted, yet they contain some form of toxicity (Levin, 2016; Pacheco-Torgal & Jalali, 2011). Thus, there is the need for environmental assessment of building materials so as to substitute those prone to health ramifications with more environmentally friendly ones in the delivery of sustainable building construction projects (Farahzadi et al., 2016).

In their study of building material emissions through regression, He et al. (2005) argued that emission parameters have a causal connection between the molecular structure of compounds and material properties. Their study explained that the two most widely accepted physical models for emission determination are (1) the diffusion coefficient [D] and (2) the partition coefficient [K]. Thus, many researchers use Fick's law to interpret the mass transfer inside the material due to the concentration difference. The argument by He et al. (2005) was affirmed by the research carried out by Zhang et al. (2018, p.3) – through material efficiency by separation and

---

N. B. Ansah (✉) · E. Adinyira · K. Agyekum · I. Aidoo
Department of Construction Technology and Management, College of Art and Built Environment, Kwame Nkrumah University of Science and Technology (KNUST), Kumasi, Ghana

dematerialisation. The material efficiency measurements 'include all changes that result in decreasing the number of materials used to produce one unit of economic output or to fulfil human needs'.

To be able to assess the emission characteristics of building construction materials, there is a need to determine the emission factors and the taxonomy of chemicals emitted concerning the materials' health intricacies. Analysing the search results from a bibliometric review, an overview of natural and alternative material emissions is presented in this paper with discussions and pathway for future research.

The procedure involved a detailed literature review of building construction materials to provide a summary of existing studies of particle emissions. The literature is relative to the health implications of the materials in question that apply to the built environment. The study depended on those building construction materials available on the market with emission quality deductions and consisted of three stages.

Stage 1 consisted of articles retrieved from reputable databases from a list of publications using the keywords as the benchmark. Further, articles from each database were grouped and sorted based on their relevance to the objectives of the study. In stage 2, a systematic review of the articles (peer-reviewed journals, original industry reports) and books to solicit for the data of particulate matter emissions of building construction materials and their health benefits and/or challenges was carried out. Critical reading was carried out to obtain evidence and to provide a useful evaluation of the text. Stage 3, the final stage, considered the impact of emissions on human health from literature.

37.2 Search Strategy

The scope of the literature search was within the confines of the widely academically recognised databases, namely, Google Scholar, ScienceDirect, ResearchGate and Web of Science. The literature was then sorted according to their background, for example, *Journal of Cleaner Production* (*JCP*), Multidisciplinary Digital Publishing Institute (MDPI), ResearchGate and Original Industry Reports and Letters. The studies that did not connate to the objectives of the review were removed. The literature retrieval was done using the keywords and Boolean logical operators – for example, natural and alternative building material emissions. Sources of literature that had strong affinity to the study theme were used as a foundation for the review (Ramdhani et al., 2014). Also, particulate matter emissions, factors affecting material emissions and building construction material efficiency were included in the search plan as text words. To ensure that high-quality literature was used, refereed journals and original documents were selected for the study (Wallace & Wray, 2013).

37.2.1 Frontline Literature

A total of 127 journal articles, industry reports, letters and unpublished articles were retrieved from various databases. Following Sun et al. (2020) and Chan et al. (2020) study, a scrutiny of all the articles was carried out. Repeated articles were removed and those articles with the required relevance to the study were selected. After this consideration, a total of 107 articles remained. With the need for more relevance, a comprehensive examination of abstracts, conclusions and full-text analysis was carried out. After the examination of the abstracts, conclusions and text analysis, 53 of the articles found to be relevant to the research were collected for further studies. Seven out of the 53 publications were deemed to be most relevant to the study and were captioned 'frontline literature', and they were used as the basis for the literature synthesisation. Table 37.1 shows the keywords setting used for the search. Table 37.2 shows the spread across various search results of the final literature. Table 37.3 clarifies the synthesised matrix organised by frontline literature (Fig. 37.1).

37.3 Family of Construction Materials

37.3.1 Natural Materials (Traditional)

In their study of building material emissions for the construction of classrooms, Moulton-Patterson et al. (2003) classified 'commonly used building products containing low or no recycled content' as standard or natural materials. When the working life of construction material is increased, the eco-friendliness of the content is improved (Edwards and Bennett, 2003; Hertwich et al., 2019). In their study of material efficiency, Ruuska and Häkkinen (2014, p. 267) argued that the use of natural materials supports the quality of life of the occupiers of the building. They posit that the 'natural building material that has the required emission stipulations provides a better construction option and also reduces emission'. Natural materials are found as either renewable or non-renewable. The renewable materials are those that

Table 37.1 Keywords setting used for the search

Parameter	Setting
Keywords	Natural material and particulate matter emissions
	Alternative material and particulate matter emissions
	Natural material emission factors
	Alternative material emission factors
	Particulate matter emission factors

Table 37.2 The spread of literature search relevant to the study

No.	Database/background	Initial retrieval	Final retrieval relevant to study	References
1	Google scholar	50	21	Maoeng et al. (2020), Had and Brain (2020), Milner et al. (2020), Aisyah et al. (2019), Huang et al. (2019), Keita et al. (2018), Souza and Borsato (2016), Gonçalves de Lassio and Naked Haddad (2016), He et al. (2005), Upstill-Goddard et al. (2015), Doroudiani et al. (2012), Glass (2011), Reed (2011), Ghumra et al. (2011), Holton et al. (2008), James and Yang (2005), Magee (2005), Moulton-Patterson et al. (2003), Edwards and Bennett (2003) and Bellis (1998)
2	ScienceDirect	19	11	Xia et al. (2020), Jung et al. (2019), Nwodo and Anumba (2019), Harb et al. (2018), Martínez-Rocamora et al. (2016), Wille and Boisvert-Cotulio (2015), Azari (2014), Ramesh et al. (2010), Haapio and Viitaniemi (2008), Aoki and Tanabe (2007) and Wegener et al. (2007)
3	Original industry reports and letters	16	11	Raifman et al. (2020), Greenstone and Ryan (2020), IISD (2019), Chin et al. (2019) Hertwich et al. (2019), SimaPro Library Database Manual Colophon, M. (n.d.), Meng et al. (2015), Hillman et al. (2015), United States Environmental Protection Agency (2013), Baetens et al. (2010) and Järnström (2007)
5	Multidisciplinary digital publishing institute	9	5	Shi et al. (2020) Kong et al. (2020), Sun et al. (2020), Mohajerani et al., 2019) and Ruuska and Häkkinen (2014)
6	*Journal of cleaner productions*	7	4	Cheriyan and Choi (2020), Silva et al. (2019), Khoshnava et al. (2018) and Gmelin and Seuring (2014)
7	ResearchGate	10	5	Zhang et al. (2018), Levin (2016), Farahzadi et al. (2016), Jalali (2015) and Zhang (1997)
	Total	111	57	

can be replenished after harvesting, while the non-renewable material resource is those that can only be gathered once. The theory of durability concerning material efficiency was corroborated by Lifset and Eckelman (2013) and Levin (2016). In his plenary architecture lecture, Levin (2016, p.15) explained, 'selecting natural building materials that are durable has sufficiently environmental benefits than the one that must be substituted more than once in the life of the building'. For example, when the concrete cover is doubled from 10 mm to 20 mm, the service life of

37 Bibliometric Study on Particle Emissions of Natural and Alternative Building...

Table 37.3 A synthesised matrix organised by frontline literature

Author and date	Purpose	Finding	Background
Kong et al. (2020)	To evaluate the carbon emissions during the construction process of a prefabricated concrete slab	The carbon emission from prefabricated concrete slab is 35% less than cast in place (p. 12)	MDPI
Hertwich et al. (2019)	To address the current state of knowledge in reducing emissions through material efficiency focusing on product groups: Building, vehicle and electrical	A considerable potential exists to decrease abundant emissions connected to material production used in buildings and vehicles. Evidence for emission reduction in the electrical group is limited (p. 13)	*Environmental research letters*
Farahzadi et al. (2016)	To examine the similarities in energy efficacy and emissions of using traditional materials with proposed alternative ones (p. 187)	Using eco-friendly materials in building construction reduces energy consumption and promotes emission reduction	ResearchGate
Levin (2016)	To exhibit analysis to assist design alternatives in reducing harmful impact on indoor and general environments	'Durable materials are less likely to emit contaminants into the atmosphere, will require less toxic chemicals for the maintenance and refurbishing and, by definition, will be longer lasting' (p. 17).	ResearchGate
Pacheco-Torgal and Jalali (2011)	To discuss the effects of toxic building materials	'Several building materials that comply with regulations remain toxic to human health' (p. 5)	ResearchGate
He et al. (2005)	To use emissions data to provide information on emission parameters (p. 60)	Compound molecular structure and material properties affect diffusion and subsequently affect emissions	Google scholar
Moulton-Patterson et al. (2003)	To compare emissions data of traditional materials with that from no or low recycled products (standard) and those with high recycled content, rapidly renewable and/or products containing low or no VOCs (alternative)	Many products tested emitted chemical concentrations exceeding the allowable concentration limits used in the study	Google scholar

reinforcement – 'defined as the time it takes carbonation to reach the reinforcement', Levin (2016, p. 63) – is increased by 400% but increases concrete consumption by only 5–10%. Farahzadi et al. (2016) proposed a variety of conventional construction materials: viz. standard bricks, oil paint, aluminium frames, polystyrene thermal insulation and air-filled double glazed windows.

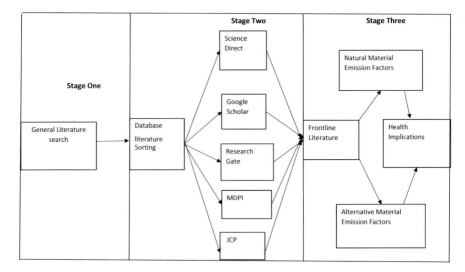

Fig. 37.1 Literature search process after Sun et al. (2020)

37.3.2 Alternative Materials

Moulton-Patterson et al. (2003) carried out an extensive study on material emissions for the construction of schools in the United States. During their study, they classified alternative materials as those with 'higher amounts of recycled content, rapidly renewable materials, and/or products containing no or low volatile organic compounds – VOCs' (p. 1). However, in their study of the assessment of alternative materials, Farahzadi et al. (2016) also established the following materials as being alternative construction materials, thus clay blocks, glass wool thermal insulation, acrylic paint, wooden frames and argon-filled double glazed windows. In his study of building ecology (design alternatives), Levin (2016, p. 7) characterised building materials as one of the essential primary criteria for classifying a building as either healthy or not. The trade-offs in using alternative building construction material are to accomplish the expected results. The optimum way of doing this is 'to select low emitting products, to condition or treat the product before installation or to ventilate the building after installation before occupancy'.

To this effect, the most prudent way of reducing environmental pollution from building construction materials and thus raise its efficiency is to use them again at the end of their useful life to prevent them from going through the processes of extraction to processing (Hertwich et al., 2019) (Table 37.4).

After Farahzadi et al. (2016)

Table 37.4 Conventional and alternative building construction materials

No	Conventional materials	Alternative materials
1	Brick	Clay block
2	Petrochemical insulation (polystyrene)	Mineral-based insulating (glass wool)
3	Synthetic and petrochemical paints (oil base)	Natural paints (acrylic)
4	Aluminium window frames	Wooden window frames
5	Double glazed 6 mm spacing (air-filled)	Double glazed 6 mm spacing (argon-filled)

37.4 Natural and Alternative Material Selection

In their study of validating a set of empirically weighted sustainability indicators for construction products, Ghumra et al. (2011) posit that pollution from the use of the construction materials was highest on the ranking of all indicators. Life cycle assessment (LCA) has the most significant attention based on the indicators' respective weightings used in their study. Even though LCA is a widely accepted criterion in the promotion of sustainable material selection, the process is intensive and difficult to handle. Upstill-Goddard et al. (2015) argue that those responsible for material selection may choose other methodology and still aim for higher performance. To this effect, Upstill-Goddard et al. (2015) suggest that LCA methodology should be carried out and promoted as a separate entity.

Khoshnava et al. (2018) conclude that the action of building material selection involves a complex challenge which usually looks at quality, performance, beauty and cost to reveal the main serviceability functions. In selecting building materials, whilst the focus is usually on the environmental impact reduction, it is imperative to consider the economic and social impacts as well. The emission characteristics which contribute to human health efficiency are considered under social impact.

37.5 Particulate Matter (PM) Emissions

Particulate matter emissions are characterised by the presence of small particles and liquids.

Depending on their size, particles can be inhaled deeper into different parts of the respiratory system causing serious health challenges.

Doroudiani et al. (2012, p. 264) investigated the toxic release from construction materials during fire. The authors reported in their study that 'particle size larger than 5μm are filtered in the upper respiratory system while the smaller ones can travel to bronchial and alveolar areas'.

In their study on cities' ambient particulate matter source contribution Karaguliana et al. (2015) and Bylone (2019) clarified that as of year 2015, PM was judged to be the core function for health effects of pollution.

37.5.1 Emissions and Human Health

Pacheco-Torgal and Jalali (2011, p. 2) substantiated that a large number of building construction materials exhibit some form of toxicity, 'thus causing several health-related problems such as asthma, itchiness, burning eyes, skin irritations or rashes, nose and throat irritation, nausea, headache, dizziness, fatigue, reproductive impairment, disruption of the endocrine system, impaired child development and birth defects, immune system suppression and cancer'. Table 37.5 shows some cancer-causing agents and their likely sources from paints used in the building industry.

The global burden of disease (GBD) reported in 2015 that air pollution is the fifth-ranked mortality factor (Burnett et al., 2018).

During a fire, significant toxic chemicals are emitted from building materials. These chemicals released are very harmful to human health (Doroudiani et al., 2012). Considering Doroudiani et al. (2012) study and GBD (2015) report, it is imperative that emissions from both the usage of materials and their combustion properties are taken into account during selection of natural and alternative materials for construction projects.

37.5.2 Factors Affecting Building Material Emissions

The detailed study carried out by Moulton-Patterson et al. (2003) outlined certain critical factors that affect the emissions from materials. Examples of these factors include:

- Quantity of material used in a particular operation.
- The assumed average ventilation rate.
- The time between completion of construction and occupancy.
- Building ventilation rate before and during occupancy.
- Age of material between manufacturing and installation.
- Storage, delivery and construction practices.

Table 37.5 Carcinogenic properties in paint

Property	Likely source
Chromates	Primers
Cadmium	Pigments
Benzene	Solvents
Nickel compounds	Pigments
Tetrachloroethylene	Organic solvent
Lead	Primers, dryers
Methylene chloride	Paint strippers

Source: Pacheco-Torgal and Jalali (2011)

These factors critically affirm that material efficiency has a larger role to play in terms of emissions. This means emission of a material is a direct function of the quantity of material and its concentration present in a product which affirms Fick's law.

37.6 Material Efficiency

Holton et al. (2008) emphasise the role of responsible material sourcing leading to enhanced material selection through material efficiency to provide an avenue to unlock opportunities to improve competitiveness.

In this perspective, Glass (2011) argued that responsible sourcing in industrial procurement practice is a challenge to selecting sustainable efficient materials. The study explains that for selecting an efficient construction material, there is the need to be proactive rather than reactive. 'The construction industry's fragmented supply network is a fundamental problem' (Glass, 2011, p. 169). Zhang et al. (2018) established that eco-friendly indicators emphasise the environmental reduction of resource use. The study by Ruuska and Häkkinen (2014), Edwards and Bennett (2003) and Hertwich et al. (2019) corroborates to the assertion that the durability and longevity of a material contribute to its eco-friendliness and thus its efficiency.

37.7 Conclusion and Future Research Direction

Motivated by building material emissions and their accompanying health complexness, this study concentrated on the need for insight into particle emissions of both natural and alternative building materials as directions for sustainable built environment achievement. The literature was sorted and categorised to help understand the pathway to the study of material emissions in the academic fraternity. A list of the important databases with their brief research engagement helped to outline the scope of this review. Durability and longevity of construction materials appear to be very significant in the study of construction material selection as corroborated by the study carried out by Ghumra et al. (2011).

The problem of selecting good quality material with emission-free health challenges stems from the fact that the built environment professionals have little clue of knowing about the toxicity of building materials. Throughout the review, literature has been quiet on the emissions of PM_{10} and $PM_{2.5}$ released from the use of both natural and alternative materials. This calls for concern, and a more significant research direction is required towards the study of this phenomenon to yield the required data that is essential for analysing the health complexities accompanying the use of these construction materials. Alternatively, the review confirms several studies on emissions relative to carbon and volatile organic compounds on building materials.

The study has been limited because to ascertain a more in-depth theoretical account, detailed analysis and discovery are paramount to bring forwards the boundary of emissions between natural and alternative materials.

Future research pathway(s) should provide more information on particulate matter emissions (PM_{10}, $PM_{2.5}$) of building construction materials with various quantitative emission factors of both natural and alternative materials.

Responsible material sourcing has also been found as an enhanced methodology to promote the mitigation of challenges associated with material selection. To this effect, even though life cycle assessment methodology is cumbersome, it provides a better inclusion in responsible sourcing (ethical management of sustainability issues within the construction supply chain) to demonstrate transparency with regard to the materials within a particular product, and thus aids in their selection.

When this is done, the confidence for material selection will be expanded to enable the built environment professional to make bold and knowledgeable decisions in this direction.

References

Aisyah, S., Rohana, S., Herberd, A., Zainuddin, A., Armi, M. (2019). Exposure of particulate matter 2.5 (PM2.5) on lung function performance of construction workers. In *AIP Conference Proceedings 2124* (020030). AIP Publishing, pp. 1–8.

Aoki, T., & Tanabe, S. (2007). Generation of sub-micron particles and secondary pollutants from building materials by ozone reaction. *Atmospheric Environment, 41*(15), 3139–3150. https://doi.org/10.1016/j.atmosenv.2006.07.053.

Azari, R. (2014). Integrated energy and environmental life cycle assessment of office building envelopes. *Energy and Buildings, 82*, 156–162. https://doi.org/10.1016/j.enbuild.2014.06.041.

Baetens, R., Jelle, B. P., & Gustavsen, A. (2010). Phase change materials for building applications: A state-of-the-art review Phase Change Materials for Building Applications. *SINTEF, 42*, 1361–1368.

Bellis, L. D. E. (1998). *Material Emission Rates Literature Review, and the Impact of Indoor Air Temperature and Relative Humidity, 33*(5), 261–277.

Burnett, R., Chen, H., Szyszkowicz, M., Fann, N., Hub-bell, B., Pope, C. A., et al. (2018). Global estimates of mortality associated with long-term exposure to outdoor fine particulate matter. *Proceedings of the National Academy of Sciences, 115*(38), 9592–9597.

Bylone, M. (2019). Healthy work environment 101. *AACN Advanced Critical Care, 22*(1), 19–21. https://doi.org/10.1097/NCI.0b013e3182049053.

Chan, A. P. C., Tetteh, M. O., & Nani, G. (2020). Drivers for international construction joint ventures adoption: A systematic literature review. *International Journal of Construction Management, 10*(2), 1–13. https://doi.org/10.1080/15623599.2020.1734417.

Cheriyan, D., & Choi, J. (2020). A review of research on particulate matter pollution in the construction industry. *Journal of Cleaner Production, 254*. https://doi.org/10.1016/j.jclepro.2020.120077.

Chin, K., Laguerre, A., Ramasubramanian, P., Pleshkov, D., Stephens, B., & G. E. (2019). Emerging investigator series: Primary emissions, ozone reactivity, and byproduct emissions from building insulation materials. *Environmental Science: Processes and Impacts, 21*(8), 1255–1267. https://doi.org/10.1039/c9em00024k.

Doroudiani, S., Doroudiani, B. and Doroudiani, Z. (2012) Materials that release toxic fumes during fire. *Toxicity of Building Materials*. Woodhead Publishing Limited. doi: https://doi.org/10.1533/9780857096357.241.

Edwards, S., & Bennett, P. (2003). Construction products and life-cycle thinking. *Industry and Environment, 26*(2–3), 57–61.

Farahzadi, L., Bakhitari, A. L., Gutierrez, U. R., and Azemati, H. (2016). Assessment of alternative building materials in the exterior walls for reduction of operational energy and CO2 emissions assessment of alternative building materials in the exterior walls for reduction of operational energy and CO_2 emissions. *International Journal of Engineering and Advanced Technology (IJEAT)*, (September), pp. 0–7.

Ghumra, S., Glass, J., Frost, M., Watkins, M., Mundy, J. (2011). Validating a set of empirically weighted sustainability indicators for construction products Shamir. Materials and Energy Assessment in Ceequal Transport Project, *Proceedings in the Institution of Civil Engineers*, pp. 153–164.

Glass, J. (2011). Briefing: Responsible sourcing of construction products. In *Proceedings of the Institution of Civil Engineers: Engineering Sustainability*, pp. 167–170. doi: https://doi.org/10.1680/ensu.1000011.

Gmelin, H., & Seuring, S. (2014). Determinants of a sustainable new product development. *Journal of Cleaner Production, 69*, 1–9. https://doi.org/10.1016/j.jclepro.2014.01.053.

Gonçalves de Lassio, J. G., & Naked Haddad, A. (2016). Life cycle assessment of building construction materials: Case study for a housing complex TT - Evaluación de ciclo de Vida de materiales de edificaciones: Estudio de Caso en complejo de viviendas. *Revista de la construcción, 15*(2), 69–77. https://doi.org/10.4067/S0718-915X2016000200007.

Greenstone, M. and Ryan, N. (2020). *Continuous Emissions Monitoring Systems (CEMS) in India March 2020 Impact Evaluation Report 111* (March).

Haapio, A., & Viitaniemi, P. (2008). A critical review of building environmental assessment tools. *Environmental Impact Assessment Review, 28*(7), 469–482. https://doi.org/10.1016/j.eiar.2008.01.002.

Had, L. and Brain, A. (2020). *Concrete solutions that lower both emissions and air pollution air quality and climate change intertwine in unexpected ways; a concrete example.*

Harb, P., Locoge, N., & Thevenet, F. (2018). Emissions and treatment of VOCs emitted from wood-based construction materials: Impact on indoor air quality. *Chemical Engineering Journal, 354*(August), 641–652. https://doi.org/10.1016/j.cej.2018.08.085.

He, G., Yang, X., & Shaw, C. Y. (2005). Material emission parameters obtained through regression. *Indoor and Built Environment, 14*(1), 59–68. https://doi.org/10.1177/1420326X05050347.

Hertwich, Edgar G., Ali, Saleem, Ciacci, Luca, Fishman, Tomer, Heeren, Niko, Masanet, Eric (2019). Material efficiency strategies to reducing greenhouse gas emissions associated with buildings, vehicles, and electronics - A review. *Environmental Research Letters*.

Holton, I., Glass, J., & Price, A. (2008). Developing a successful sector sustainability strategy: Six lessons from the UK construction products industry. *Corporate Social Responsibility and Environmental Management, 15*(1), 29–42. https://doi.org/10.1002/csr.

Huang, Kun-chih, Tsay, Shyan-Yaw, Lin, Fang- Ming and Chung, J.-W. (2019). Efficiency and performance tests of the sorptive building materials that reduce indoor formaldehyde concentrations. *PLoS ONE*.

Hillman, K., Dangaard, A., Ola, E., Jonsson, D., and Fluck, L. (2015). Climate benefits of material recycling, inventory of average greenhouse gas emissions for Denmark, Norway and Sweden. http://dx.doi.org/10.6027/TN2015-547. Available at www.norden.org/nordpub

International Institute of Sustainable Development [IISD]. (2019). Carbon Accounting gaps, Emission, Omissions. *Report*, (April), pp. 1–74.

Jalali, S. (2015). Toxicity of building materials: A key issue in sustainable construction. (September 2011). doi: https://doi.org/10.1080/19397038.2011.569583.

James, J. P., & Yang, X. (2005). Emissions of volatile organic compounds from several green and non-green building materials: A comparison. *Indoor and Built Environment, 14*(1), 69–74. https://doi.org/10.1177/1420326X05050504.

Järnström, H. (2007). *Reference values for building material emissions and indoor air quality in residential buildings*. Finland: VTT Publications.

Jung, S., Kang, H., Sung, S., & H. T. (2019). Health risk assessment for occupants as a decision-making tool to quantify the environmental effects of particulate matter in construction projects. *Building and Environment, 161*, 1–16.

Karaguliana, F., Belis, C. A., Dora, C. F. C., Prüss-Ustün, A. M., Bonjour, S., Adair-Rohani, H., & Amann, M. (2015). Contributions to cities' ambient particulate matter (PM): A systematic review of local source contributions at global level. *Atmospheric Environment, 120*, 475–483.

Keita, Sekou, Liousse, Cathy, Yboué, Veronique, Dominutti, Pamela, Assamoi, Eric-Michel, Doumbia, Madina, Bahino Julien, Guinot, B. (2018). Particle and VOC emission factor measurements for anthropogenic sources in West Africa. *Atmospheric Chemistry and Physics*.

Khoshnava, S. M., Rostami, R., Valipour, A., Ismail, M., & Rahmat, A. R. (2018). Rank of green building material criteria based on the three pillars of sustainability using the hybrid multi criteria decision making method. *Journal of Cleaner Production, 173*, 82–99. https://doi.org/10.1016/j.jclepro.2016.10.066.

Kong, A., Kang, H., He, S., Li, N., & Wang, W. (2020). Study on the carbon emissions in the whole construction process of prefabricated floor slab. *Applied Sciences (Switzerland), 10*(7). https://doi.org/10.3390/app10072326.

Levin, H. (2016). Building ecology: An architect's perspective -- plenary lecture. (November 2014).

Lifset, R., & Eckelman, M. (2013). 'Material efficiency in a multi-material world', *philosophical transactions of the Royal Society A: Mathematical. Physical and Engineering Sciences, 371*(1986). https://doi.org/10.1098/rsta.2012.0002.

Magee, R. (2005). A material emission database for 90 target VOCs NRC Publications Archive (NPArC) Archives des publications du CNRC (NPArC) A Material emission database for 90 target VOCs. Won, D. Y.; Magee, R. J.; Yang, W.; Lusztyk, E.; Nong, G.; Shaw, C. (January).

Maoeng, M., Edoun, E. I., & Mbohwa, C. (2020). Sustainable development practices in the south African construction industry: A review of related literature. *Applied Sciences, 10*, 1418–1426.

Martínez-Rocamora, A., Solís-Guzmán, J., & Marrero, M. (2016). LCA databases focused on construction materials: A review. *Renewable and Sustainable Energy Reviews, 58*, 563–570. https://doi.org/10.1016/j.rser.2015.12.243.

Meng, J., Liu, J., & Tao, S. (2015). Tracing primary PM2.5 emissions via Chinese supply chains. *Environmental Research Letters, 11*, 1–13.

Milner, J., Hamilton, I., & Woodcock, J. (2020). Health benefits of policies to reduce carbon emissions. *BMJ, 368*, 6–11. https://doi.org/10.1136/bmj.l6758.

Mohajerani, A., Siu-Qun, H., Mehdi, M., Arulrajah, A., Horpibulsuk, S., Kadir, M., & Aeslina Farshid, A. T. R. (2019). Amazing types, properties, and applications of fibres in construction materials. *Materials, 12*(16), 1–45. https://doi.org/10.3390/ma12162513.

Moulton-Patterson, L., Peace, C., & Leary, M. (2003). Building material emissions study. *Integrated Waste Management Board, 22*, 1–328.

Nwodo, M. N., & Anumba, C. J. (2019). A review of life cycle assessment of buildings using a systematic approach. *Building and Environment, 162*(July), 100. https://doi.org/10.1016/j.buildenv.2019.106290.

Pacheco-Torgal, F., & Jalali, S. (2011). Toxicity of building materials: A key issue in sustainable construction. *International Journal of Sustainable Engineering, 4*(3), 281–287. https://doi.org/10.1080/19397038.2011.569583.

Raifman, Matthew, Russell, Armistead, Skipper, Nash, Kinney, P. (2020) 'Quantifying the health impacts of eliminating air pollution emissions in the City of Boston', Environmental Research Letters 24 1–19.

Ramdhani, A., Ramdhani, M., & Amin, A. (2014). Writing a literature review research paper: A step-by-step approach. *International Journal of Basic and Applied Science, 3*(01), 47–56.

Ramesh, T., Prakash, R., & Shukla, K. K. (2010). Life cycle energy analysis of buildings: An overview. *Energy and Buildings, 42*(10), 1592–1600. https://doi.org/10.1016/j.enbuild.2010.05.007.

Reed, C. H. (2011). *Reference material to improve reliability of building product VOC emissions testing.* (September), pp. 30–33.

Ruuska, A., & Häkkinen, T. (2014). Material efficiency of building construction. *Buildings, 4*(3), 266–294. https://doi.org/10.3390/buildings4030266.

Shi, Qingwei, Gao, Jingxin, Wang, Xia, Ren, Hong, Cai, Weiguang, and Wei, H. (2020). *Temporal and spatial variability of carbon emission intensity of urban residential buildings: Testing the effect of economics and geographic location in China.* Sustainability, Switzerland, pp. 1–23.

Silva, R. V., de Brito, J., & Dhir, R. K. (2019). Use of recycled aggregates arising from construction and demolition waste in new construction applications. *Journal of Cleaner Production, 236*, 117629. https://doi.org/10.1016/j.jclepro.2019.117629.

SimaPro Library Database Manual Colophon, M (n.d.).

Souza, V., & Borsato, M. (2016). Sustainable design and its interfaces: An overview Vitor de Souza * and Milton Borsato. *International Journal of Agile Systems and Management, 9*(3), 183–211.

Sun, J., Zhou, Z., Huang, J., & Guoxing, L. (2020). A bibliometric analysis of the impacts of air pollution on children. *International Journal of Environmental Research and Public Health, 12*, 1–11. https://doi.org/10.3390/su12072695.

United States Environmental Protection Agency. (2013). *Recommended procedures for development of emissions factors and use of the WebFIRE Database.*

Upstill-Goddard, J., Glass, J., Dainty, A., Nicholson, I. (2015). Analysis of responsible sourcing performance in BES 6001 certificates. In *Proceedings of the Institution of Civil Engineers: Engineering Sustainability*, pp. 71–81.

Wallace, M., & Wray, A. (2013). *Critical Reading and writing for postgraduates, educate~.* New Delhi: Sage.

Wegener, A., Sleeswijk, O., van Lauran, G., Jeroen, S., & Jaap, H. M. (2007). Normalisation in product life cycle assessment: An LCA of the global and European economic systems in the year 2000. *Science of the Total Environment, 390*(1), 227–240.

Wille, K., & Boisvert-Cotulio, C. (2015). Material efficiency in the design of ultra-high performance concrete. *Construction and Building Materials, 86*, 33–43. https://doi.org/10.1016/j.conbuildmat.2015.03.087.

Xia, B., Ding, T., & Xiao, J. (2020). Life cycle assessment of concrete structures with reuse and recycling strategies: A novel framework and case study. *Waste Management, 105*, 268–278. https://doi.org/10.1016/j.wasman.2020.02.015.

Zhang, C., Chen, W. and Ruth, M. (2018). Measuring material efficiency: A review of the historical evolution of indicators. *Methodologies and Findings.* (April). doi: https://doi.org/10.1016/j.resconrec.2018.01.028.

Zhang, J. J. (1997). A review of volatile organics emission data for building materials and furnishings. doi: https://doi.org/10.4224/20338000.

Part IX
Sustainable Transport

Chapter 38
Social Acceptance and Societal Readiness to EVs

Khalid Kamal Abdelgader Mohamed, Henok K. Wolde, and Salma M. S. Alarefi

38.1 Introduction

The transport sector is among the major contributors of greenhouse gas emissions with emissions expected to reach 50% by 2030 (Egbue & Long, 2012). The latter has motived the shift towards electrification of transpiration (Lieven et al., 2011). Despite the advances in the technology, the uptake of EVs by the public has been modest due to concerns over the sustainability of the fuel, cost, safety and reliability as well as the maturity of the technology (Lieven et al., 2011). Researchers have been engaged with enhancing the sustainability of plugin EVs. For example, renewable power-assisted EV charging stations have been emerging as a greener charging (Lewis et al., 2012). Innovative technologies like carbon capture and sequestration to reduce emission levels have also been investigated (Klemeš et al., 2012; Krupa et al., 2012), that is, in addition to the well-documented investigations of the life cycle analysis of EV batteries and the range, maintenance, charging infrastructures and initial cost of EV (Jensen et al., 2014; Krupa et al., 2012; Krupa et al., 2014; Offer et al., 2010).

Inevitably, the public awareness and attitude towards electrification of transpiration have a vital role in stimulating interest in the technology. In fact, a study conducted in (Lewis et al., 2012) reported that 10% of the participants claim to have concerns about the reliability of EVs. In relation to the performance of EVs, however, understanding the correlation between the fuel and mileage seems to be a great discouraging factor. Unlike combustion engine cars which illustrate a visual estimation of fuel consumption, such indication is not available in EVs. The latter could lead to anxiety issues affecting the safety of the driver (Klemeš et al., 2012).

K. K. A. Mohamed · H. K. Wolde · S. M. S. Alarefi (✉)
School of Electronics and Electrical Engineering, University of Leeds, Leeds, UK
e-mail: S.Alarefi@leeds.ac.uk

Arguably, the low uptake could be accredited to the acknowledged trade-off between storage capacity and cost of EVs. This remains under investigation for improved fuel economy and expected to vanish as the technology matures. In relation to that, when deciding on a car, however, it is argued that consumers seem to overlook the long-term financial benefit (Krupa et al., 2014). This is interesting, given that the fuel economy of EVs (i.e. in terms fuel per mile) is 50% less than that of a traditional gasoline-fuelled vehicle (Jensen et al., 2014).

In 2016, the electric stock market for plugin hybrid and battery-based EVs has reached two million (from a couple of hundred thousand in 2010) (Erdem, 2011). According to the International Energy Agency (IEA), the global EV population is projected to reach 130 million by 2030 (Hawkins et al., 2013). As well as fuel sustainability, the availability of accessible public charging stations contributes to modest uptake levels (Lieven et al., 2011). In terms of fuel flexibility, the uptake in urban areas continues to be challenged by the poor charging infrastructure both at residential level and workplaces (Lieven et al., 2011; Lane & Potter, 2007). In the UK, for example, home charging is not possible for most households with limited if any parking spaces (Krupa et al., 2012). In Leeds, for instance, currently there are 20 EV charging stations that are capable of charging 88 EVs (Offer et al., 2010). In fact, Leeds City Council is promoting and enforcing the use of EVs in the city. The council predicts wide spread of EVs for work commute by 2030 (Krause et al., 2013). The city council has 44 EVs with plans to replace existing 50 vehicles by cleaner and greener zero-emission Nissan vans. In total, the city council will have 95 EVs more than any country authority in 2018 (Kenneth & Reid, 2008).

Despite advances in EV technologies which continue to emerge as a promising alternative for sustainable urbanisation, the uptake in the city of Leeds is modest and continues to be challenged by the public attitude towards the technology. It is with no doubt that public engagement and interest in the uptake are vital to motivate the sustainable transition towards electrifications of the transportations. This study therefore investigates the attitude of the people of Leeds towards the uptake of EVs. To set the scene, literature related to the common factors which impacts the social acceptance of EVs is briefly reviewed based on studies carried out in the UK. The methodology of the qualitative survey is outlined. Before concluding, the implications of results are briefly discussed and supported by outline of recommendation and future directions.

38.2 Literature Review

The global public attitude towards adoption of EV remains to be controversial. In fact, the environmental sustainability of EV fuel continues to be challenged by environmental enthusiasts and influencers with concerns over pollution shifting (Egbue & Long, 2012; Lieven et al., 2011). Not to mention the high initial and maintenance cost and long payback times of EVs. From an economic sustainability prospective, the economic viability of investing in EV does not compare with other sustainable socioeconomic investment choices such as a residential grid-connected solar PV system (Lewis et al., 2012; Klemeš et al., 2012).

According to the National Grid, 43% of the UK households lack the front parking space required for home charging (Krupa et al., 2012). The social acceptance and readiness for the transition depend on different factors such as finical, vehicle performance and charging infrastructure (Jensen et al., 2014; Erdem, 2011). The financial concerns are related to the price of the EV (Graham-Rowe et al., 2012), the battery cost (Jensen et al., 2014) and the poor understanding of the fuel and maintenance cost (Lewis et al., 2012). Whereas concerns over vehicle performance are related to the safety, range, battery life and reliability (Lieven et al., 2011; Turrentine & Kurani, 2007), the nature of the power supply and the availability of public and residential charging contribute to consumer's great concerns over the charging infrastructure (Lieven et al., 2011; Lane & Potter, 2007; Sierzchula et al., 2014). It is obvious that public awareness of EV technologies is a function of age, income, level of education and environmental awareness (Zhang et al., 2013). The interest of environmentally aware and highly educated consumers is well defined in existing studies (Lieven et al., 2011; Jensen et al., 2014; Nayum & Klöckner, 2014).

According to (Klemeš et al., 2012), following a week trail, UK drivers reported their unsatisfaction with the charge duration and appearance of the car. Furthermore, the limited range of the vehicles and their performance are major issues in their acceptance by the society and prolong the readiness to decide and own these vehicles (Lieven et al., 2011).

Indeed, EVs have great potential to compete with traditional vehicles if the associated infrastructures are made widely available (i.e. as that of existing filling stations for gasoline/diesel cars) (Erdem et al., 2010). As well as fast charging technologies for workspace, the comfort of home charging, overnight charging in particular, would then be a flexibility factor that could further motivate the transition towards electrification of transportation (Dagsvik et al., 2002). While the technology continues to progress, improving the sustainability of the supply through existing mature technologies (i.e. energy diversifications through renewables) is likely to influence sustainability informed consumers and aid with the transition. Interestingly, a Swedish study found that personality and attitude of consumers' had an influence on the decision-making of car owner's (Kishi & Satoh, 2005). Similarly, a study in (Jansson, 2011) reported that participants believed that driving a hybrid vehicle reflected their personalities and values. It is reported that better educated, environmental consciousness considers the adoption of the niche technology as a status symbol and that owners feel respected for driving EV (Heffner et al., 2007).

Given the abovementioned, the social acceptance of EVs is discouraged by many factors including the availability of charging infrastructure. It is evident that the availability of home charging increases consumers' willingness to purchase EVs. The environmental benefits associated with EVs such as zero emissions persuaded users to overlook the shortcomings of EVs such as appearance or performance issues. There is however a gap on understanding the role of renewable powered charging stations in the stimulation of the public interest and engagement with EVs.

Examples of key forces that continue to challenge public acceptance of the EVs are critically analysed through PESTE and summarised in Table 38.1.

Table 38.1 PESTE analysis of the key forces that challenge public acceptance of EVs and avenues for opportunities

	Threats	Opportunities
Political	• Risk of green washing activity to meet the net 2050 zero-emission targets. • Post Brexit uncertainties impact on batteries sourced from EU due to lack of active battery plants in the U.	• Innovative legislation and incentives to support individual and organisational take-ups. • Political intervention can accelerate technological adoption in some societies (Rogers, 1995).
Economic	• Increasing production cost for UK manufacturing companies. • High competitions with emerging economies (e.g. China).	Extra government funding to pave the road to zero strategy for • **Manufacturer** – Help mitigate difficult market condition, increased cost and low uptake. • **Research and development** – Aid with the development of cost-effective EVs and, sustainable charging infrastructure to eradicate consumers concerns over the technology. • **Individuals** – Widespread instalments of charging points in new-build homes, accessible and inclusive tax-free loans and incentives to further encourage uptake of EVs.
Societal	• Early adopters are often better educated, are more literate and have higher social status. • High prices limit access and hence adoption (Element Energy, 2009). • Poses alternative on three levels: Symbolic, organisational and behavioural basis (Hard & Jamison, 1997). • A car is more than just a method of transport; for some, it is a status symbol or an expansion of ones' own physical space (Latour, 2005). • Early take-up of EVs in the UK is gendered (Brady, 2010; Heffner et al., 2007a). • Vehicle symbolism price and economy (Scharff, 1992).	Sustainability education to communicate the need for sustainable transportation and benefits it can add to the health and wellbeing of the society (e.g. creation of jobs and training opportunities, promotion of green societies and economies) at • **Early learning** – For less car-dependent society overturn. • **Social and public media** – Encourage and support sustainable sociotechnical system. • **Universities and in the workplace** – Help to eradicate stereotype about EV symbol.

(continued)

Table 38.1 (continued)

	Threats	Opportunities
Technology	• Risks to pedestrians due to low auditable of EV engine at low speeds. • Complexity relationship with technology. • Risk associated with stereotyping its application due to limited travel range. Historically EVs were deemed to be suitable for women (i.e. housewives and mothers) (U.K. Government, 2020). • Uptake mainly motivated by Technophilia. • Risks of low consumers' tolerance of technical deficiencies for an emerging technology of EVs.	Opportunities for mutual shaping of the technology as well as the society through technology infusion and technology diffusion
Environment	• Environmental consciousness concern over the sustainability of the fuel.	Sustainable charging fuel (through renewable-assisted charging stations)
Legal	Modest uptake makes it difficult to anticipate various legal aspects of the technology (socially, environmentally and economically)	Avenues for the UK to lead and pioneer in the legislations of sustainable societies

38.3 Research Methodology

38.3.1 Survey

The survey was conducted on different demographics of the community based on age and education levels. The qualitative survey was in the form of a questionnaire containing 11 closed-ended multiple choice questions.

The questions were categorised into three sections: the first section is related to the participant's demography which includes age and education level, the second section studies the attitude towards the UK government policy and the understanding of the role of EVs in carbon emission reduction, and the third section investigated the public willing to invest on EVs. To gain a deeper understanding of motivating factor that would stimulate the uptake of EVs, the response towards renewably assisted EV charging station as well as the automation of EVs was investigated. The survey was distributed through Leeds social media groups (i.e. Facebook, LinkedIn, WhatsApp) and by emails to the student and staff community at the University of Leeds, and the responses were collected for analysis through Google Forms.

38.3.2 Data and Quality Control

The survey data was gathered from 69 participants based in the city of Leeds and the data were checked for anomalies. Participants were encouraged to give their honest opinions through open-ended follow-up questions that promoted the participants to share their thoughts, to further explore the public's attitude towards EVs and electric vehicle charging infrastructure. The questionnaire was designed to be simple, included minimal technical information and did not include vague questions.

38.4 Results and Analysis

To explore current perception of the energy sources implored to supply charging stations, a survey was designed and distributed. The educational level composition is shown in Fig. 38.1.

As expected, all participants seem to be aware of the impact of carbon emissions. In fact, more than two-thirds of the participants are in favour of the 2050 ban of internal combustion engine vehicles. Evidently, there seems to be great correlation between the participants' educational background and appreciation of the significance of the matter. Whereas two-thirds of the population is made up of participants with postgraduate qualifications, the remaining minority is claimed by participants of primary level of education. Interestingly, more than two-thirds of the participants acknowledge the role of electrifying transportation to reduce the levels of carbon emissions. In relation to the latter, it was also seen that when the participants were asked if usage of EVs could reduce carbon emissions, about a third of the population seem to be hesitant or rather uncertain about that. In relation to the latter, it was also seen that when the participants were asked if usage of EVs could reduce carbon emissions, whereas a third of the population seem to be hesitant or rather uncertain, a fifth of the responders (i.e. 7%) are in strong disagreement. Figure 38.2 illustrates the percentages of the responses to the role of electrifications in emission reduction categorised in terms of educational background.

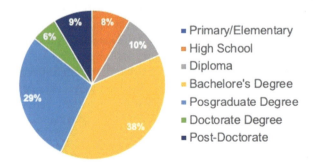

Fig. 38.1 Educational level composition of the data sample

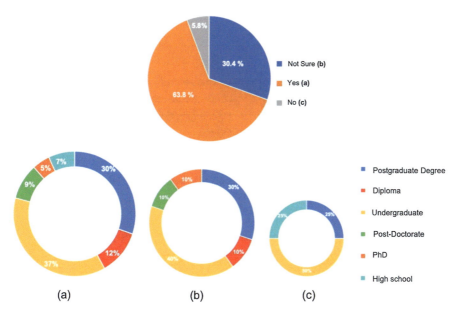

Fig. 38.2 Evaluation of the participants' acknowledgement of the role of EVs in reducing emissions. (**a**) Educational background of participants in agreement, (**b**) educational background of participants who were not sure, (c) educational background of participants in disagreement

Having established the participants' attitude towards electrification of transportation, the participants' interest in purchasing an electric car was further investigated. In response to that, around half of the participants showed interest in investing in EVs. As well as being perceived as better alternative to conventional cars (48.7%) and zero-emission contributions (33.3%), the large population of the participants are motivated by the fact that electricity is cheaper than fossil fuel (3%), potential for automation (5%) and better tax benefits (3%). Figure 38.3 indicates in percentages the factors that discourage them from buying an electric car. Table 38.2 further analyses the responses of the third majority in Fig. 38.3.

The carbon emitting nature of EV fuels (i.e. fossil fuel-generated electricity) seems to be a great discouraging factor. Of the one-quarter population of participants who claimed to be less likely to buy an EV, a third were discouraged by the fact that that EVs are charged with partially fossil-fuelled energy sources. Among the other factors, concerns about the charging times, frequency of charging, EV range and cost of EVs seem to influence the modest uptake. In relation to that, around 20% of the participants who are not willing to invest in EVs claim to prefer conventional cars over EV mainly being deterred by their market cost.

Having established an insight into the understanding and appreciation of the concept of electrification of transportation, the survey further examines the response to the technology with enhanced sustainability. In that regard, the participants were asked if renewable-powered EV charging stations would influence their attitude towards EVs. Moreover, more than four-fifths of the participants agreed that

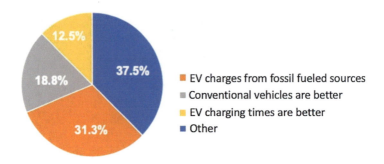

Fig. 38.3 Proportion of participants that are less likely to invest in an electric car

sustainable EV fuel would positively impact their attitude towards the uptake of EV, and results are shown in Fig. 38.4.

It is evident that the great majority (more than four-fifths of participants) appreciate the role of EVs in emission reduction and the usage of EVs clearly; however, this needs to be supported by informing the public about the portfolio of energy being used to power their charging stations. Distributed renewably powered off-grid charging stations are a direct solution for this as the power generation methodology can be clearly seen by the public. Meanwhile, a minority of 12% responded by conditions for their attitude change which revolved around increasing EV mileage and reducing their environmental impact which hints at a possibility of misinformation in the public perception of EVs. The results of this question have no correlation to educational background of the sample used, meaning the effect can be and will be seen across all backgrounds. In addition, the shift in answers from a two-third majority in favour of EV usage to over 80% as seen in Fig. 38.4 shows a substantial positive shift in attitude towards using EVs.

EVs provide an opportunity for automation using self-driving cars which is not a tangible possibility for conventional engine cars in addition to their ability to be charged from homes. This can be used to supplement distancing measures in response to the COVID-19 pandemic. Participants' appreciation of the matter was examined. Majority of over 60% agreed that the comfort of charging from home makes EV usage a safer option.

Further, the participants were invited to share their views on the role of automated EVs in supporting local communities in health crisis such the current crisis imposed by the breakdown of the novel virus, and results are indicated in Fig. 38.5. In that regard, the most majority (two-thirds) acknowledged the role of EV automation in supporting communities. The latter is accredited to the nature automation to do with the lack of human intervention. This indeed indicates that public opinion on EVs can be further examined through sustainability educations (i.e. environmental, socioeconomic benefits). A third disagreed however with mixed educational background levels matching the sample composition, so there was no direct correlation between these factors. About a tenth claim that EV could be safer based on conditions tabulated in Table 38.3.

Table 38.2 Analysis of the mutually inclusive responses by participants who are less like to invest on an EV

	Cost (50%)	Technical (30%)	Environmental (30%)	Social (10%)
Educational level	• Undergraduate (42.8%) • Postgraduate (~28.5%) • Post-doctorate (~28.5%)	• Undergraduate (20%) • Postgraduate (20%) • Post-doctorate (60%)	• Undergraduate (50%) • Postgraduate degree (50%)	• Undergraduate (50%) • Diploma (50%)
Age group	• 18–24 (31%) • 25–34 (23%) • 35–44 (8%) • 55+ (31%)	• 18–24 (40%) • 25–34 (20%) • 35–44 (20%) • 55+ (20%)	• 25–34 (33.3%) • 35–44 (33.3%) • 45–54 (33.3%)	• 25–34 (100%)
Comments	• Conventional cars are cheaper • EVs are much more expensive • Financial restraints	• Charging points are rare and slow where I live • Frequency of charging EVs is greater than that of refuelling traditional cars • Lithium batteries (lifetime and end-of-life recovery routes)	• Potentially use fossil fuel power sources • I don't want to buy/own a car, even electric. I'll rent one if needed, be but I'm using public transport and plan to buy an electric bike one day. • EVs not charged from renewable energy sources	• Unlikely to drive generally • Can't charge at home (no garage)

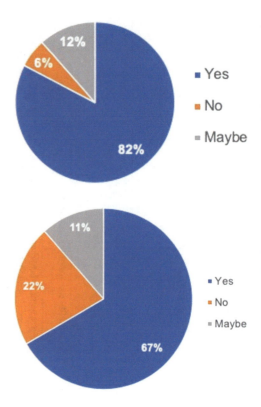

Fig. 38.4 Evaluations of responders' attitude to EV if powered through renewable-assisted charging stations

Fig. 38.5 Sample response to automation technology as a percentage, in answer to the question 'Could automated EV technology support local communities in health crises?'

Table 38.3 Analysis of the responses that participants claim to condition the safety of EV in controlling the spread of the coronavirus

EV as sustainable alternative	Interest in buying an EV	Comments (EV automation might work if…)
Agree (fuel sustainability)	Very unlikely	Control systems are foolproof and fail safe
Agree (life cycle of EVs)	Very unlikely	As a replacement transport service and for remote delivery of medicines, food and to collect specimens or swap physical objects (meals, fruits, vegetables etc.) between neighbours – In and out of health crisis
Agree	Very likely	If there are accessible places to charge the EV
Agree (concernsover cost)	Very unlikely	Self-driving EVs? They could be useful for home deliveries … They would need autonomous pick-and-place robotics to remove each customer's goods and deposit them outside their home
Agree	Very likely	I do not think the level of automation needed, which isn't already in place, would have an impact in the timeframe needed
Agree	Likely	Local businesses receive help to modernise (i.e. petrol stations)

38.5 Discussion

38.5.1 Implications

The context of the survey was to pin down public opinion by extrapolating from a moderate sample size. Overall, the public appears to have a moderately informed opinion and positive stance towards reducing emissions contributed by the transport sector, but the motivation seems to vary. There is a clear evidence of the public support to a sustainable shifting from fossil-fuelled vehicles towards EVs.

Sceptics over the introduction of EVs originated by concerns over its cost, which is predicted to drop as the technology matures. EV range seems to stimulate discussions over the viability of the technology. The latter is an issue that is closely tied to cost-effectiveness of existing long-range EVs. Interestingly, participants that are not certain about the role of EVs in emission reduction seem to be from groups with no higher education experience, which implies simply a lack of public information on the topic.

Overall, EVs are perceived as a better alternative to conventional cars due to their technical specifications and potential for automation. Considering contact restriction imposed by the breakout of the novel COVID-19 pandemic, the latter seems to be most favourable with the majority proportion of participants interested in opting in and supporting the technology. The introduction of renewable powered-assisted EV charging station seems to have sparked great interest in the uptake of EVs.

Post-COVID-19 carbon savings made by reduced transportation activities are at great threat (U.K. Government, 2020). That is because the constraints of the 'new normal' lifestyle imposed by the social distancing restrictions to stop the spread of coronavirus have seen an increase in the use of personal cars over public transport. Electrifying the transportation system (EV, e-bikes, e-scoters) could provide a responsible and sustainable transport alternative (Budd & Ison, 2020).

38.5.2 Recommendations

Evidently, the sustainability of EV technologies (i.e. fuel, batteries) will continue to pose great challenges in simulating interest and accelerating the acceptance of such technological shift. This necessitates the need for better communication channels to communicate the implications of the technological advancement to the public. As well as educating the public, public engagement will also introduce a sense of acknowledgement, which would in turn motivate the uptake of the technology. The following recommendations are equally pointed at policymakers at different levels (e.g. from city level to national level) and educational institutes:

1. Increased public awareness of the benefits of revolutionising the energy sector through energy diversification and its impact on greening of transportation.

2. Enhanced public understanding of potential health and safety benefits that is to be made possible by advancement of EV technology to help eradicate negative stereotypes on EVs.
3. Improved visibility of charging infrastructure through city-level public channels.
4. As well as environmental benefits, improved public knowledge of potential avenues for economic growth through widespread adoption of the technology.

38.5.3 Further Research

For the research to provide tangible insight into public attitude, a bigger sample size must be used. This would make the use of data extrapolation methods viable. In this study, the small sample size meant that using statistical techniques (e.g. standard deviation for correlation analysis) runs the risk of exaggeration of trends. Further work should use a bigger distribution infrastructure with the use of more grouping questions to differentiate more demographics based on both societal and economical basis, which could help to provide precise data for policymakers. Finally, adopting an incentivised approach to the dissemination of future surveys could encourage more participation from members of the public.

38.6 Conclusion

By and large, the output of the survey showed great public appreciation of the role EVs can play in reducing emissions, with nearly two-thirds of the participants in favour of the sustainable alternative. Evidently, participants are more likely to opt-in for the technology if the charging power is generated through means of renewable energy sources.

Considering the automation potential of EVs and light of the current social contact restrictions introduced by COVID-19, participants acknowledged the potential for automated EVs to support local communities during public health crisis. In that regard, motivated by the need to reduce contact through social distancing measures, participants showed appreciations to the benefit EV offers in relation to home charging.

The outcomes of the survey highlight the importance of improving the sustainability of the power supply for EV charging stations in order to enhance the public perception of the technology and hence increase the uptake. In the interim, however, the reported results create an avenue for policymakers and regulating bodies to consider planning the zero-emission transition for the sector.

References

Brady, J. (2010). *Electric Car Cultures: An ethnography of the everyday use of electric vehicles in the UK*. Masters thesis, Durham University. Retrieved from http://etheses.dur.ac.uk/690/

Budd, L., & Ison, S. (2020). Responsible transport: A post-COVID agenda for transport policy and practice. *Transportation Research Interdisciplinary Perspectives, 6*, 100151.

Dagsvik, J. K., Wennemo, T., Wetterwald, D. G., & Aaberge, R. (2002). Potential demand for alternative fuel vehicles. *Transp. Res. Part B Methodol., 36*(4), 361–384.

Egbue, O., & Long, S. (2012). Barriers to widespread adoption of electric vehicles: An analysis of consumer attitudes and perceptions. *Energy Policy, 48*(2012), 717–729.

Element Energy. (2009). Strategies for the uptake of electric vehicles and associated infrastructure implications—Report for the committee on climate change October 2009.

Erdem, C. (2011). Corrigendum to 'identifying the factors affecting the willingness to pay for fuel-efficient vehicles in Turkey: A case of hybrids'. *Energy Policy, 39*(8), 4673.

Erdem, C., Şentürk, I., & Şimşek, T. (2010). Identifying the factors affecting the willingness to pay for fuel- efficient vehicles in Turkey: A case of hybrids. *Energy Policy, 38*(6), 3038–3043.

Graham-Rowe, E., et al. (2012). Mainstream consumers driving plug-in battery-electric and plug-in hybrid electric cars: A qualitative analysis of responses and evaluations. *Transp Res Part A Policy Pract., 46*(1), 140–153.

Hard, M., & Jamison, A. (1997). Alternative cars: The contrasting stories of steam and diesel automotive engines. *Technology in Society., 19*(2), 145–160.

Hawkins, T. R., Singh, B., Majeau-Bettez, G., & Strømman, A. H. (2013). Comparative environmental life cycle assessment of conventional and electric vehicles. *Journal of Industrial Ecology, 17*(1), 53–64.

Heffner, R., Kurani, K. S., & Turrentine, T. S. (2007a). Symbolism in California's early market for hybrid electric vehicles. *Transportation Research Part D, 12*(6), 396–413.

Heffner, R. R., Kurani, K. S., & Turrentine, T. S. (2007). Symbolism in California's early market for hybrid electric vehicles. *Transport Research Part D Transportational Environment., 12*(6), 396–413.

Jansson, J. (2011). Consumer eco-innovation adoption: Assessing attitudinal factors and perceived product characteristics. *Business Strategy and the Environment, 20*(3), 192–210.

Jensen, A. F., Cherchi, E., & de Dios Ortúzar, J. (2014). A long panel survey to elicit variation in preferences and attitudes in the choice of electric vehicles. *Transportation (Amst)., 41*(5), 973–993.

Kenneth, S., & Reid, R. (2008). *Driving plug-in hybrid electric vehicles*.

Kishi, K., & Satoh, K. (2005). Evaluation of willingness to buy a low-pollution car in Japan. *Journal of Eastern Asia Society Transportation Studies, 6*.

Klemeš, J. J., Varbanov, P. S., & Huisingh, D. (2012). Recent cleaner production advances in process monitoring and optimisation. *Journal of Clean Production., 34*, 1–8.

Krause, R. M., Carley, S. R., Lane, B. W., & Graham, J. D. (2013). Perception and reality: Public knowledge of plug-in electric vehicles in 21 U.S. cities. *Energy Policy, 63*(2013), 433–440.

Krupa, J. S., Chatterjee, S., Eldridge, E., Rizzo, D. M., Eppstein, M. J. (2012). Evolutionary feature selection for classification: A plug-in hybrid vehicle adoption application. *GECCO'12 - Proc. 14th Int. Conf. Genet. Evol. Comput.* pp. 1111–1118.

Krupa, J. S., et al. (2014). Analysis of a consumer survey on plug-in hybrid electric vehicles. *Transp Res Part A Policy Pract., 64*, 14–34.

Lane, B., & Potter, S. (2007). The adoption of cleaner vehicles in the UK: Exploring the consumer attitude- action gap. *Journal of Cleaner Production, 15*(11–12), 1085–1092.

Latour, B. (2005). *Re-assembling the social: An introduction to actor-network theory*. Oxford: Oxford University Press.

Lewis, A. M., Kelly, J. C., Keoleian, G. A. (2012). Evaluating the life cycle greenhouse gas emissions from a lightweight plug-in hybrid electric vehicle in a regional context. In *2012 IEEE international symposium on sustainable systems and technology (ISSST)*, pp. 1–6.

Lieven, T., Mühlmeier, S., Henkel, S., & Waller, J. F. (2011). Who will buy electric cars? An empirical study in Germany. *Transp Res Part D Transp Environ., 16*(3), 236–243.

Nayum, A., & Klöckner, C. A. (2014). A comprehensive socio-psychological approach to car type choice. *Journal of Environmental Psychology, 40*, 401–411.

Offer, G. J., Howey, D., Contestabile, M., Clague, R., & Brandon, N. P. (2010). Comparative analysis of battery electric, hydrogen fuel cell and hybrid vehicles in a future sustainable road transport system. *Energy Policy, 38*(1), 24–29.

Rogers, E. M. (1995). *Diffusion of innovations* (4th ed.). New York: Free Press.

Scharff, V. (1992). *Gender, electricity and automobility in the Car and the City: The automobile, the built environment, and daily urban life* edited by Wachs, M and Crawford, M 1992 (University of Michigan Press, Ann Arbor) pp 75–85.

Sierzchula, W., Bakker, S., Maat, K., & Van Wee, B. (2014). The influence of financial incentives and other socio-economic factors on electric vehicle adoption. *Energy Policy, 68*, 183–194.

Turrentine, T. S., & Kurani, K. S. (2007). Car buyers and fuel economy? *Energy Policy, 35*(2), 1213–1223.

U.K. Government. (2020). *Our Plan To Rebuild: The UK Government's COVID-19 Recovery Strategy*. Retrieved July 24, 2020, from https://assets.publishing.service.gov.uk/government/uploads/system/uploads/attachment_data/file/884760/Our_plan_to_rebuild_The_UK_Government_s_COVID-19_recovery_strategy.pdf.

Zhang, X., Wang, K., Hao, Y., Fan, J. L., & Wei, Y. M. (2013). The impact of government policy on preference for NEVs: The evidence from China. *Energy Policy, 61*(2013), 382–393.

Chapter 39
Transport and Waste: Killing Two Birds with One Stone—The Sustainable Energy

Hina Akram, Shoaib Hussain, and Talib E. Butt

39.1 Introduction

The notion that any human activity should be sustainable probably has been accepted as a basic premise, though this realization should have been much faster than it has been. Among the various factors influencing the environment of our planet, exponential increase in human population has always been at the top of the tiers. Two of the main anthropogenic influences are transport and waste. Among several energy-consuming sectors, the transport is the prominent one. Energy consumption for both travel and freight movement has rapidly been growing. It is estimated that transportation energy consumption would increase by nearly 40% between 2018 and 2050. Thereby, this sector is placing considerably an additional strain on fossil fuel reserves (García-Olivares et al., 2018). On the other hand, the production of municipal solid waste (MSW) – constituted as quotidian waste – and commercial waste (CW) is continuously on the rise as areas become more industrialized and urbanized. It was estimated that the average annual global waste produced in 2012 was 1.3 billion tons and is predicted to still increase by almost 60%, i.e., 2.2 billion by 2025 (Hoornweg & Bhada-Tata, 2012). Therefore, waste-to-energy can be a sustainable waste management option by employing incineration technologies.

H. Akram (✉)
College of Earth & Environmental Science, University of the Punjab, Lahore, Pakistan

Department of Life Sciences, University of the Punjab, Lahore, Pakistan

Department of Mathematics and Applied sciences, Minhaj University, Lahore, Pakistan

S. Hussain
College of Earth & Environmental Science, University of the Punjab, Lahore, Pakistan

T. E. Butt
Faculty of Engineering & Environment, Northumbria University, Newcastle upon Tyne, UK

From the above description regarding transport and waste, it can be seen that the energy is a common denominator; however, the former requires the energy to operate, while the latter would yield energy if incinerated. Thus, the output of energy in waste incineration can be channeled in transport sector, thus killing two birds (transport and waste) with one stone (energy). The coupling of the two sectors (i.e., transport and waste) via energy is what this study focuses on. However, no evidence has been found in the review of literature regarding the idea of tying the two sectors (waste and transport) together via the thread of energy. It is also necessary to establish the pragmatism of this venture in various modes of the transport, that is, in transport sectors, modes are classified into land, air, and water. This study aims to outline the potential to employ waste-to-energy option to run transport sector.

39.2 Transport

Transport modes are essential components of transport systems since they are the means by which mobility is supported. Modes can be grouped into three broad categories based on the medium they exploit: land, water, and air (Fig. 39.1). Each mode has its own requirements and features and is adapted to serve the specific demands of freight and passenger traffic (United States Department of the Interior, 2006).

Land transportation involves moving passengers and freight with vehicles over a prepared surface and further expands into roadways and railways. Air transportation is the movement of passengers and freight by any conveyance that can sustain controlled flight. Water transportation concerns the movement of passengers and freight over water masses, from oceans to rivers (Rodrigue, 2020a, 2020b).

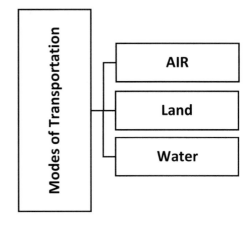

Fig. 39.1 Modes of transport

39.3 Waste and Energy

Around the world, waste generation rates are rising. In 2016, the worlds' cities generated 2.01 billion tons of solid waste, amounting to a footprint of 0.74 kilograms per person per day. As of 2018, around 20.01 bn metric tons of municipal solid waste was produced every year (World Bank, 2018). The 1.3 billion tons of waste produced per year is continuing to rise with the increase of the global population. It is presented in Fig. 39.2 that most of waste residing on our planet is being generated from Asian and European regions of the world. Central Asia, South and East Asia, Europe, and Pacific regions are the main contributors of the waste generated around the world (World Bank, 2018).

The majority of all types of refuse is still disposed of to landfill, regardless of whether the waste can be reused or not (Bovea & Powell, 2016). In addition to this, the method disregards the potential to harness the remaining potential energy within the litter, and so the rate of resource depletion will be increasing as we continue to extract energy from new reserves of nonrenewables, despite being equipped with an array of materials that can be used in energy recovery. By waste-to-energy practice in incinerators, the life of global fuel reserves could be prolonged (Cohen-Rosenthal & Musnikow, 2017). To the face of this potential, Fig. 39.3 shows that the primary source of energy is abstracted from nonrenewables, in particular coal, oil, and gas – all three are fossil fuels. Figure 39.3 also reflects the high-level worldwide dependency on unsustainable fuels.

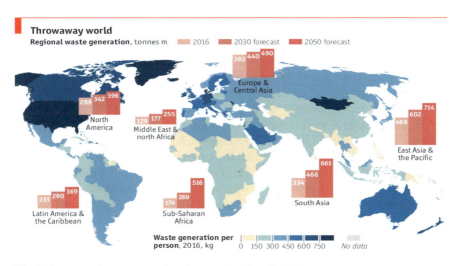

Fig. 39.2 Regional waste generation (Source: World Bank, 2018)

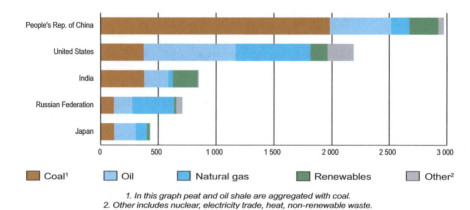

Fig. 39.3 Graph showing the top five energy consumers by region, and proportion of energy sources used (International Energy Agency, 2017)

39.4 Innovative Insights

Due to escalating demand of energy in transport sector, switching transport to renewable energy systems has become a priority, for both passenger and freight transport. This means a substitute for energy generation is required to reduce the strain on fossil fuel. On the other hand, elevation in the quantity of waste due to ever-increasing commodities will result in an increase in landfilling, as the majority of municipal solid waste (MSW) and commercial waste (CW) is not currently reprocessed effectively. However, it may be possible to use waste more sustainably by employing waste-to-energy option directly into maritime transport (Fig. 39.4).

Unlike fossil fuels and other mainstream energy sources, the calorific values of MSW and commercial waste tend to be lower, yet their accessibility is what makes them desirable in this context. MSW consists of the typical domestic waste produced in a household environment, such as paper, plastic, cardboard, food, etc. (Hoornweg & Bhada-Tata, 2012). All materials possess a calorific value (CV) which equates to the amount of energy released from a set amount of the said material after complete combustion (Lupa et al., 2011). Typical MSW's net CVs range from 4 to 35 MJ/kg from organic materials to certain plastics (World Energy Council [WEC], 2016). Although these values are lower than that of fossil fuels, they can still be beneficial in partially powering systems on a maritime, such as heating arrangements, lighting, and domestic electrical outlets. The idea behind targeting the maritime transport is that the incinerator requirement for waste incineration is so large in size (Figs. 39.5 and 39.6 show an example) that only gigantic maritime vessels would be able to accommodate it.

The production of waste is continuous; thus, the source of it will be plentiful. This would also reduce energy wastage due to line losses because the energy

Fig. 39.4 Van diagram of waste-to-energy and transport

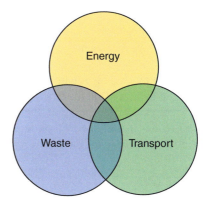

Fig. 39.5 A typical incinerator

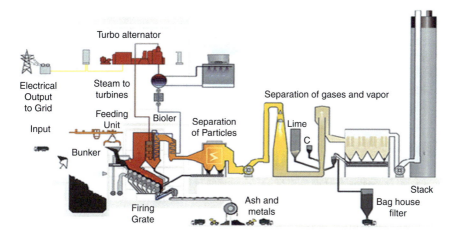

Fig. 39.6 Schematic of waste incineration process

generated at the source through incineration of the waste on board of the vessels will be consumed within the vessels as oppose to terrestrial waste-to-energy scenario in which incinerator plants can be miles away from the city to which the energy would be supplied through grid and power line. In addition to generating energy for transport, the process also reduces landfilling by reusing the waste that would be managed in this way, and the strain placed on fossil fuel reserves could be mitigated by a small extent, depending on the volume and type of waste used. This technology can render it possible to look at waste products to partially fund certain sectors, particularly due to the increasingly excessive amount that is being produced globally.

39.5 Concluding Remarks

This paper considers two main sectors from the anthroposphere, which are transport and waste. Initially, the two sectors are individually discussed, in the context of energy. It is established that one sector can generate energy (i.e., waste) via incineration and the other sector consumes energy (i.e., transport). This paper innovatively connects both sectors via energy being a common denominator. There are three basic modes of transport, i.e., land, air, and water. Though waste-to-energy on an onboard incinerator in transport is a juvenile concept, its practicality seems to be more feasible in maritime transport rather than the other two modes.

This conceptual research paves a path on a number of fronts. Some examples of which are as follows: numerical estimation models can be developed to establish the potential of unlocking the advantages of this innovative idea, at local, regional, national, continental, and international levels. Some technical studies may also be required in terms of typical, fossil fuel-driven power engines being replaced by waste incineration engines not only to propel the maritime vessels but also to operate the onboard built environment functions. In environmental sustainability context, the carbon footprint control technologies such as carbon capture and storage (CCS) can be applied. This idea has a great potential to enhance environmental sustainability and contribute to SDGs and climate change agenda.

In summary, the idea of the amalgamation of the three sectors, i.e., transport waste and energy, has potential multidimensional benefits which can be reaped by further research.

References

Bovea, M. D., & Powell, J. C. (2016). Developments in life cycle assessment applied to evaluate the environmental performance of construction and demolition wastes. *Waste Management, 50*, 151–172.

Cohen-Rosenthal, E., & Musnikow, J. (2017). *Eco-industrial strategies: Unleashing synergy between economic development and the environment*. New York: Routledge.

García-Olivares, A., Solé, J., & Osychenko, O. (2018). Transportation in a 100% renewable energy system. *Energy Conversion and Management, 158*, 266–285.

Hoornweg, D. & Bhada-Tata, P. (2012). What a waste: A global review of solid waste management.

International Energy Agency. (2017). Key world energy statistics. *Top five countries by total primary energy supply (TPES)*, p. 42.

Lupa, C. J., Ricketts, L. J., Sweetman, A., & Herbert, B. M. (2011). The use of commercial and industrial waste in energy recovery systems–a UK preliminary study. *Waste Management, 31*(8), 1759–1764.

Rodrigue, J. P. (2020b). The distribution network of Amazon and the footprint of freight digitalization. *Journal of Transport Geography, 88*, 102825.

Rodrigue, J.-P. (2020a). *Geography of transportation* (fifth ed., 456 pages). New York: Routledge.

United States Department of the Interior. (2006). *Major Roads of the United States.*

The World Bank. (2018). What a Waste: An Updated Look into the Future of Solid Waste Management, Website: https://www.worldbank.org/en/news/immersive-story/2018/09/20/what-a-waste-an-updated-look-into-the-future-of-solid-waste-management, Downloaded: March 2021, The World Bank Group.

World Energy Council [WEC]. (2016). *Waste as a Fuel* (p. 8). New York: World Energy Resources.

Index

A
Academic fraternity, 503
Academic knowledge, 278
Accountability, 328
ACHASM Contribution to "Covid-19" "Construction Return to Work" Documentation' media release, 167
Achieved sustainability, 5
Active drug trafficking, 485
Advanced Engineering Project Management, 159
Advanced recovery fees (ARF), 471
Adversarial contracted teams, 302
Adversarial/cooperative attitudes
    Axelrod's test, 268
    commercial pressures, 270
    competitive bidding, 266
    competitive tendering, 266
    complexities of human nature, 268
    Constructing the Team Report, 266
    contractors, 269
    cooperation *vs.* defection, 265
    forgiving strategy, 268
    framework agreement, 272
    game scores by group, 273
    Latham report, 267
    partnering, 269, 270, 272
    project management framework group, 273
    Quantity Surveyors, 274
    Re-Thinking Construction in 1998, 267
    supply chain transparency, 267, 271
    Tit for Tat approach, 268
    unpaired' design, 272
AEC industry, 441

Aerogenerators, 186
Agile methodology, 375
Agriculture adaptation, 483
Air permeability, 51, 53, 54
Air quality device, 141
Air transportation, 526
Alternative construction materials, 500, 501
Alternative dehumidification, 56, 57
Alternative qualitative research methods, 200
Amalgamation, 530
American Society of Safety Professionals (ASSP), 237, 239, 243
American sustainable universities, 163
Analytical/creative thinking, 353
Andragogy, 236, 237, 243–245
Anthropogenic actions, 9
Anthropogenic influences, 525
Anthroposphere, 530
Architectural design, 447
Architectural technology (AT), 432
Architecture, engineering and construction (AEC), 249, 250, 256, 257, 259–261, 431
Armed conflict, 484–485
Asbestos-containing materials, 171
Asset life cycle components, 153
Asset management (AM), 154
    definition, 395
    digital twins, 406
    implementation benefits, 398
    interoperability, 396
    SEAM, 395 (*see also* Smart enterprise asset management (SEAM))
    whole lifecycle, 407

Asset whole lifecycle, 407
Association for Project Management (APM), 186, 407
Asymmetric negotiation, 269
Attic insulation, 64
ATTMA TSL2 technical standard, 52
Australian Constructor's Association (ACA), 193
Automated compliance, 416, 426
Automated EVs, 518, 520
Autonomation, 443
Availability error, 352
Axelrod's test, 268

B
Bamako Convention, 465
Basel Action Network (BAN), 465
Basel Convention, 465
Basel Total Ban, 465
Baseline risk assessment (BRA), 204, 207
Battery system combinations, 73
Beaumont Morgan, 182
Behavioural economic theory, 314, 322
Behaviours and Potential Solutions (BS), 332
Beneficial index value (BIV), 253, 254
BES6001, 251, 252
Bibliometric analysis
    ability, 444
    BIM, 443
    building sustainability, 442–443
    lean construction, 443, 444
    relationships, 455
    systematic review, 444
Bibliometrix R-package, 445
BIM asset management Kitemark, 415
BIM authoring tools, 436
BIM compliance strategy, 437
BIM education, 431, 438
BIM framework, 428
BIM-Lean, 445–447
BIM-Lean synergy, 442, 453, 455
BIM-Lean-Sustainability, 444, 451, 453
BIM pedagogy, 432
BIM principles, 438
BIM-specific software, 438
BIM-specific tools, 435
BIM-Sustainability, 445, 449
Blacklist contractors, 269
Bloemfontein campus, 109
Bodies of knowledge, 297
Bottom-up approaches, 3
Bottom-up societies, 5

Breakfast room game, 183
British Standards, 88
Brundtland Report, 442
Building construction material emissions
    alternative materials, 500
    carbon and volatile organic compounds, 503
    emission characteristics, 496
    emission quality deduction stages, 496
    environmental assessment, 495
    frontline literature, 497
    health complexness, 503
    literature review, 496
    material efficiency, 496, 503
    material selection, 501
    natural materials, 497, 499
    physical models, 495
    PM emissions, 501–503
    quantitative emission factors, 504
    regression, 495
    reputable databases, 496
    research, 495
    search strategy, 496–498
Building documentation, 416
Building information modelling (BIM), 387, 388
    advantages, 251, 441
    AM integration, 415
    barriers to implementing and integrating responsible sourcing, 257–260
    benefits to implementing and integrating responsible sourcing, 254–256
    BES6001, 251, 252
    cause-effect relationships, 416
    consumers, 250
    data analysis, 253, 254
    data collection, 252, 253
    facilities management, 415
    focus, 415
    frameworks, 250
    implementation, 416
    LCA, 252
    life cycle stages, 415
    local sourcing, 251
    participant and organisational demographics, 254
    PRISMA, 252
    recommendations, 261
    sustainable construction, 441
    sustainable globe, 443
    technological importance, 443
    transformative information technology, 442
    value-action gap, 250

Index

whole design, 443
working environments, 251
Building life cycle management, 428
Building management systems, 411
Building material selection, 501
Building Surveying, 432–433
Built environment
  agencies and departments, 95
  construction activities, 95
  financial incentives, 102
  Ghana, 93
  MWH, 95
  settlements, 96
  sustainable construction, 94
Built environment professionals, 503
Built environment-related SEAM, 408
Business cartels, 328
Business case, 283
Business model, 35
Business organizations, 7

C

Calorific value (CV), 528
Campus sustainability execution, 106
Campus sustainability self-reporting framework, 114, 115
Canal negotiating local topography, 170
Cancer-causing agents, 502
Candidate and Professional Construction Health and Safety Agents (CHSAs), 167
Carbon capture and storage (CCS), 530
Carbon dioxide (CO_2), 138, 139, 145, 151
Carbon emissions, 59
Carbon energy, 85
Care home construction project, 356, 357
Central University of Technology (CUT), 108, 110
Chemical reactions, 478
China Belt costs, 278
Chinese residents, 478
Chlorofluorocarbons (CFCs), 4
Churchill Mansions, digital twin solution
  phase 1 goals, 425
  phase 2 advantages, 425, 426
  phase 2 goals, 427
  phase 3 advantages, 427
  plans, 424
  prototype, 425
  Scale Fast, 426
  Sitedesk, 422
  start small, 423
  think big, 427

CIBSE test reference, 81
Circular business models
  adoption, 33
  built environment, 35
  collaborations, 37
  construction supply chain, 34, 37
  sharing platforms, 37
  value identification, 35
Circular construction supply chains transitioning
  business model, 35, 37
  CE, 34
  circular business models, 37
  circular material production, 37
  collaborations, 41, 42
  cost minimisation strategies, 39
  critical success factors, 42
  early engagements, 41
  economic incentives, 41
  French participants, 41
  initial codes, 38
  limitations, 43
  linear economy, 35
  minimised recycling costs, 39
  positive communication, 41
  qualitative methodology, 37
  secondary resources availability, 42
  semi-structured interviews, 37, 38
  supply business models, 37
  thematic analysis approach, 38–40
Circular construction techniques, 35
Circular design models, 35
Circular economy (CE)
  construction supply chain, 34
  definition, 33
  implementation, 34
  key challenges, 34
  production and consumption activities, 35
  strategies, 35
  sustainability benefits, 33–34
  sustainable economic model, 35
  sustainable economic paradigm, 33
Circular educational economy, 470
Circular material production, 37
City of Leeds, 512, 516
City services, 371
Client and economic influences, 205
Climate change, 478
  diversity loss, 4
  environmental crisis, 3
  flooding, 6
  impacts, 9, 151
  prediction, 59
  projection, 9

Cloud-based file sharing platforms, 435
CO_2 emissions, 11
Coca cultivation, 483, 489
Code for Sustainable Homes (CSH), 71
Cognitive benefits, 317
Cognitive fit theory, 315–317, 322
Cognitive legitimacy, 27
Cognitive load, 316
Collaboration, 437, 479
Collaboration and information management process, 438
Collaboration catalyst, 317
Collaborative BIM environment, 438
Collaborative design initiatives, 91
Collaborative design process, 89
Collaborative learning, 471
Collusion, 327–335
Collusive activities, 327
Commercial pressures, 270
Commercial waste (CW), 525, 528
Common data environment (CDE), 422, 431, 434
Communication, 342
Communication barriers, 396
Communication plans, 342–344
Communication theory, 337
Community-based participatory research (CBPR), 386
Competition and Markets Authority (CMA), 333, 334
Competitive bidding, 266, 272
Competitive tendering, 266
Competitiveness, 503
Complementarities, 329
Complex networks, 34
Compliance, 417, 420, 421, 425
Compliant environmental conditions, 57
Comprehensive Buildings Safety legislation, 417
Computational modelling, 54
Computational thermal model, 52
Concrete experience, 337, 345
Confidential information, 195
Conflicts of interest, 330
Connected assets, 407
Constructing circular buildings, 41
Constructing the Team Report, 266
Construction, 13
Construction accidents, 209, 211, 212
Construction activities, 163
Construction and engineering education, 431
Construction contractor
    actions and motivations, 17

    interviews, 13
    public sector projects, 13
    social value measurement, 13, 19
    SV quantification, 16
    UK, 17
Construction contracts, 269
Construction design and processes, 205
Construction games and simulations, 186
Construction industry, 15, 217, 218, 220, 228, 237, 265–270, 274, 327–335, 441, 453
    climate change, 93
    economic development, 93
    natural environment impacts, 163
Construction Industry Development Board (CIDB), 203
Construction Industry Institute, 269
Construction professional, 187, 265, 271, 272
Construction project work, 176
Construction Regulations, 203, 204, 213, 214
Construction site, 235, 237–240, 244
Constructivism, 17
Consumer capitalism, 5
Contemporary issues, 309
    in PM
        defining success, 299, 300
        deterministic planning, 301
        false performance reporting, 302, 303
        managing risk and uncertainty, 302
        methodologies, 301
        planning fallacy, 300
        portfolio and programme management, 303, 304
        stakeholder management, 303
    selection, 299
Contractors, 269
Contractors' business models, 18
Conventional construction materials, 501
Conventional construction supply chain, 42
Cooperation *vs.* defection, 265
Corona 19 virus transmissions, 7
Corporate social responsibility (CSR), 249, 254, 256, 257
    business responsibility, 14
    construction organisations, 23
    criteria, 26
    definition, 23
    economic responsibilities, 14
    global business survey, 14
    holistic expectations, 26
    obligations, 175
    organisational, 13, 26
    philanthropic business priority, 14

stakeholder, 16
SV debates, 15
COVID-19, 71, 82, 274, 278, 291, 331
COVID-19 national lockdown, 89
COVID-19 pandemic, 138, 429, 521, 522
COVID-19 restrictions, 108
Covid-19 virus, 477
Critical decision immersive simulations, 318
Critical infrastructure, 480
Critical plant assets, 411
Critical realism approach, 195
Critical thinking, 438
Crop mapping, 483
Crop selection model, 491, 492
CSR activities, 30, 31
CSR movement and agenda, 25
CSR pyramid, 14
CSR requirements, 186
CT sensors, 54
Cultural relationship, 480
Cultural shift, 71
Customers' environmental consciousness, 87
CUT's internal experts, 110
Cutting-edge technologies, 441

D
Daily Safe Task Instructions (DSTIs), 209
Data visualisation, 353
Database-supported VR/BIM-based Communication and Simulation (DVBCS), 387
Decentralisation, 475
Decision making, 314, 315, 317, 318, 322, 323
Decision-making errors, 352
Decision-making process, 138
Decision-making tool, 317
Deconstruction supply chains, 41
Deforestation, 484
Dehumidification demand, 50, 52, 54, 55, 57
Dehumidification systems, 49, 56
Demand side response (DSR), 72, 81, 82
Demographical information, 253
Demolition firms, 42
Department of Construction Management's Facebook posts
  analysis, 166–170, 172
  construction health and safety, 166
  current reality, 165
  environment-related posts, 166, 167
  findings, 166
  impacts, 165

non-environmental posts, 166, 172
public relations, 164
research method, 165
social media, 165
summary, 166
topical issues, 173
Department of Higher Education and Training (DHET), 112
Descriptive statistical tool, 253
Descriptive statistics, 206
Design visualisation, 387
Detect collusive bidding, 328
Deterministic, 298, 301, 303, 304, 309, 310
DHET 'motivation, 115
Digital systems, 396
Digital tools, 416
Digital twins
  AM, 406
  digital representation, 406
  integrated, 418
  IoT, 406
  multiple, 406
Digitisation, 429
Distributed energy resources (DERs)
  batteries, 73
  dwellings (see ZP dwellings)
  effectiveness, 72
  FiT, 72
  PV, 72
  SC, 73
  smart home technologies, 73
  SS, 73
Distributed renewably powered off-grid charging stations, 518
Distribution network operator (DNO), 47
Diversity, 478
DMAIC process, 352
Domestic hot water (DHW), 74
Domestic waste, 528
Dutch construction industry, 329

E
East Suffolk Council (ESC), 88
Eastern Cape Department of Roads and Public Works' (ECDRPW) projects, 203, 206–208, 210, 214
Easy-to-measure-and-communicate financial metrics, 19
Eccentric cost, 282
Eco-friendly, 121, 124, 133, 134
EcoLab Architecture Hackathon, 89

EcoLab project
  challenges, 86
  collaborative design initiatives, 91
  collaborative environment, 88, 89
  consumer, 86–88
  digital skills and innovation, 85
  Hackathon, 89, 90
  implementation, 86
  national planning polices, 85
  planning permission, 85
  targets, 85
Ecological systems, 218
Economic and environmental sustainability, 9
Economic growth, 10
Economic incentives, 43
Economy 7 tariff, 81
Ecosystem, 396
Ecosystem services, 408
Education scheme, 335
Elderly-specific user requirements, 390
Electric stock market, 512
Electrical electronic equipment (EEE), 464
Electrical substation
  categories, 48
  criteria, 48
  critical infrastructure assets, 47
  PD, 48
  subsidiary station, 48
Electrical vehicles (EVs)
  city council, 512
  competence, 513
  environmental benefits, 513
  environmental sustainability, 512
  fuel economy, 512
  life cycle analysis, 511
  niche technology, 513
  performance, 511, 513
  public awareness, 513
  reliability, 511
  social acceptance, 512, 513
  storage capacity and cost, 512
  technological advancement, 512
Electricity grid, 81
Electricity North West Limited (ENWL)
  air permeability test, 50
  dehumidification, 49, 50
  deployment, 51
  DNO, 47
  environmental control, 47
  extraction rate, 50
  low humidity environment, 48
  objective, 47
  psychrometric chart, 49
  RH, 49
  substation population, 48
  temperature effect, 49
Electricity Supply Board (ESB), 192
Electrification, 517
Electrification of transportation, 517, 521
Electronic waste (e-waste)
  business, 463
  educational institutions, 462
  electronic and electrical devices, 461
  environmental impacts, 461
  environmental policies and legislation, 461
  health hazards, 463
  internal generation, 462
  metals, 462
  sources, 461
  workers, 463
Emerging trends, 309
Emotional engagement, 177
Employee satisfaction, 137, 146
Employers' Onus, 236–238
End-of-life concept, 35
Energy consumption, 57, 137, 525
Energy efficiency multiplier (EEM), 61
Energy performance, 51
Energy wastage, 528
Engineering and Environmental Science, 445
Engineering concepts, 51
Engineering curriculum, 431
Environmental and energy monitoring campaign
  computational thermal model, 52
  dehumidification demand, 52
  dehumidifiers, 52
  Pendleton substation, 53
  specific humidity, 52
  temperature and RH sensors, 51
Environmental and social problems, 484
Environmental conditions, 55, 57
Environmental consciousness, 513
Environmental control management analysis
  case studies, 51
  commissioned substations, 56
  dehumidification demand, 54–56
  effectiveness, 54
  heating demand, 54
  interventions, 53, 57
  payback analysis, 55
  substation population, 51
  substations, 57
  thermal models, 56
  weighted average environmental conditions, 54

Environmental degradation, 163
Environmental design impacts, 385
Environmental management practices, 163
Environmental monitoring, 57
Environmental sustainability, 4, 530
Environment-related posts, 164, 172
EPR policy approach, 464, 471
e-procurement services, 411
Ergonomics problems, 205, 209, 211
Ethical issues, 195
Ethical practice, 327–335
EU legislation, 67
European Cooperation in Science and Technology (COST), 282
European Union (EU), 59, 464
EV automation, 518, 522
Evidence mapping, 375
Evidence-based research (EBR), 386
EVs in emission reduction, 521
EVs public/social acceptance study
    analysis, 516
    automation, 518
    carbon emissions impacts, 516
    carbon emissions reduction, 517
    carbon emitting nature, 517
    coronavirus spread control, 520
    data and quality control, 516
    electrification, 517
    emission reductions, 522
    future research, 522
    implications, 521
    participants' appreciation, 518
    PESTE analysis, 514–515
    portfolio of energy, 518
    recommendations, 521, 522
    UK government policy, 515
    zero-emission contributions, 517
E-waste dumping, 470, 471
E-waste management, 466, 468
    aim, 462
    Basel Convention, 465
    educational institutions, 471
    EPR approach, 471
    EUP Directive 2002/32/EC, 465
    limitations, 465
    opportunities, 470
    policy proposal, 466
    proposed policy barriers, 466, 467
    recycling/reuse, 464
    regulatory framework, 464
    research gap, 472
    RoHS Directive 2002/95/EC, 464
    stakeholder recommendations, 471, 472
    survey, 468
    WEEE Directive 2002/96/EC, 464

E-waste mismanagement, 471
Excavation techniques, 195
Exchange programmes, 471
Executive Instrument (EI), 95
Experienced and competent contractors, 199
Experiential construction project management education, 183
Experiential treatment, 318
Explanatory approach, 298
Exploratory sequential design, 338
Extended producer responsibility (EPR), 464, 471
External and compelling motivation, 363
Externally refurbish (EWI), 126
Extinction Rebellion, 4

F

Facebook page administrators, 173
Facebook page statistics, 173
Facility management (FM), 156
Fact-based natural science approach, 16
Factoring influencing upscaling, PV systems
    funding challenges, 111, 112
    limited commitment, 112, 113
    limited knowledge/appreciation, 113, 114
    limited motivation and incentives, 115
    limited schemes, 114, 115
    limited SOs, 114
    unavailability/limited space challenges, 112
Faculty of engineering, 111
Federation of Concrete Manufacturers (FCM-Participant 5), 41
Feed-in tariff (FiT), 72
FF&E Furniture Project, 357, 358
Fick's law, 503
Fire compliance, 418, 428
First-time environments, 282
Follow-on deductive approach, 338
Forming relationships with stakeholders, 345
Fossil fuel-driven power engines, 530
Fossil fuels, 521, 528
Fragmented supply network, 503
Frontline literature, 497, 499
Fuel sustainability, 512
Fuzzy concept, 372

G

Game production interactions and exchanges, 184, 185
Game skills/competencies, 184
Gantt charts, 181, 319

Gantt chart-type modelling
  case study 1, 355
  case study 2, 356, 357
  case study 3, 357, 358
  case study 4, 359, 360
  constraint findings, 361, 362
  coordination, 364
  critical path identification, 354
  feasible solutions, 363
  hypotheses, 364
  key planning tool, 353
  mapping 4, 361
  potential gains, 363
  process implementation, 363
  visualisation, 353, 363
Generalisability, 31
Gesture recognition, 391
Ghanaian construction, 94
Ghanaian sustainable construction analysis
  actions, 102
  barriers, 97–99, 101
  climate change, 101
  conceptual approach, 94
  drivers, 97, 100–101
  implementation, 96, 97, 101
  literature review, 94, 95
  materials selection, 94
  socio-economic development, 99
GHG emissions, 151, 152
GIS software, 483
Global businesses, 398
Global E-waste Monitor, 461
Global warming, 4
Google translate application, 244
Government Advisory Committee on Climate Change, 88
Government's austerity measures, 31
Green Building Council publication, 397
Green buildings, 449
Green Campus Initiative, 107
Green cities, 477
Green construction, 442
Greenhouse gas (GHG), 151, 441, 511
Grid electricity demand, 73
Grid substation, 48
Gross domestic product (GDP), 93, 278
Group project, 432

H
Habitat fragmentation, 478
Hard and soft paradigms
  PM, 314, 315
Hard skills, 344
Hard systems model, 315
Hazard identification and risk assessment (HIRA), 204, 205, 209
Head-in-the-sand approach, 11
Health ageing, 385
Health and safety (H&S), 217
  actions on ECDRPW's projects, 207, 208
  construction accidents, 209, 211, 212
  Construction Regulations, 203, 213, 214
  ECDRPW's projects, 203
  ergonomics problems, 205, 209, 211
  factors contribute to exposure to hazards on construction sites, 209, 211
  factors in accidents, 205, 206
  legislation, 204, 205
  legislative requirements, 203
  multi-stakeholder approach, 213
  public sector projects, 213, 214
  research method and sample stratum, 206
  in the South African construction, 203
  training on ECDRPW's projects, 209, 210
Health and Safety Authority (HSA), 192, 237
Health and Safety Executive (HSE), 193
Health and safety management plan, 194
Health and safety management systems (H&SMS), 218, 220, 228
Health and wellbeing, 9
Healthcare delivery, 386
Healthcare facilities design
  communication and simulation, 387
  COVID-19 pandemic response, 392
  deficiencies, 387
  design, 387
  provision, 385
  requirements and preferences, 388, 392
  user engagement, 386
  VR, 386
  vulnerability, 392
Healthcare industry, 385
Healthy indoor environment, 137
Heating degree days (HDD), 80
Heating demand, 54, 56
Heuristic learning, 471
Higher education institutions (HEIs), 107
High-fidelity mixed-reality (MR) system, 391, 392
Hirschman's Hiding Hand Theory (HHP), 291
Hoftsader's Law, 271
Housing, 59
HSA's Code of Practice
  client duties, 193
  compulsory elements, 200

contractor duties, 193
designer duties, 193
effectiveness, 200
evaluation, 196
excavation techniques, 194
impacts, 199
interviews, 199
knowledge and awareness, 195
safety improvement strategies, 198
service strikes, 197
trenchless technologies, 199
using as-built drawings, 194
HSE (2014) and ACA (2014) guidelines, 197
Human cognitive, 345
Human relationship skills, 9
Human society, 7
Humidity controls, 57
Humidity environments partial discharge, 48
HVAC systems, 139
Hybrid ventilation systems, 140, 146
Hydra Suite laboratory, 318, 319, 321
Hypothetical possibilities, 338

I
Ill-considered procurement strategies, 364
Illegal transboundary movements, 466
Illicit crops, 483
Illicit crops management in Colombia
  aerial fumigation, 484
  agroclimatic characteristics, 489
  cassava cultivation yields, 491
  crop conditions, 491
  crop planning alternatives, 485
  data, 484
  effective law enforcement, 484
  investment, 492
  Landsat satellite images, 485
  mapping, 484
  mathematical programming model, 488
  NDVI processing, 485, 489, 493
  parameters, 493
  Sierra Nevada de Santa Marta, 485
  simulation model, 486
  statistical models, 484
  vegetation cover, 493
  vegetation index, 485
Immediate accident circumstances and shaping factors, 205
Impact assessment, 236, 242
Improvement methodologies, 351, 354
Improvement process constraints, 354
Indoor air quality (IAQ), 138, 146, 147

Indoor CO_2 concentration, 138
Inductive research, 338
Industrial and 'big business' sector, 6
Industrial procurement practice, 503
Industrialisation, 93
Industry sectors, 7
Industry-wide BIM adoption, 431
Inequality, 29
Informal management structure, 7
Information transmission, 8
Innovation, 364, 371
Innovative digital technologies, 385
Innovative Lego-based construction project management game/simulation, 175
Innovative sculptures and captivating mosaics, 172
Innovative technologies, 412, 511
Institution's PV portfolio (P1), 113
Insulated partition, 53
Integrated BIM asset management model
  drivers, 416, 417
  process-interaction flow, 417, 418
Integrated digital twin, 418
Integrity Department (INT), 329
Intensive agricultural system, 484
Interactive VR communication systems
  BIM, 388
  development workflow, 388
  elderly friendly functionality, 390, 391
  FBX file formatting, 388
  hand gestures, 389
  HCI, 389
  OOP, 389
  testing, 389
  thematic analysis, 390
  two-stage approach, 388
  virtual building design, 391, 392
Interconnected multidisciplinary system, 408
Interest-based negotiation, 310
Intermediate impossible, 178
Internal combustion engine vehicles, 516
International Asset management, 154
International Council for Research and Innovation in Building and Construction (CIB), 96
International Energy Agency (IEA), 512
Internet, 278
Internet of Things (IoT), 410–411
Interoperability, 395
Inter-professional group projects, 432
Inter-professional studies (IPS), 432, 438
Inter-project learning, 344, 346
Intervention groups, 322

Intuitive functions, 391
Invisible price, 7
Irish Council for Social Housing, 59
Iron principles, 288–290
Iron triangle, 278, 290–292, 349
   adaptions, 280
   adopted, 288
   business case, 283
   constraints, 286–288
   elements, 287
   factors, 280
   flexible model/approach, 285, 286
   limitations, 284
   measures of project success, 278
   performance data of case studies, 285, 286
   popularity, 291
   principles, 289
   problem-solving method, 290
   project life cycle, 283
   quality, 280
   scope creep, 284
   simplicity, 281
   TPM, 284
Irrational decision-making, 351
ISO 55001:2014 (BSi, nd), 399
IT Software Procurement Project, 359, 360

J
Journal of Cleaner Production (JCP), 496
Just-in-time flow (JIT), 443

K
Kinaesthetic learners, 242, 245
Kolb's learning theory, 337, 344, 345

L
LA practices, 30
Laboratory e-waste management policy
   effectiveness, 469
   e-waste disposal, 468
   financial constraints, 468
   illegal transboundary movements, 473
   informative learning activities, 470
   laboratory session, 468
   learning experience, 469
   quantitative research method, 467
   ratings, 469
   responses, 467, 468
   social acceptance, 467, 470
   space allocation, 468
   survey, 467–469
Laboratory e-waste management system, 472
Lamella properties, 129, 130, 132, 134
Land transportation, 526
Landscape Urbanism, 479
Language barriers, 244
LAs' geographical remit, 31
Latent value, 350, 353, 356, 361
Latham report, 267
Lean construction, 443, 444
Lean thinking, 444, 446
Lean-sustainability, 445, 449, 451
Learning construction PM techniques, 178
Learning style, 237, 239
Learning theory, 344
Learning tools, 345
Learning transfer, 338, 344, 345
Leeds construction scene, 272
Legislation, 14
Legitimacy theory, 24, 26, 27
Lego-based simulations
   bottom-up *vs.* top-down costing, 182
   construction and logistics team, 183
   design decisions, 179
   design issues and choices, 179
   illusory real, 178
   learning experience, 179
   learning outcomes, 178
   model experiences, 177
   PM, 178, 182
   Race model categories, 181
   solving problems, 178
   teaching session, 182
Lego serious play
   branded facilitation methodology, 177
   illusions, 177
   sensation, 177
Life asset management, 160
Life cycle analysis (LCA), 252, 501, 504
Life cycle-oriented built environment
   management, 154
Life skills, 178
Light detection and ranging (LiDAR), 407
Li-ion battery system, 73
Linear economy, 34
Literature review, 330–332
Literature synthetisation, 497
Lived experience, 304
Local authorities (LA), 131
   funding, 24
   government policy, 24
   Institute of Fiscal Studies, 24
   LGA, 24

Index 543

public sector organizations, 23
ramifications, 24
spending cuts, 24
Local Government Association (LGA), 24
Low and zero carbon (LZC), 87
Low carbon building construction knowledge, 88, 90
Low-carbon targets, 72
Lower socio-economic communities, 29

M

Maintenance activity, 57
Making Learning Happen (book), 177
Management of Technological Risk, 158
Management philanthropic values, 15
Manchester City Council Construction Skills Network, 186
Mapei, 122
Maritime transport, 528
Market competition, 398
Markets, 327
Material efficiency, 503
Mathematical programming model, 488
Mean score (MS), 206, 207, 209
    DSTIs, 209
    H&S induction, 209
    HIRA, 209
    ranging, 207
Mechanical and electrical (M&E), 62, 65
Mechanical ventilation systems, 137
Mechanical ventilation with heat recovery (MVHR), 65
Medium-sized organisations, 254
Megaproject management
    academic knowledge, 278
    audiences, 278
    characterised, 277
    complexity, 282, 288
    COVID-19, 278
    cross-case analysis, 284
    destructive and dictated view, 277
    eccentric cost, 282
    evaluation models, 277
    evolving and dynamic systems, 279
    exponential benefits, 283
    extreme time delays, 282
    HHP, 291
    iron principles, 288–290
    iron triangle (*see* Iron triangle)
    methodology, 284
    multi-trillion-dollar global delivery model, 277
    organisational structure, 281
    originality/value, 285
    project performance evaluation, 279
    research limitations, 285
    risk management (*see* Risk management)
    stakeholders, 278
    subjectivity, 281
    substantial stakeholder involvement, 281
    TPM theory, 287, 288
    uncertainty, 282, 288
    unique first-time environments, 282
Megaproject pathologies, 279
Mental map, 353
Metaphysic, 477
Migration crisis, 8
Minimal viable product (MVP), 421
Ministry of Justice, 489
Mixed method approach, 338
Monnit TM wireless sensor system, 51
MoSCoW analysis, 182, 186
Multicriteria analysis, 155
Multidisciplinary Digital Publishing Institute (MDPI), 496
Multiplayer functionality, 391, 392
Multiple digital twins
    advantages, 407
    APM, 407
    beneficiaries, 408
    BIM integration, 406
    connected assets, 407
    data analysis, 407
    lifecycle management, 407
    real-world systems, 406
    3D model scanning, 407
Multi-stakeholder approach, 213
Multi-trillion-dollar global delivery model, 277
Municipal solid waste (MSW), 525, 527, 528

N

Narrative analysis, 28
Natural building materials (traditional), 497, 499
Natural ecosystems, 476
Natural ventilation, 137, 139, 140, 146
    building type and location, 141
    carbon dioxide concentration, 145
    carbon dioxide level, 144, 146
    IAQ, 142, 143, 146
    office building, 170
    openable windows, 146
    quantitative research designs, 140
    research findings, 146
    SBS survey, 144, 145

Naturally ventilated buildings, 138, 139
NDVI processing, 485
Nearly zero-energy buildings (nZEB)
  advantages and disadvantages, 67–68
  analysis, 62, 63
  barriers and challenges, 60
  cost-effectiveness, 67
  EEM, 61
  energy efficient technologies, 61
  energy performance, 60
  EU legislation, 60
  implementation barriers, 61
  internal plastering, 66
  M&E services, 65
  materials, 64
  measurement and verification, 61
  preliminaries, 66
  requirements, 67
  research method, 61, 62
  sequential mixed method approach, 60
  short-term expenditures, 61
  social housing, 60
  standards and regulations, 60
  subcontractors, 65, 66
  superstructure, 64
  time, 66
Net Zero Carbon substation, 57
Network diagrams, 352
Network planning, 364
New normal' lifestyle, 521
New Urbanism, 479
Non-government organisations (NGOs), 164
Non-immersive technologies, 391
Non-industrial indoor environment, 138
Non-traditional, 121
Non-traditional dwellings
  Airey properties, 131
  BRE, 122
  Lamella properties, 132
  limitations, 134
  literature review, 122, 127–130
  Mapei, 122
  non-costs, 131
  non-specialist parties, 131
  philosophical worldviews, 123
  post-war, 133
  professional experience, 130
  qualitative and quantitative mixed methods, 122
  questionnaire survey, 124, 125, 127
  research aim, 124
  research designs, 123
  research methods, 123
  themes, 126
Non-traditional Houses, 128
Non-traditional properties, 121
Normalized difference vegetation index (NDVI)
  processing, 485
  satellite images, 485
  vegetation cover, 489, 493
Normative project management, 337
Northern Ireland, 65
Numerical estimation models, 530
nZEB guidelines, 65
nZEB standards, 64, 65

O

Object-based parametric modeling, 443
Objected-oriented programming (OOP), 389
Objectivism, 17
Objectivity-based methodology, 252
Object-oriented modelling, 387
Occupational Health and Safety Act, 204
Office environment, 137, 146
Office of Fair Trading (OFT), 327
Off-site construction, 35
Open tendering, 266
Open-air incineration, 463
Operational hydrology, 95
Operational life cycle phases, 159
Operational management, 397
Operational tools, 352
Operationalisation, 105
Opinion on Collusive Practices Overseas and UK comparison (OPO), 332
Opinion on the Ethical Stance (OES), 332
Organisation for Economic Co-operation and Development (OECD), 442
Organisation's planning, 352
Organisational CSR, 14, 26
Organisational resilience, 219
Originating influences, 205
Out-of-use equipment, 466

P

Paradigm shift, 351
Partial discharges (PD), 48
Particle emissions, 495, 496, 503
Particulate matter (PM) emissions
  affecting factors, 502, 503
  characteristics, 501
  health challenges, 501
  source contribution, 501
Partnering, 269, 270
Partnering management approach, 270

Index   545

Payback analysis, 55
Peak demand shift tariff, 82
Pedagogy, 236, 237
Pendleton, 53, 56
Performance, 214
    analysing, UK-based case studies, 279
    in construction, 284
    developing, 292
    evaluation, 279, 281, 282
    factors, 290
    megaprojects, 277–279, 284
    success rate, 285
Performance gap, 51, 56, 57
Personal liberty, 5
Personal protective equipment (PPE), 205, 463
Personalia-related posts, 173
PESTE analysis, 514–515
Photovoltaic (PV), 72
    adoption, 105
    deployment influencing factors, 106
    greenness variation, 105
    operational innovation, 105
    renewable energy technology, 106
Plan visualisation, 354
Planning fallacy, 300, 309
Plan-task-coordination, 364
Plastering, 66
Plastic reusable bags, 6
Plastics consumption, 6
Policy makers, 87
Pollution, 495
Population, 8
Positive communication strategies, 42
Post-COVID-19 carbon savings, 521
Post-Covid-19 world, 475
Postgraduate course Asset Management, 157
Postgraduate education in sustainable asset management
    BIM-related skills, 158
    engineering and management course, 160
    engineering environment, 160
    learner feedback, 158
    middle-level engineering asset, 157
    online course, 157
    physical infrastructure assets, 157
    project management, 159
    risk management teaching, 159
    RQ, 159
    strategic asset management principles, 157
    support, 157
    sustainability orientation, 159
    sustainable engineering and built environment, 158
    themes, 157
Post project learning, 338
Post project reviews
    learning tool, 345
    project lifecycle, 337
    utilisation, 343
Power purchase agreement (PPA), 107
Practitioner development, 314
Pragmatism, 526
PRC frame structure, 121
Precision Medicine Initiatives, 386
Predetermined communication plans, 342
Preferred Reporting Items for Systematic Reviews and Meta-Analyses (PRISMA) flow chart, 221, 252
Primary schools, 175, 176, 186
Principal contractors (PCs), 204
Prisoner's dilemma, 272, 273
Private rented sector (PRS), 87
Problem-based learning (PBL)
    AEC, 432
    BIM process, 433, 437
    CAD drawing, 436
    CDE, 434
    communication and collaboration, 435
    data collection and analysis methods, 433
    direct supervision, 437
    feasibility report development, 437
    IPS, 432
    objectives, 432
    teaching BIM, 437
    technical challenges, 438
Problem-solving, 315, 316, 323
Process improvements, 352
Procurement, 20, 265
Pro-environmental behaviours, 91
Professional ethics, 330
Programme management, 303, 304
    applicability and validity, 375
    decision-making, 376
    fundamental goals, 375
    methodologies, 376
    modular project solutions, 376
    project strategy, 376, 381
    research, 375
    scientifically based evidence, 375
    smart cities' challenges, 376–380
    strategic goals, 375
    sustainability, 376
Project BIM compliant, 436
Project complexity, 314
Project conceptualisation, 314
Project environment, 307

Project learning, 337, 338
Project life cycle, 283
Project management (PM), 205
   analysis, 308–309
   APM, 297
   application, 315
   approaches, 364
   behavioural economic theory, 314
   bodies of knowledge, 297
   cognitive fit theory, 315–317
   concepts, 376
   contemporary issues, 299–304
   data visualisation, 365
   decision-making, 314
   despite decades, 313
   deterministic approaches, 298
   development, 313
   discipline, 314
   experience, 176
   explanatory approach, 298
   hard and soft paradigms, 314, 315
   knowledge, 298
   limitation, 323
   lived experience, 298
   methodologies, 277, 313, 317, 318
   multidimensional view, 350
   multi-disciplinary nature, 317
   opportunity, 176
   organisations, 349
   perceptions, 350
   PMI, 297
   practitioners, 176
   problem-solving, 322
   professional body approaches, 297
   project managers, 351
   quality management principle, 297
   rationale and outputs, 342, 343
   reflective professionals, 298
   research method, 318, 319
   research review, 317, 318
   rethinking, 306, 309
   skills and competencies, 176
   stakeholder management, 298
   tools implementation, 352
   trained technicians, 298
   value, 350
   visualisation tools, 315–317, 350, 351
Project Management Institute, 351
Project management office (PMO), 304
Project Management theory, 337
Project Manager, 65, 365
Project professionals, 338, 339, 344, 346
Project stakeholders, 270
Project strategy, 375
Public construction projects, 327
Public engagement, 512, 521
Public relations, 164
Public sector, 329
Public sector organisations, 23
Public sector projects, 213, 214
The Public Services (Social Value) Act (2012) (SVA)
   additional requirements, 15
   advantages, 29
   contractor CSR, 23
   enactment and ramifications, 28
   government review, 25
   LAs, 23, 28, 29
   legal requirement, 25
   legislation, 15
   legislative duty, 25
   narrative analysis, 28
   procurement, 31
   requirements, 18, 25, 30
   role, 15
Purposive sampling technique, 108
PV and battery combination, 73, 76, 81, 82
PV deployment framework, 110
PV electricity, 76
PV procurement, 112
PV self-consumption, 72
PV system, 73
PV systems deployment in SAUs study
   barriers, 106, 107
   case study design approach, 108
   comparative analysis, 107
   comprehensive analysis, 106
   CUT, 108, 110, 111
   factoring influencing upscaling (*see* Factoring influencing upscaling, PV systems)
   government, 108
   organisational perspective, 106
   payback period, 106
   place limitations, 107
   PPA contracts, 107
   purposive sampling technique, 108, 109
   SD implementation, 107
   SD operationalisation, 107
   SP challenges, 107
   themes, 109
   transcribed verbatim, 109
   two-step approach, 108
Pygmalion effect, 337, 345

Q

Qualitative data collection, 331
Qualitative information, 339
Qualitative research, 28, 330

Index

Quality in Sustainability (QiS), 398, 409
Quality of Effectiveness (QoE), 398, 409
Quality of Service (QoS), 409
Quality of urban life, 476
Quantitative and qualitative researchers, 444
Quantitative research, 330
Quantitative study, 310
Quantity Surveying and Project Management courses, 433
Quantity surveying sector, 329
Quantity surveyor, 64, 65, 274
Questionnaires interviews, 330

R

Race model categories, 181
Race model for experiential learning, 181
Radio frequency identification (RFID), 155
Reactive, 164
Realistic construction project choices, 178
Real-time performance data, 412
Reconciliation ecology, 478
Redevelopment, 130, 133
Reflective professionals, 298
Regional waste generation, 527
Regulatory compliance, 417
Relationship building, 343
Relative humidity (RH), 49, 56
Reliable evidenced-based information, 396
Remote sensing methods, 483
Renewable energy, 106, 108, 114, 115
Renewable power-assisted EV charging stations, 511
Renewable-powered EV charging stations, 513, 518, 520, 521
Replacement thermostats, 53, 54
Republic of South Africa (RSA), 204
Resilience
    ability to anticipate, 227
    ability to learn, 227, 228
    ability to monitor, 226, 227
    ability to respond response, 226
    concept, 218, 219, 222
    definitions, 219, 222–226
    development dialogues, 218
    dimensions, 222–226
    disaster risk reduction, 219
    ecological systems, 218
    ecosystem, 218
    engineering, 218
    H&S management, 219
    international, national and local policy, 218
    modulus, 218

    organisational resilience, 219
    PRISMA flow chart, 221
Resilience engineering (RE), 219, 222
Respiratory movement amplitude, 139
Responsibility, 343
Responsible material sourcing, 504
Responsible sourcing (RS), 249–254, 256, 257, 259–261
Restriction of Hazardous Substances (RoHS) Directive 2002/95/EC, 464
Rethinking, 309
    PM, 308
Re-Thinking Construction in 1998, 267
Rhetoric questionnaire's questions, 346
Rigid insulation boards, 53
Risk management, 205, 291
    demands, 292
    employed, 278
    profile, 288
    and quality assurance, 277
    research stream, 279
    in uncertainty, 285, 290, 292
Risk mitigation, 338
Risk perception, 338
Roof covering, 53, 57
Royal Society's geoengineering report, 10
Rules of Conduct, 329

S

Safety communication, 235
Safety improvement, 197
Safety management, 219, 220
Safety managers (SMs), 238
Safety, Health and Welfare at Work (Construction) Regulations 2013 (HSA, 2013), 193
Sales of electric cars, 5
SBS symptoms, 146
Scepticism, 466
Sceptics, 521
Scientific mapping, 445
Scoping review, 375
SEAM ecosystem
    complementary technologies, 408
    digital ecosystem framework, 408
    joined-up digital strategy, 408, 409
    re-usable and interoperable data, 411
    smart building use case, 409
SEEDS 2019 Conference Dinner, 170
Self-administered questionnaire, 206
Self-consumption (SC), 72, 73, 82
Self-fulfilling prophecy, 345

Self-sufficiency (SS), 73
Semi-structured interview approach, 196
Semi-formal interviews, 330
Semi-structured interview approach, 195
Semi-structured interviews, 28
Sensor drift, 56
Sequential mixed-method approach, 192
Sequential selection strategy, 67, 200
Service life of reinforcement, 498–499
Service strikes, 196
Shared digital model (Revit), 89
Shifting PV production, 73
Sick building syndrome (SBS), 139, 140, 144, 145
Site managers (SMs)
    contractors, 206, 213
    functions, 203
    self-administered questionnaire, 206
Six Sigma (SS), 351
Smart and interoperable approach, 397
Smart asset management
    business advantage, 421
    digital twin, 418
    digital vision, 421
    functionalities, 420, 421
    integrated digital twin, 421
    MVP, 421
    solution, 420
    three-phased approach, 422
    use case, 422, 423
    vehicle analogy, 422
Smart BIM-enabled asset management ecosystem, 419
Smart cities' project development model
    benefit realisation, 381
    bottom-up approach, 381
    description, 382
    methodologies and strategy, 381
    top-down approach, 381
Smart citizens, 373
Smart city
    aspects, 373
    categories, 373
    challenges, 372
    definition, 371
    human dimension, 374
    initiatives, 375, 381
    institutional factors, 374
    objectives, 373
    programme management, 375
    project development, 372
    responsive and resilient environment, 372
    technologies/infrastructure, 373
    urban development, 372

Smart data management, 412
Smart energy system, 81
Smart enterprise asset management (SEAM)
    business need, 398
    conflicting terminologies, 397
    cutting-edge technologies, 397
    digital twins, 399, 406–408
    ecosystem, 396
    ecosystem of partners, 410, 411
    ecosystemic approach, 398
    enhanced asset analysis, 397
    integrated modules, 409
    lifecycle management, 399, 406
    macro-level methodology, 398
    performance possibilities, 397
    technical capabilities, 398
Smart government building, 396
SMART systems, 85
Smart technologies, 376
SMART technologies, 91
Social acceptance, 513
Social and cultural behaviour, 9
Social distancing, 522
Social housing
    construction professionals, 62
    construction projects, 60, 63, 67, 68
    development scheme, 62
    foundations, 59
    in Ireland, 61, 62
    M&E, 63
    and nZEB, 60
    provision, 59
Social inequality, 7, 26
Social media, 164, 165
Social process, 314
Social Responsibilities of the Businessman (book), 26
Social science analysis method, 17
Social sustainability
    achievement barriers, 8
    company performance, 10
    consideration, 4
    importance, 9, 10
    international change, 4
    organisations, 11
    problems, 5–8
    requirement, 7
Social system, 307, 345
Social value (SV), 186
    interpretations, 27
    procurement criterion, 13
    subjective concept, 13
    SVA, 15
Social Value Act, 13

Index

Socioeconomic areas, 477
Socio-environmental activities, 249
Sodium-nickel-chloride (NaNiCl), 73
Soft skills, 177, 344, 345
South African universities (SAUs), 106
South East Macclesfield, 54, 56
Space heating, 76, 82
Space standards, 85
Spatial problem representations, 316
Spectroradiometer field, 484
Speed drawing, 89
Spoiling publicity, 7
Stakeholder engagement, 343
Stakeholder management, 159, 298, 303
Stakeholder satisfaction, 9, 299
Start Small, 421
Statutory nZEB legislation, 61
Steel purlins, 130
Stockholm Resilience Centre Report, 10
Strategic asset management approach, 159
Strategic decision-making, 317
Strategic thinking, 352
Structherm, 122
Structural strengthen (SEWI), 126
Subcontractors, 65, 66
Suburbanism, 475
Suffolk Climate Change partnership, 88
Suffolk County Council, 86
Superstructure, 64
Supply business models, 37
Supply chain analysis, 267
Supply chains, 34, 250, 251, 254, 256, 257, 259, 260
    ethical, 249
    manufacturing, 250
    resilient, 250
Supply side barriers, 87
Sustainability, 3, 10
Sustainability achievement, 8
Sustainability agenda, 4
Sustainability and facilities management, 156
Sustainability and life skills, 175
Sustainability crisis, 4
Sustainability educations, 518
Sustainability evaluation, 16
Sustainability informed consumers, 513
Sustainability of EVs, 511, 512, 521, 522
Sustainability offices (SOs), 114
Sustainability practices, 101
Sustainability Tracking, Assessment and Rating System (STARS), 114
Sustainable asset management process
    asset register, 155

    BIM, 156
    environmental factors, 154
    interrelated and interacting elements, 155
    management principles, 157
    postgraduate education (*see* Postgraduate education in sustainable asset management)
    stakeholders recognition, 154
    strategic approach, 156
    sustainable construction, 155
Sustainable building construction projects, 495
Sustainable building practices, 115
Sustainable built environment achievement, 503
Sustainable companies, 10
Sustainable construction, 88, 442
    advantages, 94
    aim, 96
    built environment, 93
    CIB, 96
    definition, 96
    practice desire, 94
    sustainability principles applications, 96
Sustainable crop alternatives, 489
Sustainable crop planning alternatives, 485
Sustainable design implementation, 86
Sustainable development (SD), 97, 106, 442
Sustainable Development Goals (SDG), 4, 153, 477
Sustainable education, 462, 466
Sustainable energy, 172
Sustainable EV fuel, 518
Sustainable e-waste management, 465
Sustainable procurement (SP), 107
Sustainable societies, 3
Sustainable socioeconomic investment choices, 512–513
Sustainable urban development and management, 159
Sustainable urban environment, 152, 160
SV measurement and communication
    agreement, 15
    business operations, 18
    construction contractors, 16
    construction professionals empowerment, 16
    CSR, 15
    financial and nonfinancial metrics, 19
    frameworks and models, 16
    imaginative and adventurous pathways, 16
    industrial survey, 18
    interviews analysis, 18
    monetary-derived method, 16

SV measurement and communication (*cont.*)
  narrative analysis, 17
  objective requirements, 18
  ontological positions, 17
  semi-structured interviews, 17
  stakeholder understandings and perceptions, 19
  subjective, 16
  third-party measurement tool, 18
SV practices, 28
System of systems, 373
System thermodynamics, 11
Systematic review, 375

T
Teaching BIM, 437
Team working, 314
Technological entities management, 398
Telephone interviews, 332
Teleportation functions, 391
Tender requirements, 43
Terrestrial waste-to-energy scenario, 530
Thematic coding method, 350
Theme Matrix
  findings, 344
  interpersonal intelligence, 340
  similarities, 340
  themes, 340, 341
  usage, 340
Theory for practice, 298
Theory of Constraints (TOC), 351–353
Theory of Programme Management, 300
Three-phase asset management, 406
Time-of-use (TOU), 72
Tit for Tat approach, 268
Toolbox talks (TBTs)
  andragogy, 236
  ASSP, 237, 239, 243
  authority of person, 242
  benefits, 240
  characteristics of trainer, 242
  construction site, 235
  conversations, 241
  costs, 238
  creation of booklets, 240
  data collection, 240
  delivery methods, 245
  design and delivery, 239, 244
  discussion alone/discussion paired with demonstration, 241
  drives inconsistency, 243
  duration and frequency, 240, 243
  Employers' Onus, 236–238
  feedback session, 239
  H & S training, 238
  kinaesthetic learners, 242, 245
  language and learning barriers, 241–242
  language barriers, 244
  learning style, 237, 239
  literature review, 239
  management, 241
  materials, 235
  monitoring, 245
  opportunities for managers, 237
  optimisation, 242
  pedagogy, 236
  perception, 240
  post-completion, 239
  post-talk discussion, 242
  researcher observations, 235
  safety anecdotes, 236
  safety communication, 235
  safety programs in construction, 235
  self-protective answers, 240
  SM, 238
  stickiness, 244
  structure and delivery, 239
  supervise, 241
  teaching style, 243
  train the trainer, 243
  types, 243
  worker's performance and safety, 244
  worker's safety, 236
Top-down approach, 3
Top-down society, 5
Total cost of ownership (TCO), 399
Total grid consumption, 77
TOU tariffs, 81
Toxic chemicals, 502
Toyota Production System (TPS), 443
Traditional project management (TPM) theory, 287, 288
Traditional supply chains, 34
Trained technicians, 298
Transport investment projects (TIPs), 287, 289
Transport modes, 526, 530
Transportation, 39
Transportation energy consumption, 525
Trinity, 445–447, 451, 455, 456
Trust, 270

U
UK building regulations, 71, 125
UK construction, 19, 266

Index

UK construction industry, 17, 23, 31
UK construction sector, 330
UK parliament passed legislation, 71
UK population, 475
UK's Association for Project Management (APM), 297
UK's net zero emissions target, 71, 82
UN's Sustainable Development Goals, 480
UN's vision, 477
Uncertainty, 277–279, 281–283, 285–292
Underbidding and business case exaggerations, 302
Underground utility services
 data-driven planning, 191
 governmental statutory organisations, 192
 HSA, 192
 importance, 191
 inherent feature, 191
 operative inexperience, 191
 strikes, 199
 work practices, 192 (*see also* HSA's Code of Practice)
Unethical practices, 330
United Kingdom (UK), 23, 192
United Nation's New Urban Agenda, 477
University of South Africa (UNISA), 107
University of Technology (UoT), 108
Unsustainable design and construction, 94
Unsustainable fuels, 527
Urban dwellers, 478
Urban forest, 476
 air pollutants impact reduction, 478
 biodiversity, 478
 deployment, 477
 dwellers, 478
 fresh air, 478
 healthy, 478
 positive effects, 479
 research, 476
 robust and resilient, 480
 strategic planning and legislation, 479
 sustainable creation, 479
 vulnerability, 479
Urban living environment, 151
Urban sustainability management issues
 asset life cycle, 153
 building construction materials, 153
 built environment, 153, 154
 construction sector, 153
 GHG emissions, 152, 153
Urban utopias, 475
Urbanisation, 93, 476
US Environmental Protection Agency, 145
User engagement
 critical criterion, 386

healthcare, 386
mode, 386
technologies and strategies, 386
traditional methods, 386
VR technologies, 387
Utility network infrastructure, 192

V

Value
 creation, 351
 debate, 351
 definition, 349
 misinterpretation, 350
 project stakeholder's benefits, 349
 project success, 349
 satisfactory needs, 349
 success and realisation, 354
Value analysis, 352
Value creation, 305, 314
Value generation, 305
Value Management (VM), 351
Value planning, 352
Value stream, 352
Ventilation points, 54
Verbalising and assessing stages, 181
Viable urban futures, 480
Victorian slums, 476
Visible price, 7
Vision 2020, 112
Visual representation, 315, 353
Visualisation, 315–317
 anticipation, 322
 controlled environment, 321
 number of problems, 321
 PM, 323
 problems and solutions, 321
 research hypotheses, 318
 research method, 318–320
 research question, 318
VM structure, 352
VR simulation, 392
VR technologies
 BIM CAVE, 387
 BIM-VIS, 387
 DVBCS, 387
 semi-immersed VR environment, 387

W

Waste Electrical and Electronic Equipment Directive 2002/96/EC, 462
Waste Electrical Electronic Equipment (WEEE) Directive 2002/96/EC, 464
Waste incineration, 526, 528, 529

Waste-to-energy (WtE)
  fossil fuels, 528
  global fuel reserves, 527
  incinerator, 530
  maritime transport, 528
  renewable energy systems, 528
  sustainable waste management, 525
  transport sector, 526
Water consumption, 85
Water pump manufacturing project, 355
Water transportation, 526
Web-based information, 164
Weighted average score (WAS), 253
Western democracy, 6
Whilst government policy, 87
Whole lifecycle, 412
Wildlife, Ecology and Conservation Science, 91
Windermere, 53, 56
Work breakdown structure (WBS), 352, 354
Work environment satisfaction, 137
Workplace factors, 205

World Business Council for Sustainable Development (WBCSD), 442
World Green Building Council, 153

Z
Zero-Carbon Homes, 71
Zero-emission transition, 523
ZP dwellings
  analysis, 81
  Covid-19 lockdown impacts, 78–81
  data sources, 74, 75
  DSR experimentation and evaluation, 81
  electricity balance, 76, 77
  electricity consumption, 74
  household characteristics, 74, 75
  PV panels, 74
  SC, 74
  SS, 74
  statistical analysis, 74
  total energy consumption, 76

Printed in the United States
by Baker & Taylor Publisher Services